3 8008 00402 3346

AF616166

WITHDRAWN

International Petroleum Encyclopedia 2003

PennWell Corporation
1421 South Sheridan Road
Tulsa, Oklahoma 74112-6600 USA

800.752.9764
+1.918.831.9421
sales@pennwell.com
www.pennwellbooks.com
www.pennwell.com

Library of Congress Cataloging-in-Publication Data Pending

ISBN 0-87814-893-0

Printed in the United States of America
1 2 3 4 5 07 06 05 04 03

contents

46
43
37
44-45
48
21
14-15
18
17
26
23
38
30
32-33

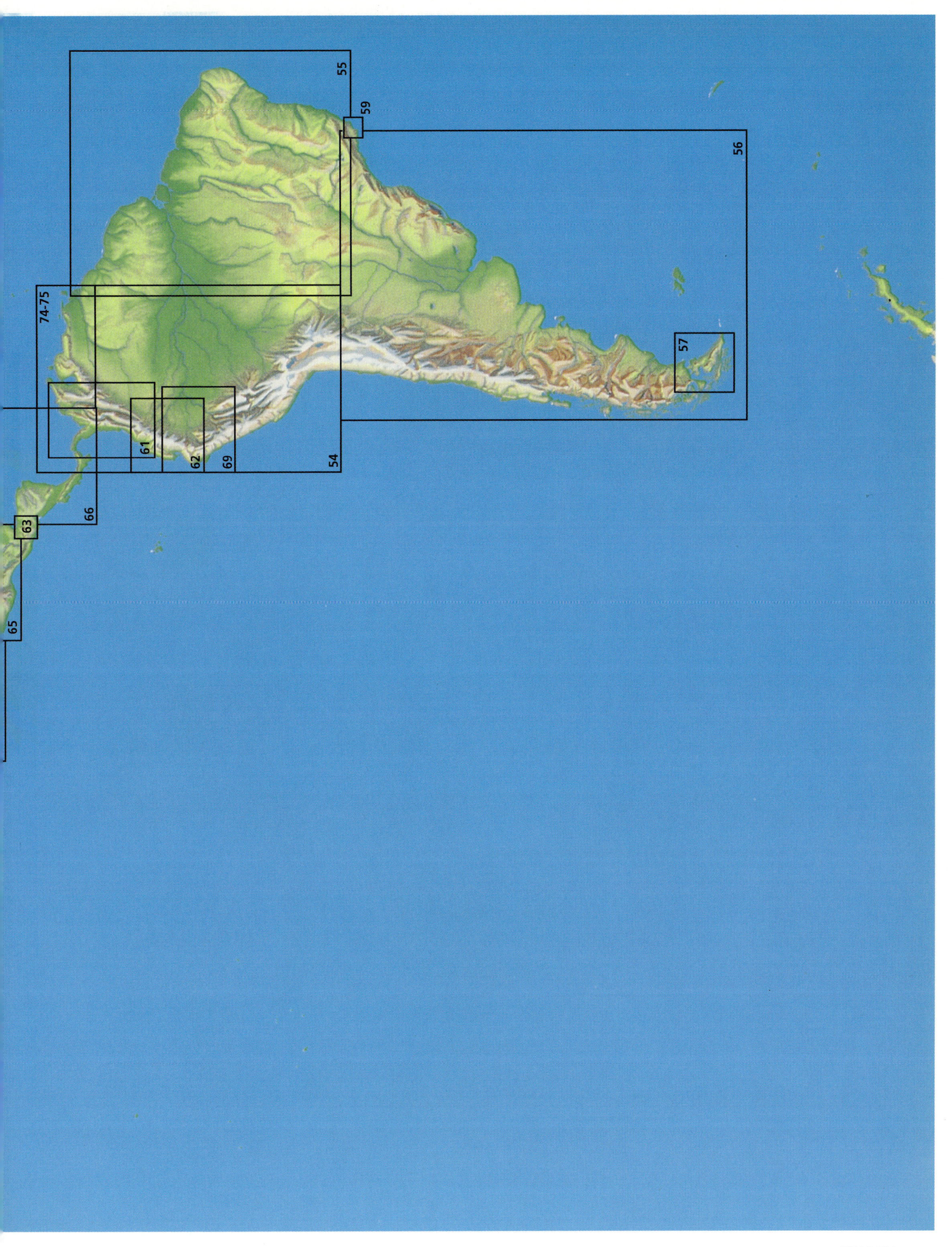
55
59
56
74-75
57
61
62
69
54
66
63
65

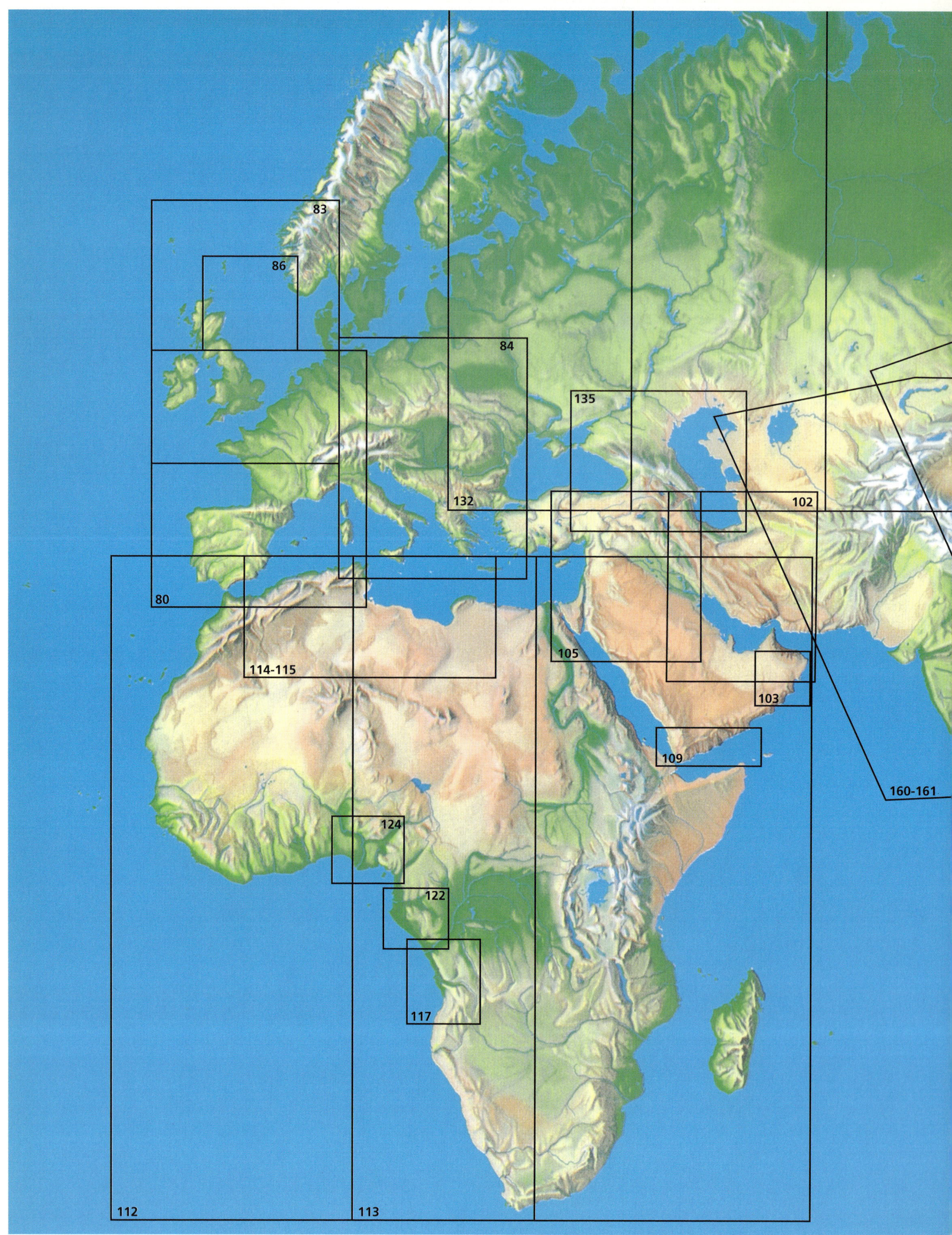
83
86
84
135
132
102
80
105
114-115
103
109
160-161
124
122
117
112
113

133
133
154
165
152
151
168-169
158
158

Editor
Rebecca L. Busby

Art Director
Charles Thomas

Production
Heather Skeith

Lead Illustrator
Kay Wayne

International Petroleum Encyclopedia
is published annually by:
PennWell Corp., 1421 S. Sheridan Rd.,
Box 1260, Tulsa, OK 74101, USA
Phone: 918-835-3161; Fax: 918-831-9555

Chairman
Frank T. Lauinger

President and Chief Executive Officer
Robert F. Biolchini

Senior Vice-President,
Finance and Administration,
Chief Financial Officer
Thomas L. Stone

Publisher
Bob Smock

Managing Editor
Marla Patterson

Advertising Production Coordinator
Mary McGee

IPE ESSAY

Security of Supply – It's time to put energy policy at the heart of foreign policy

BY SIR RICHARD GIORDANO KBE
CHAIRMAN – BG GROUP

AT A TIME when we in the oil and gas industries have the luxury of gazing out over the prospect of several decades of growth in demand for our products, the tone of the energy debate at present is unusually gloomy. And understandably so.

A series of dramatic, interlinked global and political developments ranging from the events of 9/11 to economic downturn have led many politicians and policy-makers — justifiably or not — to question for the first time in many years our industries' ability to continue meeting demand and ensuring uninterrupted supply to our markets.

The audacity, scale, and level of violence explicit in the September 11 attacks still haunt us all, prompting us to question whether our own and our industries' security is robust; and, even though subsequent al-Qaeda attacks in Bali and Kenya have been upon 'softer' targets, they have demonstrated the random nature of the new terror we are faced with and a sharpening of terrorists' focus upon any individuals or organizations associated with the US, the UK, or other 'developed' Western nations and their allies.

Uncertainties about conflict in Iraq — unresolved, as I write — have added to the gloom. Naturally, the emergence of a more democratic Iraq would be welcome, but even that outcome would be not without its difficulties for global oil and gas markets, given the huge reserves it would unlock and the consequent adjustments to the balance of power among producer countries. The inevitable price spikes in oil markets as a result of the turmoil around Iraq over the past year have also added to the jitteriness of governments and decision-makers.

It may also be the case that we have so far underestimated the impact that the choking off of oil supplies from Venezuela as of December 2002 could have on energy policy. One can imagine oil and gas-importing countries concluding that, if political unrest can disrupt production from the country with the largest oil reserves in the Western hemisphere, there may well be a need for more strategic stocks and more intervention in the market by policy-makers.

This gloom has been set against a backdrop of a persistent and wide-ranging economic slowdown, lingering concerns about California's brownouts, and fears that we may be witnessing the demise of the nuclear industry. All of these elements have combined, producing political uncertainty that, in turn, has pushed security of energy supply concerns rapidly up the agenda.

In this essay, as chairman of an international, integrated gas major, I want to focus principally on the prospects for natural gas, though many of my conclusions will apply to oil too; but, before looking for solutions to the challenge of guaranteeing secure supplies, I would first like to underscore the compelling reasons why — despite the current atmosphere of gloom — we in the gas industry really do have cause for optimism.

First of all, the US Energy Information Agency (EIA) forecasts compound annual growth rates for gas consumption of 3.2% in the years to 2020 — the strongest of all fuels and stronger even than oil at 2.2%. The International Energy Agency (IEA), in its *World Energy Outlook 2002*, forecasts a doubling of primary gas consumption between now and 2030, with natural gas's share of the world energy market rising from 23% to 28% over the same period.

But such numbers, however accurate or inaccurate they prove to be, are insufficiently eloquent to describe the true prospects for natural gas over the coming decades. They do not articulate the reasons why this will be inevitably one of the real growth stories of the first half of this century.

This consolidation of and expansion from the strong position gas has built up in the global fuel mix will be boosted by a series of factors, including:

- A boom in trading, as the import requirements of the main industrialized nations surge;
- A recognition that the economics of LNG are strengthening and the scale of LNG activity growing fast — perhaps by as much as 60% in the next 5 years and 100% over the decade; and
- The desire of some of the world's largest nations to cut pollution and improve air quality in their most polluted cities, together with increasing international pressure on countries to meet ever more demanding climate change targets, will hasten the switch from fuels such as coal, oil, and wood to natural gas with its lower carbon emissions.

In terms of trade, imports will account for virtually all of the growth in gas consumption in the US, northeast Asia, and Europe in the years to 2020, and about two-thirds of additional gas production will cross borders. This represents an international gas supply trade that is more than double existing levels and will take the share of trade in gas production globally to around 45%.[1] This figure compares with 52% for oil, and it represents a massive transformation for a commodity once seen as all but restricted to intra-regional trade. If trade in natural gas is not quite at a level that could be described as inter-regional, it is certainly increasingly cross-border, and there is little doubt that inter-regional trade, though limited now, will grow strongly.

Of course, however ambitious the industry is in terms of fixed infrastructure, it is the case that growth and trading flexibility on the scale outlined would not be possible without the surge of confidence that has flowed through the LNG industry in recent years. This is a confidence that shows every sign of being sustained. BG Group has played an important part in contributing to that confidence in LNG by achieving historically low costs at its Atlantic LNG plant in Trinidad — technology that will be replicated and developed in Egypt and elsewhere.

BG has achieved historically low costs at its Atlantic LNG operation in Trinidad.

This has enabled LNG to compete with pipeline gas in US markets. More broadly, LNG can contribute in forthcoming years to easing some concerns about gas-price volatility. It is unlikely ever to displace pipeline gas and one can never eliminate volatility from any commodity market, but LNG will act as a 'price-taker,' capping the marginal price of gas bought and sold. It is also significant that, whereas LNG used to belong only in the realm of long-term contracts, there have been signs in recent years of a form of spot market in LNG cargoes emerging. An expansion in LNG trading is likely to create more opportunities for speculative trading of this kind, and that in turn should add a suppleness and a flexibility to gas trading generally.

"In a growing global market for natural gas and one in which inter-regional trade will be one of the methods of meeting demand, LNG must play a significant and increasing role."

The strength of this trend is evident also in the growth in the market for LNG carriers. Nineteen are due to enter service this year — the highest number on record. With still more orders possible, it could be that the world fleet will top 200 this year.[2] Both of these trends are welcome developments and clear signs that LNG can play an important role in easing security of supply concerns.

There has been some focus of late on cancelled LNG projects, as though this represented some kind of stalling of growth in LNG expansion. However, nothing could be further from the truth. These cancellations represent the normal commercial fallout from competing proposals and, in fact, indicate the highly competitive nature of an industry in which several players are typically competing to build each export and import terminal.

It is quite clear that, in a growing global market for natural gas and one in which inter-regional trade will be one of the methods of meeting demand, LNG must play a significant and increasing role.

The increasing focus on environmental concerns is another factor that will help natural gas to prosper. In the past, environmental policy-makers have tended to lump together all hydrocarbons and treat them as though they have an identical impact and emit the same amount of greenhouse gases. Nothing could be further from the truth. On combustion, natural gas produces 22% less CO_2 equivalent emissions than oil and 40% less than coal. In fact, the UK is on course to meet tough Kyoto emissions targets almost entirely on the back of a major shift from coal-fired power generation to gas-fired stations during the 1990s. And consumer nations will increasingly turn to gas as a source of power generation. More than 60% of the growth in gas demand between now and 2030 will be burned in gas-fired power plants.[3]

What each of these developments underlines is the fact that gas is increasingly the fuel of choice across the world because — probably in order of priority — it is convenient to use, plentiful, and clean.

The recent story of the extraordinary expansion of natural gas markets is in large part due to a two-stage process: of governments liberalizing their energy markets, then private sector initiatives taking center stage to create fast-growing, competitive, international energy trade.

We observed earlier that much of the growth in gas demand in the US, Europe, and northeast Asia will come from imports. What this latest phase in the developments of gas trade signals is an end to the historic perception that security of energy supply can be constructed only on the back of indigenous energy resources. Increasingly, there is an awareness that safeguarding the *supply* of energy imports is the key to security — and that fuel and energy are no different to a whole range of essential commodity and product imports we have come to rely upon.

As John Mitchell says, in one of the most cogent analyses in recent years of the prospects for the oil and gas industries: "The energy security policy concerns of 30 years ago have been reversed. The development of global investment and supply gives cheaper security for importers than policies focused on reducing trade ... Policies to reduce the economic risk of energy trade by subsidising domestic investments would limit the security and flexibility which the global system provides. Such subsidies and market distortions impose costs on the country which adopts them."[4]

BG's Hibiscus platform, North Coast Marine Area, Trinidad, began gas production in 2002.

There are two principal lessons to be drawn from this. First, while it was essential for governments to liberalize and cut loose their energy industries, it has been the subsequent dynamism of the competitive, international energy markets that has been the true motor to growth and the factor that has increasingly underpinned security of supply. Second, countries that seek to focus on indigenous supply rather than on international energy trade — except in circumstances of exceptional indigenous resource — risk putting themselves and their industries at a competitive disadvantage.

Ironically, one of the early spin-offs of

IPE Essay Guest Author

Sir Richard Giordano KBE
Chairman — BG Group

Sir Richard Giordano KBE is chairman and non-executive director of BG Group. He was appointed chairman of British Gas plc in January 1994, having been a non-executive director since December 1993. He is non-executive deputy chairman of Rio Tinto plc and a non-executive director of Georgia-Pacific Corporation Inc.

After working as a lawyer with Shearman & Sterling, the international law firm based in the US, he was chairman and chief executive of BOC Group plc from 1985 to 1991 and chairman from 1994 to 1996. He has also previously served on the boards of National Power plc and Reuters plc and has been non-executive deputy chairman of Grand Metropolitan plc.

BG Group is an international, integrated gas 'major' with assets in around 20 countries across the world, ranging from the US and the UK to South America, continental Europe, North Africa, the Caspian, and the Far East.

The company was part of the former British Gas. Its center of gravity is now firmly in international exploration and production but it is active throughout the gas-chain, from discovery to delivery to customers. It is listed in London and on the New York Stock Exchange.

liberalization — particularly in the US and the UK, where market opening was most comprehensive — was the virtual demise of government energy policies.

In the US, as long ago as the Reagan era, there was a general perception that the President and his administration had all but concluded that there was no longer any need for energy policy *per se*. Fuel was plentiful and the market would deliver. And, in the UK, active energy policy in the 1980s and 1990s involved cutting energy industries free — not second-guessing them.

"Countries that seek to focus on indigenous supply rather than on international energy trade risk putting themselves and their industries at a competitive disadvantage."

However, more recently, although the need for a vital international trade in energy is largely a given, two apparent areas of conflict have emerged and now stand at the center of the debate about the future of energy policy. They are:

- The demands of security of energy supply versus the motor of liberalization — can these two elements co-exist?
- And, regulation and intervention versus free markets — what is the correct balance?

In the UK and Europe in particular, the argument is increasingly frequently heard that markets can — and do — deliver lower prices but they are inimical to systems which require supply back-up to cope with the possibility of technical or political disruption. Markets abhor slack in the system but security of supply demands it, the claim goes. Similarly, regulation is essential to prevent market dominance and abuse, but at what point does intervention become addictive, unnecessary, and an impediment to dynamic competition?

Whilst these are important questions and, undoubtedly, balances need to be struck, there is a danger that the pendulum may be swinging back, that government or institutional intervention will become fashionable once again, and that some of the real benefits delivered by the market will be put at risk. And, if one is looking for the best example of these tensions between liberalization and intervention, the European Commission is probably the best case in point.

On the one hand, it is to be applauded for creating through active liberalizing measures across the European Union what will be the biggest liberalized gas market in the world; on the other hand, its policy-makers show increasing signs of seeking to intervene in the market — for example, by pushing for a pan-European system of energy taxes and by seeking to gain powers that would enable bureaucrats to control and direct the use of oil and gas stocks.

There are two broad points that emerge from these two areas of tension. First, our industry recognizes that there is a need for government and/or regulator intervention where there is a threat of an incumbent or newly established company taking advantage of a monopoly position. However, what governments, regulators, and policy-makers in general must resist is a temptation to believe that security of energy supply requirements supersede the demands of competitive energy markets to such an extent that a return to large-scale central planning of energy provision might be timely.

The scale of international energy trade is such now that any government — or institution like the European Commission — that believes it can take back the levers of control and run an energy policy more efficiently than the market is deluding itself. Yes, there may be a case for frameworks that help, say, proven renewable or energy efficiency technologies to reach a position at which they become genuinely competitive and capable of standing alone. Yes, there is a case for using economic instruments — such as carbon trading — to guide the market in the direction of lower carbon fuels. But any attempt to 'fix' the fuel mix in a given economy will end up with that economy bearing potentially crippling extra costs and putting itself at an economic disadvantage compared to its competitors.

"Regulation is essential to prevent market dominance and abuse, but at what point does intervention become addictive, unnecessary, and an impediment to dynamic competition?"

The second broad point to emerge relates to the kind of regulation we face in the energy industries. It is widely acknowledged in principle that the best kind of regulation is 'light touch' regulation. Ideally, a regulatory authority creates a framework designed to foster competition and to remove the threat of incumbent and monopoly power. After setting that regulatory framework, it then steps back and allows the market to function, intervening at later stages only when a new anti-competitive threat emerges.

But, if that is the theory, in practice the opposite increasingly appears to be the case. In fact, the tendency is for regulators' offices, once established, to expand seemingly inexorably. That, in turn, inclines them towards delving ever deeper into the workings of their industries. Before long, they are seeking to micro-manage markets, putting at risk the benefits gained from true competition.

In Brazil and the US, there have been recent examples of the constructive approach to regulation. The Brazilian regulator has been working with industry to enable the gas industry to expand and has shown real pragmatism in his approach, while the Hackberry decision in the US has seen the FERC decide against regulation of new LNG terminals on the basis that the wholesale market these terminals will be supplying is already sufficiently competitive.

In contrast, as well as some of the European Commission excesses alluded to above, Ofgem in Great Britain has created a largely sound regulatory framework but now risks disrupting competition by seeking to delve ever deeper into the workings of the market — in our view often needlessly — as well as seeking more and more jurisdiction. Still more alarming have been fresh powers taken by Northern Ireland's regulator — powers which actually risk cutting the value of assets in the province and deterring future investment.

Now this is not the oil and gas industry asking to be left alone and being resistant to regulation. However, what is undeniably the case is that flawed or downright poor regulation can — and does — give liberalization and competitive markets a bad name. The best example, of course, is California and the brownouts and blackouts of 2000-2001.

The opponents of liberalization were quick to begin laughing up their sleeves and to claim this phenomenon as proof that open markets cannot work. How could they work, if a state that was the epitome of the developed world could not secure sufficient power to keep the lights on, the air-conditioning whirring, and the PCs humming.

Governments working together can help Caspian gas access European markets.

The reality was that California was a victim of poor regulation — and, to a degree, of power-price manipulation by some rogue traders. It was not a victim of liberalization. A system which put a highly restrictive ceiling on the increases in tariffs consumers could pay, which had environmental policies which enabled any new power station proposal to be easily blocked, in a state in which demand for power rose by 32% in 5 years — that is a system inevitably heading for a fall.

As Lawrence Makovich and Daniel Yergin of CERA concluded: "The common diagnosis of California's electricity power debacle is wrong. The state is not suffering from deregulation. It is afflicted by a strange mutant ailment — partial deregulation and now partial reregulation — that has produced a flawed market.

"California designed a market that disconnected customers from prices and, at the same time, made it neither profitable nor possible to build a new power plant. The result is a serious power shortage."[5]

"The EU-Russia Energy Dialogue has proved to be a difficult process and there is much work still to be done. We may not yet see the light at the end of the Russian tunnel, but at least it appears that we are entering the tunnel."

BG Group's own chief executive, Frank Chapman, summed it up still more pithily: "To those who say to me, 'What about California?' I say: the US — alongside the UK — is an excellent example of a liberalised gas market. The Californian power market is not. What you've got is not liberalisation but, as we've witnessed, a recipe for disaster."[6]

Now this is not to say that there is no role for governments in energy policy or in the energy debate. In fact, because of the developments I outlined at the top of this essay, this is something of a watershed period. It is time now to demonstrate how energy markets can best guarantee security of energy supply and, to achieve this, we need governments — or, more accurately, governments working together — to join forces with private companies to address the key strategic challenges before us. In short, we need governments to begin to place energy policy at the heart of foreign policy.

Too often in the past, governments have run energy policies and foreign policies in isolation, and it has been mere coincidence if there was overlap. However, as we seek to unlock and bring to market at economic prices the vast remaining oil and gas reserves that can ease supply concerns — and there is now little doubt that reserves sufficient to meet up to 100 years of global demand are out there — governments need to have the inherent challenges foremost in their minds as they frame their diplomatic strategies and policies.

Now this is not to urge governments or groups of governments working together to adjust their principled positions towards nations that can become major suppliers of oil and gas. If a regime is beyond the pale in terms of repression or human rights, then that regime remains beyond the pale — whether or not they have reserves.

What I am suggesting rather is that governments approach diplomatic negotiations with energy concerns at the heart of their thinking. Questions which Ministers and Prime Ministers and senior officials should be asking could include:

- What can we do to open up markets and convince would-be producer nations that our security of energy supply can mean security of revenues for them?
- As we embark upon trade talks or similar negotiations, what can we do to try to urge potential producers to create liquid, transparent markets, which will contribute to the smooth operation of international energy trading?
- How can we link discussions about energy prospects with other potentially beneficial outcomes?
- Given that much of the growth of the oil and — in particular — gas industries will be dependent on new infrastructure, what can be done to foster a climate in which private companies will feel secure — and indeed incentivized — to invest in large-scale infrastructure projects?

These are a few obvious starting points, but specific examples of where diplomatic support to resolve specific *impasses* will perhaps give an even clearer picture of the way in which governments can work with industry to help create a security of supply framework based on solid foundations.

"Diplomatic efforts may be capable of speeding up the transition to gas in India and China. Such a shift represents probably the single largest realistic action that could be taken to reduce the impact of climate change."

Russia is, of course, the most obvious case in point. It is not simply the 33,000 billion cubic meters (bcm) of gas reserves[7] (1.2 tcf) Russia holds and its position as the country with the largest gas reserves in the world that make it central to the debate; the failure to access some of the largest reserves available to it and the stranglehold it retains on much of the gas reserves in the Caspian have combined to place it at the center of industry and diplomatic efforts to find solutions.

The failure to establish production-sharing agreements involving private 'supermajor' and 'major' players in the industry and the inability to secure sound legal and contractual foundations that will enable these companies to operate confidently have already consumed much time and effort and, to date, brought few real results.

However, praise should be given to the European Commission for the lead it is showing in seeking to hasten reform within the Russian system. The EU-Russia Energy Dialogue has proved to be a difficult process and there is much work still to be done but, ironically, 'unelected' politicians such as Trade Commissioner Mario Monti and Energy and Transport Commissioner Loyola de Palacio have blazed a trail. As I write, there are glimmers of hope and prospects for progress. We may not yet see the light at the end of the Russian tunnel, but at least it appears that we are entering the tunnel.

The Pacific LNG project can supply Bolivian gas to US west coast markets.

"Encouraging Pacific LNG — a project to take gas from Bolivia over the Andes and, as LNG, into US markets on the west coast — could help eliminate the 'coca' economy and enable Bolivian standards of living to rise on the back of legitimate trade."

I should like to see elected politicians with particularly strong relationships with President Putin — such as UK Prime Minister Tony Blair and US President George Bush — bolster this European effort with supportive initiatives of their own. Last year, Prime Minister Blair and President Bush agreed to launch a US-UK energy dialogue. It is essential that that dialogue, which is set to report initial findings in the first half of this year, identifies reform of the Russian gas network and export markets as a high-priority target of diplomatic efforts.

As well as enabling more Russian reserves to be developed, the EU — backed by the US and the UK — should be pushing for direct access to European markets for Caspian gas. The *quid pro quo* for Russia would be freedom of access to European downstream markets.

Already, some of the best informed analysts in the industry are forecasting major reforms within the Russian gas system. For example, Jonathan Stern predicts: "While Russian gas deliveries to Europe will increase substantially up to 2010 in absolute terms, and perhaps also as a percentage of European gas demand, Gazprom's monopoly over gas exports to Europe will undoubtedly be broken at some stage during the decade. While Gazprom will remain an extremely powerful export presence, it is entirely possible that as many as six other companies will have significant quantities of Russian gas to export to Europe in competition both with one another and with other suppliers."[8]

What these and other similar analyses demonstrate is that hopes of reform to the Russian system are far from pie-in-the-sky. Indeed, they show how timely a sustained diplomatic push could be and indicate the huge impact such an effort could have in increasing security of supply.

Other areas where diplomatic efforts could be critical to bringing about major gas developments — and associated social benefits — include India and China, South America, and Iran.

BG Group is heavily involved in seeking to expand some of India's gas markets, as well as being the supplier of compressed natural gas for Natural Gas Vehicles in Mumbai state — a contribution to tackling severe air-quality problems. Both India and China are keen to expand their use of natural gas and, over time, reduce coal, wood — and cow-dung! — burning, which has and continues to produce appalling levels of atmospheric pollution. Yet, despite the Indian and Chinese desire to improve their air quality, current IEA forecasts still suggest coal use will increase to 2030, though more slowly than gas and oil, and that India and China will account for two-thirds of that increased demand.

Diplomatic efforts — and perhaps even practical support — may be capable of speeding up this transition to gas in India and China. And a major incentive for some global diplomatic pressure here lies in the huge impact such a shift could have for tackling global warming. It represents probably the single largest realistic action that could be taken to reduce the impact of climate change.

A further incentive for stronger diplomatic efforts to help developing countries to exploit their natural resources lies in the social and security spin-offs that can emerge. For example, the US authorities would be well advised to put more strategic weight into encouraging Pacific LNG — a project to take gas from Bolivia over the Andes and, as LNG, into US markets on the west coast. Such a move could help eliminate the 'coca' economy with all of its pernicious implications from the country and enable standards of living in the country to rise on the back of legitimate trade.

In relation to Iran, US and European perspectives on how best to encourage the regime to reform and to reject support for terrorism currently differ wildly. But the European Commission is currently placing a great deal of diplomatic effort behind building a productive dialogue. With member state backing, the Commission is offering Iran trade and energy opportunities but is requiring in return respect for human rights and assurances that it will cease fund-

US support for Bolivia's gas economy could enable standards of living in the country to rise.

ing terrorism. This may prove to be a hard deal to cut but, given the choice between hope and active diplomacy or despair and withdrawal, surely the former is preferable every time.

The key to achieving wide-ranging results from this 'glorious' coalition of developed consumer nations, developing countries with natural resources, and private companies lies in our stripping away extraneous issues and seeking constantly to focus on the mutual benefits that can be the product.

As I suggested earlier: the developed world's security of energy supply is the developing world's security of revenues. As Jonathan Stern puts it: "This need for revenue is likely to be a source of competition for market share in the future ... Moreover, given the rigidities of gas exports compared with oil — the pipelines and the markets are where they are, and export volumes cannot be reoriented to other destinations — any threat to use gas as a commercial or political weapon risks depriving exporters of any long-term export future."[10]

If that is the deal between consumer and producer nations, the oil and gas companies' contribution will be of an increasingly challenging nature. It will focus on corporate governance and it will mean governments, institutions, NGOs and, indeed, consumers pressing ever more insistently for assurances that company activities are socially progressive, environmentally sound, and lend themselves to furthering development in new producer nations. There will be pressure to eliminate any hint of asset-stripping, to stop lining the pockets of repressive leaders, and to focus on 'guaranteeing' raising the living standards of the poor in these countries.

"The EC is offering Iran trade and energy opportunities but is requiring in return respect for human rights and assurances that it will cease funding terrorism. This may prove to be a hard deal to cut but, given the choice between hope and active diplomacy or despair and withdrawal, surely the former is preferable every time.

Some of these goals oil and gas companies can deliver single-handedly. The more ambitious targets of 'guaranteeing' higher standards of living in host nations will require wide-ranging and imaginative solutions, frameworks, and support from governments and bodies such as G8, the WTO, the EU, and others.

But oil and gas companies need to be aware that commitment to more active corporate governance is non-negotiable. The genie, if you like, is out of the bottle. And it will not float back in.

Already the World Bank has launched a review of the extractive industries. More recently, global financier George Soros has seized the imagination of many governments, NGOs, and other organizations with his 'publish what you pay' proposal — effectively a call for host nations to reveal how they are deploying the revenues they are receiving from foreign investors but a challenge also to our industry to find means of being absolutely transparent.

Yet, this is not something we should be fearful about. We in the industry must stand up and make our case eloquently and consistently for diplomatic back-up *en route* to security of supply delivery, for energy policy at the heart of foreign policy. We must press our case for light-touch regulation and regimes that encourage investment. We must stress the need for minimal intervention in competitive markets. But we must rise to the corporate governance challenge too.

It is time for us all — private companies, governments, policy-makers, and leaders — to be bold and imaginative. And, if we are, our goals of delivery to shareholders, security of supply for consumer nations, security of revenues for producers, and higher standards of living in developing nations are there for the taking. It won't be easy but let us begin that journey hopefully.

IPE

IPE ESSAY FOOTNOTES

[1] Figures and analysis from John V. Mitchell, *Renewing Energy Security, Sustainable Development Programme, The Royal Institute of International Affairs*, July 2002, page 9.

[2] *Lloyd's List*, 4 February, 2003, page 2.

[3] IEA: http://www.worldenergyoutlook.com/weo/pubs/weo2002/WEO2002_1sum.pdf

[4] *The New Economy of Oil, Impacts on Business, Geopolitics and Society*, John Mitchell with Koji Morita, Norman Selley, and Jonathan Stern, *Energy and Environment Programme, Royal Institute of International Affairs*, 2001, page 277.

[5] *California in the Dark, CERA News, 19/3/01*, by Lawrence Makovich, senior director of Cambridge Energy Research Associates, and Daniel Yergin, chairman.

[6] Address to the Society of British Gas Industries and the Institute of Gas Engineers, May 1, 2001.

[7] Wood Mackenzie presentation to International Association of Oil and Gas Producers, December 2002.

[8] *Security of European Natural Gas Supplies, The impact of import dependence and liberalization*, Jonathan Stern, *The Royal Institute of International Affairs, Sustainable Development Programme*, July 2002, page 18.

[9] http://www.worldenergyoutlook.com/weo/pubs/weo2002/WEO2002_1sum.pdf, page 3.

[10] *Security of European Natural Gas Supplies, The impact of import dependence and liberalization*, Jonathan Stern, *The Royal Institute of International Affairs, Sustainable Development Programme*, July 2002, page 19.

IPE CHRONOLOGY

2002 in Brief

JANUARY

Markets

OPEC cuts of 1.5 million b/d take effect

Non-OPEC exporters pledge cuts of 462,500 b/d

Exploration/Development

New oil reserves found in UK North Sea's Buzzard, largest amount in decade

Apache wildcat discovers oil, gas in Egypt

Production

NW Shelf begins Echo-Yodel gas production 3 months early

Malaysia's Angsi field starts 6 months early, to reach 65,000 b/d, 450 MMscfd

Algeria's Hassi Berkine begins oil production 2 months early

North Sea Hanze producing 31,500 b/d

Gulf of Mexico Crosby oil field starts producing 20,000 b/d

Natural Gas

Timetable slips for massive Saudi gas deals

GdF signs Egypt's 2nd major LNG deal

S. Korea plans 1.7 mtpy LNG terminal

Shell purchases LNG expansion capacity at El Paso's US terminal in Georgia

Pipelines

Iran-Turkey gas line inaugurated

Gas line approved from Timor Sea to Australia

"Definition of the [gas pipeline] environmental issues is a job for northerners, not for southern or international environmentalists who make a comfortable living on the expresso and gore-tex circuit opposing development." — Nellie Cournoyea, Mackenzie Valley Aboriginal Pipeline Corp. Chair

Government

US approves FPSOs in Gulf of Mexico, announces deepwater royalty relief

Companies

PanCanadian to purchase Alberta Energy, forming EnCana

FEBRUARY

Exploration/Development

Gas field discovered in Ninilchik prospect on Alaska's Kenai peninsula

Oil found in Carnarvon basin off Australia

Ebano prospect holds oil off Equatorial Guinea near Ceiba field

Perenco discovers oil off Gabon near Gombe field, tested at 6,300 b/d

Venezuela launches exploratory drilling in Orinoco Delta

Production

Syria's Desgas project reaches full capacity of 450 MMcfd of associated gas

Malaysia's Larut oil/gas field onstream, to peak at 30,000 b/d and 35 MMcfd

Oil production resumes from Kuwait facilities damaged by explosion

Ivory Coast Espoir field starts oil production

"I disagree with those who claim that the Enron collapse sounds the death knell for competition in energy markets or justifies nationwide reimposition of traditional cost-based regulation of electricity." — FERC Chairman Pat Wood

Pipelines

Blue Stream gas lines laid in Black Sea waters over 7,000 ft deep

PetroChina to build 2,600-mile gas pipeline from Tarim basin to Shanghai

Subsea 2,000-mile gas line planned from Papua New Guinea to Australia

Three pipeline systems will serve deepwater Gulf of Mexico oil discoveries

Refining/Petrochemicals

Borouge JV starts shipments from UAE's Ruwais petrochemical complex

Government

Pakistan to privatize its biggest oil distribution company

US contracts royalty-in-kind oil for SPR

MARCH

Markets

OPEC extends quota restrictions, non-OPEC exporters to maintain cuts

Exploration/Development

Appraisal well flows 4,360 b/d of oil off Trinidad in Angostura field

West Patricia oil find off Malaysia confirmed to be commercial

Production

Jade field in UK North Sea starts up, to peak at 200 MMcfd and 16,000 b/d

Iran's South Pars gas development Phases 2-3 onstream, to reach 2 bcfd

Venezuela's Zuata Sweet syncrude reaches first production, to peak at 180,000 b/d

Natural Gas

GdF orders world's first diesel electric LNG tanker, to carry 74,000 cu m

Japan to buy 3 mtpy of LNG from Australia's proposed Darwin terminal

Shell, Marathon to build LNG regasification terminals in Baja California, Mexico

GTL demo plant under construction in OK

"The deepwater has become so heated. We believe there are a number of countries in West Africa that have many undeveloped areas both on their shelf and onshore." — Joe Bruso, Sovereign Oil & Gas Co. Pres.

Pipelines

Bulgaria and Greece agree on 697,000 b/d Russian oil export line to bypass Bosporus

Marathon plans 675-km North Sea gas line connecting Brae, Norway's Heimdal

Ladyfern gas sales pipeline onstream in BC

Retail

ExxonMobil seeks to market refined products in Taiwan

Government

Angola and Congo (Brazzaville) agree on maritime border, allowing exploration

California MTBE ban delayed

APRIL

Exploration/Development

Elon oil discovery off Equatorial Guinea extends Rio Muni basin play

Gendalo appraisal well flows 30 MMcfd and 2,200 b/d condensate off Indonesia

Oil discovered in Tahiti prospect in deepwater Gulf of Mexico

Statoil finds oil on Staer structure near Norne in Norwegian North Sea

Ranggas appraisal well flows 8,158 b/d and 6.4 MMcfd off Indonesia

Bintang development off Malaysia to yield 1 tcf, mostly for domestic power

Shell to develop Goldeneye gas condensate project off Scotland

White Rose development off eastern Canada will proceed, 100,000 b/d FPSO planned

"Latin America seems to be sliding backwards after being for so many years a 'poster child' for reform and progress." — Michelle Michot Foss, Inst. for Energy, Law, & Enterprise Exec. Director

Production

KingKong-Yosemite in Gulf of Mexico begins production at 140 MMcfd

Egypt's Ras Kanayes lease starts producing at 2,130 b/d, 17.8 MMcfd

Algeria's Hassi Berkine 4th oil production train starts up at 75,000 b/d

Hannay field onstream at 15,000 b/d in UK North Sea

Natural Gas

Oil export pipeline planned from Kazakhstan's Alibekmola field

Petrochemicals/Retail

China's Sinopec to build 450 gasoline stations with foreign companies in 2002

Giant ammonia-urea fertilizer complex under contract for Oman

Government

Iraq halts oil-for-food exports for 1 month

Venezuela's Chavez is briefly ousted

Companies

Shell to buy Enterprise Oil

MAY

Markets/Trading

Nymex crude futures settle above $29/bbl

Russia exporting record crude volumes

US energy trading probe expands

Exploration/Development

Natural gas discovered off Sicily in Panda prospect, constrained flow of 19 MMcfd

Canadian Mackenzie Delta well finds gas, restricted flow at 30 MMcfd

Condensate, light oil found in Norway's North Sea near Gullfaks, Statfjord

Thick net oil, gas pay discovered in Ranggas appraisal well off Indonesia

China discovers 21-tcf gas field in Ih Ju Meng area of Inner Mongolia

Petrobras sets world record water depth for gas manifold installation in 6,184 ft

US Gulf Atlantis discovery's reserves increased to 575 million boe from 300

World's 1st expandable tubular technology application completed in Texas

Burlington to develop China's Sichuan gas

Natural Gas

Norway's Snohvit LNG project approved

El Paso unveils LNG tanker concept with onboard regasification

Pipelines/Transport

US gas lines approved to serve West Coast power plants

Shuttle tanker designed for deepwater Gulf of Mexico crude

"We have proven we can develop oil and gas throughout the world without harming the environment. When people take an objective look at the industry, they will see that." -- Mark A. Rubin, SPE Exec. Director

Government

Russia, Kazakhstan agree on Caspian Sea ownership rights

Australia, East Timor agree on Bayu-Undan gas revenues split

USGS releases new NPR-A resource estimates, ANWR still attractive

JUNE

Markets/Trading

OPEC leaves quota unchanged but is producing 1-1.5 million b/d excess

Norway, Russia officially end restraint while Mexico, Oman continue cuts

US energy traders admit round-trip deals

"We are competing with Venezuela, Mexico, Algeria, and Saudi Arabia. None of them are signatories to Kyoto. As easy as you turn on oil sands investment, you can turn it off." — Eric Newell, Syncrude Canada Chairman, CEO

Exploration/Development

Indonesian appraisal well adds 1.07 tcf to Senoro-Toili Block gas reserves

Deepwater gas well's open flow potential estimated at 90 MMcfd off Egypt

Abang oil discovery off Equatorial Guinea to be developed with 5 other fields

Three gas fields off Viet Nam declared commercial, to be developed by Unocal

Sanha gas-condensate fields off Angola to be developed with 5 platforms

Norway offers 11 companies licenses on 32 blocks in 17th round

Kazakhstan's oil supergiant Kashagan declared commercial, to be developed

Production

Gibson, Plato fields begin producing 22,265 b/d off Australia in Carnarvon basin

Boomvang starts up in US Gulf, to peak at 160 MMcfd and 32,000 b/d

Lost Ark in Gulf of Mexico flowing 28 MMcfd, to reach 40

Natural Gas

OxyPet gets Enron's stake in UAE's Dolphin gas project

Qatar LNG plant to get 2 more trains, largest ever to be built

Shell to buy up to 3.7 mtpy of LNG from NW Shelf in 2004-09

Government

Russia and Ukraine resolve gas transport dispute, easing Russian exports

JULY

Markets/Trading

Platt's includes two more North Sea crudes in its Brent valuation

Exploration/Development

UK North Sea Buzzard oil discovery to begin development of 400 million bbl

Etame's 1st well flows 7,630 b/d off Gabon

Woolybutt oil field off NW Australia to get FPSO

Drilling begins on Camisea gas development in Peru

Redhawk gas field to be developed in 5,300 ft water, Kerr-McGee's deepest

Denmark approves development of Nini, Cecilie fields' 65 million bbl

Apache finds oil, gas onshore and off Egypt

Gulf of Mexico West Cameron discovery flows at 22 MMcfd

Canyon Station 500 MMcfd production platform commissioned in US Gulf

"That the Dolphin [gas] project came about was not just a dream based on politics. There was also a real need for it." — Fereidun Fesharaki, FACTS Pres.

Production

Thailand's N. Pailin producing 230 MMcfd of gas/condensate, could yield 380

Kazakhstan's Karachaganak will increase condensate production to 7 mtpy

Qinghuangdao adds 20,000 b/d with 2 more platforms in China's Bohai Gulf

Natural Gas

Egypt's West Delta Deep gas development, LNG project approved

GdF approves 2nd Fos LNG terminal

Norway's Snohvit LNG project orders barge, storage/loading facilities

UK to buy Qatar LNG for 25 years

Pipelines

JV formed to develop western China's gas, build 2,500-mile West-East pipeline

Companies

Dynegy buys Enron's 16,600-mile Northern Natural gas pipeline system

AUGUST

Markets/Trading

Nymex crude futures close above $30/bbl, highest since Feb. 2001

Exploration/Development

Large oil field discovered off Brazil could hold 600 million boe

Caister Murdoch development well flows 148 MMscfd in UK North Sea

FPSO ordered for Bijupira-Salema field off Brazil, to produce 70,000 b/d

Pemex to develop Ku-Maloob-Zaap production of 800,000 b/d by 2011

Arthit structure off Thailand is commercial, holding 1.5 tcf

Libya approves Repsol development of Murzuq basin's A field

Production facilities ordered for Russian Sakhalin I project

Bongkot gas field off Thailand to get 13th platform

Argentina's Neuquen basin well flows 516 b/d of condensate, 8.5 MMcfd of gas

Natural Gas

China to buy 3.3 mtpy of NW Shelf LNG, plus stake in Australian gas project

Indonesia begins 100 MMcfd of gas deliveries from S.Natuna to Malaysia

Pipelines

Tidelands' dual gas, propane-butane US export lines to Mexico approved

"What has become known as the 'Trinidad' model for development of a diversified gas market is being emulated from western Australia to Mozambique." -- Gregory McGuire, NGC Manager of Strategic Planning

Refining/Petrochemicals

Trinidad approves 224,000 b/d refinery, one of world's largest

Thailand plans 250,000 tpy ethylene plant at Rayong complex

Rayong Olefins secures LPG, NGL feedstock made from indigenous Thai gas

Companies

US approves merger of Conoco, Phillips

SEPTEMBER

Markets

OPEC leaves quotas unchanged

Iraq exports jump to 1.9 million b/d

Exploration/Development

BP finds 1 tcf of gas in Iron Horse field off Trinidad, near Amherstia, Cassia

Ultradeepwater (6,628 ft) Plutao discovery off Angola in Block 31 flows 5,357 b/d

US Gulf deepwater Redhawk gas field to get world's 1st cell spar production platform

Aparo appraisal suggests joint development with Bonga field off Nigeria

Gulf of Mexico K2 field's reserves increased to 100 million bbl

Yoho development off Nigeria begins, field holds 400 million bbl

Azerbaijan's ACG Phase 2 construction sanctioned, to develop 1.6 billion bbl

Ivanhoe to develop China's Sichuan gas

"The solution [to worker shortages] will be a greater international workforce. The technical staff will become more nationally diverse and global, as companies grow and today's workforce ages." -- Janeen Judah, ChevronTexaco Technical Support Manager-Latin America

Natural Gas

Hibiscus field off Trinidad produces 1st gas into new Atlantic LNG Train 2

Iran's 280-tcf South Pars gas project contracts Phases 9-10

Indonesia's Tangguh LNG project to supply China with 2.6 mtpy

Pipelines

Australia's 455-mile pipeline carries gas to Tasmania's 120-MW power plant

Work begins on 1,110-mile BTC oil line from Caspian Sea to Turkey

Triton 275 MMcfd, 41-mile line planned for western Gulf of Mexico

World Bank funds Chad-Cameroon 665-mile oil export pipeline, terminal

Refining/Marketing

Cogen plant planned for St.John, NB, refinery

OCTOBER

Markets/Trading

Terrorists attack oil tanker off Yemen

US traders admit falsifying price data

US oil stocks fall to 2-decade low

Exploration/Development

Discovery well flows 2,500 b/d, 6.6 MMcfd off Viet Nam in Cuu Long basin

Brazil's Jubarte, Roncador, Barracuda, and Caratinga fields to get FPSOs

Calder field platform installed off UK, to develop 294 bcf at peak 145 MMscfd

Kalamkas well flows 2,300 b/d in Kazakh Caspian sector near giant Kashagan

Sanha gas-condensate field off Angola to get world's 1st newbuild LPG FPSO

Production

Tullich, Maclure fields begin producing 19,500 b/d in UK North Sea

Thailand's 5 new production licenses to add 100 MMcfd of gas, 15,000 b/d of oil

Canada's MacKay River SAGD oil sands facility operating, to reach 30,000 b/d

Natural Gas

Statoil to operate Iran's 280-tcf South Pars gas project Phases 6-8

Tanker ordered for Norwegian LNG

Pipelines/Transport

US Gulf Redhawk field to get 86-mile, 330 MMcfd gathering line

Georgian Black Sea Poti oil terminal complete, to handle initial 40,000 b/d

"The business world's main motive for social responsibility is long-term, educated self-interest, not altruism. Companies act responsibly in large measure because they can do well by doing good." — Olav Fjell, Statoil Pres., CEO

Refining/Petrochemicals

ExxonMobil to build 160,000 b/d refinery, 800,000 tpy ethylene plant in China

Government

Brazil elects leftist Pres. Lula da Silva

Court says Cameroon owns Bakassi Peninsula, Nigeria refuses to leave

NOVEMBER

Exploration/Development

Kikeh discovery off Malaysia totals 400-700 million bbl, one of SE Asia's largest

Deepwater gas discovery off Egypt, 3rd reported by Apache

Jubarte starts early oil production off Brazil

Drilling stopped at Kazakh Tengiz oil field

Gas recycle facilities installed in Timor Sea's Bayu-Undan field

UK approves North Sea Blake Flank development, to produce 20 million bbl

"In the wider context of corporate governance, transparency is the key to restoring trust — showing people that we have nothing to hide, and therefore they have nothing to fear." — John Browne, BP Chief Exec.

Production

Camden Hills well producing gas in record 7,209 ft of water in Gulf of Mexico

Rang Dong field off Viet Nam producing 65,000 b/d with 1st oil from new platforms

Natural Gas

BG to build Italy's 2nd LNG terminal, at Brindisi, to take 3 mtpy from Egypt

Qatar's Ras Laffan contracts 4th LNG train, to produce 4.7 mtpy in 2005

Statoil acquires one-third capacity at US LNG regasification terminal

Pipelines/Transport

Spain deals with fuel oil spill from sunken tanker off Galicia

Russia to build Murmansk Arctic oil port to boost tanker exports

Interests in North Sea's 675-km, 900 MMcfd Symphony line total 1.65 bcfd

Work begins on 375 MMcfd line from Texas to Mexican power complex

Refining/Petrochemicals

Iran to build 300,000 tpy polyethylene unit

France's Lavera complex to become star site for integrated refining/petrochemicals

China to build world-scale petrochemicals complex producing 2.3 mtpy

DECEMBER

Markets/Trading

Nymex crude futures rise to $32.72, highest since Nov. 2000

OPEC increases quotas by 1.3 million b/d but urges stricter compliance

Exploration/Development

Reserves of 5 tcf discovered off India in Krishna Godavari basin

Iraq cancels Russian West Qurna development contract

Shenzi, Vortex wells discover oil, gas in Atwater Foldbelt area of US Gulf

Apache reports 4th oil discovery off Egypt

Production

Alaska's Kuparuk River field producing 16,000 b/d from 3 Palm wells

Murdoch K field flows 204 MMscfd in UK North Sea's CMS area holding 18 tcf

Natural Gas

China to build Guangdong LNG terminal, regasification plant, and pipeline

NW Shelf building 4th LNG train with 4.2 mtpy capacity, 5th train being designed

ChevronTexaco applies to build 800 MMcfd offshore LNG terminal in US Gulf

ExxonMobil to site UK LNG terminal

"I do not believe that Lula's administration [in Brazil] would disregard the importance of private capital (national or foreign) to develop our energy sources." — Jean-Paul Terra Prates, Grupo Expetro Exec. Director

Pipelines

Argentina's Southern Cross line begins gas flow to Uruguay, could take 180 bcfd

Bream 200 MMscfd gas line starts up in Australia's Bass Strait

Pipeline integration allows Russia to boost oil exports from Croatia's Omisalj port

Government

General strike in Venezuela cripples oil production, refining, and exports

Russia sells 5.9% of Lukoil

US streamlines refinery clean air rules

IPE

NORTH AMERICA

United States

UNITED STATES

CAPITAL: WASHINGTON, DC
MONETARY UNIT: DOLLAR
REFINING CAPACITY: 16,623,301 B/D
OIL PRODUCTION: 5.77 MILLION B/D
OIL RESERVES: 22.446 BILLION BBL
GAS RESERVES: 183.46 TCF

The US economy grew by an estimated 2.4% in 2002, well above 2001's level of 0.3%, and growth of 2.8% was anticipated for 2003. A stimulus package was passed in March, and interest rates reached their lowest levels in more than 40 years. Amid mixed economic signals, however, the economy was still vulnerable to a second recessionary dip.

The spectre of war in Iraq continued to loom, along with escalating conflict in the Middle East and the lingering threat of terrorist attacks. These uncertainties aggravated the volatility of oil prices and stock markets. In December a general strike in Venezuela – one of America's biggest oil suppliers – virtually shut down oil exports, boosting US prices to 2-year highs around year-end.

Petroleum markets

As hostilities heated up during 2002 between the US and Iraq, Saudi Arabia repeatedly insisted it would unilaterally make up any shortfall in global oil supply in the event of a US-led attack. In addition to Saudi Arabia's output of more than 8 million b/d, the country could tap surplus capacity of 2-2.5 million b/d within 30-90 days, said the US Energy Information Administration (EIA).

However, the simultaneous loss of Iraq's production and Venezuela's pre-strike output of about 3 million b/d could not be replaced without drawing down world oil inventories, which were already tight. Crude oil stocks in the US fell to their lowest level in more than 20 years, said EIA in October, approaching the point where regional supply disruptions could develop. Global oil stocks also fell to uncomfortably low levels, reported the International Energy Agency in September, although product inventories remained within a comfortable range.

Oil prices recover

US oil prices rose gradually throughout most of 2002 as the economy improved, relations with Iraq worsened, and the Organization of Petroleum Exporting Countries (OPEC) restrained production. In May, for the first time since the September 2001 terrorist attacks, prices settled above $29/bbl (Nymex near-month crude futures) and in August climbed above $30, capping a rise of nine straight sessions. During October, the $2-$4/bbl "war premium" was trimmed, and prices closed below $30/bbl.

But late in the year, when the strike strangled Venezuela's oil exports during December, prices rose sharply and hit a 2-year high of more than $33/bbl. At an emergency meeting in January 2003, OPEC boosted its production quota by 6.5%, or 1.5 million b/d, bringing total official output to 24.5 million b/d. During November and December, Saudi Arabia had increased its exports to the US, said EIA; the arrival of this oil cushioned the loss of short-haul Venezuelan supplies and prevented further shortfalls and price increases.

OECD COMMERCIAL OIL STOCKS

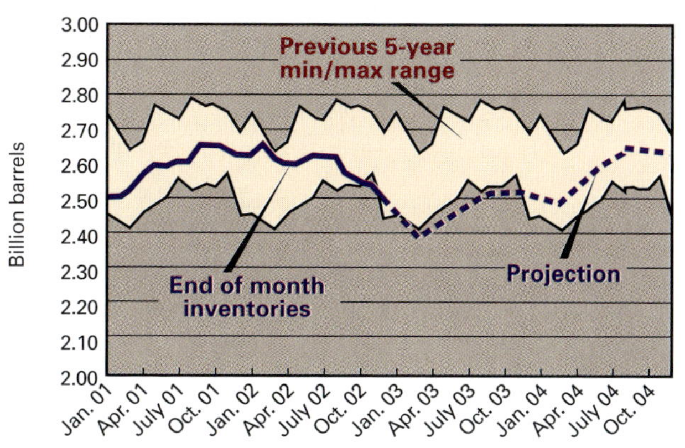

Source: History EIA; Projections Short Term Energy Outlook, January 2003

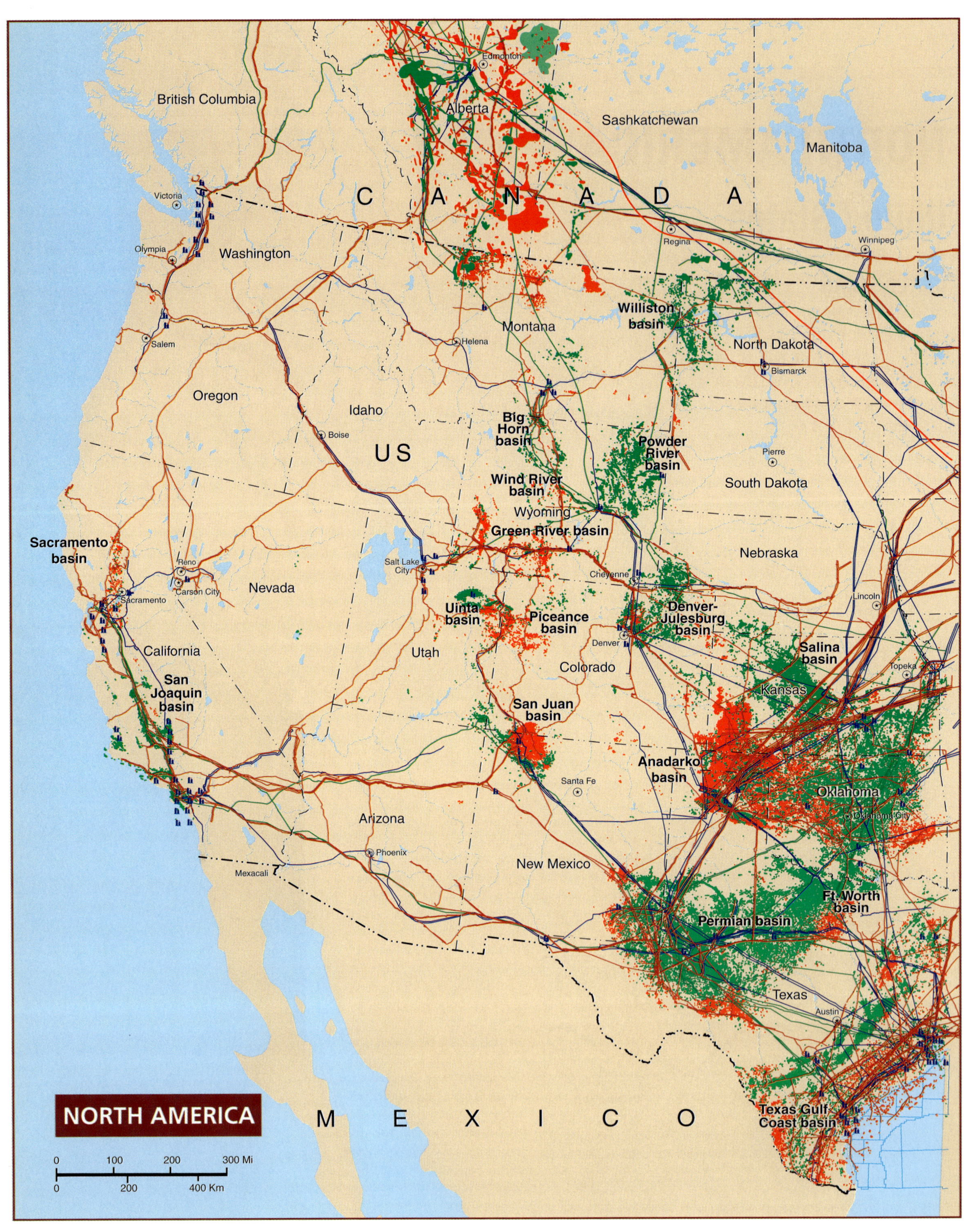

British Columbia
Alberta
Edmonton
Sashkatchewan
Manitoba
C A N A D A
Victoria
Olympia
Washington
Regina
Winnipeg
Williston basin
Montana
Helena
North Dakota
Bismarck
Salem
Oregon
Idaho
Boise
US
Big Horn basin
Powder River basin
Pierre
South Dakota
Wind River basin
Wyoming
Green River basin
Sacramento basin
Nebraska
Reno
Carson City
Nevada
Salt Lake City
Cheyenne
Sacramento
Uinta basin
Piceance basin
Denver-Julesburg basin
Lincoln
Denver
California
Utah
Colorado
Salina basin
Topeka
San Joaquin basin
Kansas
San Juan basin
Anadarko basin
Santa Fe
Oklahoma
Oklahoma City
Arizona
Phoenix
Mexacali
New Mexico
Ft. Worth basin
Permian basin
Texas
Austin
NORTH AMERICA
M E X I C O
Texas Gulf Coast basin
0 100 200 300 Mi
0 200 400 Km

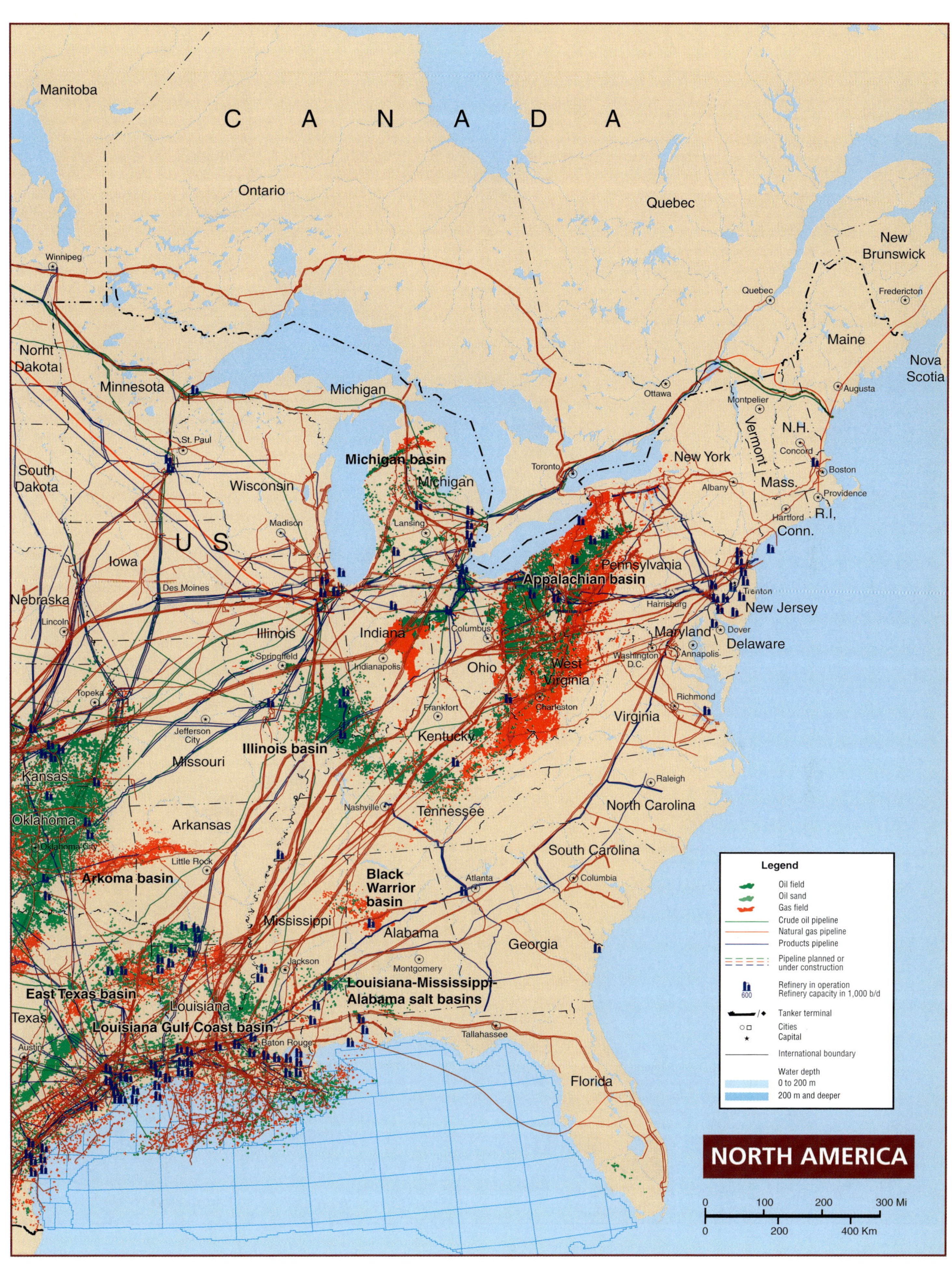

Manitoba
CANADA
Ontario
Quebec
New Brunswick
Nova Scotia
Maine
Winnipeg
Quebec
Fredericton
Norht Dakota
Minnesota
Michigan
Ottawa
Augusta
Montpelier
Vermont
N.H.
Concord
St. Paul
Michigan basin
Michigan
Toronto
New York
Boston
South Dakota
Wisconsin
Albany
Mass.
Providence
R.I.
Hartford
Conn.
Madison
Lansing
US
Iowa
Pennsylvania
Appalachian basin
Des Moines
Trenton
Nebraska
Harrisburg
New Jersey
Lincoln
Illinois
Indiana
Columbus
Maryland
Dover
Delaware
Springfield
Indianapolis
Ohio
Washington D.C.
Annapolis
West Virginia
Topeka
Charleston
Richmond
Frankfort
Virginia
Jefferson City
Illinois basin
Kentucky
Missouri
Kansas
Raleigh
Nashville
Tennessee
North Carolina
Oklahoma
Oklahoma City
Arkansas
South Carolina
Little Rock
Columbia
Arkoma basin
Black Warrior basin
Atlanta
Mississippi
Alabama
Georgia
Jackson
Montgomery
Louisiana-Mississippi-Alabama salt basins
East Texas basin
Louisiana
Texas
Louisiana Gulf Coast basin
Tallahassee
Austin
Baton Rouge
Florida
Legend
Oil field
Oil sand
Gas field
Crude oil pipeline
Natural gas pipeline
Products pipeline
Pipeline planned or under construction
Refinery in operation
Refinery capacity in 1,000 b/d
600
Tanker terminal
Cities
Capital
International boundary
Water depth
0 to 200 m
200 m and deeper
NORTH AMERICA
0
100
200
300 Mi
0
200
400 Km

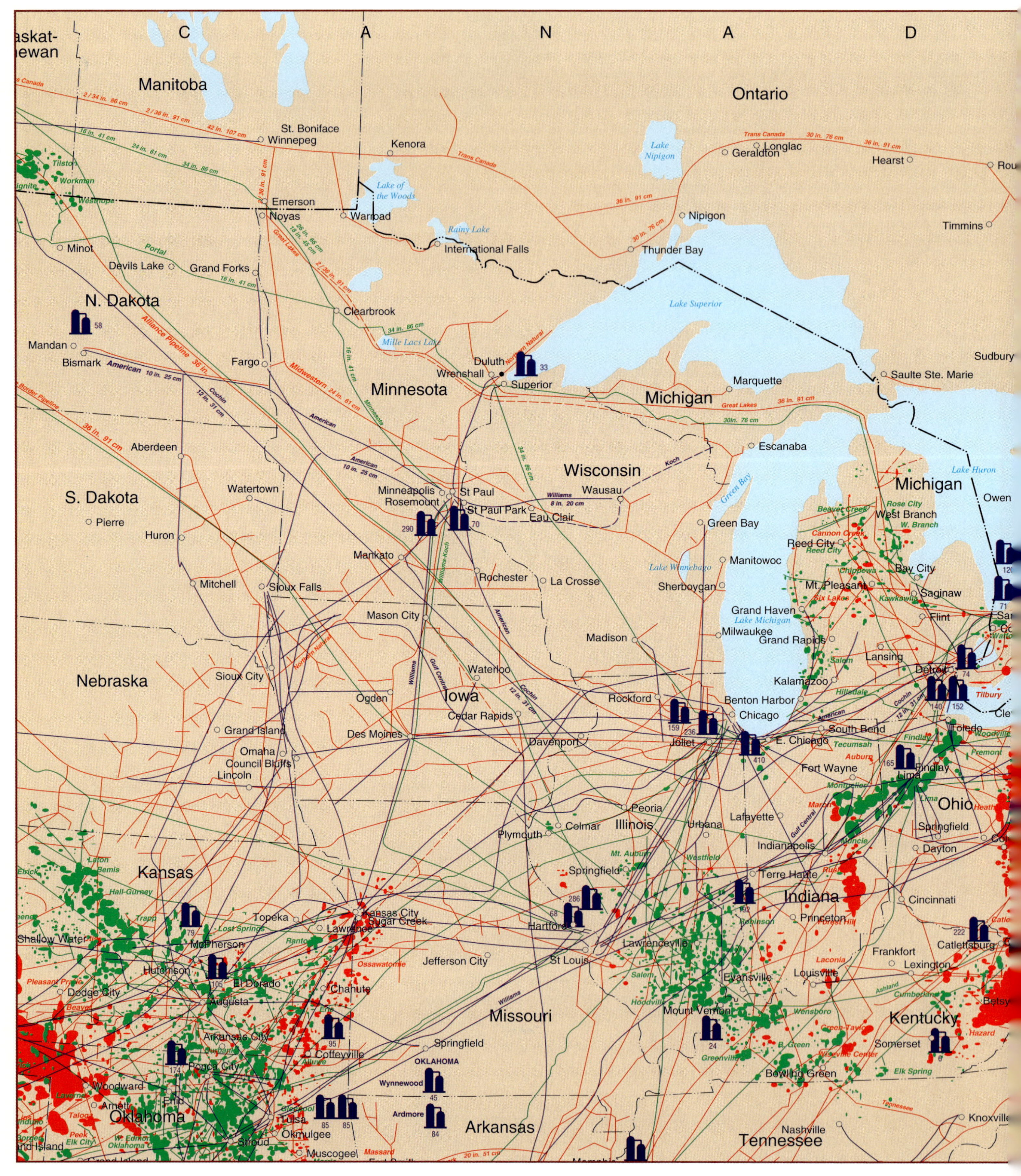

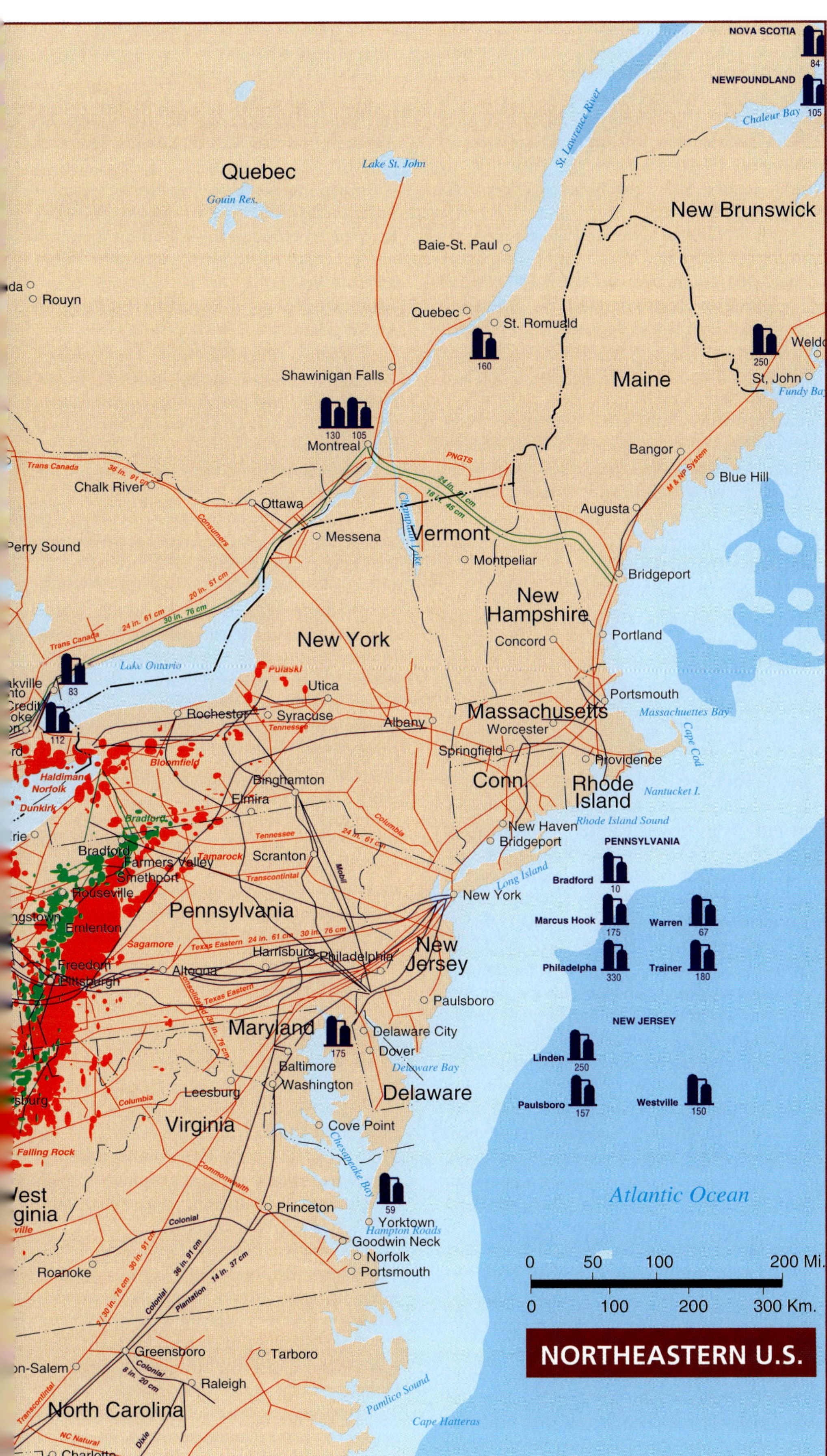

Short-term outlook

Although risks remain high for price volatility and an oil price spike in 2003, the market overall will be relatively balanced, EIA predicted, given certain fragile assumptions. Gradual restoration of Venezuela's full capacity, supplemental production from other OPEC countries, and uninterrupted exports from Iraq could lead to slowly declining oil prices in 2003 and into 2004.

Refining margins in the US, which slumped to a low point in mid-2002, should strengthen during 2003-04 as motor gasoline demand rises and the cost of producing gasoline increases, due in part to substitution of ethanol for methyl tertiary butyl ether (MTBE), said EIA. Higher refining margins in 2003 were also predicted by Prudential Securities Inc., due to reduced product imports from Venezuela, tighter product inventories, and improving demand.

US gasoline pump prices are projected to rise to an average $1.50/gal in 2003, then decline slightly in 2004, EIA said. Heating fuel prices surged early in the 2002-03 winter, and seasonal increases in household fuel costs vs. the previous winter were expected for natural gas (34%), heating oil (43%), and propane (20%).

Although natural gas spot market prices exceeded $5/MMBtu in late 2002, gas wellhead prices averaged only $2.90/MMBtu throughout the year, compared to $4 in 2001. EIA projects that this price will increase by about $1/MMBtu in 2003, reaching an average gas wellhead price of $3.90/MMBtu, and will climb to $4.12/MMBtu in 2004.

Underground working-gas storage volumes, which dropped rapidly in late 2002, ended the year 19-20% lower than the previous year and 5% below the last 5-year average. EIA anticipates continued low volumes of stored gas throughout 2003. Canadian gas storage levels also were low.

US oil demand during 2002 showed only a slight year-to-year gain over 2001, about 0.3%, but this small net increase masked individual fuel use patterns and a late-year surge. EIA expects strong demand growth of 3%/yr in all major fuel categories in 2003-04. World oil demand will also recover, increasing by 1.3 million b/d in 2003, with almost half of this growth coming from the US. EIA projects global demand to expand another 1.4 million b/d in 2004. America's crude oil production in 2003 is expected to fall by about 3% to 5.62 million b/d and by 2% in 2004, said EIA, following fairly stable production levels during 1999-2002.

Solid growth is expected for US natural gas demand, with increases of 4.7% in 2003 and 2.7% in 2004. Gas demand declined in 2002 by 1.7%, mainly due to industrial weakness, but winter 2002-03 consumption was estimated to be 8.7% higher than the past winter. Dry natural gas production appeared to be flat in 2002 but might have been overestimated, said EIA, which anticipates production of 522.3 billion cubic meters, or bcm (19.45 tcf) in 2003.

Simmons & Co. International (Houston) said in December that gas market fundamentals are improving. Gas production will likely decline 1.5% in 2003, which, coupled with power generation demand, should tighten the gas market and increase prices to the point where industrial demand growth is discouraged.

Net gas imports will increase in 2003-04 after falling in 2002, said EIA. However, Canadian analysts predicted that Canada's rising domestic gas demand and declining production will reduce gas exports to the US by 2% in 2003. Salomon Smith Barney Inc. (New York) estimated that US imports of Canadian gas decreased 3% in 2002 and could drop another 4% in 2003.

US reserves

Year-end 2002 data from *Oil & Gas Journal* list US estimated proved reserves of 22.446 billion bbl of oil and 4.9267 trillion cubic meters (183.46 tcf) of natural gas, as of January 1, 2003.

In 2001, US crude oil reserves increased by almost 2%, reported the EIA in October. Discoveries totaled 2.565 billion bbl, and reserve additions exceeded production by 21%. New field discoveries reached their highest level since Alaska's Prudhoe Bay reserves were added in the 1970s. Most of the reserves increase came from frontier areas of the deepwater Gulf of Mexico and Alaska, including Thunder Horse field in more than 1,800 m of water (6,000 ft), which is expected to be the gulf's largest field.

Proved reserves of dry natural gas jumped by 3.4% in 2001 to 22.758 tcf, said EIA, and gas reserves exceeded output by 31%. Most of the gas reserve additions came from Pinedale field in Wyoming, the Lobo trend and Barnett shale areas in Texas, and Wattenberg field in Colorado. EIA said that improved technology helped producers realize dramatic production gains in these fields. Coalbed methane accounted for 9.6% of proved dry gas reserves. However, proven reserves of natural gas liquids declined by 4.5%.

Longer term outlook

A major influence on US energy markets through 2025 will be the availability of adequate natural gas supplies at competitive prices, said EIA. To meet rising demand, America will depend on new, large-volume gas supply projects such as deepwater offshore wells, LNG import facilities, Canada's Mackenzie Delta pipeline, and an Alaskan pipeline. LNG imports are expected to increase at 8.6%/yr, reaching 22.3 bcm (830 bcf) in 2020.

The US will also depend increasingly on imported oil, which could supply 65-70% of total petroleum demand by 2025, even if consumption is moderated by efficiency improvements as expected. By comparison, net imports accounted for only 37% of US oil demand in 1980, 42% in 1990, and 55% in 2001.

According to the American Gas Association (AGA), nontraditional sources such as Alaskan gas and LNG will likely account for a significantly larger share of America's future gas supply mix.

Global outlook

World oil demand will grow by only 1.04 million b/d in 2003, according to the International Energy Agency (IEA), but world energy demand will expand briskly over the longer term through 2030, by 1.7%/yr. Abundant resources are available to meet this growth, including coal and renewables, but oil and gas would fuel most of the increase. IEA predicted oil demand growth of 1.6%/yr and a doubling in consumption of natural gas, the fastest-growing fuel at 2.4%/yr, with power generation accounting for 60% of the increase. IEA said that North America will become a net importer for 26% of its gas by 2030.

Looking farther out, toward 2050, the World Energy Council said that natural gas supplies in North America and other key markets will flatten and then decline unless global prices rise significantly and the US opens more federal land to exploration.

An early peak in world oil production was predicted by Douglas-Westwood Ltd. (Canterbury, UK) in mid-2002. The firm claimed that world oil reserves are being drawn down at an unprecedented rate, with supplies likely to become constrained by 2010 even without demand growth. A Princeton University analysis that was published in November suggested an even earlier world production peak, around 2005.

Energy policy

In his January 2002 State of the Union address, President Bush identified Iran and Iraq as part of an axis of evil that supports terrorism. In April he met with Saudi Crown Prince Abdullah to discuss Middle East conflicts, and Saudi Arabia denied that it would use oil exports as a tool to express displeasure with US policy.

President Bush agreed with Russian President Putin in May to encourage US investment in Russia's oil and gas sector as part of a new energy partnership, which supports improvements to Russia's export infrastructure. The first direct shipment of Russian crude oil to the US, about 2 million bbl of Urals blend, arrived in July. After another meeting in November, Russia announced plans to consider building an Arctic port to export oil to the US.

The government decided in May to buy back oil and gas drilling leases on the Florida coast and in the Everglades because of environmental concerns. A similar buyback plan was approved in July by the US House of Representatives for leases off the California coast.

The long-awaited pipeline safety bill was passed by the US Congress in November, the first major rewrite in more than 20 years. US industry praised the Pipeline Safety Improvement Act of 2002, which gives federal and state agencies broad new powers to regulate and enforce safety standards. Its provisions require inspection of urban pipelines within 5 years and other lines within 10 years. The law also expands protection for whistleblowers and strengthens the national One-Call system, which deters third-party damage.

In late 2002 Congress abandoned a last-ditch effort to pass energy legislation before 2003, crushing hopes for financial incentives to build an Alaskan natural gas pipeline from the North Slope to the lower 48 states. A new bill could eventually contain oil and gas production incentives, as well as an ethanol mandate.

National security

In late 2002 the Bush Administration and Congress created the Department of Homeland Security, the most extensive reorganization of the federal government since the 1940s. The new department will pull together 170,000 federal employees from 22 agencies, including an unspecified number from the Department of Energy.

The law was endorsed by the American Petroleum Institute (API). API and other industry trade groups successfully averted incorporation of tough new security rules covering refineries and chemical plants that would have been enforced by the US Environmental Protection Agency (EPA). This plan, when proposed in mid-2002, was blasted by several associations and dropped from the homeland security legislation.

Environmental issues

New Source Review

Air pollution control guidelines for power plants and refineries were updated by EPA in November. The National Petrochemical & Refiners Association (NPRA) and other industry trade groups praised the changes to the New Source Review provision of the Clean Air Act, which should encourage the kind of large investments US refiners will need to comply with tighter clean fuel standards.

The proposed rules would draw a brighter line between routine maintenance, including repairs and replacement, and major modifications such as plant expansion, which trigger permit review. Refiners and power generators had endorsed reform, saying that the existing program was seriously flawed and has actually discouraged projects that would reduce emissions and improve energy efficiency.

Diesel sulfur

Low-sulfur diesel regulations continued to attract attention from refiners seeking to avoid possible retail supply disruptions when the program begins in mid-2006. API in December reiterated concerns that potential diesel supply problems are not being addressed. In November EPA presented a report from its clean diesel review panel that updated information on technical advances being made to comply with the rule. The panel's work was endorsed by API and NPRA.

In August EPA regulations were approved to penalize manufacturers of diesel engines that exceed pollutant limits taking effect in October. The rules are part of a plan to reduce diesel truck and bus emissions by 90% by 2007. In September EPA extended air quality regulations to large engines and off-road vehicles, which are growing sources of pollution.

Oil pipelines could have difficulty adjusting operations to the 15-ppm rule for ultra-low sulfur diesel, based on tests conducted on two US product pipelines that were reported in late 2002.

Coalbed methane

Three US government agencies endorsed in early 2003 a dramatic expansion of coalbed methane production on public lands in Montana and Wyoming. Their report advocates an environmental impact statement for the two states that would allow more drilling under new industry-supported safeguards that comply with clean air and water rules.

The report, which is subject to public comment, was released by the Department of the Interior's Bureau of Land Management, the Department of Agriculture's Forest Service, and EPA. Previously, a draft EPA report said that hydraulic fracturing of coalbed methane wells does not contaminate underground drinking water sources.

Producers want to drill 39,000 new gas wells in northeastern Wyoming, with about 24,000 of these on federal land. About 670 bcm of natural gas (25 tcf) might be recoverable from coalbed seams in the state's Powder River basin, said the Bureau of Land Management. The coalbed methane industry is also interested in southeastern Montana, which currently has only operating 250 wells.

Greenhouse gases

In February the administration announced a plan to link reductions in carbon emissions to US economic output, focusing on emissions intensity per unit of gross domestic product (GDP) rather than on fixed targets. Its goal is to reduce the American economy's greenhouse-gas intensity by 18% by 2012. Emissions of carbon dioxide, the primary greenhouse gas, are targeted for reduction by the Kyoto Protocol, an international treaty that the US has refused to sign.

Several federal agencies collaborated on a greenhouse gas emissions plan that would encourage industry to reduce carbon dioxide emissions voluntarily. The Departments of Energy, Commerce, and Agriculture and the EPA recommended in July that President Bush create a transferable credit system for emission reductions.

In mid-2002 President Bush unveiled his Clear Skies initiative, which would promote a market-based approach to meeting emission standards for nitrogen oxides and sul-

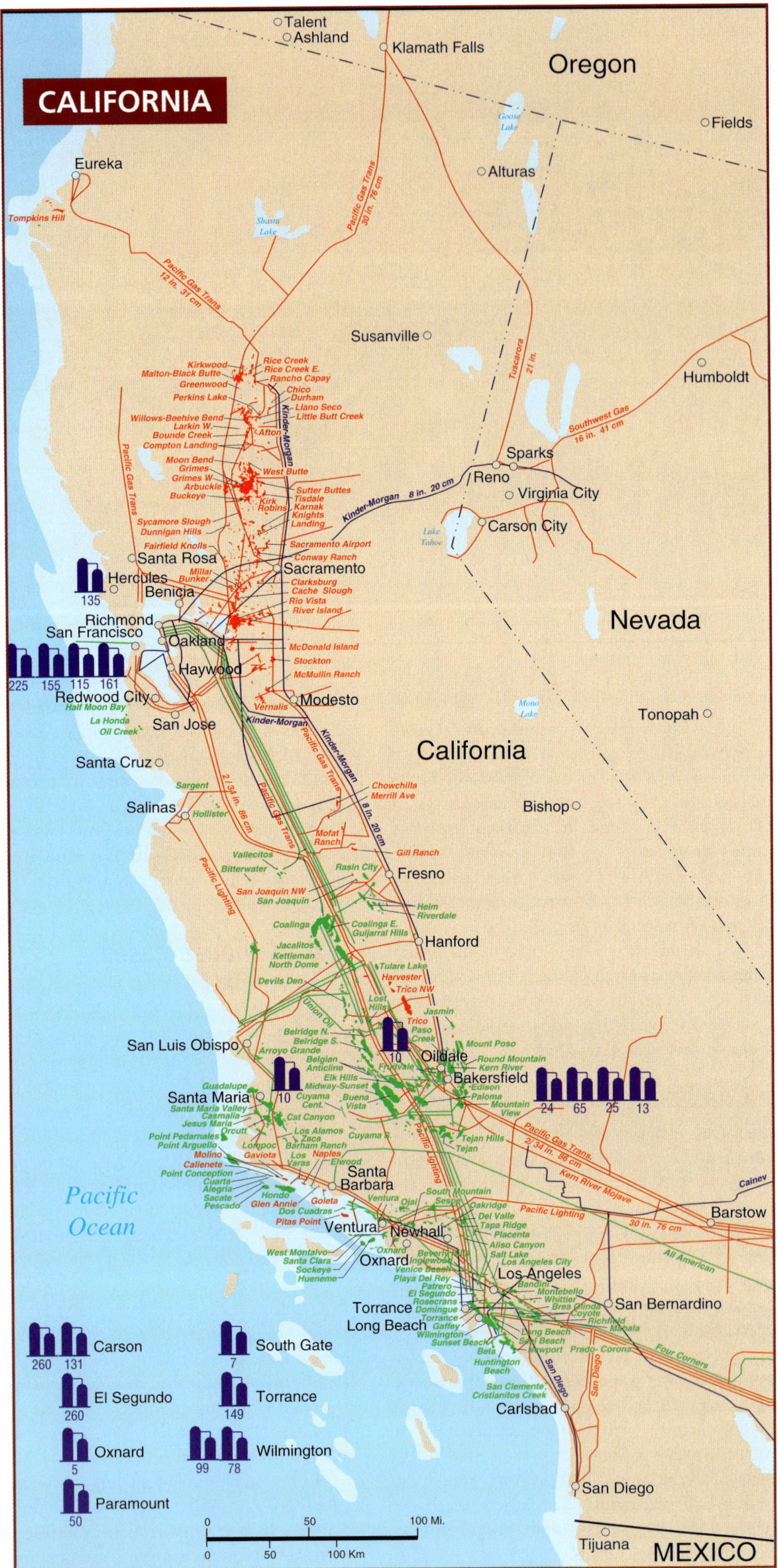

fur as well as mercury, but would not affect emissions of carbon dioxide. The Bush proposal, which promises to simplify and strengthen the Clean Air Act without crippling American industry, would use market incentives to reduce power plant emissions 70% by 2018.

In early 2003 a bipartisan initiative in Congress proposed a comprehensive cap on carbon dioxide emissions. The legislation would permit a nationwide market-based trading program to achieve the reductions. Some American utilities, pressured not only by environmentalists and lawmakers but also by investors, were taking independent steps to mitigate their greenhouse gas emissions.

Global prices for carbon dioxide emission credits were estimated by Natsource LLC (New York) and Global Change Strategies International Inc. (Ottawa). Their analysts said in September that a tonne of carbon dioxide equivalent will fetch a bit more than $5 until 2005, when the Kyoto Protocol could take effect, and reach an average of $11/tonne by 2010.

Oil producers could boost production and reduce greenhouse gas emissions by expanding enhanced recovery methods, such as carbon dioxide injection. The Interstate Oil and Gas Compact Commission was working with the US Department of Energy to study potential uses for carbon dioxide sequestration within the oil industry.

Methane emissions were reduced by 1.1 bcm (42 bcf) in 2001 by industry participants in EPA's Natural Gas STAR program, surpassing their goal and saving an estimated $126 million worth of gas.

Emergency reserves

President Bush refused in early 2003 to tap oil from the Department of Energy's Strategic Petroleum Reserve (SPR) to make up the Venezuelan deficit. The SPR held 592 million bbl of oil as of November 2002, the largest volume of crude ever stored since emergency stockpiling began in 1977. The SPR was on track to reach full capacity of 700 million bbl in 2005.

Oil producers began bidding in early 2002 to supply crude oil volumes to the SPR instead of making cash royalty payments. Under the royalty-in-kind program, the crude is delivered to market centers and exchanged by the government for oil suited for SPR use.

The US Minerals Management Service (MMS) awarded the first such contracts in February to four Gulf of Mexico producers for about 60,000 b/d of royalty oil, and the reserve's fill rate was then accelerated to

100,000 b/d. MMS officials predicted that royalty-in-kind oil volumes dedicated to the SPR will increase to 130,000 b/d in 2003. As part of this program, 285,000 bbl of Russian oil was earmarked for the SPR in October.

The government's 2 million bbl Northeast Home Heating Oil Reserve will be managed by the same three companies as the previous winter under new leases to be signed by the Energy Department. Instead of an option to extend the leases by 1 year, the new contracts will offer 4-year extensions. The companies operate storage terminals in several Northeast US locations.

Energy merchants

After the implosion of Enron Corp. in late 2001, the US Department of Justice launched a criminal probe, and a wave of corporate scandals spread throughout and beyond the energy industry. Integrated energy firms were accused of manipulating markets during California's power crisis, and in May energy marketers and traders were defending "round-trip" deals. These deals involve the simultaneous sale and purchase with the same parties at the same price, a procedure that inflates trading volumes without moving any energy.

In the midst of regulatory investigations into financial reporting practices and an onslaught of litigation, energy merchants shed or scaled back their trading operations and divested key assets. Dynegy Inc. (Houston) unveiled in July a plan to improve liquidity and reduce debt, including selling off its Northern Natural Gas pipeline unit. According to Standard & Poor's Rating Service, US energy merchants will need $90 billion in refinancing by 2006.

Several companies exited the energy trading business, including Dynegy, El Paso Corp. (Houston), Aquila Inc. (Kansas City), and CMS Energy Corp. (Dearborn, MI, US), while others such as Willams Cos. (Tulsa), Reliant Resources (Houston), Mirant (Atlanta), and Calpine Corp. (San Jose, CA, US) reduced their work forces. Many of these firms also fell off the list of the world's top 50 energy companies as ranked by Petroleum Finance Co. (Washington, DC), and the list became dominated by non-US companies.

In September the Federal Energy Regulatory Commission (FERC) said that an El Paso unit withheld natural gas from California during the power crisis, and in November an earlier settlement between California and Williams Cos. Inc. (Tulsa) was publicized, revealing apparent collusion with a power plant operator to extend generator maintenance. Natural gas traders from at least five companies confessed to giving false prices to industry publishers that compile indexes.

To alleviate this situation, a coalition of 31 large energy companies unveiled voluntary guidelines in November that focus on governance, valuation, credit risk management, and disclosure of energy trading activities. The guidelines intend to increase transparency and win back investor confidence.

Now that US natural gas and power markets have become intertwined, the road ahead remains bumpy for both, said Energy Security Analysis Inc. (Boston) in December. For the gas market, the key to moving forward will be to recognize the inextricable link that has been forged with the power industry. The problems of liquidity, volatility, credit, and transparency that are facing regional electricity markets will continue to have a direct impact on gas demand and pricing in local and national gas markets.

Power industry analysts in early 2003 anticipated a poor outlook for the coming year, given falling electricity prices and an overcapacity of generation. However, some brokers expressed cautious optimism that the worst was over for the power market as a whole, even if some companies continue to deteriorate. The longer term market has remained surprisingly resilient, they said. Merrill Lynch reported that stock prices for many battered energy merchants had begun to recover in early 2003.

Pipeline planning

Early in the year as gas prices slid and demand stagnated, industry interest declined in building pipelines, especially to California, and some of the capacity to be completed in 2002, estimated at 2.3 bcm/yr (230 MMcfd), was quietly put on the back burner. Still, many other projects went forward, and the industry remained optimistic about growing demand. As in 2001, pipeline projects emphasized adding compression to existing systems.

Northeast

Maritimes & Northeast Pipeline applied to FERC for an expansion that would double its gas line capacity from offshore Nova Scotia, Canada, to the northeastern US. The expansion to 7.8 bcm/yr (800 MMcfd) would be completed in 2004. Maritimes & Northeast Pipeline is owned by affiliates of Duke Energy Corp. (Charlotte), ExxonMobil Corp. (Irving, TX, US) and Emera Inc. (Halifax).

Service began on two natural gas pipeline expansions that meet northeastern power plant demand, said Williams in November. The second phase of the MarketLink expansion brings about 1.3 bcm/yr (130 MMcfd) into the Transco system, and the Leidy east expansion added the same volume. With a planned third expansion, Williams would increase its Transco rated design capacity to about 74 bcm/yr (7.5 bcfd).

A 56-km (35-mile), 24-inch extension of the Iroquois Pipeline Operating Co. (Shelton, CT, US) Eastchester gas line will connect Long Island to New York City. The extension, to be complete in spring 2003, is part of Iroquois's plans to supply 2.3 bcm/yr (230 MMcfd) of western Canadian gas to the New York metropolitan area. Onshore facilities were also being built.

More than 1 million bbl of winter peak capacity will be added to its northeastern LPG delivery system in 2003, said Teppco Partners LP in November. Plans include construction of three pump stations on the Ohio-Pennsylvania LPG common carrier line and an additional 3.5 million bbl of brine containment in Texas. US LPG demand could increase by 2005 to more than 67 million tonnes/yr (tpy), said Purvin & Gertz (Houston) in June.

Southeast, Mid-Atlantic

The Patriot pipeline extension, 151 km (94 miles) of 24-inch pipe, will introduce natural gas to parts of southwest Virginia and also reach North Carolina. Duke Energy said that southeastern gas markets exceed the capacity of the project. FERC approved the project in November.

Capacity doubling was planned for a 306-km (190-mile) segment of the Plantation pipeline, which transports refined products throughout the southeast, said operator Kinder Morgan Energy Partners LP (Houston). The 8-inch line from Bremen, GA, to Knoxville, TN, will be replaced with 20-inch line to reach 90,000 b/d capacity.

Additional gasoline and other liquid fuels will also be transported to Knoxville by a new 16-inch line, said Colonial Pipeline Co. (Atlanta) in September. The expansion, beginning with construction of a 69-km (43-mile) section in mid-2003, ultimately will increase deliveries to 180,000 b/d of products.

The proposed Seafarer system gas pipeline system will transport 10 bcm/yr (1 bcfd) from a planned LNG terminal at Grand Bahamas to southern Florida. New open seasons for the 261-km (162-mile), 26-inch system were held in mid-year by developer El Paso.

The Gulfstream gas pipeline system connecting southern Alabama and Mississippi to central Florida incorporated several unusual features at its compressor Station 100, said Willbros Engineers Inc. (Panama) in November. The special design was needed to move 11.1 bcm/yr of gas (1.130 bscfd) from a single station without liquid condensation in the cold gulf pipeline segment.

Midwest

A 229-km (142-mile), 36-inch natural gas system was being built by CMS-operated Guardian Pipeline to carry up to 7.4 bcm/yr (750 MMcfd) from Illinois to Wisconsin, with service to begin in November.

A new gas system in northern Illinois, including 28 miles of 36-inch line, entered service in the spring carrying 3.7 bcm/yr (380 MMcfd), said Horizon Pipeline Co. LLC.

The 42-km (26-mile), 30-inch Westleg gas pipeline will parallel an existing lateral and provide up to 2.2 bcm/yr (220 MMcfd) to northern Illinois and southeastern Wisconsin, said El Paso unit ANR Pipeline Co. in September. About a third of the gas could go to a planned 600-MW power plant. The line should be complete in late 2004.

Rocky Mountains

The Jonah natural gas gathering and transportation system in Wyoming will be expanded, increasing the Pinedale field lateral's capacity to 2.5 bcm/yr from 0.54 bcm/yr (to 250 MMcfd from 55 MMcfd), said Teppco in April.

Gas from Cheyenne, CO, would be transported to Midwestern markets by expanding the Advantage gas pipeline if Kinder Morgan's open season is successful.

Western region

A pipeline expansion that will double capacity to western markets, to 8.88 bcm/yr (906 MMcfd), was approved by FERC in July, said Kern River Gas Transmission Co. (Salt Lake City). The 1,150-km (716-mile), $1.2 billion extension was permitted in half the normal time under FERC's new prefiling process. Kern River began construction in September, with service planned for May 2003.

Two gas pipelines were approved by FERC in April to serve western power plants. Kern River Gas will build a lateral near Los Angeles to fuel a 650-MW station near Victorville, and Northwest Pipeline Corp. will build a gas line to serve a 720-MW plant in Grays Harbor County, WA.

Northern California's gas pipeline system capacity was increased by 1.8 bcm/yr (180 MMcfd), said Pacific Gas & Electric Co. (San Francisco), which completed 22.7 km (14.1 miles) of new 42-inch line in September. The project increases firm volumes on the system's Redwood Path, which transports Canadian gas from Oregon.

Mexico connections

Dual pipelines to deliver natural gas and propane-butane from Eagle Pass, TX, to Piedras Negras, Mexico, were being built mid-year by Tidelands Oil & Gas Corp. (US) subsidiary Reef International LLC. Several transportation and sales delivery agreements for both the liquids and the natural gas were in place.

In August Reef began directional drilling of the 12-inch gas pipeline and bilateral 6-inch liquids crossings, 1.5 m (5 ft) apart, under the Rio Grande River. The crossing was expected to be completed by year-end. Work on a terminal in Eagle Pass was also under way.

Tidelands was also planning another international pipeline crossing, a 30-inch natural gas trunkline at El Paso, TX, which will deliver natural gas to Juarez, Mexico. After permitting, Reef anticipated starting construction at year-end 2002.

Construction began in November on the Monterrey pipeline designed to carry 3.68 bcm/yr (375 MMcfd) into Mexico, said Kinder Morgan, which has a 15-year contract for all of the capacity. The 30-inch, 150-km (95-mile) line from Starr County, TX, will connect to a 1,000-MW power plant and should be completed by mid-2003. The pipeline could be expandable or the flow reversed to export gas to the US.

A 3.08 bcm/yr (314 MMcfd) gas line was proposed from south Texas to Hidalgo County by El Paso unit Tennessee Gas Pipeline Co. in April. Construction of the 15-km (9.3-mile), 30-inch lateral would feed power generators in Mexico.

Exploration and production

Financing of exploration projects and property acquisitions by independent E&P companies were difficult in late 2002, due to tight equity markets. Also, rig insurance

FIELDS WITH TRENTON-BLACK RIVER AS PRIMARY PAY

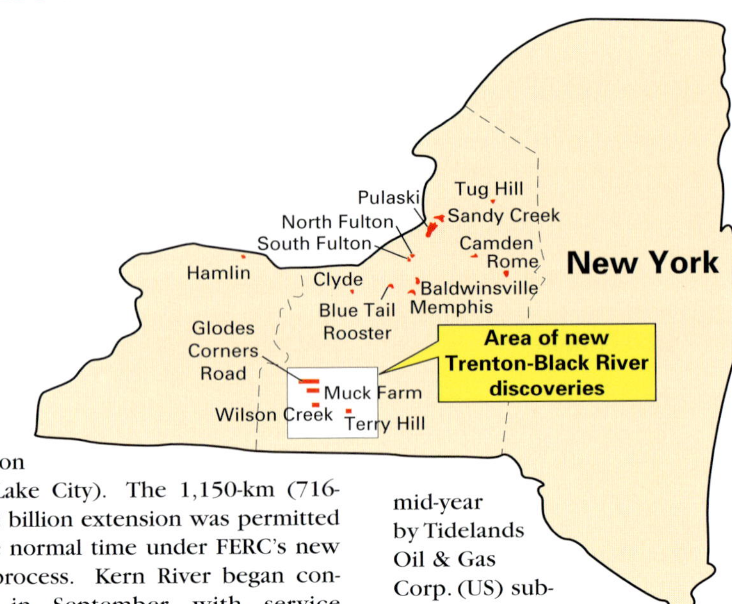

DEVONIAN FIELDS IN NEW YORK

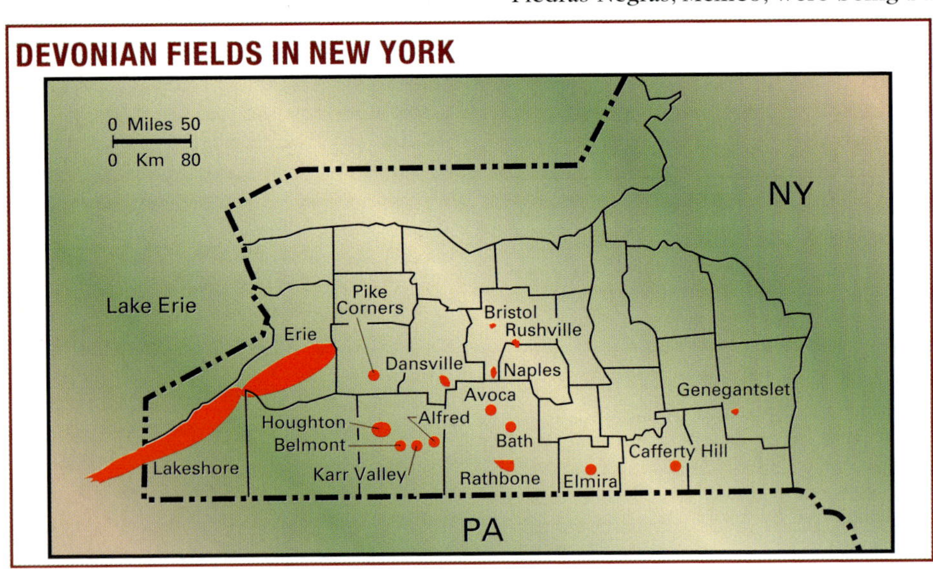

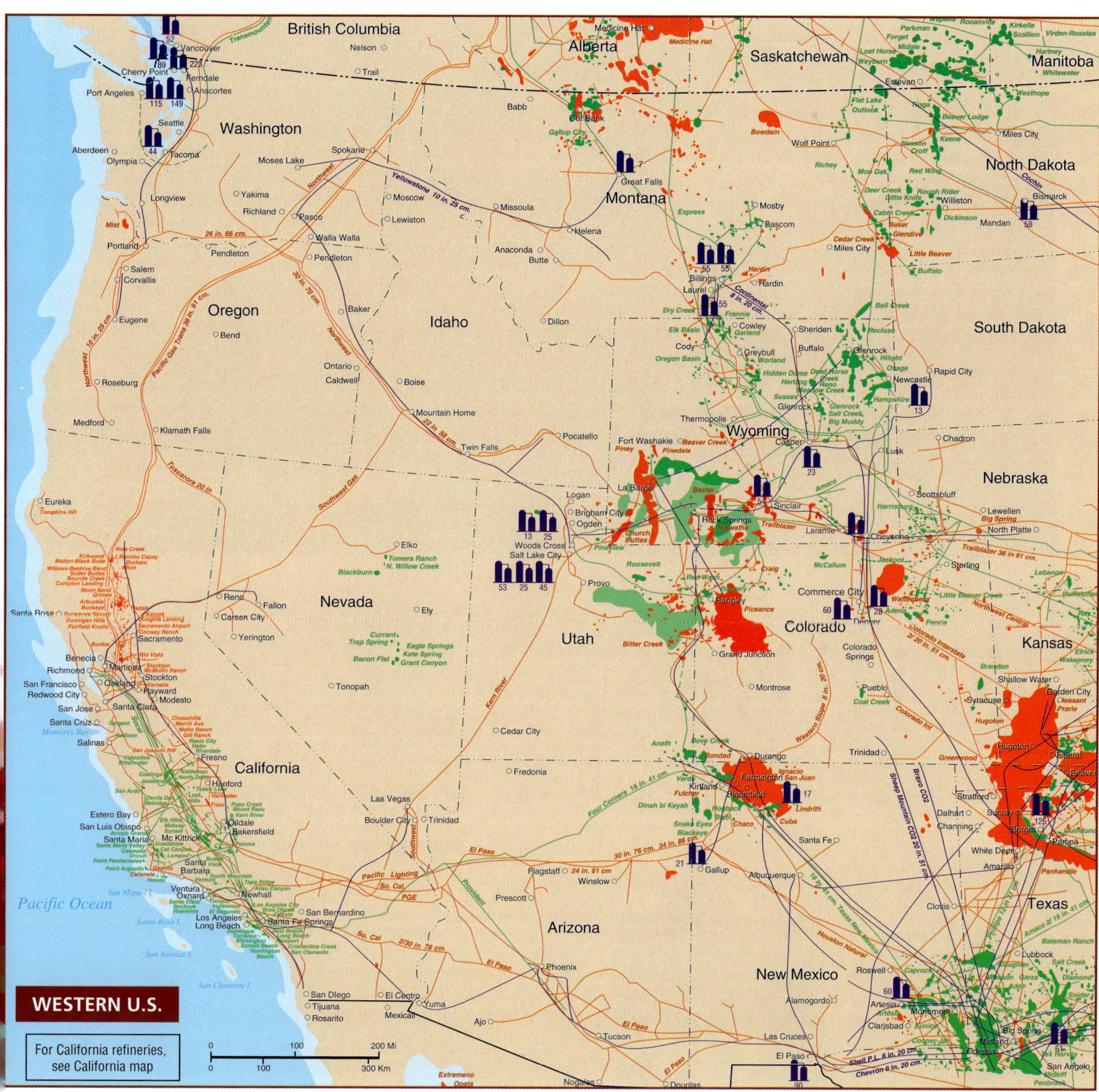

premiums increased dramatically, due to underwriting losses, low interest rates, declining equity markets, and other factors.

Debate continued over industry access to the Rocky Mountains, where current oil and gas supply scenarios are too narrow because they focus on resources on federal lands, said Rand Corp. (Santa Monica, CA) in March. Natural gas prices in the Rocky region fell faster than elsewhere during 2002 due to increased coalbed methane drilling and pipeline bottlenecks.

Trenton-Black River

New York state's Trenton-Black River play and deeper horizons of the Appalachian basin could be much larger than previously indicated. A model of the Ordovician pay zone was developed by Direct Geochemical Services (Golden, CO) and Pyron Consulting (Pottstown, PA), applying their own methodology called integrated exploration technology.

The model of Glodes Corner Road field is applicable to the search for natural gas in the Trenton-Black River reservoir. The methodology successfully identified the best wells in the field and optimal spots for more development.

The same integrated methodology was applied to the New York Devonian black shale, the two companies reported in early 2003. Development of the Upper and Middle Devonian shale/siltstone pay of the Appalachian basin has been inhibited by limited historic data and the complexity of fractured reser-

voirs, despite numerous oil and gas shows.

Modeling of Genegantslet field in Chanango County, NY, involved integration of subsurface mapping techniques, production history, and surface soil geochemistry, as well as analysis of petrophysical, remote sensing, and fracture trace data. The model could be more applicable to gas exploration in the Hamilton (Marcellus) reservoir.

Subsurface mapping, supported by geochemical data, revealed a strong correlation between the best producing wells and thinning of the interval. The presence of fractures alone is not enough reason to produce a shale reservoir in this area.

Other E&P

Enormous shallow gas accumulations in a Montana play could be economically developed with new reverse circulation drilling technology, combined with carbon dioxide fracturing, said K2 Energy Corp. (Calgary) in March. In extended flow tests at four wells, the technology yielded commercial gas flows from the Bow Island formation that were 57-100% higher than rotary air drilled wells.

The results represent a major kick-start to the area's shallow gas play. Stabilized rates exceeding 8 million cu m/d (300 MMcfd) could be achieved with multizone stimulations, allowing capital cost recovery within a year of start-up.

An unexpected oil flow was found in Texas's Glen Rose reservoir, and the play was growing in September, said Saxet Energy Ltd. (Houston) and The Exploration Co. (San Antonio). Since the discovery well began producing in March, 268,000 bbl of sweet, light oil had been produced, along with fresh water. Six wells were yielding a combined 3,670 b/d in mid-year. Comanche-Halsell field in the Maverick-McKnight basin area will undergo another 800 sq km (300 sq mi) of 3D data to be processed by year-end.

Natural gas was discovered in eastern Texas's Bossier play area and in northern Louisiana's Ansley prospect, reported Anadarko

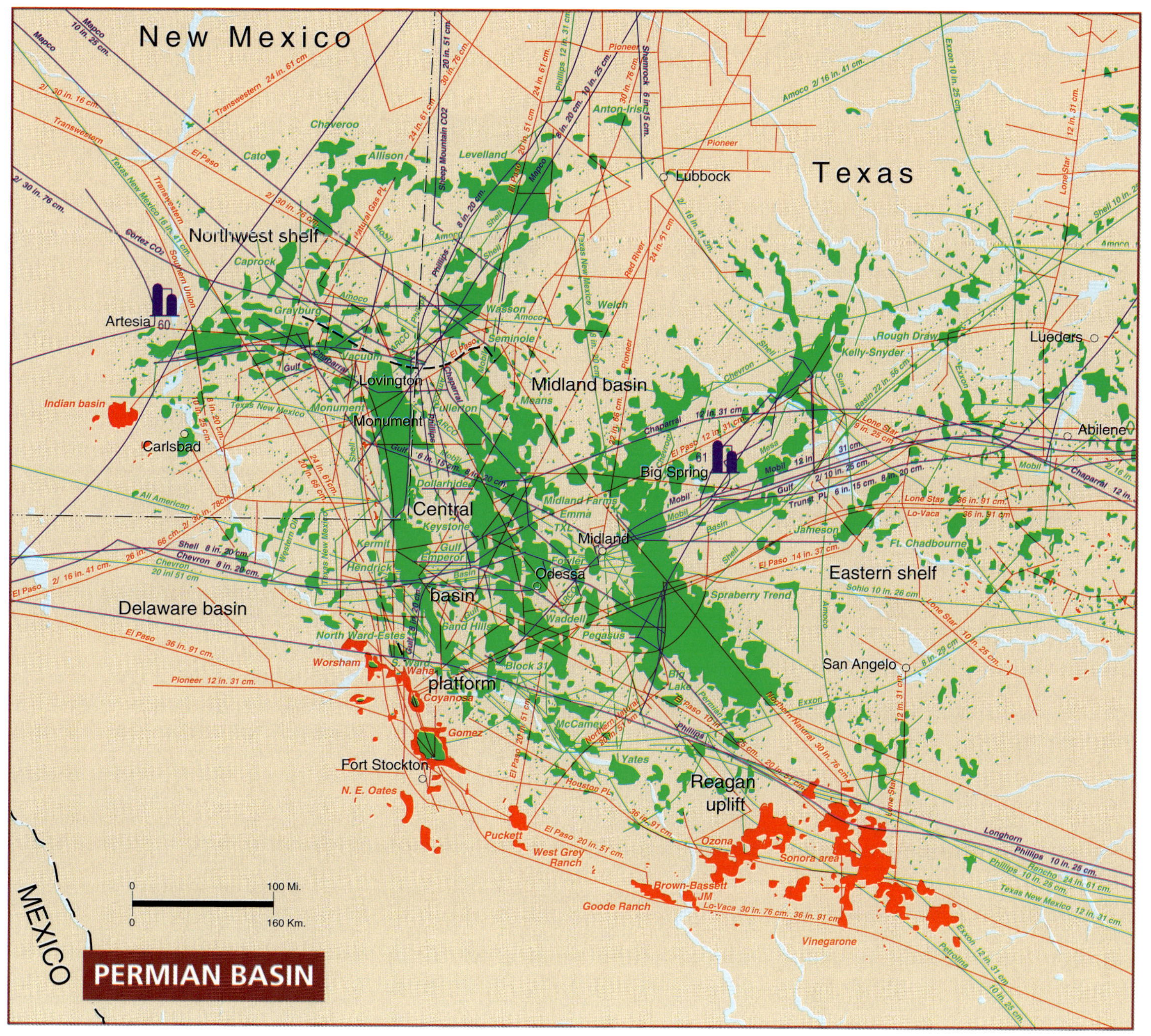

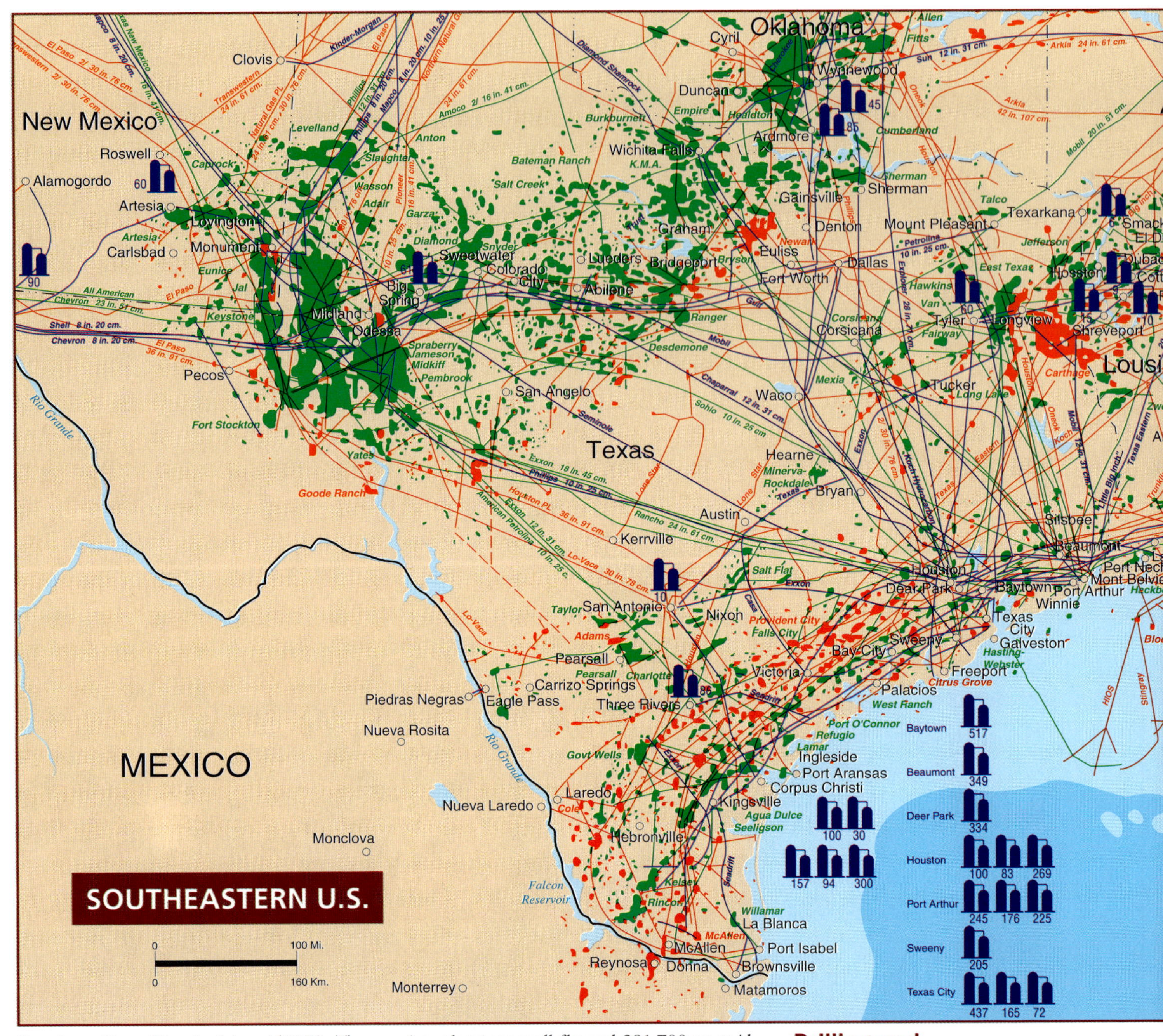

Petroleum Corp. (Houston) in mid-2002. The Gregory A-1 well and the Davis Bros. 11-1 well each flowed initially 200,000 cu m/d (8 MMcfd) into sales pipelines. Additional drilling of both discoveries was planned.

A substantial gas shale resource was being exploited by Devon Energy Corp. (Oklahoma City), which expects to recover an initial 8% of about 1.48 bcm/sq km (142.5 bcf/sq mi) of original gas in place from the Mississippian Barnett shale in the Fort Worth basin. Net reserves of 56 bcme (2.1 tcfe) have been booked in the overpressured, unconventional Barnett gas play. An additional 8-10% could be recovered by restimulating wells and halving well spacing.

In Liberty County, TX, the Hankamer No. 1 exploratory well flowed 281,700 cu m/d of natural gas (10.49 MMscfd) and 772 b/d of oil, said Carrizo Oil & Gas Inc. (Houston) in November. Sales should begin by year-end, along with spudding of another well.

Investigations were under way to determine the cause of a November explosion and fire at a development well being drilled at Tapia Canyon heavy oil field in Los Angeles County. The blast killed one and seriously injured a second person.

Coalbed gas could revitalize the Cherokee and Forest City basins in eastern Kansas and western Missouri, said the Kansas Geological Survey in late 2002. Coalbed gas from Middle Pennsylvanian rocks there is an emerging energy play.

Drilling and production technology

The world's first application of expandable tubing was completed in September, announced Shell Exploration & Production Co. (Houston), which calls the technology MonoDiameter. Shell drilled and completed a Vicksburg gas well using the technology, which eliminates the telescoping effect in current well design, allowing operators to slim down the top of the well while increasing the diameter at total depth. The well was drilled with a series of 9-5/8 inch liners run through each other. To complete the well, a conventional casing string was run through the nested liners, and the inner cas-

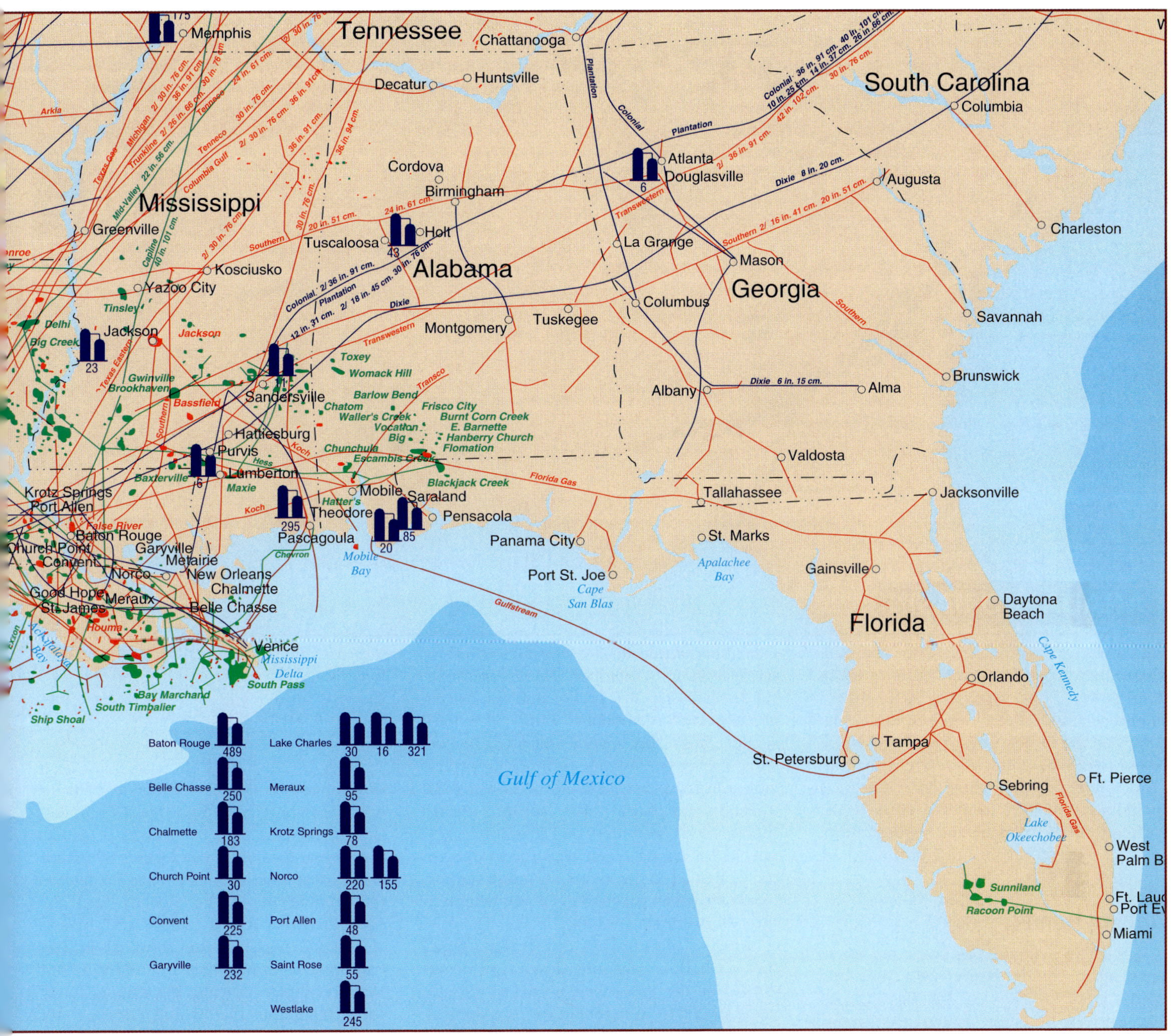

ing string served as the production casing.

Shell's success is indicative of the industry's rapid advances in expandable tubular technology, said *Oil & Gas Journal* in early 2003. Companies often have used expandable liners to solve unexpected drilling problems. Until recently, however, the concept has not been part of normal well design. Engineers working with the technology think that by 2005 expandable casings will be considered production quality or the last string required in a well. Shell plans the first offshore MonoDiameter well in the Gulf of Mexico in 2003. Expandable technology promises to cut material usage, costs, and environmental effects of drilling.

Another area of rapid progress anticipated for 2003 is the use of surface blowout preventers (BOPs) on floating drilling equipment in deep waters. First used off Indonesia, surface BOPs reduce wellbore size, allowing the use of smaller and cheaper rigs. Extending the capability to drill in harsher environments will require advances in well design, risers, seafloor disconnects, and rig equipment. Along with expandable tubular technology, surface BOPs for floating drilling operations in deep water will yield very meaningful gains to lower well costs.

ChevronTexaco Corp. (San Francisco) successfully drilled a deep, hot wildcat in southern Texas using a conventional diesel oil-base fluid without hole problems and at high rates of penetration.

Deepwater production

As the trend for lighter tension-leg platforms and spar designs continues, operators will be able to produce from very deep water with dry trees that facilitate downhole well remediation, predicted *Oil & Gas Journal* in early 2003. Advances in flow assurance enable operators to develop deepwater fields with ever-longer flowlines. The advances, many still under development, include new chemicals and electrified lines for preventing deposition of hydrates, asphaltene, scale, and paraffin.

Several systems have emerged for controlling sand in deepwater reservoirs,

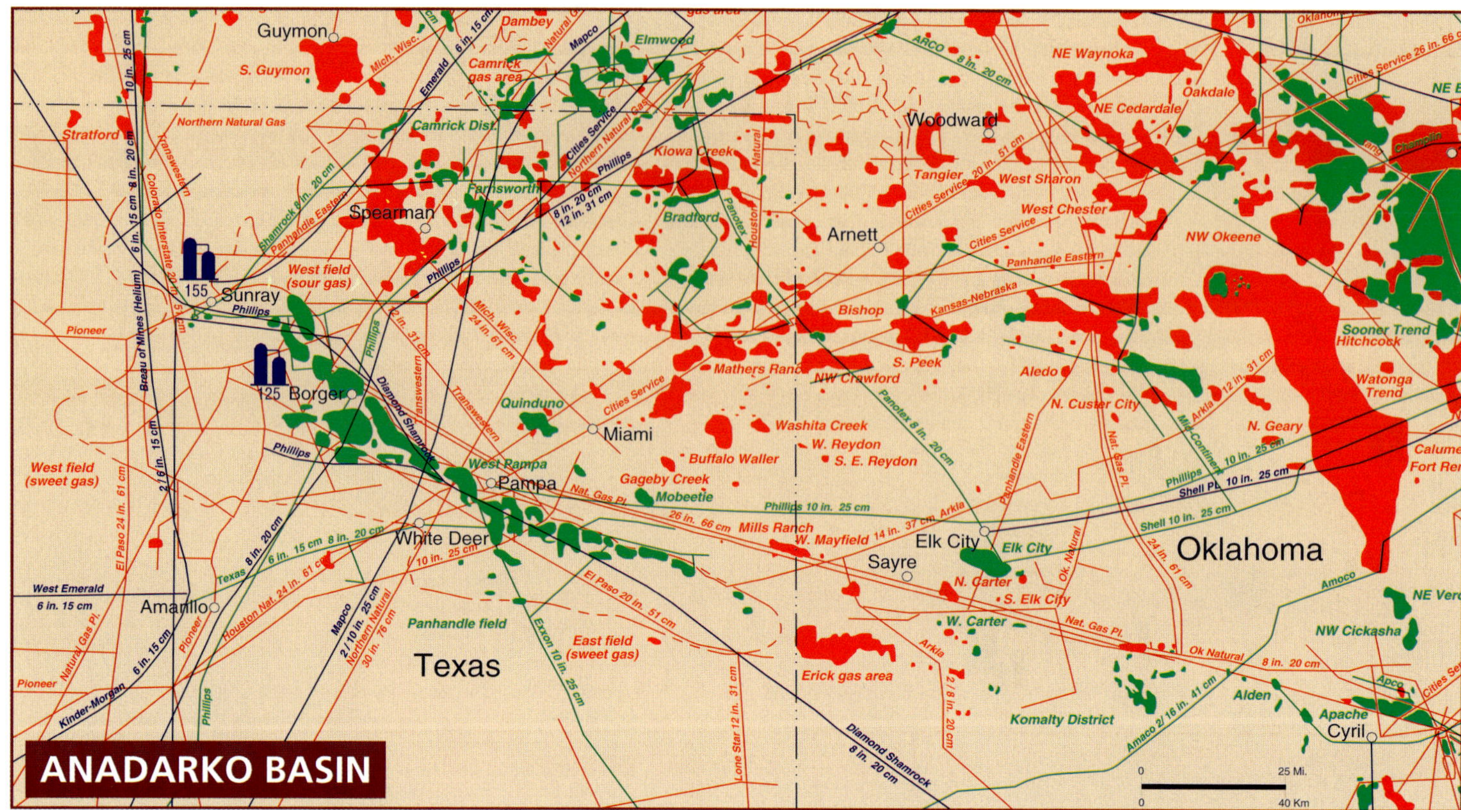

which is especially important for the long horizontal holes frequently employed in deep waters. Also, advances in downhole monitoring and control are helping operators working in deep water and elsewhere to improve scheduling of well remediation.

Other important technologies for 2003 include production from marginal wells, carbon dioxide injection and sequestration, tight gas development, and utilization of stranded gas.

Drilling trends

US drilling activity has come full circle once again, falling nearly 43% from July 2001's peak to a low of 738 rigs in April 2002, driven by falling natural gas prices. However, with winter 2002-03 gas demand appearing very favorable, the drilling market was poised to boost activity in 2003 above mid-2001's levels and could reach the limit of existing rig capacity, prompting more new-builds, said *Oil & Gas Journal* in September.

In addition to this gas-driven up-cycle, activity should increase as relationships continue to strengthen between operators and contract drillers. Contractors were trying to maintain rigs and crews and even grow their fleet, despite 2002's soft market, and producers were focusing on partnerships with selected key contractors and suppliers.

The drilling industry is well-positioned for profitable growth as economic recovery gains momentum in 2003, predicted industry analysts in September. North American rig forecasts are 20% higher, and gas prices are expected to increase while oil prices remain high for several months before trending lower. Volatile commodity prices and industry activity should be considered normal and built into future plans. Independent producers said that 2002's relatively low US rig count was due in part to the lack of good prospects, although tighter North American gas markets could brighten the independents' outlook.

Analysts reported in October that operators' cash flow was still down 25% from 2001, but the gas-driven recovery should begin in 2003. To maintain current domestic production, at least 900 rigs would have to be drilling, said Lehman Bros. Inc., which predicted gas production declines of 5% in 2002 and 3% in 2003.

Oil field services demand should rebound throughout 2003 following a sluggish fourth quarter 2002, according to Gerson Lehrman Group Inc. (New York), which started a quarterly survey gauging oil field service capital expenditure trends. In the initial late-2002 survey, 20 drilling and asset managers expected their share of company capital budgets to increase by 29% in 2003 to a mean $68 million.

LNG developments

US LNG imports are expected to increase at 8.6%/yr, reaching 22.3 bcm (830 bcf) in 2020, according to the US EIA. ChevronTexaco and El Paso in late 2002 proposed offshore LNG projects in the Gulf of Mexico.

Early in 2002 all of the available expansion capacity at the Elba Island, GA, LNG terminal was purchased by Royal Dutch/Shell Group (Netherlands) from El Paso. Under the 30-year deal, Shell will receive 89 million cu m of storage (3.3 bcf) and 3.5 bcm/yr (360 MMcfd) of sendout capacity. Shell also confirmed that it will build a 12.7 bcm/yr (1.3 bcfd) regasification terminal in Baja California to supply gas in California and Mexico. El Paso was also developing an offshore technique that would eliminate the need for land-based terminals.

Plans for another expansion at CMS Trunkline LNG's Lake Charles, LA, terminal included a fourth cryogenic storage tank with 140,000 cu m capacity, a second marine unloading dock, and additional equipment to boost sendout to 11.8 bcm/yr from 6.2 bcm/yr (to 1.2 bcfd from 630 MMcfd).

Earlier Trunkline had sold to BG Group plc (UK) unit BG LNG Services Inc. the firm rights to all of its uncommitted vaporization and storage capacity, about 130 million cu m (5 bcf), which increases to 169 million cu m (6.3 bcf) after another contract runs out in mid-2005. Also, after the terminal's expansion, BG would gain additional storage and sendout capacity during 2005-2023.

A new LNG receiving and regasification

terminal in Hackberry, LA, was planned by a Dynegy unit. The proposed 14.7 bcm/yr (1.5 bcfd) facility would be built on the site of an existing terminal by year-end 2006, bringing the US total number of terminals to five.

Three new security zones were created for LNG carriers in Boston Harbor by the US Coast Guard in October. The zoning will likely increase congestion, but was needed to protect the carriers and surrounding areas from sabotage or accidents.

Gas storage

A new salt cavern facility called the Southern Pines Energy Center was approved in October, announced SG Resources Mississippi LLC. Two caverns, each with capacity of 215 million cu m (8 bcf), will provide 322 million cu m (12 bcf) of storage and support withdrawal rates up to 32 million cu m/d (1.2 bcfd) and injection rates up to 16 million cu m/d (0.6 bcfd). The facility should begin operating in 2004. SG Resources Louisiana LLC was also moving forward with another two-cavern facility totaling 430 million cu m of storage (16 bcf).

Expansion of the Petal salt dome facility in Mississippi was completed in mid-2002, along with a 100-km (60-mile) takeaway pipeline, said an El Paso unit. The dome now has 232 million cu m (8.65 bcf) of working gas capacity and will supply gas to Southern Co. (Atlanta) under a 20-year contract.

An open season for a new gas storage project in the Rocky Mountains attracted bids totaling more than 698 million cu m (26 bcf) and was fully subscribed. Kinder Morgan's proposed Cheyenne Market Center in Colorado, which will have 1.03 million cu m/d (38.4 MMcfd) of injection capability and 1.676 million cu m/d (62.4 MMcfd) of withdrawal deliverability, should open in mid-2004.

Underground storage tank testing company Tanknology-NDE International Inc. (Austin,TX) was fined in July for giving false information to federal regulators.

Petrochemicals, gas downstream

The world's largest naphtha steam cracker was inaugurated in Texas in June, with a capacity of 920,000 tonnes. BASF Fina Petrochemicals LP (Germany) and partners also will build an olefins complex at the Port Arthur site, to start up by year-end 2003.

A cryogenic gas processing train with capacity of 9.4 million cu m/d (350 MMscfd) will be added to the Neptune plant in Louisiana by Technip-Coflexip Group (Paris) under contract to a joint venture of Enterprise Products Partners LP and Marathon Oil Co. (both of Houston). The train should begin operating in late 2003.

A gas-to-liquids (GTL) demonstration plant was being constructed by ConocoPhillips (Houston) in Oklahoma to convert 100,000 cu m/d of natural gas (4 MMcfd) into 400 b/d of diesel, kerosene, and other products. The facility was to be finished by year-end. Another GTL plant will be built near Tulsa to demonstrate an ultra-clean fuels production process, said Marathon Oil and Syntroleum Corp. (Tulsa). The GTL plant should be finished in mid-2003.

Huntsman Corp. (Salt Lake City) permanently idled 15% of its total ethylene oxide and ethylene glycol production, due to weak demand and inadequate margins.

Refining outlook

Clean fuels regulations will be the strongest influence on refining technology in 2003, said *Oil & Gas Journal*. New rules require gasoline sulfur levels averaging 30 ppm by 2005 and diesel sulfur levels of 15 ppm by 2006. As US and European sulfur standards toughen, the most important area for development is diesel desulfurization. Current methods are expensive, and gasoline desulfurization processes don't work as well with diesel, which has sulfur species that are more difficult to treat.

Because requirements for ultra-low sulfur diesel take effect in the US in mid-2006, timing is important, and refiners could start deciding on capital expenditures for these units in 2003. Catalyst manufacturers are researching fluid cracking catalysts able to remove sulfur more efficiently.

US refiners are watching their European counterparts, which produce more diesel and some of which already make 10-ppm sulfur diesel. Refiners also are looking for ways to increase reliability by improving plant software, field instruments, and control systems and by using digital plant architectures to raise plant uptimes and profits.

Another trend in 2003 will be conversion of MTBE units if the US mandate for oxygen in reformulated gasoline disappears. Anticipating phaseout of MTBE and an accompanying ethanol mandate, some California refiners already blend ethanol. Given potentially idle MTBE capacity, refiners could seek an inexpensive technology to convert MTBE units to oligomerization or dimerization capacity.

ChevronTexaco will switch in May 2003 from MTBE to ethanol in gasoline blends sold in southern California. In mid-2002, Chevron began selling a lower-emissions fuel for diesel engines in southern California in limited quantities. The fuel reduces emissions of nitrogen oxides by 14% and particulate matter by 63% without substantial engine modifications.

Low refining margins through most of 2002 made refiners reluctant to invest in plant improvements and capacity additions. Production of reformulated gasoline is behind schedule, which raises the prospect of price spikes in 2003. The petrochemical business could remain distressed, due to planned ethylene capacity additions, especially in the Middle East, which could suppress prices on a global basis. Supply of propylene is also growing strongly, largely from refinery fluid catalytic cracking units.

During the second half of 2002, US downstream fundamentals gradually strengthened, and annual product demand, which was close to flat, is expected to grow 1.7% in 2003, according to Merrill Lynch. Factors improving the downstream outlook included discounts on heavy sour crude, strong gasoline demand, growing distillate demand, and a gradual tightening of product inventories. The forecast assumes a normal winter and a stronger economic recovery in late 2002 through 2003.

Other refining developments

Crudes processed in US refineries continued to decline in quality, though less rapidly than before, according to consultant Edward J. Swain's analysis of EIA data. Gravities decreased by 0.07 degrees API per year over the past 10 years, while the sulfur content of the lower quality crudes increased moderately, about 0.032 wt%/yr.

Production of low-sulfur fuels could increase refiners' demand for hydrogen, leading Praxair Inc. (Tonawanda, NY, US) to effectively double its hydrogen pipeline capacity in May. Praxair was also planning to build two Gulf Coast hydrogen plants.

Premcor Inc. shut down its refinery in Hartford, IL, in October. The facility had a capacity of 70,000 bbl/calendar day (b/cd). Premcor is one of the US's largest independent refiners and marketers of unbranded vehicle fuels and heating oil. In late 2002 Premcor agreed to buy Williams' 175,000 b/cd refinery in Memphis, TN.

Upgrading of ConocoPhillips's 89,000 b/cd Ferndale, WA, refinery was under way to comply with EPA clean gasoline regulations. Plans include installation of a catalytic gasoline desulfurizer unit using Phillips's S Zorb sulfur removal technology, as well as revamping an existing gasoline splitter. Work should be done before 2004.

A sulfur recovery plant was planned for the 175,000 b/cd Marcus Hook, PA, refinery, said Sunoco Inc. (Philadelphia), which had agreed with state regulators to reduce gas flaring. Also, a sulfur processing unit was being installed at Flint Hills Resources' 300,000 b/cd refining complex in Texas, allowing production of fuels with 80% lower sulfur levels.

The Flying J refinery in Salt Lake City reduced electrical consumption by 5-6% and lowered purchased natural gas use by 35-40% by implementing recommendations from the US Energy Department's best practices program. The improvements resulted in cost savings of $900,000 in the first year.

Industry business

The ConocoPhillips merger was completed in August, creating the third-largest US-based integrated oil firm. Some refineries, gasoline outlets, and natural gas production facilities will be divested.

Anadarko announced in October that it will acquire Howell Corp. (Houston) in a move to expand its Wyoming oil production. Pennzoil-Quaker State Co. was purchased by Royal Dutch/Shell in a deal announced in April. And in March, Williams sold its Kern River gas system to MidAmerican Energy Holdings Co. (Des Moines, IA).

These deals and other consolidation eroded the OGJ200 list to only 176 firms. Also, the list now includes only US-headquartered companies, which excluded several Canadian and European giants.

IPE

NORTH AMERICA

Gulf of Mexico: Record Water Depth for Gas Production

Natural gas began to flow in September into the Canyon Station production-handling platform from two deepwater wells in the eastern gulf's Mississippi Canyon, setting a world record water depth for natural gas production. The platform's first gas, flowing through multiphase gathering lines, came from wells in water depths of 1,500-2,200 m (5,000-7,200 ft) in Aconcagua and King's Peak fields.

The production platform was to reach full capacity of 13.4 million cu m/d (500 MMcfd) when nine more wells begin producing in late 2002. Production from federal leases totaling 140 sq km (35,000 acres, or 55 sq mi) is dedicated to the facility, owned by Williams.

The natural gas flows through the 92-km (57-mile) Canyon Express system, one of the longest multiphase gathering lines in the world and reportedly the gulf's deepest. Operation of the system marks the first time production from such deepwater fields is being commingled in a single flowline and the first use of subsea multiphase meters to determine the production from individual wells.

Canyon Express is owned by five working partners, which also own reserves in fields tied into the system. Aconcagua field is operated by TotalFinaElf SA (France) and King's Peak by BP plc (UK). The third field feeding production to the platform is Camden Hills, operated by Marathon Oil. Subsea wells completed in 2,197 m of water (7,209 ft) at Camden Hills are the Gulf of Mexico's deepest water producers.

The fixed-leg Canyon Station platform, commissioned in July, stands in 90 m of water (300 ft) on East Main Pass Block 261 and uses a new methanol recovery process to provide flow assurance for the deepwater wells. Williams Field Services operates all subsea well monitoring, flow control, and chemical injection functions from the platform, which treats and processes gas and gas liquids for delivery onshore.

Other Mississippi Canyon E&P

One of the Gulf of Mexico's largest oil fields, Thunder Horse (formerly Crazy Horse), will be developed using a semisubmersible platform in about 1,800 m of water (6,000 ft). Advanced drilling and mooring equipment will be delivered in early 2003 as part of BP's ultra-deep project involving Thunder Horse and three oil and gas fields nearby.

The Matterhorn deepwater development will include a monocolumn tension-leg-platform (mini-TLP), the fourth one in the Gulf of Mexico and the first dry-tree monocolumn TLP. TotalFinaElf planned to install the column in 864 m of water (2,835 ft) in mid-2003.

Development drilling of Devils Tower will be conducted from the truss spar platform in 1,700 m of water (5,600 ft) on Mississippi Canyon Block 773 beginning in 2003. Part of the Devils Tower project is the 20-inch Mountaineer oil pipeline system being built in shallow water by Horizon Offshore Inc. (Houston) for Williams.

The three-well Crosby development began production of 20,000 b/d of oil in early 2002, reported Shell E&P. Crosby, in 1,340 m of water (4,400 ft) on Blocks 398 and 399, is tied back 16 km (10 miles) to Ursa TLP, also in Mississippi Canyon on Block 809 in 1,204 m of water (3,950 ft). At peak during 2002, Crosby was to reach 60,000 b/d of oil and 2.4 million cu m/d of gas (90 MMcfd), boosting Ursa's total production to 170,000 b/d.

Na Kika development

Shell and BP made progress on their $1.26 billion Na Kika project to develop widely spaced oil and gas fields in the ultra-deepwater Mississippi Canyon area. The semisubmersible production and development system was being fabricated in South Korea and will be transported to the site in 2003, when first production is expected. Mating the semi's topsides structure to its hull required a superlift technique using strand jacks instead of cranes.

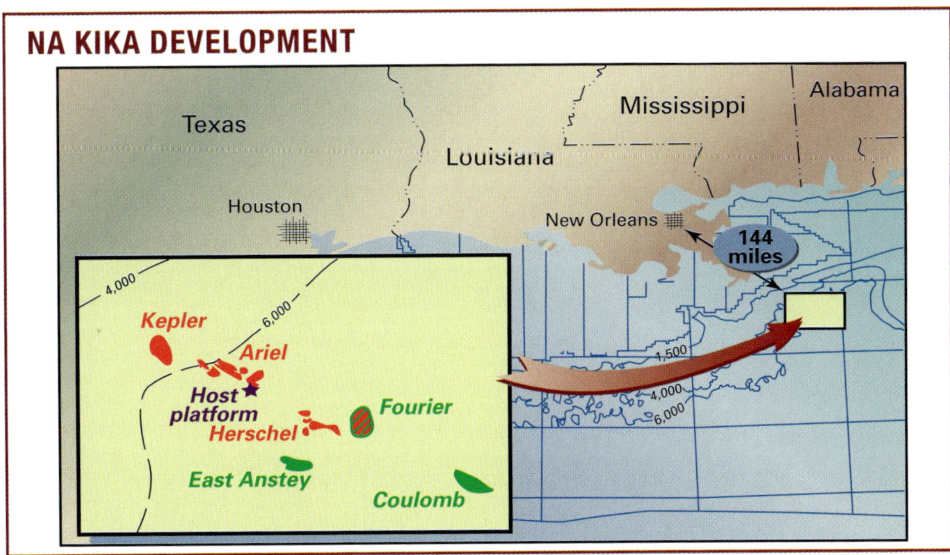

The semisubmersible, in a first for the Gulf of Mexico, will be permanently moored 225 km (140 miles) southeast of New Orleans on Mississippi Canyon Block 474, where it will serve as a host facility for six surrounding independent fields:Ariel (Block 429), Fourier (522), Herschel (520), Kepler (383), East Anstey (607), and Coulomb (657), which is scheduled for 2005 development. Peak rates from the fields are expected to reach 11.4 million cu m/d of gas (425 MMcfd) and 110,000 b/d of 25-29 degree gravity oil, with ultimate recovery expected to exceed 300 million boe.

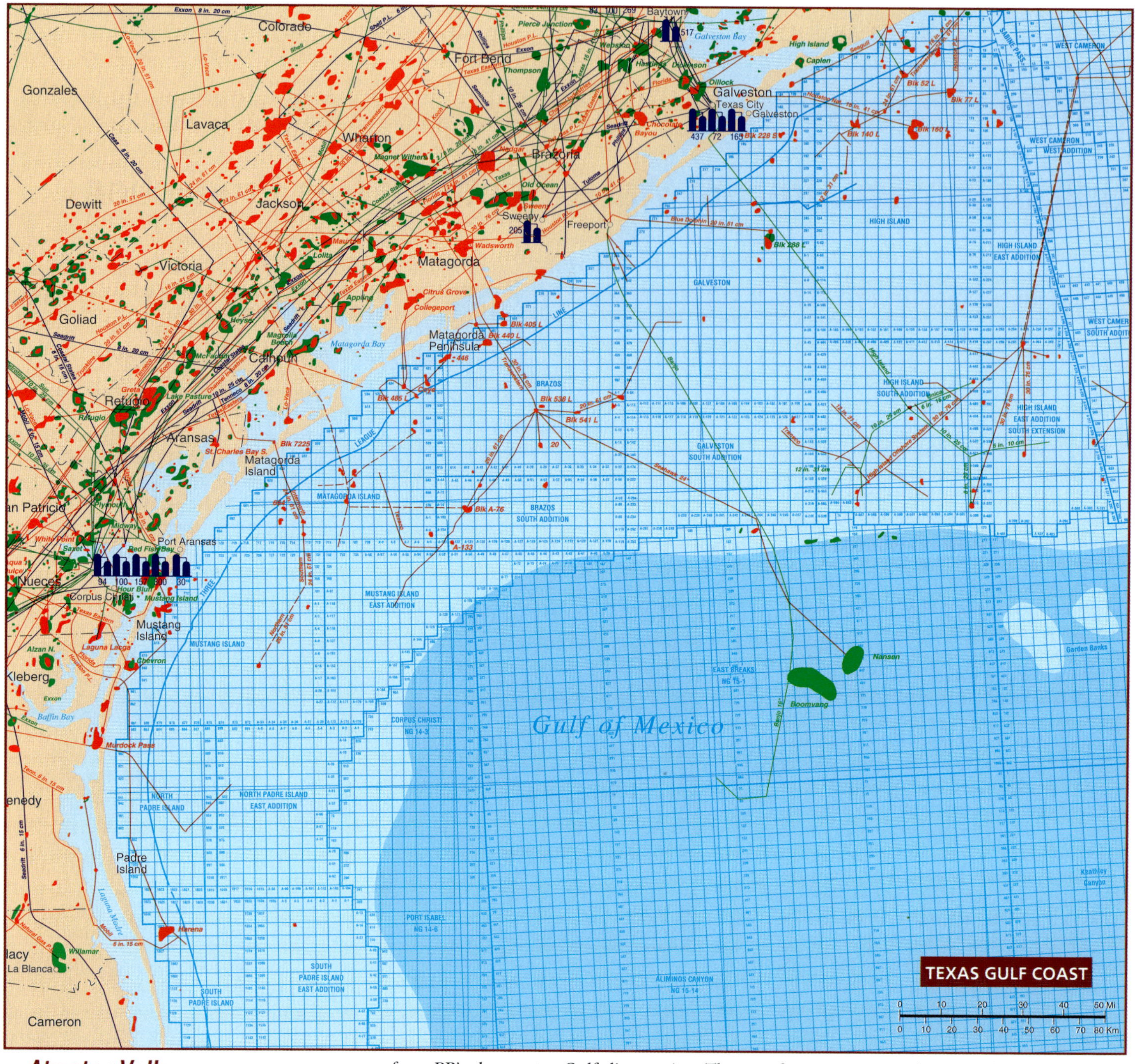

Atwater Valley

Just south of Mississippi Canyon, natural gas was discovered on the Vortex prospect on Atwater Valley Block 261, said BHP Billiton Ltd. (Melbourne) in December. The Deepwater Millennium spudded the Vortex-1 exploratory well in water depths of 2,543 m (8,344 ft). A sidetrack was also drilled 880 m (2,900 ft) west of Vortex-1 to a total depth of 5,892 m (19,330 ft).

Mardi Gras oil pipelines

Three new crude oil pipeline systems — Caesar, Cameron Highway Offshore, and Proteus-Endymion — will carry production from BP's deepwater Gulf discoveries. The large-diameter lines represent the third component of BP's proposed Mardi Gras Transportation System. Caesar will span 190 km (120 miles) and link Holstein, Mad Dog, and Atlantis fields with Ship Shoal Block 332. With its main portion of 28-inch diameter, Caesar initially will carry 450,000 b/d, the first pipeline of its size to be installed in gulf waters deeper than 1,500 m (5,000 ft).

Reportedly, Mad Dog was on schedule to produce first oil by year-end 2004, and Atlantis field's reserves were increased to 575 million boe from 300, BP said in May. Production will begin from Atlantis in 2005 from a moored semisubmersible with capacities of 150,000 b/d and 4.8 million cu m/d (180 MMscfd).

The 610-km (380-mile) Cameron Highway Offshore pipeline will transport production from Ship Shoal Block 332 to Texas landfall, delivering as much as 500,000 b/d of oil from the southern Green Canyon and western gulf areas. El Paso will participate in building and using Cameron. The Caesar and Cameron systems will start up in 2004, when Holstein field begins production.

The third system will serve Mississippi Canyon's Thunder Horse field. The 28-inch Proteus oil pipeline's deepwater portion will extend 110 km (70 miles) from the platform to a booster station to be installed on South Pass Block 89. There Proteus will connect to the Endymion oil pipeline, a 140-km (90-mile), 30-inch system extending to onshore Louisiana and connecting with storage and offloading facilities.

Combined, Proteus and Endymion will have an initial capacity of 420,000 b/d. Both segments are slated for completion in early 2005, in line with Thunder Horse's intended start-up.

Garden Banks

In the western gulf, natural gas from the deepwater Garden Banks area will be delivered through the new Triton pipeline, to be built by Shell/Enterprise Products Partners venture. The 16-inch, 7.38 million cu m/d (275 MMcfd) pipeline will extend 66 km (41 miles) from Gunnison field to a connection with a Stingray Pipeline Co. natural gas pipeline to shore. Operation could begin by late 2003.

Gunnison, in 305 m of water (3,150 ft), was being developed with a truss spar platform that can serve as a hub for future development and third-party production. The truss spar, scheduled for late 2003 delivery, is being designed to produce 40,000 b/d of oil and 5.4 million cu m/d of gas (200 MMcfd). The flowline, risers, and jumper were ordered in late 2002. Kerr-McGee Corp. (Oklahoma City) operates Gunnison on Blocks 667, 668, and 669, where it was the area's first discovery.

Also in Garden Banks, the Red Hawk natural gas field will be developed in more than 1,600 m of water (5,300 ft) using the world's first cell spar production platform, said Kerr-McGee in July. Red Hawk, with estimated proven reserves of more than 6.7 bcm (250 bcf) on Block 877, could begin producing by mid-2004 and will be Kerr-McGee's deepest water development to date. Drilling should begin in 2003, resulting in initial production of 3.2 million cu m/d (120 MMcfd) and ultimate capacity of 8 million cu m/d (300 MMcfd).

A new 16-inch, 138-km (86-mile) gas gathering pipeline will built to carry up to 3.2 bcm/yr of gas (330 MMcfd), originating in deep water at Red Hawk and connecting to the ANR pipeline system at Vermilion Block 397. Service is scheduled for mid-2004.

Magnolia field, also in Garden Banks in 1,400 m of water (4,700 ft), will be developed using a TLP with eight production riser systems, a record depth for this type of floating structure. Installation was planned for 2004, said ConocoPhillips in September. Magnolia, on Blocks 783 and 784, should produce 150 million boe.

Serrano field's subsea production system became the third to produce via the Auger TLP, said Shell E&P in early 2002. The Serrano wells, in 1,040 m of water (3,400 ft), were flowing 1.4 million cu m/d of gas (52.5 MMcfd) and 4,625 b/d of oil from Blocks 516 and 472 and were expected to peak at 4.3 million cu m/d (160 MMcfd).

K2 DEEPWATER SUBSALT APPRAISAL WELL

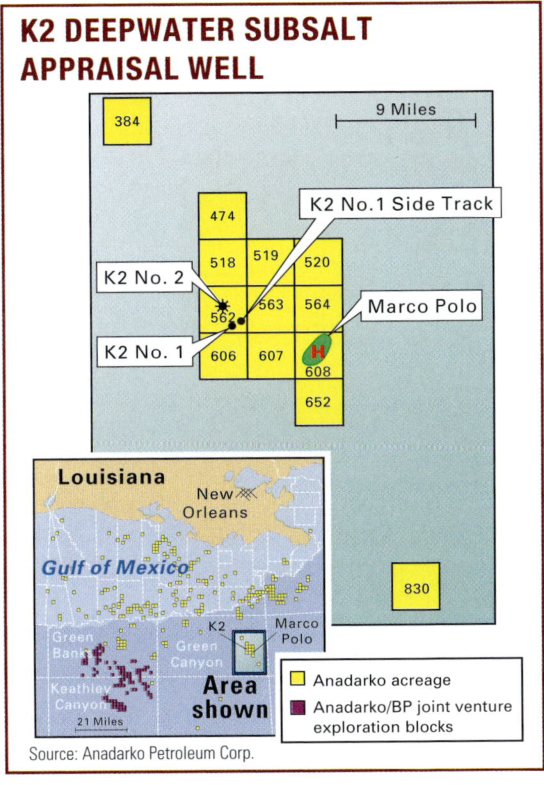

Source: Anadarko Petroleum Corp.

Green Canyon

The deepwater Shenzi well on Green Canyon Block 654 found hydrocarbons in the Western Atwater Foldbelt area 200 km (125 miles) off the Louisiana coast, said BHP Billiton in late 2002. Appraisal drilling will be necessary to determine the size of the find and the significance of the well.

Shenzi, which encountered a gross hydrocarbon column of 142 m (465 ft) with 43 m (140 ft) of net pay, is on the same block where oil and gas were discovered previously at Mad Dog, Atlantis, and Neptune. The Western Atwater Foldbelt is set to become a core producing area for BHP starting in late 2004.

Oil was found on the Tahiti prospect of Block 640, a significant discovery, said ChevronTexaco in April. The well was drilled in 1,224 m of water (4,017 ft) using an ultra-deepwater drillship.

Development was under way at Marco Polo, with four wells drilled during 2002 on Block 608. A TLP able to produce 100,000 b/d of oil and 6.7 million cu m/d of natural gas (250 MMcfd) will be installed in late 2003 in 1,300 m of water (4,300 ft), which could set a record. Production is targeted for early 2004, said Anadarko in September. Marco Polo was Anadarko's first deepwater discovery.

Nearby at K2 field, estimated reserves were increased to more than 100 million bbl of oil by a confirmation well on Block 562 in 1,200 m of water (3,900 ft). K2 is operated by the Agip Petroleum unit of Eni SpA (Italy), which spudded the well in April.

Gas production began in March from the King Kong-Yosemite project, which flowed at 3.8 million cu m/d (140 MMcfd) and was expected to reach platform capacity of 4 million cu m/d (150 MMcfd) with the addition of a third well, said Mariner Energy Inc. (Houston). The wells are on Green Canyon Blocks 472, 473, and 516 in 1,200 m of water (3,800 ft).

Walker Ridge

Just south of Green Canyon, the Cascade prospect was drilled in 2,500 m of water (8,200 ft) on Walker Ridge Block 206 and found hydrocarbons, said BHP Billiton and partners in June. Also on Walker Ridge, a wildcat was being drilled in August by ConocoPhillips in 2,028 m of water (6,654 ft) on Block 285.

East Breaks

At Boomvang field in 1,052 m of water (3,453 ft), the first of three subsea wells was producing 1.3 million cu m/d of gas (50 MMcfd), said Kerr-McGee in July. Boomvang should produce at its peak 4.3 million cu m/d (160 MMcfd) and 32,000 b/d of oil by mid-2003. The field, on Blocks 642, 643, and 688, holds reserves estimated at 70-100 million boe. Boomvang was developed using a truss spar that is a twin to the one at nearby Nansen field, which was developed with the world's first truss spar.

Kerr-McGee began production earlier in 2002 at Nansen, also in East Breaks, in 1,122 m of water (3,680 ft) on Blocks 602 and 646. Output should peak by year-end at 40,000 b/d and 2.1 million cu m/d (80 MMcfd), and the Nansen spar can accommodate more gas from satellite fields. The field's reserves are an estimated 140-180 million boe.

Navajo gas field on Block 690 in 1,200-m (4,000-ft) water, developed by Kerr-McGee as a subsea tieback to the Nansen spar, began production in mid-year. Along with Kerr-McGee's West Navajo and Northwest Navajo

Texas
Louisiana
Calcasieu
Lake Charles
Jefferson Davis
Jennings
Acadia
Church Point
Lafayette
St. Martin
Iberville
Ascension
Livingston
Port Allen
Baton Rouge
St. John The Baptist
St. James
Convent
Reserve
Assumption
Napoleonville
Cameron
Hackberry
Big Lake
Black Bayou
Sabine Lake
Calcasieu Lake
Grand Lake
Mud Lake
Holly Beach
Vermilion
Abbeville
New Iberia
IBERIA
Weeks Island
Avery Island
Vermilion Bay
Cote Blanche Island
West Cole Blanche Bay
White Lake
Lac Blanc
Buck Point
N. Fresh Water Bayou
Recan Island
St. Mary
St. Martin
Morgan City
Patterson
Baitman Lake
Lake Sand
Houma
Terrebonne
Bayou Penchant
Lake Docade
Lapeyrose
Mosquite Bayou
Elaine
Lake Pelto
Lake Barre
Atchafalaya Bay
Tiger Shoal
Rabbit Island
West Cameron 65
WEST CAMERON
SABINE PASS
Blk 45
Blk 149
Blk 65
Blk 180
Blk 63
Blk 192
Blk 89
THREE
LINE
EAST CAMERON
Vermilion 84
Blk 71
Blk 76
235
SOUTH MARSH ISLAND NORTH ADDITION
S. Marsh Is. 257
S. Marsh Is. 255
Blk 131
Blk 0
Blk 23
Blk 100
Blk 32
Blk 28
Blk 27
WEST CAMERON WEST ADDITION
West Cameron 264
Blk 260
28
Blk 164
VERMILION
SOUTH MARSH ISLAND
Blk 38
Blk 48
Blk 66
136
EUGENE ISLAND
Blk 175
Blk 188
Blk 198
Blk 205
Blk 107
SHIP SHOAL
Blk 169
SOUTH PELTO
Blk 172
Blk 208
HIGH ISLAND EAST ADDITION
Blk 464
West Cameron 479
494
WEST CAMERON SOUTH ADDITION
A-279
VERMILION SOUTH ADDITION
Blk 222
Blk 276
Blk 266
294
310
Blk 330
West Cameron 551
HIGH ISLAND EAST ADDITION SOUTH ADDITION
EAST CAMERON SOUTH ADDITION
SOUTH MARSH ISLAND SOUTH ADDITION
EUGENE ISLAND SOUTH ADDITION
SHIP SHOAL SOUTH ADDITION
Ship Shoal 320
East Cameron 359
Vermilion 380
Vermilion 397
Auger Garden Banks
224
236
Serrano
Oragano
GARDEN BANKS NG 15-2
GREEN CANYON NG 15-3
Magnolia

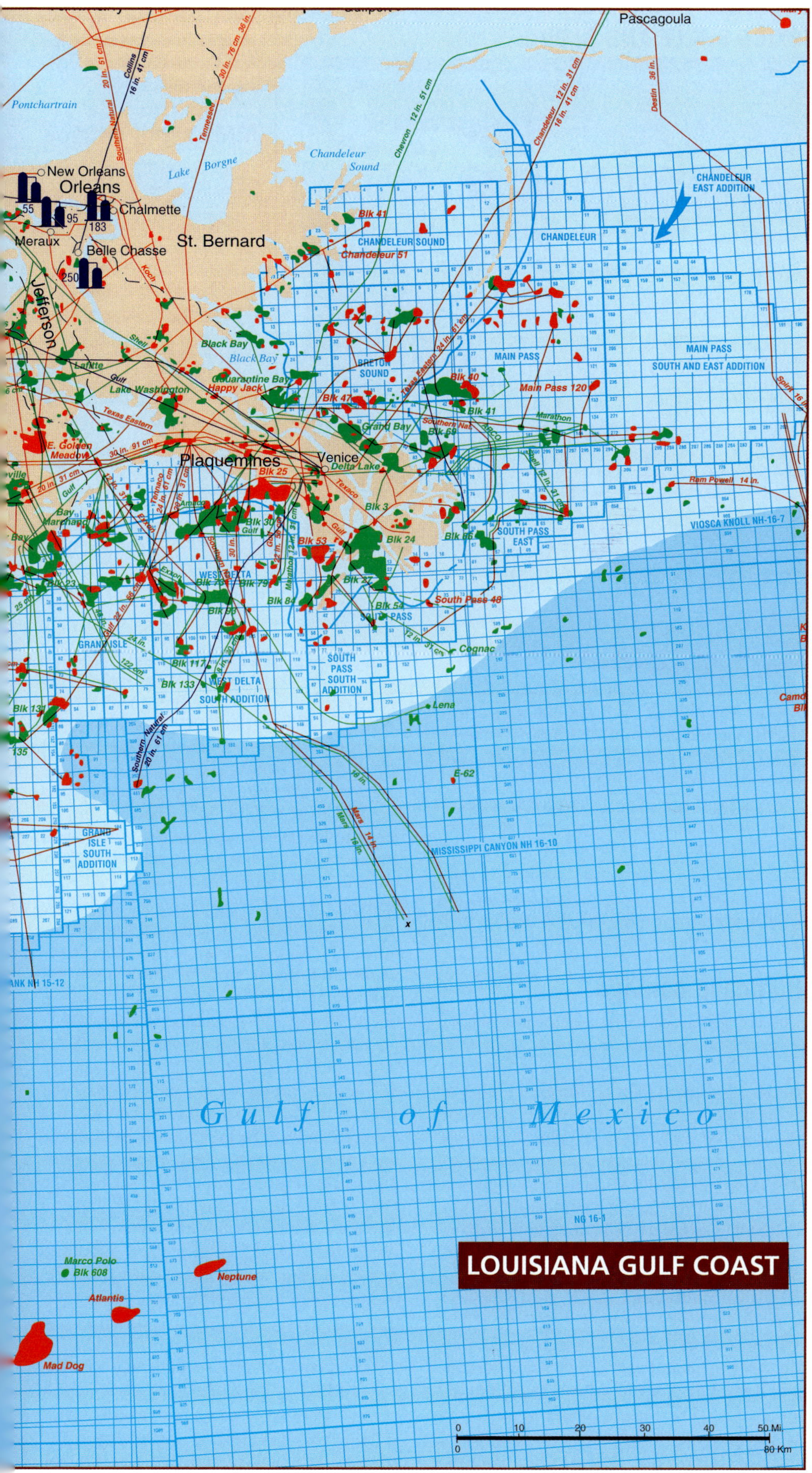

prospects on Blocks 689 and 646, Navajo field reserves total 20-30 million boe, of which 60% is gas.

Production began in June at Lost Ark field on Block 421 in 820 m of water (2,700 ft). A single well flowed at 752,000 cu m/d (28 MMcfd), which will be ramped up to 1.1 million cu m/d (40 MMcfd), said Noble Energy (Houston). Fuel injection technology on its diesel-powered rigs reduces nitrogen oxides emissions by 30% and trims fuel consumption and costs.

Falcon field on East Breaks will be developed with an 8.1 million cu m/d (300 MMcfd) platform, to be installed in 120 m of water (390 ft) on Mustang Island Block 103, said an El Paso unit in May. The Falcon Nest platform, to be completed in early 2003, will connect to an offshore gathering system via 32 km (20 miles) of 18-inch pipeline. El Paso partners Pioneer Natural Resources Co. (Irving, TX, US) and Mariner Energy, Inc. (Houston) will dedicate nine blocks to the platform in addition to Falcon field's blocks.

Alaminos Canyon

Just south of East Breaks, a greenfield prospect called Great White was discovered in September on Alaminos Canyon Block 857 in about 2,500 m of water (8,000 ft). Also in Alaminos Canyon, Unocal Corp. (El Segundo, CA, US) drilled the deepest water exploration well in 9,743 ft, and Shell Offshore drilled the deepest water discovery well, in 2,323 m (7,620 ft).

By 2003 Unocal's Trident ultra-deepwater discovery will be confirmed by a second appraisal well on Block 947, about 3 km (2 miles) away. The Trident prospect covers seven Alaminos Canyon blocks.

Shallow water activities

Two wells discovered natural gas in the West Delta and Main Pass areas of the central Gulf of Mexico, reported Pogo Producing Co. (Houston) in late 2002. West Delta Block 54 No. 1 was drilled to 3,902 m subsea (12,804 ft) and found a 10-m (33-ft) hydrocarbon column in its primary objective below 3,800 m (12,600 ft).

The well was drilled on a 1.4 sq km structure (350 acre, or 0.5 sq mi) and flowed about 80,500 million cu m/d of natural gas (3 MMcfd) and 730 b/d of oil on test. The well will begin production by mid-2003 from Pogo's South Pass Block 24 facilities.

Pogo's second discovery well, Main Pass Block 62 No. 3, found 12 m (40 ft) of gas pay on a previously untested "G-pod" structure with an area of 0.8 sq km (200 acre, or 0.3 sq mi) in the northwest portion of

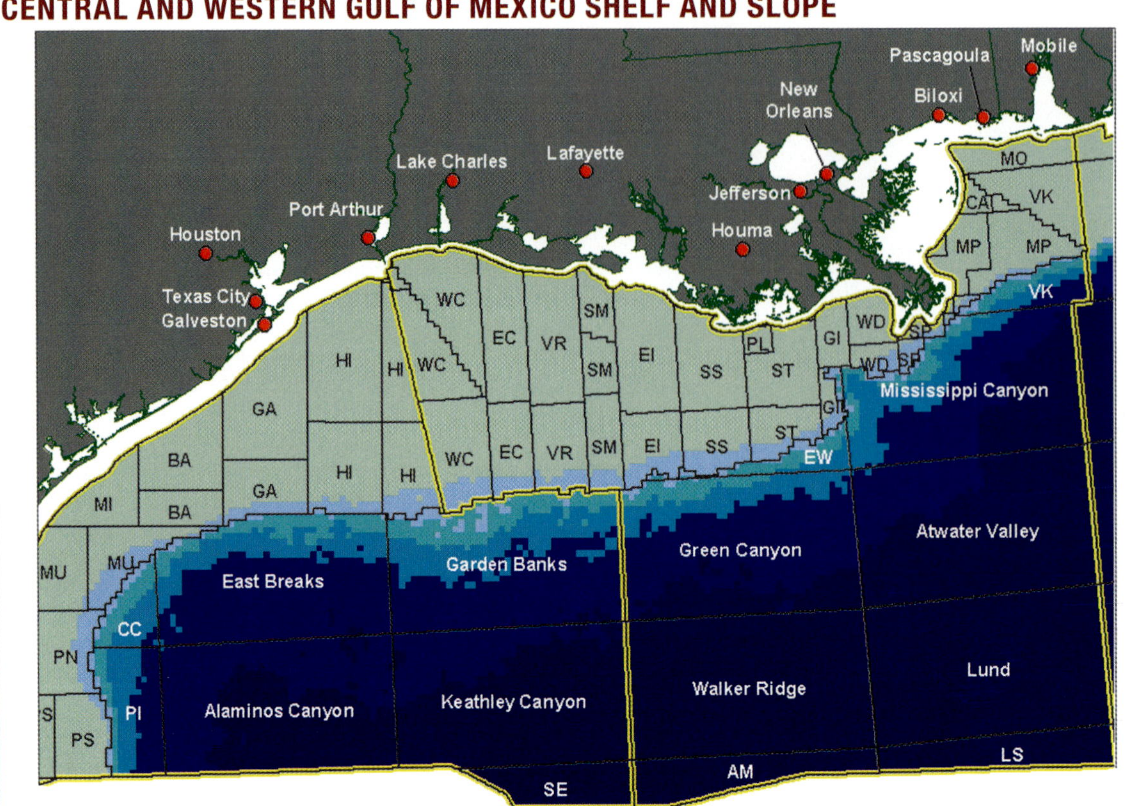

CENTRAL AND WESTERN GULF OF MEXICO SHELF AND SLOPE

Blocks 61/62 field.

A new 5-km (3-mile) flowline will connect the well to Main Pass Blocks 61/62 production facilities. This well and a previous gas discovery at the "E-pod" on Block 61's northeast corner should begin producing through the A-platform facilities by midyear 2003. Pogo had two more exploratory wells being drilled on other prospects in the central gulf's Main Pass area.

In West Cameron, natural gas was discovered in July on Block 347 by Remington Oil & Gas Corp. (Dallas) and partners. The discovery well in 25 m of water (82 ft) flowed 591,000 cu m/d (22 MMcfd) and will begin producing in 2003.

Another West Cameron discovery on Block 45 started production in December, said Stone Energy Corp. (Lafayette, LA, US). The single well, drilled to 5,012 m (16,444 ft), cut 18 net m (59 ft) of gas-productive sand and flowed 677,000 cu m/d of gas (25.2 MMcfd) and 125 b/d of oil. The well produces from the West Cameron 45 A platform 1,200 m (4,000 ft) to the west.

Production began in mid-2002 from wells in three other shallow areas: East Cameron Block 184, producing 1 million cu m/d of gas (36 MMcfd) and 1,050 b/d of oil; South Timbalier Block 274, 270,000 cu m/d (10 MMcfd) and 1,056 b/d of oil; and Eugene Island Block 397, 1,000 b/d of oil and 12 million cu m/d (444 Mcfd).

The Gryphon 2 well should begin producing natural gas from High Island 52 field in early 2003. The completed well was being tested in late 2002.

Apache Corp. (Houston) agreed in early 2003 to purchase shallow water gulf assets, mainly gas producers, from BP. The assets' net share of output was about 71,000 boe/d. The transaction makes Apache the largest acreage holder on the continental shelf.

Hurricane damage

Central Gulf of Mexico oil production of 14.4 million bbl was shut in for almost a month by Hurricane Lili and Tropical Storm Isidore, which blew through during September and October. Several rigs were damaged by Lili, reported their owners in October. Rowan International Inc. (Houston) lost its Rowan-Houston drilling rig, which was severed from its legs and sank about 530 m (1,750 ft) northwest of its prestorm location on Ship Shoal Block 207. The hull, which had been 19 m (63 ft) above the water, was found resting on the bottom in 32 m (105 ft) and appeared severely damaged.

The Ocean Lexington semisubmersible rig became separated from its moorings and drifted 72 km (45 miles), grounding in an estimated 11 m of water (35 ft) off Louisiana, said Diamond Offshore Drilling Inc. (Houston). Also, leg and hull damage was sustained by the Noble John Sandifer cantilever jack-up in the Eugene Island area.

MMS's gulf outlook

The US Minerals Management Service reported that the deepwater Gulf of Mexico had 3,227 active leases with 2,253 approved applications to drill and 27 active platforms in water over 300 m deep (1,000 ft), as of late October. Deepwater production in 2002 should average 1.0-1.1 million b/d of oil and 33-39 bcm/yr of gas (3.4-4.0 bcfd).

MMS also predicted that leases on the gulf's outer continental shelf will be producing 2-2.47 million b/d of oil and 107.5-222.7 bcm/yr of gas (10.97-16.39 bcfd) by the end of 2006. Production from water deeper than 300 m (1,000 ft) will play a greater role, accounting for up to 77% of oil and 26% of gas on a daily basis.

The deepwater gulf is an expanding frontier that will be of increasing importance to US energy supplies, the MMS said in May, noting that that 59% of all gulf oil production comes from deep water. One of the newest and most remarkable trends is the increasing practice of using subsea completions. The MMS reported in April that deepwater gulf production in 2001 again hit record levels, with an estimated 335 million bbl of oil and 31.7 bcm of gas (1.18 tcf).

The use of FPSOs in the gulf will no longer be barred, as a result of an MMS environmental ruling in January 2002 that approved their operation in central and western areas. Producers must now examine the economics of FPSO projects versus pipelines or other transport. Also, since FPSOs store only liquids, producers have to find a use for associated gas, which can't be reinjected. MMS might allow gas-fired power generators offshore, with the electricity cabled in, or a temporary gas reinjection scheme that would eventually produce the gas.

An economic analysis indicated that the use of FPSOs to produce a dry-gas reservoir is technically feasible, said Fluor Corp. (Aliso Viejo, CA, US) in December. The analysis also determined that shipping by

CNG would have a higher return on investment than pipeline or LNG transport.

Gulf of Mexico well logs are now available faster, thanks to the MMS's new web-based system. Orders are completed in less than 30 minutes, and any number of logs can be ordered. Previously a month was required to send out a maximum of about 125 logs. During the system's first 2 months of operation, 180 CDs containing over 33,000 well log images were delivered.

Lease sales

MMS's August lease sale (184) of western gulf prospects generated less money for more tracts than its previous sale, with interests divided almost equally between shallow and deep waters. Final high bids in the second phase totaled nearly $148.6 million for 316 leases awarded. Earlier, the central gulf sale drew $363.2 million for 506 tracts, also lower than its predecessor. In August the MMS released environmental assessment details of upcoming lease sales in 2003 and 2005. Whale protection measures took effect in August, but the softened rules will not cost as much as had been feared.

The MMS's royalty-relief incentives for new deepwater drilling leases, which succeeded in spurring exploration and development, are being extended to some shallow water drilling. The next central gulf lease sale (185), set for March 2003, will include a royalty relief incentive to drill for deep gas deposits in the shallow water shelf in waters less than 200 m (660 ft). Royalty relief will apply to the first 540 million cu m (20 bcf) of production, if it begins within 5 years of award, from wells drilled into new gas reservoirs 4,600 m (15,000 ft) below sea level.

Undiscovered deep gas resources on the near-shore shelf could be in the range of 130-540 bcm (5-20 tcf), said MMS in November. Royalty relief incentives could add another 6.7-16 bcm of production (250-600 bcf) during 2003-07. Additional royalty relief will apply in deeper waters as well:

- In 400-799 m (1,312-2,621 ft), 5 million boe per lease
- In 800-1,599 m (2,625-5,246 ft), 9 million boe per lease
- In 1,600 m (5,249 ft) or greater, 12 million boe per lease.

The proposed Sale 185 encompasses 4,411 available central gulf blocks covering 93,500 sq km (23.1 million acres, or 36,100 sq mi). The blocks are 5-340 km (3-210 miles) offshore in water depths ranging from 4 m to more than 3,425 m (from 13 ft to more than 11,237 ft). Undiscovered, economically recoverable hydrocarbons in this area are estimated at 270-650 million bbl of oil and 42.7-88.6 bcm of natural gas (1.59-3.3 tcf).

Offshore LNG proposals

ChevronTexaco in late 2002 applied for a license to construct and operate a deepwater port in the Gulf of Mexico off Louisiana where an LNG terminal would be sited. The development, called Port Pelican, would consist of an LNG ship receiving terminal, LNG storage and regasification facilities, and an interconnection to existing pipeline infrastructure for delivery as regasified natural gas into the US interstate pipeline network.

Phase I, to begin operating by 2006, would be constructed as an offshore facility initially designed to process 21 million cu m/d (800 MMscfd) of natural gas. Phase II would expand the terminal to accommodate a total of 43 million cu m/d (1.6 bcfd). Port Pelican's design codes and standards would meet or exceed accepted industry practice.

Also late in the year, an El Paso unit applied for a license to build and operate its EP Energy Bridge, an offshore deepwater mooring buoy for LNG carriers. The system, to begin operating by 2005, would be built on West Cameron Block 603 about 187 km (116 miles) off Louisiana in 90 m of water (300 ft).

The concept features a floating mooring buoy and LNG tankers that can regasify their cargoes onboard and deliver 11-13 million cu m/d (400-500 MMcfd) into pipelines from far offshore. El Paso plans to construct a submersible, offshore buoy and riser system that includes 13 km (8 miles) of 20-inch pipeline connecting to two existing subsea pipeline systems to deliver natural gas to the main US pipeline grid. Three LNG carriers are under contract, and El Paso has an option for a fourth vessel.

IPE

NORTH AMERICA

Alaska: National Reserve Holds Huge Resource

For the first time in more than 20 years, the US Geological Survey (USGS) in mid-2002 reported resource estimates for the National Petroleum Reserve in Alaska (NPR-A), with startling results — over 4 times as much oil and 7 times as much natural gas as previously thought could be contained in the federal area's 91,100 sq km (22.5 million acres, or 35,200 sq mi).

The reserve lies along the western half of Alaska's north coast, including part of the North Slope area currently being leased and opposite from the Arctic National Wildlife Refuge (ANWR) on the coast's eastern end. Critics of ANWR leasing cited the USGS findings as further evidence that ANWR's coastal plain need not be explored.

Oil and gas industry interest in NPR-A lands remains strong, as seen in the North Slope lease sales held in May and October 2002. The May sale covering 12,000 sq km (3 million acres, or 5,000 sq mi) generated high bids of $63.76 million for 60 tracts.

In early 2003 the Bush Administration planned to offer more oil and gas leases on Alaska's North Slope, including some acreage in the northwestern section of the NPR-A. The US General Accounting Office urged regulators to ensure that producers are financially prepared to clean up any damage before new exploration begins.

USGS results

The new USGS data estimates the NPR-A's economically recoverable oil resource at 1.3-5.6 billion bbl, assuming market prices of $22-$30/bbl. Technically recoverable oil could total 5.9-13.2 billion bbl, with a mean value of 9.3 billion bbl, compared with the 1980 mean estimate of 2.1 billion bbl. A large proportion of the undiscovered oil is thought to occur in the northern third of the NPRA in moderate-size accumulations.

Natural gas resources were estimated at 1.05-2.69 bcm (39.1-83.2 tcf), with a mean value of 1.60 bcm (59.7 tcf), versus 1980's mean of 0.23 bcm (8.5 tcf). Most of this gas is thought to lie in the central and southern NPR-A. However, the economic viability of the gas depends on the availability of a pipeline to transport the product to the lower 48 states.

The latest USGS study's methodology was similar to that used in its 1998 assessment of ANWR's resources. That assessment area covered 6,100 sq km (1.5 million acres, or 2,300 sq mi), including federal lands within the 1002 area and adjacent state offshore and native lands; the ANWR assessment did not cover the entire refuge. According to the 1998 data, the ANWR 1002 area could contain 4.3-11.8 billion bbl of recoverable oil, with a mean value of 7.7 billion bbl.

Although NPR-A holds much promise for future oil and gas development, the location of ANWR's reserves might make exploration more attractive to industry, said USGS. Its economic analysis, which accounts for proximity to infrastructure, is particularly important in an area as large as the NPR-A, because some of the oil may be distant from existing facilities.

When market prices are below $35/ bbl, USGS concluded, the ANWR 1002 area contains more economically recoverable oil than NPR-A, and if prices exceed $35/bbl, the two areas have nearly equal volumes of economically recoverable oil.

Gas pipeline prospects

The US Congress in late 2002 abandoned a last-ditch effort to pass energy legislation before 2003, crushing hopes for financial incentives to build an Alaskan natural gas pipeline from the North Slope to the lower 48 states.

In November BP urged Congress to consider pipeline legislation as soon as possible to enable gas to flow within 10 years, which will be needed to keep pace with US

NPR-A RESOURCE ASSESSMENTS

	Oil F_{95}	Oil Mean	Oil F_{05}	Gas F_{95}	Gas Mean	Gas F_{05}
	Billion bbl			Tcf		
1980 assessment[1]	0.3	2.1	5.4	1.8	8.5	20.4
2002 assessment[2]	5.9	9.3	13.2	39.1	59.7	83.2

Note: F_{95} = 95% probability; F_{05} =- 5% probability.

[1]Includes entire NPR-A, as native selections had not been made. Does not include state offshore areas. Reported gas resources are total gas (nonassociated and associated). [2]Includes only the federal part of NPR-A; native lands and adjacent state offshore areas are excluded. Gas resources are nonassociated gas only.

Source: US Geological Survey

NPR-A VS. ANWR OIL POTENTIAL[1]

	Oil F_{95}	Oil Mean	Oil F_{05}	Size of area, million acres
	Billion bbl			
Entire area[2]				
ANWR 1002 area	5.7	10.4	16.0	1.9
NPR-A	6.7	10.6	15.0	24.2
Federal area				
ANWR 1002 area	4.3	7.7	11.8	1.5
NPR-A	5.9	9.3	13.2	22.5

Note: F_{95} = 95% probability; F_{05} =- 5% probability.

[1]Technically recoverable oil. [2]Includes federal and native lands and state offshore areas.

Source: US Geological Survey

ANWR VS. NPR-A OIL ECONOMICS

Market price, $/bbl	Economically recoverable oil, billion bbl — NPR-A	ANWR
15	0	0
—	—	—
20	0	3.2
21	0.4	4.0
22	1.3	4.4
23	2.8	5.0
24	3.1	5.2
25	3.7	5.6
26	4.0	5.8
27	4.8	6.0
28	5.1	6.2
29	5.4	6.3
30	5.6	6.3
35	6.4	6.6
40	6.9	6.8

Source: US Geological Survey

demand. Without a new gas supply, a gap of 150 bcm/yr (15 bcfd) or more is likely within 10 years, said BP, and Alaskan gas can close a third of this gap.

Producers earlier had assessed the cost, technology, regulatory, and environmental issues involved in building a 44 bcm/yr (4.5 bcfd) pipeline, expandable to 55 bcm/yr (5.6 bcfd). The study estimated that the toll to US markets would be \$64.20/cu m (\$2.39/Mcf) under either a southern route, which passes through Alaska's interior before entering Canada, or a northern route, which follows the Alaskan-Canadian coast.

A second pipeline, being promoted by Canadian producers that hold Mackenzie Delta gas reserves, is slated for completion within 8 years. However, BP said that gas demand will be robust enough for both projects. Duke Energy's Canadian unit said that the Mackenzie Valley pipeline would be built first, but that both lines would be constructed and would need US market support.

Instead of a controversial floor price that has been introduced in earlier legislative proposals, BP supports a "hybrid" plan, which includes a production tax credit of up to 52¢/MMBtu for Alaska gas transported to market. The credit would phase out to zero when prices are \$1.35/MMBtu.

The plan also calls for US government loan guarantees covering up to 80% of the project's capital cost and accelerated depreciation of the pipeline asset to 7 years from 15. Finally, BP would like to see a 35% tax credit for a gas treatment plant connected with the project.

Pipeline unharmed by quake

The Trans-Alaska oil pipeline was shut down briefly but escaped serious damage when an earthquake of 7.9 magnitude rocked central Alaska in early November. As a precaution, North Slope producers shut in all but 5% of their 1 million b/d of oil production, said BP, but no damage to BP facili-

ALASKA'S NORTH SLOPE

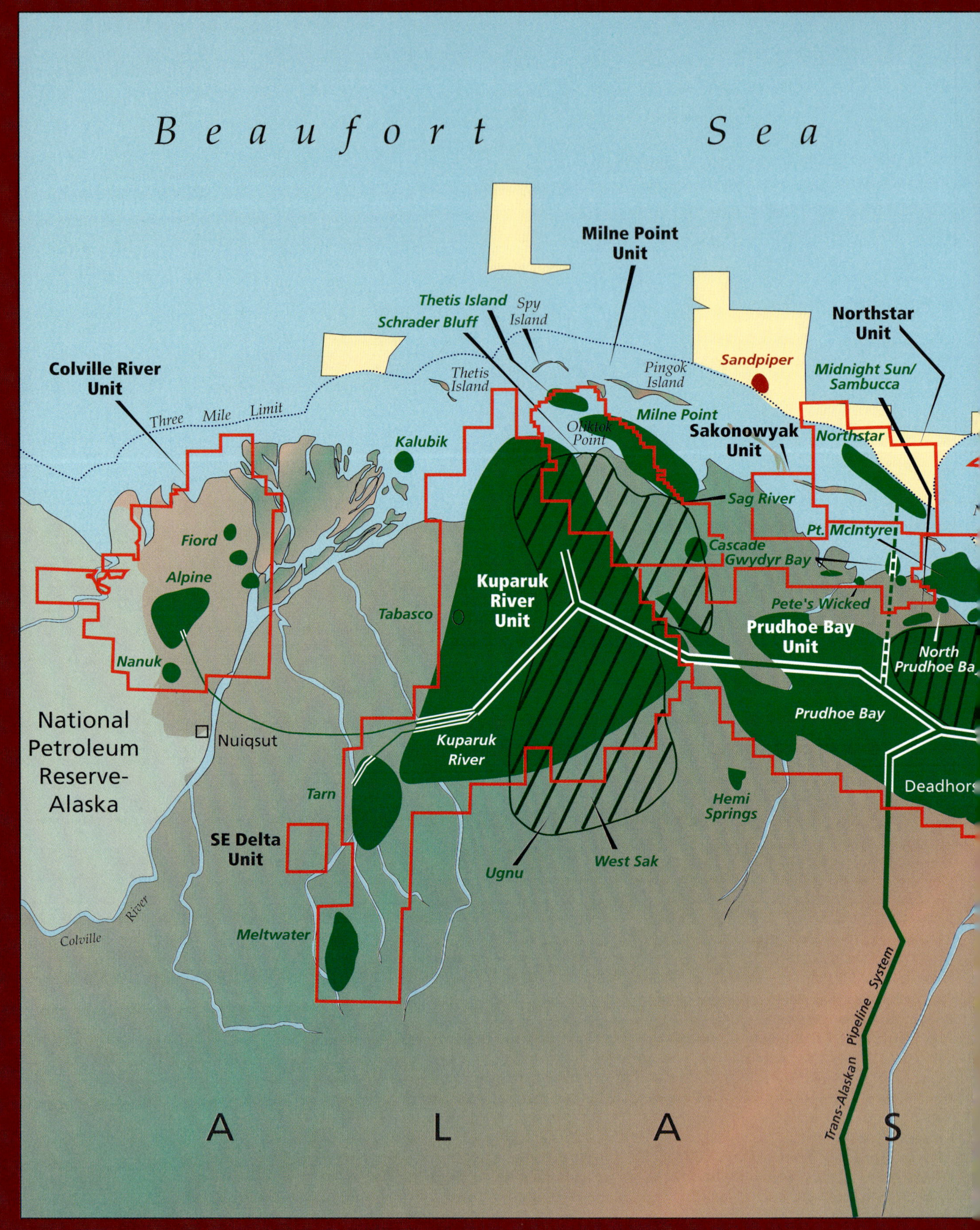

Area shown
Alaska
Oil fields
Gas fields
Oil pipeline
Proposed pipeline
Active OCS leases
Unit blocks
McCovey Unit
Cross Island
West Beach
Niakuk
Eider
Sag Delta
Sag Delta North
Duck Island Unit
Endicott
McClure Islands
Point Brower
Liberty
Stockton Island
Hammerhead
Kuvlum
Lisburne
Tigvariak Island
Maguire Islands
Bullen Point
Flaxman Islands
Flaxman Island
Point Thomson
Badami
Mikkelson
Badami Unit
Point Thomson Unit
Stinson
Sourdough
Sagavanirktok River
Staines River
Yukon Gold
Canning River
SluggerUnit
K
A
Arctic Coastal Plain - 1002 area
Arctic National Wildlife Refuge

ties or operations was reported. A refinery at Kenai also was unaffected.

The quake reportedly triggered a detection system that shut down the pipeline, but Alyeska Pipeline Service Co. (Anchorage) detected no leaks, and within 3 days the line was returned to service. Temporary supports were placed under some parts of the pipeline where support structures had been displaced, and two sections were pressure-tested before beginning the 6-hour restart.

Exploration

Alaska offered North Slope and Beaufort Sea areas for exploration in October 2002 under the state's competitive sales of 7-year oil and gas leases. The North Slope sale's 1,225 tracts totaled 69,200 sq km (5.1 million acres, or 26,700 sq mi) lying between ANWR to the east and the NPR-A to the west. The Beaufort Sea sale offered 516 tracts totaling 6,100 sq km (1.5 million acres, or 2,300 sq mi) of tide and submerged lands from Barter Island off ANWR to Tangent Point off NPR-A.

Earlier in the year, Alaska announced its leasing program, which provides the predictable schedule desired by industry. Each year through 2007, the state will hold a sale for each geographical area. In addition to the North Slope and Beaufort Sea sales held in October 2002, the program offered tracts in the Cook Inlet and North Slope Foothills in May 2002 and will do so again in May 2003.

Alaska also streamlined its permitting process in 2002 with new regulations that clarify the roles and authorities of participants and establish a more timely, predictable review. The change was made by the state's Coastal Policy Council with input from many stakeholders. The regulations were being incorporated into the Alaska Coastal Management Program.

Working interest and operatorship in 10 state leases covering 57 sq km (14,000 acres, or 22 sq mi) on Alaska's North Slope were acquired by Pioneer Natural Resources in October. The offshore leases lie in shallow waters of 1-3 m (5-10 ft) between the Kuparuk River unit, which holds an estimated 2.5 billion bbl of recoverable oil, and Thetis Island. During 2002-03, Pioneer plans to drill up to three wells to test the area, which could be prospective for oil in the same sands as the Kuparuk River.

Cook Inlet E&P

Forest Oil

Shallow water Redoubt Shoal field began producing oil in December 2002, said Forest Oil Corp. (Denver). Flow rates exceeded 4,200 b/d from Redoubt Nos. 3 and 4. The oil was transported to Forest's facilities at West McArthur River field. Redoubt Shoal production is expected to stay at or near this rate until new onshore facilities at Kustatan are completed and ramp up to 15,000 b/d in 2003.

Redoubt No. 4, drilled to a total measured depth of 6,158 m (20,203 ft), was the

ALASKA OIL AND GAS LEASING PROGRAM

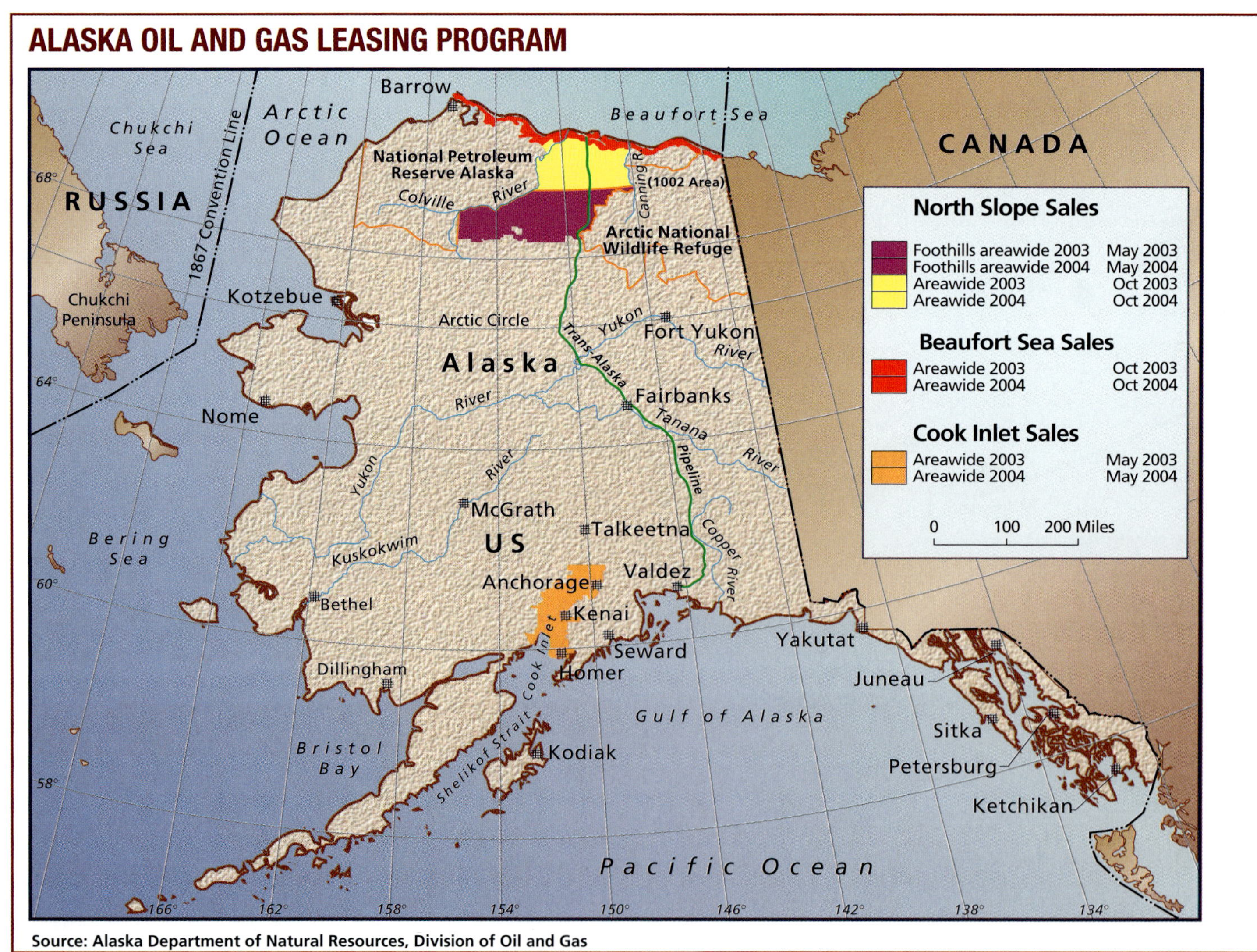

Source: Alaska Department of Natural Resources, Division of Oil and Gas

deepest deviated well ever drilled in Cook Inlet, Forest contends. The well logged 70 m (229 ft) of net oil pay, it extended the deepest known oil column by about 15 m (50 ft), and it logged 180 m (589 ft) of net gas pay in multiple shallow sands. Redoubt Shoal's recoverable oil is estimated to be at least 100 million bbl, and reserves ultimately could reach 220 million bbl, said Forest, which plans to resume drilling after finalizing completion work.

Forest estimates that Cook Inlet holds 30-100 million bbl of oil and 1.3-30 bcm of gas (50-1,100 bcf) yet to be found and developed. Other exploration prospects have been identified, including Corsair, Raptor, Sabre, Tutna, and Valkyrie, within Forest's Cook Inlet lease area, which covers about 1,100 sq km (270,000 acres, or 420 sq mi). Some of these prospects could be drilled in 2003.

Other Cook Inlet prospects

In Cook Inlet, like any maturing oil province, major oil and gas companies have tended to pull back E&D efforts while independents have move in to fill the void. For example, Escopeta-BBI Inc. (Houston), the third largest state leaseholder in the inlet, anticipates that it will start to drill in mid-2003. The company, which was seeking financing, has two major prospects in Cook Inlet: East Kitchen, north of the East Forelands area of the Kenai Peninsula, and Kitchen, just off East Forelands.

Aurora Gas, LLC (Houston) in late 2002 was acquiring holdings from ConocoPhillips and Anadarko in the Moquawkie area on the inlet's west side. The area, which covers about 170 sq km (43,000 acres, or 67 sq mi), includes the Lone Creek No. 1 discovery well and would bring Aurora's total area to about 400 net sq km (100,000 acres, or 160 sq mi) in the inlet. Aurora finished two reentry wells on its Nicolai Creek acreage in late 2020 and plans a new well there in 2003.

COOK INLET OIL, GAS OPERATIONS

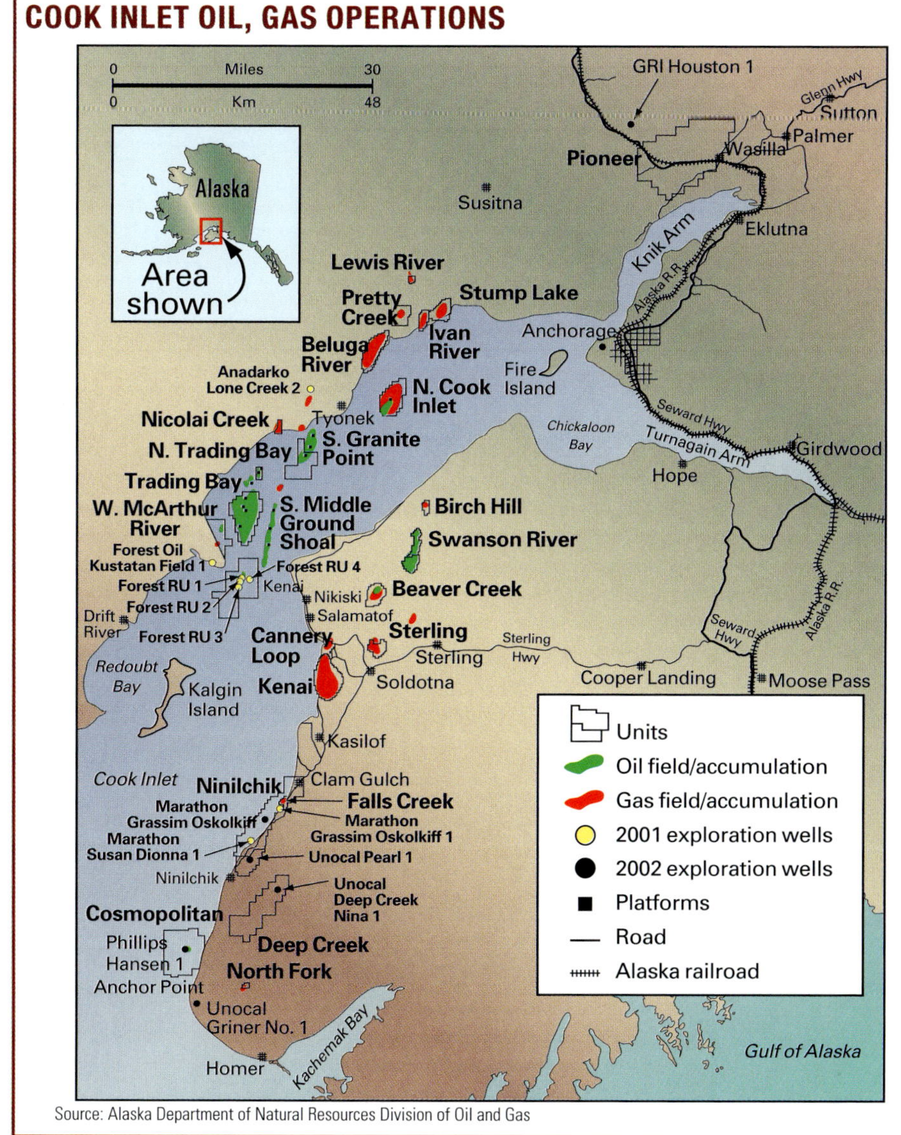

Source: Alaska Department of Natural Resources Division of Oil and Gas

XTO Energy Inc. (Fort Worth, TX) holds interest in Middle Ground Shoal field, where its target is underlying oil. The field has a relatively low production decline rate of 7%/yr and could keep producing for the next 20-25 years. A deeper Jurassic horizon under the field's current production reservoir offers the chance of additional production. In late 2002 XTO was drilling the C13-13 LN, its first well to test the east flank. The field has 450 million bbl of oil in place, with 120 million bbl produced so far. Production averaged 4,385 b/d in late 2002, with 3,837 b/d net to XTO.

In addition to independents, a few majors have clung to the Cook Inlet as well. Marathon, which remains one of the largest natural gas producers there, has acquired acreage that includes five prospects and two recent natural gas discoveries at Sterling Deep and Wolf Lake field within the Kenai National Wildlife Refuge.

Marathon and Unocal discovered a new gas reservoir in early 2002 in the Ninilchik Exploration Unit on the Kenai Peninsula. The Grassim Oskolkoff No. 1 well flowed 301,000 cu m/d (11.2 MMcfd) on test from one zone. The discovery was followed by two successful exploration wells. Grassim Oskolkoff No. 2 tested at a combined flow rate of 320,000 cu m/d (11.9 MMcfd) from three zones, and Falls Creek No. 1 RD flowed from a single zone at a rate of 183,000 cu m/d (6.8 MMcfd) from pay at 2,656 m (8,714 ft).

Ninilchik is estimated to contain more than 2.4 bcm (90 bcf) of gross proven reserves. Marathon and Unocal are planning a 53-km, 1.2 bcm/yr (33-mile, 120 MMcfd), 12-inch gas pipeline to connect these supplies with the existing system. The Kenai-Kachemak pipeline would serve commercial and utility markets in south-central Alaska. Design work and permitting was under way, and first production is anticipated for late 2003.

Other E&P

Oil production began sooner than expected from Kuparuk River field's Drill Site 3S, which contains the Palm exploration wells, said ConocoPhillips and BP units in December. The site was producing 16,000 b/d of oil from three wells at Kuparuk, North America's second largest oil field.

The Greater Kuparuk Area, which includes four satellite fields, was producing about 210,000 b/d. The Palm discovery extends Kuparuk field 5 km (3 miles) to the northwest and contains an additional estimated 35 million bbl of oil reserves.

A successful four-well pilot program was completed in mid-2002 by BP Exploration (Alaska) Inc. and its team. The program involves drilling long, horizontal, multilateral sidetracks using coiled tubing. The four wells were drilled into Milne Point Schrader Bluff field's shallow, heavy-oil reserves on the North Slope.

With drilling rates routinely in excess of 250 fph, the project demonstrated that coiled tubing could quickly drill through-tubing laterals longer than 760 m (2,500 ft) into the sandstone reservoirs, delivering good incremental oil production rates at low cost.

The Polar Resolution loaded its first shipment of Alaska North Slope crude in July. The vessel, operated by a ConocoPhillips unit, is one of the newest Millennium-class, double-hulled tankers to join the Polar fleet in Alaska.

The Resolution is ConocoPhillips' second of five 140,000-dwt oil tankers to be added to the fleet through 2005. In April the third vessel, the Polar Discovery, was christened and will join the fleet in 2003. Each ship carries a maximum 1 million bbl.

IPE

NORTH AMERICA

Canada

CANADA

CAPITAL: OTTAWA

MONETARY UNIT: CANADIAN DOLLAR

REFINING CAPACITY: 1,983,450 B/CD

OIL PRODUCTION: 2.195 MILLION B/D

OIL RESERVES: 180.021 BILLION B/D

GAS RESERVES: 60.118 TCF

Canada remained one of the most important sources of US energy imports during 2002. Its economy, highly dependent on trade, suffered from slow growth worldwide and especially in the US. However, the International Monetary Fund (IMF) projected that Canada's real GDP will grow by 3.4% in 2003.

Canada's share of world oil reserves leaped in 2002, reflecting the inclusion of Alberta's oil sands resources. The previous estimate of Canada's proved oil reserves, 4.858 billion bbl, jumped by a factor of 37, to 180.021 billion bbl, said *Oil & Gas Journal.* This change gives Canada nearly 15% of the world's oil reserves and reduces OPEC's share by more than 10 percentage points. Canada's natural gas reserves in 2002 increased by less than 1%.

In late 2002 the Canadian Association of Petroleum Producers (CAPP) released its oil and gas data for the previous year, which showed that new gas reserves during 2001 replaced 106% of gas sold, reflecting drilling emphasis on gas. Saskatchewan led the provinces with a 196% replacement rate. However, reserves of crude oil and equivalent declined during 2001. CAPP said the oil and gas industry is Canada's largest private sector investor.

CAPP noted North American demand for natural gas will challenge Canada to supply its share, given rising depletion rates and a decline in drilling due to lower prices. Most gas reserves additions have come from Sable Island fields off Nova Scotia and from Ladyfern in British Columbia. Although Sable gas production is still increasing, Ladyfern will soon decline sharply.

Canadian analysts predicted that gas exports to the US in 2003 will fall by 2%, due to production declines and rising domestic demand, especially consumption by oil sands operations. During 2002, Canada exported 0.5% less gas to the US than in 2001, the first drop since 1986. At year-end, Canada's gas inventories were about 30% lower than the previous year and 19% lower than the 5-year average.

In December 2002, Canada ratified the Kyoto climate change agreement, requiring a 6% reduction in carbon dioxide emissions by 2012 from 1990 levels. Full details of implementation might not be revealed for another 1-2 years, said Lehman Bros., and

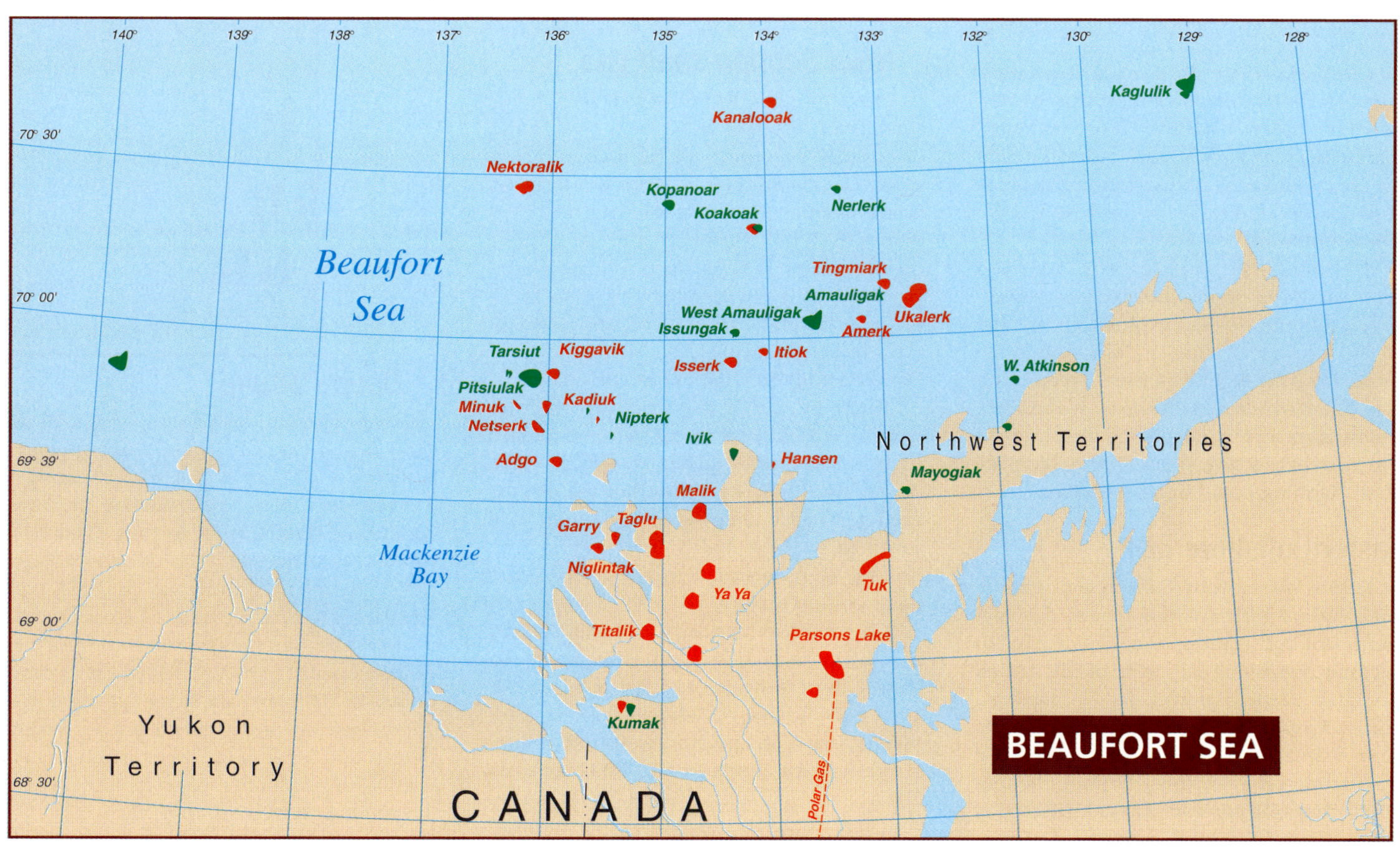

uncertainty surrounding the Kyoto protocol could delay oil sands projects. CAPP said that Kyoto will have a negative impact on economic growth.

Provincial activities

British Columbia, following a review of its 30-year-old ban on exploration off Canada's west coast, found no scientific or legal reason to maintain the moratoriums. However, many issues remain to be resolved regarding intergovernmental jurisdictions and aboriginal rights.

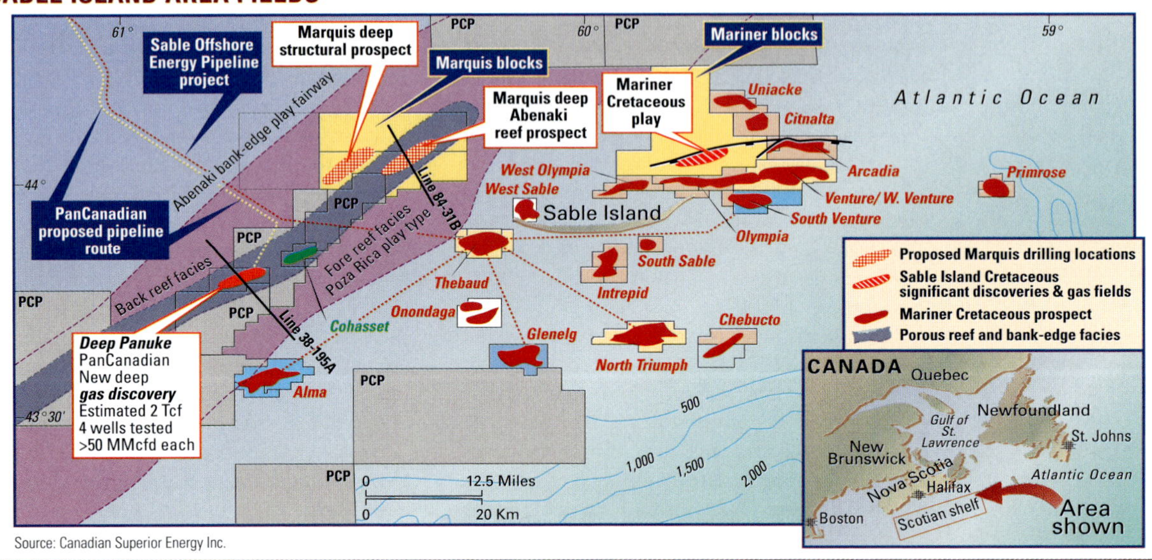

Source: Canadian Superior Energy Inc.

CAPP and Chevron Canada Resources, the offshore BC region's largest leaseholder, called the report a positive step. Previously, the Geological Survey had estimated that the area could hold undiscovered resources of 693 bcm (25.8 tcf) of natural gas and 9.8 billion bbl of oil.

A boundary dispute was resolved between Nova Scotia and Newfoundland, which should foster exploration of the 60,005 sq km (23,168 sq mi) Laurentian subbasin. Old federal permits were being converted to current licenses in the area, which contains a postulated resource of 240 bcm of gas (9 tcf) and 700 million bbl of oil. Drilling could begin in 2004.

Eastern offshore E&P

Basins off eastern Canada will experience increasing exploration and development drilling, leading the East Coast to become a growing exporter of oil, natural gas, and electric power to US markets. Within 5 years, Newfoundland will produce more than a third of Canada's conventional light crude oil, said Roger Grimes, Premier of Newfoundland and Labrador, in May 2002. The Newfoundland offshore area covers about 580,000 sq km (225,000 sq mi), with eight basins of interest.

Nova Scotia Premier John F. Hamm said his province anticipates drilling of at least another 17 deepwater wells in the next 4 years and a minimum of nine shallow water wells during the next year or two. Nova Scotia has 59 active exploration licenses, including 26 that qualify as deepwater blocks, which carry a base $6 billion in spending commitments; actual outlays should exceed that.

Nova Scotian shelf gas

Nova Scotia's Sable Island area holds established reserves of 81 bcm (3 tcf) and discovered resources of 54 bcm (2 tcf), while undiscovered resources for the whole Scotian shelf are estimated at 350 bcm (13 tcf). However, the deepwater Scotian slope could have undiscovered potential of 400-1,100 bcm of gas (15-41 tcf) and 2-5 billion bbl of oil and condensate, said the Canada-Nova Scotia Offshore Petroleum Board (Halifax) in November.

Early in 2002, the region's second major gas field was being developed, and exploration was poised to thrust into deepwater territory. Marathon Oil discovered gas in its deepwater Annapolis well south of Sable Island. More seismic and drilling were planned for 2003 to determine commerciality.

ChevronTexaco Canada began drilling the Newburn H-23 wildcat off Nova Scotia in 949 m of water (3,115 ft) in mid-2002, and Kerr-McGee Offshore Canada Ltd. planned two deepwater wells off southwestern Nova Scotia in late 2002 or early 2003.

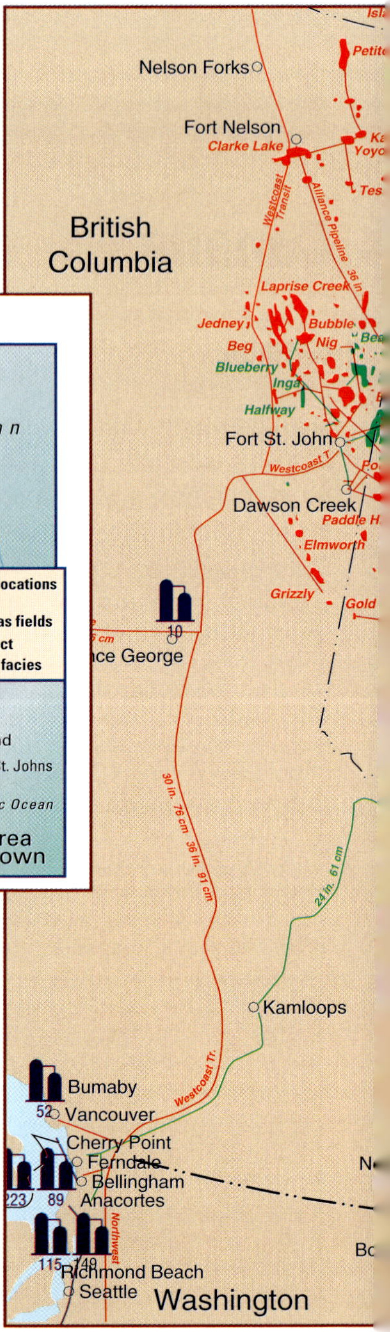

Deep Panuke

Deep Panuke field, lying about 250 km (150 miles) east of Halifax with estimated reserves of 27 bcm (1 tcf), was emerging as a world-class play and could be the most significant discovery in Atlantic Canada in more than a decade. The design phase of development was underway in 2002, with regulatory approval expected by year-end.

Deep Panuke is expected to produce 11 million cu m/d (400 MMcfd) when it comes online in 2005, with capacity expandable to 17 million cu m/d (650 MMcfd). Several delineation wells were successfully tested, each with productive capacity in excess of 1.3 million cu m/d (50 MMcfd).

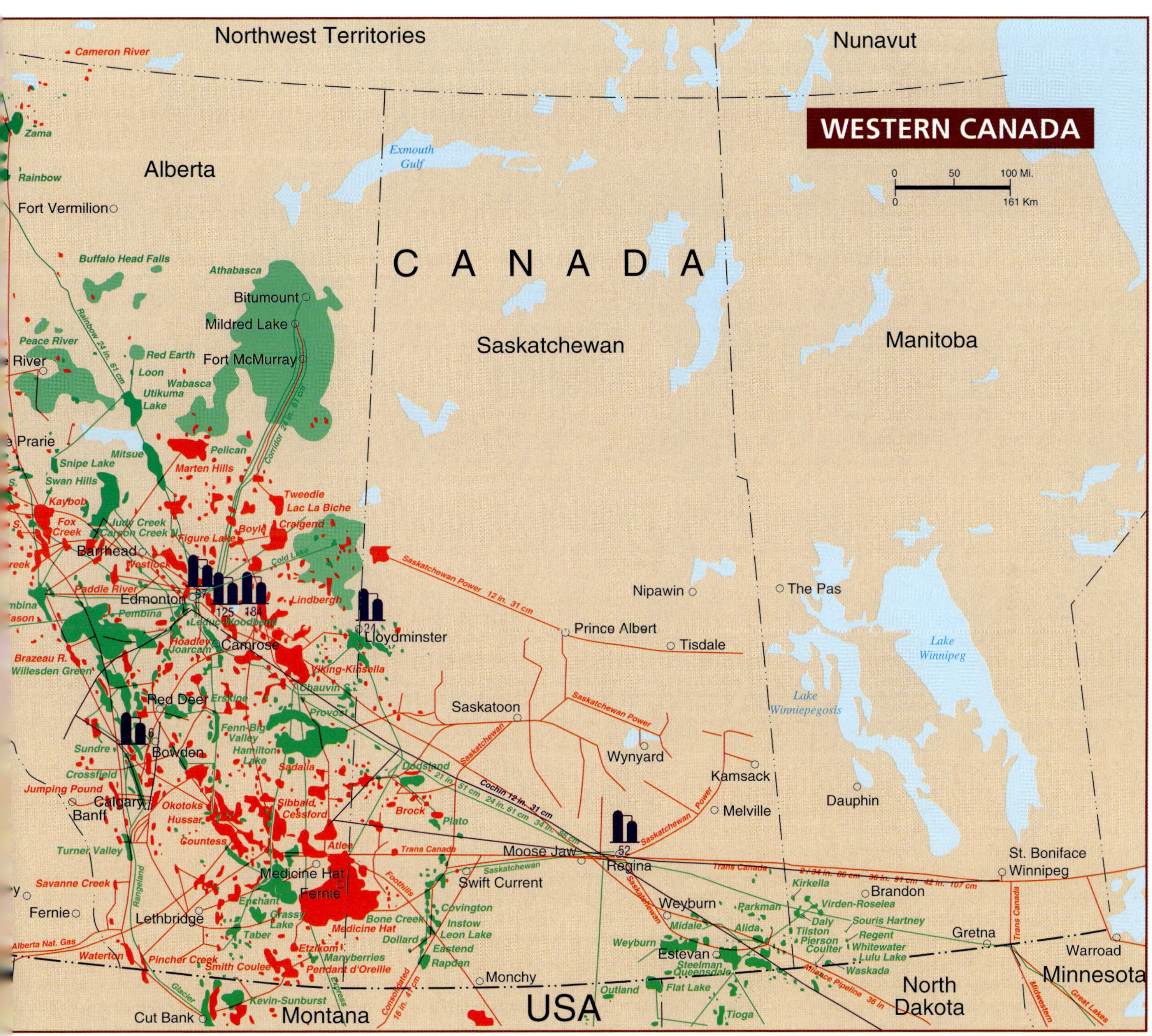

Applications to develop Deep Panuke were submitted in spring 2002 by PanCanadian Energy Corp. (Calgary), now EnCana Corp. PanCanadian anticipated spending $1.1 billion (Can.) on the project, which involves the construction of platform-based gas processing facilities offshore as well as a subsea pipeline network to transport the gas to shore.

Offshore, components will include three new bridge-linked platforms installed near the existing Panuke platform. About 179 km (111 miles) of 24-inch pipeline will carry gas from the platform complex to landfall at Goldboro, NS, where the line will tie in with the existing Maritimes & Northeast Pipeline system, which will be expanded to 8 bcm/yr (800 MMcfd) to handle the volumes.

Marquis prospect

A wildcat in the Marquis prospect, which lies 20 km (12 miles) northwest of Sable Island in water depths of less than 100 m (300 ft), was abandoned in September by Canadian Superior Energy Inc. (Calgary). The prospect is directly on trend with and analogous to the Deep Panuke Abenaki reef natural gas discovery 25 km (15 miles) to the southwest.

Although the Marquis L-35/L-35A test well failed to find commercial hydrocarbons, Canadian Superior believes the results enhanced the prospectivity of the Marquis area, where its exploration licenses cover about 450 sq km (110,000 acres, or 170 sq mi), 160 km (100 miles) offshore. An El Paso unit is partner with Canadian Superior in a joint venture for Marquis exploration.

Three industry wells are proposed for drilling to evaluate the Abenaki reef on both sides of Canadian Superior's Marquis Blocks and its Mariner Block, covering 412 sq km (101,800 acres, or 159 sq mi) in about 60 m of water (200 ft), adjacent to existing pipeline infrastructure.

Sable Island production

Gas production averaged 14-15 million

cu m/d (530-550 MMcfd) during mid-2002 at the Sable Island Offshore Energy Project (SOEP), which includes three fields off Nova Scotia. Venture, the largest SOEP field, holds 43 bcm (1.6 tcf) of proven reserves. ExxonMobil and partners put Sable reserves at 70 bcm of gas (2.6 tcf). A fourth field will go online in 2003.

Grand Banks oil

White Rose

Development of White Rose field, in the Grand Banks area's Jeanne d'Arc basin off Newfoundland and Labrador, got a green light in spring 2002, said Husky Energy Inc. and partner Petro-Canada (both of Calgary). Start-up is pegged for late 2005, with reserves estimated at 200-250 million bbl, the third and smallest development in the region. A fourth project, Hebron-Ben Nevis, which contains heavier oil than nearby fields, was deferred in February by Chevron Canada Resources.

Husky's White Rose development plan is centered on installation of an FPSO vessel with a production capacity of 100,000 b/d. Topsides facilities work was awarded to Aker Maritime Kiewit Contractors in April, and the subsea production system, including 42 km (26 miles) of piping and five manifolds, was contracted with Technip-Coflexip in September.

Plans call for 19-21 wells to recover up to 250 million bbl of oil over a 10-15 year period, with development drilling scheduled for 2003 and subsea system installation in mid-2004. White Rose production is estimated to peak at 92,000 b/d and be sustained for 4 years. The field is 350 km (220 miles) east of Newfoundland in 115-130 m of water (337-430 ft).

Terra Nova, Hibernia

Terra Nova field, about 50 km (30 miles) from Hibernia oil field, began production in January 2002 after several delays. The field achieved its 125,000 b/d capacity in 9 days from start-up. Terra Nova has 370-470 million bbl of proved and probable reserves and is expected to produce for about 15-18 years. Terra Nova's FPSO can produce up to 150,000 b/d of oil and store 960,000 bbl. The double-hulled vessel is specially reinforced to withstand icebergs.

Hibernia field, 350 km (220 miles) southeast of St. John's, Newfoundland, was averaging more than 160,000 b/d of production in mid-2002. Cumulative production was 183 million bbl, and original reserves are 884 million bbl.

As exploration matures in the Jeanne d'Arc basin, companies are eyeing deeper water prospects and expanding into new areas such as the Flemish Pass basin, east-northeast of Jeanne d'Arc, the South Whale basin, and Laurentian subbasin.

MacKenzie Delta-Beaufort Sea

Significant natural gas flows were found at the Tuk M-18 well 25 km (15 miles) south of Tuktoyaktuk in the Mackenzie Delta of Canada's Northwest Territories, announced Devon Energy and Petro-Canada in May.

The well, drilled to 3,000 m (9,850 ft), was tested at restricted rates of up to 800,000 cu m/d of natural gas (30 MMcfd). It has estimated reserve potential of 5-8 bcm (200-300 bcf) and sustained deliverability of 1.6-2.1 million cu m/d (60-80 MMcfd).

The well is part of the companies' Mackenzie Delta exploration partnership, aimed at securing pipeline space when the area's gas is tied into North American markets via a proposed pipeline along the Mackenzie Valley. Devon and Petro-Canada hold combined rights to some 4,000 sq km (1 million acres, or 1,600 sq mi) in the Mackenzie Delta — the largest landholding in the industry.

Onshore E&P

Yukon's Peel Plateau

Exploration rights in the Peel Plateau of northwest Canada were granted in early 2002 to a Hunt Oil unit, which will spend $1.16 million to explore 155 sections with an area of about 402 sq km (155 sq mi) just south of the Arctic Circle. However, a license will be required, triggering an environmental assessment. Canada's National Energy Board estimated the Peel Plateau's potential at 61.5 bcm of gas (2.29 tcf) and 21.3 million bbl of oil.

Exploration will resume in the Peel Plateau's Grandview Hills region northwest of Fort Good Hope, said Devlan Exploration Inc. (Calgary) and Vintage Petroleum (Canada) Inc. in December. With gross acreage of more than 4,000 sq km (1 million acres, or 1,600 sq mi), they plan to complete and test three wells drilled earlier that showed promise in the Devonian Bear Rock formation.

A new, deeper well is also planned, Tree River C-36, on a separate feature 9 km (6 miles) from the original Tree River B-10 well, which had earlier been drilled and cased to 1,295 m (4,249 ft). Projected total depth is 1,890 m (6,200 ft). Devlan sees Devonian, Ordovician, and Cambrian potential in the area.

Alberta gas, coalbed methane

Alberta's Wild River prospect was being explored in mid-2002 by Wiser Oil Co. (Dallas), which drilled and tested two wells and recompleted a third, all in township 57-24W5. The two new wells were drilled to 3,000 m (9,800 ft), and all three encountered one or more natural gas pay zones. Their gross combined test rate was 403,000 cu m/d (15 MMcfd). Pipelines and a new compression facility were being installed, with production to begin in 2002.

Coalbed methane was also a target in Alberta, where Trident Exploration Corp. (Calgary) farmed into the Cessford properties of Promax Energy Inc. (Calgary) in early 2002. Trident will pay for a 12-well pilot program, earning the right to begin a commercial drilling project. The Promax acreage hosts interesting shallow and deeper coal seams.

Production of coalbed methane gas from southern Alberta into sales lines began in early 2002, marking the first significant sales production in Canada, said a joint venture of EnCana and MGV Energy Inc. (Calgary). As of October, 14 wells were producing gas on an extended basis, and individual wells were each expected to produce 0.8-6.7 million cu m/d (30-250 Mcfd).

The partnership was planning a 250-well coalbed methane development on its 300-section Palliser Block, which covers about 4,000 sq km (1 million acres, or 1,600 sq mi) in southern Alberta. About 50 wells were to be completed by year-end and the rest in 2003, said the partners in October. Net reserves on the block could be, per section, 27-54 million cu m of natural gas (1-2 bcf). Outside the Palliser area, 12 of 25 proposed exploration wells were being completed and tested in central and southern Alberta.

A five-well coalbed methane pilot project at Corbett Creek was planned for late 2002 by Canscot Resources Ltd. (Calgary). Core analysis indicated more than 2.2 bcm (81 bcf) of coalbed methane in place in the two main Upper Mannville coal seams alone, of which 1.3 bcm (50 bcf) could be recoverable, including 0.94 bcm (35 bcf) net to Canscot. Methane from the pilot wells would be produced into existing processing and compression facilities. Canscot planned to start gas sales by year-end, following dewatering. If the pilot succeeds, 56 more wells could be drilled on spacing of 0.65 sq km (160 acres, or 0.25 sq mi).

Newfoundland

Oil and gas production began in mid-2002 from Garden Hill field in western Newfoundland, said Canadian Imperial Venture Corp. (St. John's, Newfoundland). The production lease was the first ever for onshore operations in the province. It covers about 130 sq km (33,000 acres, or 52 sq mi) along a prospective fairway on the Port au Port Peninsula and includes the Port au Port No. 1 discovery well.

Shares in Deer Lake Oil & Gas Inc. (St. John's, Newfoundland) will be sold to the public with provincial approval granted in September. Although the company holds oil and gas exploration rights on four western Newfoundland properties, its focus is on 292 sq km (113 sq mi) in the onshore Deer Lake basin's North Brook formation.

New Brunswick-McCully

McCully gas field appears to have been extended to the northeast, said Corridor Resources Inc. (Halifax). McCully D-48, 3 km (2 miles) northeast of the field's discovery well, cut a net 20 m (70 ft) of gas bearing sands in early 2002. As of December McCully had seven successful wells that encountered gas. The field lies within the Maritimes basin, where geological data suggest a high level of prospectivity. Stoney Creek field, 45 km (28 miles) from McCully, has two successful wells.

Once started up, McCully field will deliver an average 62,000 cu m/d of gas (2.3 MMcfd) from two wells through a short pipeline to a mill operated by Potash Corp. of Saskatchewan, which owns 50% of the wells. The gas will back out No. 2 fuel oil and will be priced based on that fuel's cost. These two wells are expected to recover combined gas of more than 270 million cu m (10 bcf) during 20 years and will sustain mill demand for 4-5 years before declining.

The two wells are expected to drain less than 8% of the original Corridor-Potash four-section area of 14.32 sq km (5.53 sq mi). The gas accumulation is undefined and appears to be much larger. As many as 13 spacing units remain undrilled. Corridor believes that three other wells drilled outside the original area in 2002 have more than doubled the potential unrisked gas-in-place figure of 13.4 bcm (500 bcf) that consulting engineers attached to the first four McCully wells. The field area is now considered to be 15 km long by 5 km at its widest point (9 by 3 miles), and some data indicate the ultimate dimensions could be 22 by 8 km (14 by 5 miles).

Corridor expects that drilling in 2003 could find enough gas to justify construction of a 50-km (30-mile) pipeline connection to the Maritimes & Northeast Pipeline. Options for further drilling include multilateral wells to develop McCully's stacked reservoirs at depths of 2,000-2,900 m (6,600-9,500 ft). Horizontal drilling is practically untried in the onshore Maritimes and may have application at McCully.

British Columbia foothills

Commercial gas was discovered in mid-2002 in the Monkman area of British Columbia's foothills by Talisman Energy Inc. (Calgary) and partners. It was the first commercial discovery to produce from the deeper, underexplored Paleozoic reservoirs in the area. The find opened up a major potential new play system, Talisman said. The discovery well, about 180 km (110 miles) northeast of Prince George, was to begin producing at least 403,000 cu m/d (15 MMcfd) in September.

The deeper play will cover a wide area and appears to contain well in excess of 27 bcm (1 tcf) of unrisked, recoverable gas. Talisman said it held the well data confidential until after the July land sale, where the company and its partners picked up 73.8 sq km of land (18,236 acres, or 28.5 sq mi) on trend with the discovery well for $12 million.

Oil sands revival

Canada's oil sands sector is expected to produce more than 50% of the nation's oil by 2010, when surface mining could be producing 1 million b/d and in-situ recovery projects another 0.7-1.2 million b/d. In mid-2002 bitumen production by the two major miners, Syncrude Canada Ltd. and Suncor Energy Inc. (both of Calgary), was close to 500,000 b/d. Suncor boosted its 2003 capital budget to $1.05 billion from $900 million in 2002, allocated mostly to Alberta oil sands.

During 2002, oil sands operators moved toward in-situ techniques to tap deeper formations in northern Alberta, although surface mining still accounted for most of the investment and production, and established mining companies also were expanding their operations.

Of an estimated 300 million bbl of bitumen, about 80% can be produced only by techniques such as steam-assisted gravity drainage (SAGD), as opposed to conventional mining. Currently, Alberta and Saskatchewan have about 30 SAGD projects, four of which are large-scale commercial projects under development.

Mildred Lake

Syncrude Canada, Canada's largest oil sands producer, was in the third phase of a program to increase production by 40% to 368,000 b/d by 2005. The upgrader at Mildred Lake mining site was being expanded and is due to start up in late 2003. The final phase of the Syncrude 21 program was scheduled to begin in 2004, including a third production train at the Aurora Mine and more expansion of the Mildred Lake upgrader. Phase 4 should increase production to 170 million bbl/yr.

Earlier in the year, Syncrude blamed maintenance issues for lower-than-expected production in 2001. The company produced 81.4 million bbl of syncrude from Athabasca oil sands bitumen in 2001, up 10% from 2000 but less than the expected 90-94 million bbl.

Muskeg River

The Athabasca Oil Sands Project, the first new, fully integrated oil sands project in 25 years, started production of bitumen in December 2002 at the Muskeg River

MACKAY RIVER PROJECT

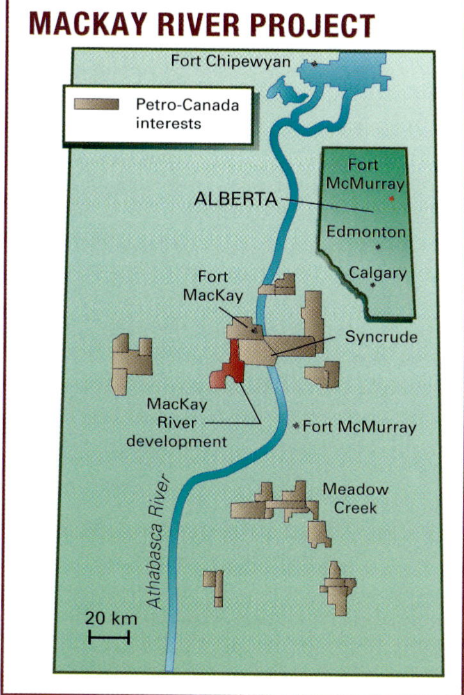

mine, 75 km (47 miles) north of Fort McMurray. The mine was to reach full production of 155,000 b/d by mid-2003, supplying the equivalent of 10% of Canada's oil needs. The project, led by Shell Canada Ltd., has been one of the world's largest construction projects in recent years.

A bitumen upgrader and modified refinery at Scotford near Edmonton are connected to the mine by two pipelines, one carrying bitumen south and the other returning diluent north. Commissioning and testing of the synthetic crude oil units were on schedule at year-end 2002, with first synthetic crude production expected early in 2003. A brief fire at the mine delayed bitumen deliveries slightly, pending the restart of Train 1, which sustained minor damage.

Additional expansion was being examined, including increasing Muskeg's output to 225,000 b/d during 2005-2010 and developing the new Jackpine Mine on the same lease. Jackpine could produce 200,000 b/d of bitumen. Shell's long-term goal is production of 530,000 b/d.

MacKay River

Petro-Canada emerged as a leader in large in-situ projects using SAGD technology. In October operations started up at its MacKay River in-situ project, where each of 25 initial well pairs is expected to produce about 1,200 b/d of bitumen by year-end.

The project was on schedule for full production of 30,000 b/d by 2004, making it the largest commercial operation of its kind in Canada. Petro-Canada hopes to convert an existing refinery to process the bitumen and was also planning a second in-situ project for Meadow Lake in the Athabasca region.

Long Lake

Another large in-situ SAGD project is Long Lake, a joint venture of OPTI Canada Inc. and Nexen, which combines SAGD with on-site upgrading. The upgrading scheme includes OPTI's OrCrude process, which produces an asphaltene by-product that will be fed to a gasification system to produce hydrogen for the hydrocracker and a syngas to fuel the SAGD steam generators. The use of syngas eliminates the variables of natural gas price and availability from project economics, said OPTI.

The Long Lake project was under regulatory review in late 2002. Construction should start in 2003 and production in 2006. The first phase is designed to produce and upgrade 70,000 bbl of 8-degree gravity bitumen to a 60,000 b/d stream of 20-degree gravity product.

Suncor projects

Production approached 225,000 b/d at Suncor Energy's Millennium mine, following an expansion, and Suncor began focusing on the four-phase Firebag in-situ SAGD project, which will develop estimated resources of 9.6 billion bbl, and the longer term Voyageur expansion.

Firebag's first phase was under construction, with each phase to contribute 35,000 b/d of production for a total increase of 140,000 b/d. The first phase will boost Suncor production to 260,000 b/d by 2005.

Tar sands evaluation

The potential for light oil underlying the Athabasca tar sands was being evaluated by Canadian geologists, who planned to reenter a 1,700-m (5,500-ft) well on the west outskirts of Fort McMurray and production-test five zones that previously had oil or gas shows.

The Athabasca tar sands is one of the world's largest oil deposits, with 1.3 trillion bbl in place across 28,000 sq km (11,000 sq mi). Although SAGD is a proven technology for producing bitumen from Athabasca tars, the technology has not been completely successful in lighter heavy oils or other formations.

Other oil sands

The Cold Lake system is Canada's largest heavy-oil gathering complex, consisting of 400,000 bbl of oil storage and more than 900 km (560 miles) of heavy blend and condensate pipelines that transport 220,000 b/d from the Cold Lake oil sands production region to crude oil market hubs at Edmonton and Hardisty, Alberta.

An EnCana unit applied for regulatory approvals to build Phases 2 and 3 of an existing SAGD in-situ project at Foster Creek in the Cold Lake area. The first phase is ramping up to 20,000 b/d after starting up at 15,000 b/d in late 2001. The second phase would push output to 65,000 b/d by 2005 and the third phase to 100,000 b/d by 2007.

Another in-situ project, at Christina Lake, was to begin steaming in mid-2002, with production to start in the fall. Phase 1 would ramp up to 10,000 b/d by 2003, followed with 40,000 b/d for Phase 2 and 70,000 b/d expected in 2008.

Petroleum pipelines

BC Gas Inc. (Vancouver) announced in early 2002 plans to build the 100,000 b/d, 517-km (321-mile), Bison bitumen pipeline from Fort McMurray to processing plants in the Edmonton area. A regulatory application was to be filed by year-end. Bison's capacity could be increased to 450,000 b/d, and heating of the line would greatly reduce the amount of diluent needed to pipe bitumen.

BC operates TransMountain Pipe Line, which moves crude oil and refined products to the West Coast. BC was also building the Corridor Pipeline to move bitumen from Athabasca oil sands to Shell's new Edmonton upgrader.

Enbridge Inc. (Calgary) was considering a 350,000 b/d heated bitumen pipeline to Edmonton with a time frame of 2005-07 and over the longer term an oil line from Alberta to the US West Coast with a capacity of 400,000 b/d.

Gas pipelines

In early 2002, Canada's Mackenzie Delta Producers Group, led by Imperial Oil Ltd. (Toronto), said it would seek regulatory approval to build a natural gas pipeline to Alberta, although the process could take 2 years before an application could be filed.

The route would pass mainly through the Northwest Territories, where aboriginal groups now support the development. The pipeline would take gas from three Mackenzie Delta fields: Taglu, with 81 bcm (3 tcf); Parsons Lake, 48 bcm (1.8 tcf); and Niglintgak, 27 bcm (1 tcf).

In June an agreement on 2003 pipeline tolls was reached with customers of Maritimes & Northeast Pipeline LP (Halifax), which will charge $0.6896/MMBtu (Can.). Maritimes planned to expand its system and add new customers.

Steel line pipe for the 63.5-km (39.5-mile) Westpass expansion was ordered in September by TransCanada PipeLines Ltd. (Calgary). The pipe, touted as the world's first commercial production and application of the X-100 (Grade 690) light-weight, high-tensile-strength pipe, will be installed on 1 km (0.6 mile) of the expansion. Westpass carries gas to British Columbia and the western US.

In February Enbridge shelved plans to build a pipeline connecting Sable Island to Quebec.

Downstream activities

A cogeneration plant was being planned in September by Irving Oil Ltd. (St. John, New Brunswick) and TransCanada Energy Ltd. (Calgary). The natural gas-fueled facility will produce steam and power for the St. John refinery, enhancing its energy efficiency and reliability while supporting New Brunswick's energy policy promoting cogeneration.

The world's first world-scale plant for production of polytrimethylene terephthalate (PTT), a thermoplastic polymer, was announced by Shell Chemicals Canada Ltd. (Montreal) and SGF Chimie (France) in a 50-50 partnership called PTT Poly Canada. The 95,000 tpy facility, which will serve the North American carpet market, should begin operation by year-end 2003 using proprietary Shell technology.

Central NGL fractionator utilization in Alberta will increase to 93% by 2010, vs. about 76% in 2000, said Hawkins Gas Consultants in October. Conversely, use of onsite fractionators to recover propane will decline from 25% to 20%.

An 80 million liter/yr fuel-grade ethanol plant was proposed for the Dauphin and Roblin region of Manitoba, said a unit of Outlook Resources Inc. (Toronto) in November.

Industry business

Canadian disclosure regulations for publicly traded oil and gas companies were being revised in late 2002, their first major overhaul in 20 years. The Alberta Securities Commission was spearheading the project to develop legislation that will likely be adopted by other agencies.

Land operations in Canada, as well as in the US lower 48 states, were closed down by WesternGeco (London) in October 2002, hinting at the economic climate that oil service and supply companies will have to weather in the short to medium term.

In April 2002 two of Canada's largest energy companies, Alberta Energy Co. Ltd. and PanCanadian Energy Corp., completed the merger that created EnCana Corp., now the world's largest independent oil and gas company in terms of enterprise value, reserves (209 bcm, or 7.8 tcf, of gas and 1.3 billion bbl of oil and liquids, equaling 2.6 billion boe), and production, with 2002 targets of 26.5 bcm/yr (2.7 bcfd) and 255,000

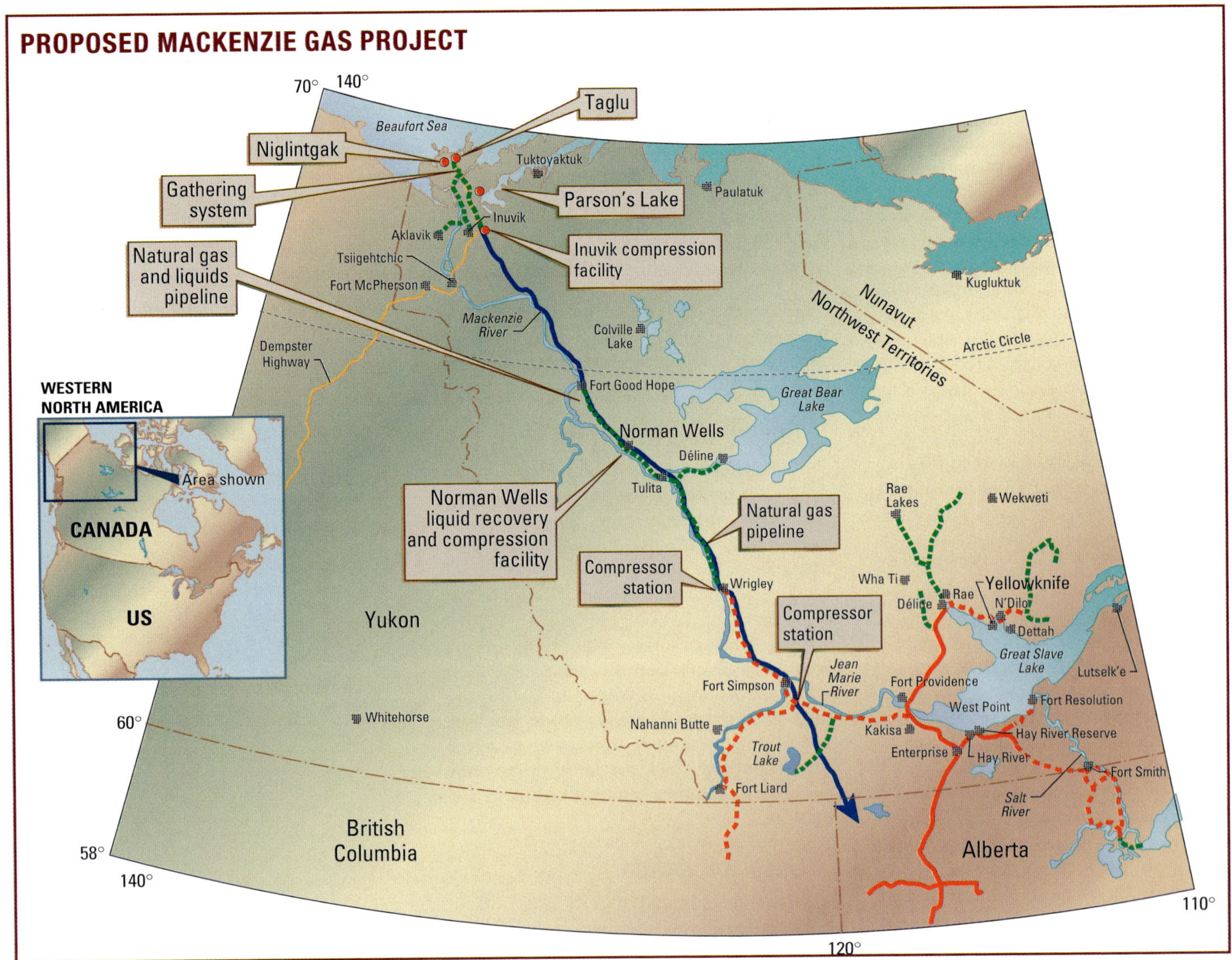

b/d of oil and liquids, or 700,000 boe/d. By 2005 EnCana hopes to increase production by 55% to 1.1 million boe/d.

EnCana is also the third largest publicly traded industrial company in Canada, with one of the largest capital investment programs ($3.8 billion in 2002) of any Canadian-headquartered company. The new company is also North America's largest independent gas producer and the largest independent gas storage operator.

During mid-2002 as gas prices languished in Alberta, EnCana evaded the glutted spot market by storing much of its natural gas. The company was developing a 1-bcm (40-bcf) storage facility that would increase its total capacity more than 40% to 3.6 bcm (135 bcf). The Countess facility, designed for peak injection of 25.5 million cu m/d (950 MMcfd) and peak withdrawal of 33.5 million cu m/d (1.25 bcfd), will use two depleted reservoirs, with the first 270 million cu m (10 bcf) of storage to become available by mid-2003.

In late 2002, EnCana sold its interest in two pipeline systems, the 2,763-km (1,717-mile) Express and the 652-km (405-mile) Cold Lake, to separate buyers, subject to regulatory approval.

Earlier in the year, Petro-Canada became the largest Canadian integrated oil and gas company when it bought the international operations of Veba Oil & Gas GmbH for $3.2 billion (Can.). The acquisition boosted Petro-Canada's production and total proved reserves by more than 70%, and production was expected to double in 5 years.

IPE

MARITIMES AND NORTHEAST PIPELINE

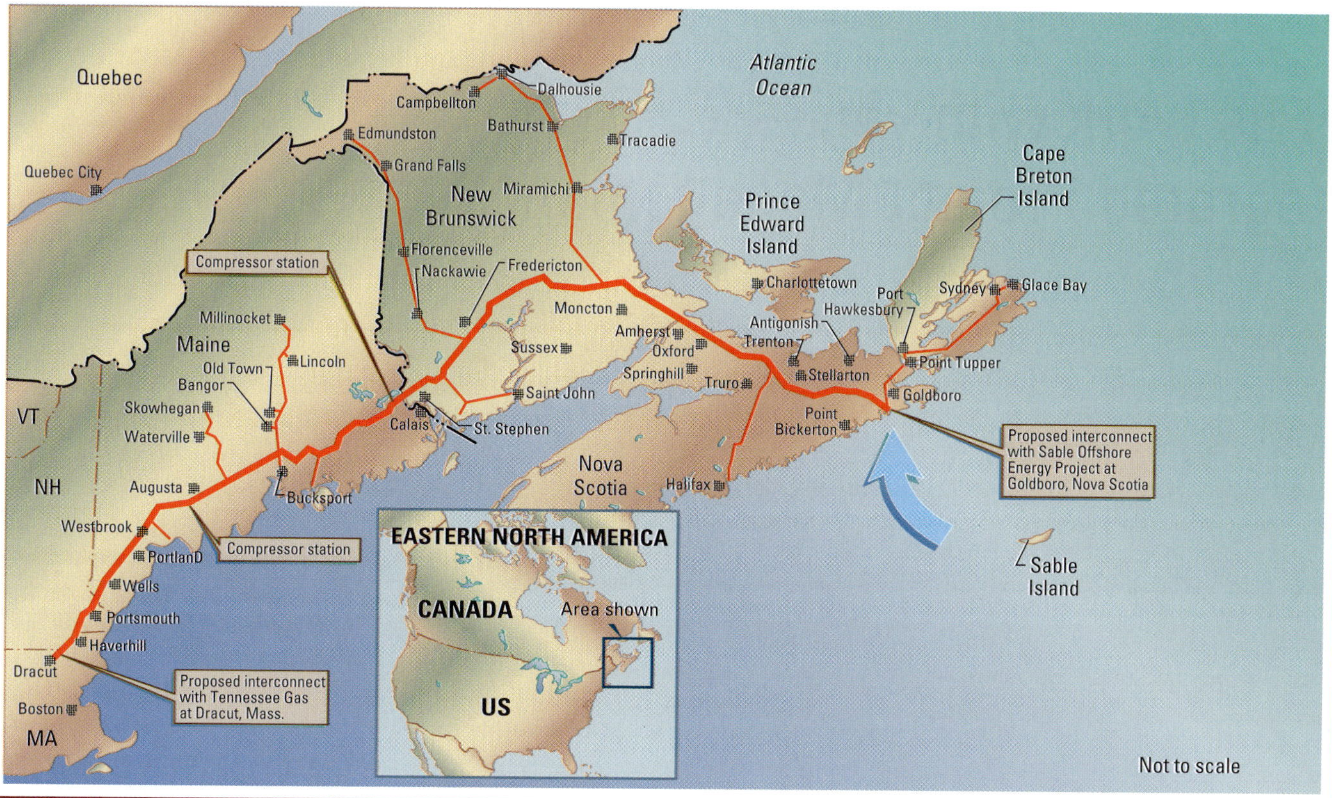

LATIN AMERICA

Argentina

CAPITAL: BUENOS AIRES
MONETARY UNIT: PESO
REFINING CAPACITY: 639,075 B/CD
OIL PRODUCTION: 750,000 B/D
OIL RESERVES: 2.87868 BILLION BBL
GAS RESERVES: 26.96 TCF

Following Argentina's abrupt economic downturn in mid-2001, the economy remained in very serious condition during 2002, undergoing its deepest recession of the post-war period. The International Monetary Fund (IMF) was working with the Argentine government to resolve the country's financial problems, but Argentina defaulted on a World Bank loan payment in late 2002. Argentina's real gross domestic product (GDP) was projected to grow only 1.0% in 2003.

The fallout from Argentina's fiscal woes staggered the oil and gas industry there. Foreign investment in Argentina all but halted, and early in 2002 Gaz de France was renegotiating its obligations. In mid-2002, the Argentine government lifted its oil export cap and reduced its diesel sales tax, which helped companies' near-term earnings outlooks.

Petrobras moves in

Brazil's Petroleo Brasileiro SA (Petrobras) seized the opportunity to shop for assets that were suddenly devalued. Petrobras acquired a controlling interest in Perez Companc SA (Buenos Aires), South America's largest independent oil and gas company, which operates and participates in various producing concessions in the Entre Lomas area in Argentina's Neuquén basin.

Petrobras also acquired full control of Petrolera Santa Fe, the Argentine subsidiary of Devon Energy Corp. (Oklahoma City, OK), which produces oil and gas from Argentina's Sierra Chata, Refugio Tupungato, Atamisqui, and El Tordillo fields and operates the concession containing El Mangrullo field, yet to enter into production. Petrobras was also considering purchasing the Argentine assets of Repsol YPF SA (Madrid, Buenos Aires).

Petrobras was planning to invest $86 million in Argentina in 2002, including $43 million to upgrade the Bahía Blanca refinery and $8 million to build a 3.5-km (2.2-mile) pipeline at the refinery. Through 2005 Petrobras committed $140 million to boost its presence in Argentina, including investments in its lubricants plant at Bahía Blanca, its network of service stations, and exploration in the Neuquén basin, where it owns interests in four blocks: Puesto Zuniga, Cerro Manrique, Puesto Gonzales, and Mata Mora.

Neuquén discoveries

The PZX-1001 well found condensate and natural gas in the Neuquén basin, said Petrobras in August. The well, drilled on the Puesto Zuniga prospect on Block CNQ-32 near Cinco Saltos in Rio Negro Province, flowed on initial tests at a rate of 516 b/d of condensate and 230,000 cu m/d (8.5 MMcfd) of gas from Jurassic Lajas at 3,650 m (12,000 ft). The company was assessing commerciality of the find and plans to drill confirmation wells.

Oil was discovered on the Neuquén basin's Cañadón Amarillo block in late 2002, reported Repsol. The Rincón Blanco X-1 well flowed 610 b/d at 800 m (2,600 ft).

Pipeline startup

The Southern Cross pipeline, Gaqsoducto Cruz del Sur (GCDS), operated by BG Group plc (UK), began transporting Argentine natural gas to Uruguay in November 2002. Primary customer is Uruguay's state electric utility.

The 193-km (120-mile), high-pressure system has the capacity to deliver 2 billion cu m (bcm)/yr (200 MMcfd). A separate 40-km (25-mile), 18-inch line known as the "Link," completed in April, connects the start of the GCDS pipeline in Argentina with the Argentine national grid.

LPG capacity

During 2001-2002, Argentina's LPG pro-

SOUTH AMERICA
ARUBA
CURACAO
GRENADA
BARBADOS
Bridgetown
TRINIDAD and TOBAGO
Port of Spain
COSTA RICA
PANAMA
Pto. Limon
Colón
Panama
Santa Marta
Barranquilla
Cartagena
Maracaibo
Caracas
Puerto la Cruz
Jose
Lechoso
Punzan
Grico
Penal-Barrackpore-Wilson
Teak
Balata-Bouvallius
Southeast Galeota
Radix
E. Queens Beach
Quarry-Coora-Quinam
Tibu
Hato
Sinco
Palmita
Silvestre
S. Silvestre
S. Cristobol
Barrancabermeja
Cantimplora
Ciudad Bolivar
Georgetown
Tambaredjo/Jossikreek
VENEZUELA
GUYANA
SURINAME
Karanambo
Volcanera
Cupiagua
Cusiana
Apiay Complex
Chichimene
Bogota
Tulua
Cali
Castilla
Andalucia
San Francisco
Dina
Tello
COLOMBIA
Tumaco
Esmeraldas
Quito
ECUADOR
Guayaquil
Amistad
Tumbes
Talara
Bayovar
Iquitos
Concordia
Manaus
Jurua
Southwest
Rio Urucu
Leste do
Uruca
Nova Olinda
B R A Z I L
Maquia
Aguaytia
Aguas Calientes
Pucallpa
Camisea
Fields
PERU
Lima
La Pampilla
Conchan
BOLIVIA
Lago de Titicaca
Pirin
La Paz
Surubi
Pto. Villarroel
Cochabamba
Sicasica
Oruro
Colpa
Santa Cruz
Tita
Rio Grande
Sucre
Corumba
Arica
Tatarenda
Camiri
Boyuibe
Monteagudo
Vuelta Grande
Caigua
Buena Vista
Sanandita
Tigre
Madre Jones
Toro
Campo Duran
Bermejo
Ramos
Lomitas Tranquitas
Rio Pescado
Caimancito
Jujuy
Salta
Antofagasta
PARAGUAY
Villa Elisa
Asuncion
0 100 200 300 400 Miles
0 200 400 600 Km
280
320
613
288
127
195
160
75
15
60
389
11
5
205
2
2
110
2
20
46
62
11
46
3
100
7
40
20
3
32
8
10°
5°
0°
5°
10°
15°
20°
80°
70°
60°

SOUTH AMERICA
0 100 200 300 400 Miles
0 200 400 600 Km
Legend
Oil field
Oil sand
Gas field
Crude oil pipeline
Natural gas pipeline
Products pipeline
Pipeline planned or under construction
Refinery in operation
Refinery capacity in 1,000 b/d
600
Tanker terminal
Cities
Capital
International boundary
Water depth
0 to 200 m
200 m and deeper
Atlantic Ocean
Georgetown
Tambaredjo/ Jossikreek
Paramaribo
Cayenne
GUYANA
SURINAME
FRENCH GUYANA
Belem
Amazon River
Sao Luis
46
Manaus
Nova Olinda
Rio Tapajos
Atum
Curima
Espada
Pescada
1-CES-8
Xareu
Fortaleza
6
Fazenda Belem
Ubarana
Agulha
Canto do Amaro
Serraria
Lorena
Guamare
Faz. Pocinhos
Estreito
BRAZIL
Recife
Maceio
Aracaju
Salvador
306
Rio Teles Pires
Rio Araguaia
Rio Tocantins
Rio Sao Francisco
Cuiaba
Brasilia
Corumba
PARAGUAY
Aquidauana
Campo Grande
140
Betim
Belo Horizonte
Campos
13
355
Duque de Caxias
Rio Grande
Parana
55°
45°
35°
10°
5°
0°
5°
10°
15°
20°

SOUTH AMERICA
CHILE
PARAGUAY
URUGUAY
ARGENTINA
Atlantic Ocean
Falkland Islands (Islas Malvinas)
Oruro
Arica
Sucre
Santa Cruz
Tita
Rio Grande
Camiri
Tatarenda
Boyuibe
Monteagudo
Vuelta Grande
Buena Vista
Caigua
Sanandita
Madre Jones
Tigre
Toro
Campo Duran
Bermejo
Ramos
Lomitas Tranquitas
Rio Pescado
Caimancito
Antofagasta
Jujuy
Salta
Metan
Corumba
Campo Grande
Aquidauana
Villa Elisa
Asuncion
Apucarana
Formosa
Resistencia
Corrientes
Paso de los Libres
Uruguaiana
Sao Gabriel
Betim
Belo Horizonte
Campos
Duque de Caxias
Rio de Janeiro
Tremembe
Sao Paulo
Maua
Cubatao
Santos
Sao Sebastiao
Irati
Curitiba
Merluza
Sao Francisco do Sul
Perimbo
Florianopolis
Laguna
Canoas
Porto Alegre
Tramandai
Rio Grande do Sul
Cordoba
San Lorenzo
Lumunta
Tupungato
Cacheuta
Mendoza
Concon
Valparaiso
Santiago
Refugio
Colorados
Barrencas
El Sosneado
San Luis
Campana
Buenos Aires
Montevideo
Dock Sud
La Plata
Las Flores
Cnl. Pringles
Concepcion
Colonia Catriel
El Medanito
Guanaco
Bahia Blanca
Punta Alta
Puerto Rosales
Plaza Huincul
Challaco
Cerro Bandera
Barda Gonzales
San Antonio
Golfo San Matias
Las Flores
Rivadavia
Comodoro Rivadavia
Anticlinal Grande
Cerro Tortuga
Cerro Dragon
Comodoro
Canadon Grande
Canadon Seco
Caleta Olivia
Pico Truncado
Santa Cruz
Rio Gallegos
Punta Arenas
Tierra Del Fuego
Rio Grande
Ushuaia
0 100 200 300 400 Miles
0 200 400 600 Km

duction capacity increased by more than a third and its natural gasoline production by nearly a quarter, following start-up of the Mega project in Neuquén province.

The project consists of a separation plant in Loma La Lata, a 600-km (370-mile) liquids pipeline from Loma La Lata to Bahía Blanca, and a fractionation plant in Bahía Blanca, along with incremental storage and shipping facilities. The increased production went into exports to Brazil.

Bolivia

CAPITAL: LA PAZ
MONETARY UNIT: BOLIVIANO
REFINING CAPACITY: 63,000 B/CD
OIL PRODUCTION: 31,000 B/D
OIL RESERVES: 440.5 MILLION BBL
GAS RESERVES: 24.0 TCF

TIERRA DEL FUEGO

The weak global economy continued to suppress growth in Bolivia during 2002, limiting the country's real GDP increase to an estimated 1.6%. In August 2002 Gonzalo Sanchez de Lozada succeeded Jorge Quiroda as Bolivia's president.

Bolivia contains South America's second-largest natural gas reserves and during 2002 continued to be largely self-sufficient in crude oil production. Bolivia's gas reserves are in part stranded due to low demand relative to their location and Bolivia's small population and limited infrastructure.

Bolivia's natural gas exports in 2002 continued to flow primarily to Brazil, with small volumes beginning to go to Argentina in mid-year. Bolivia's main hope for future economic growth hinges on increasing its natural gas exports and becoming a major energy hub for South America's Southern Cone region.

Bolivia and Brazil pledged to strengthen bilateral ties in the energy field, including a possible new gas pipeline connecting the two countries and other neighbors. Natural gas production by Petrobras at the southern San Alberto field was expected to double by 2003 to nearly 4.9 bcm/yr (500 MMcfd). However, in March 2002, bidding on an expansion of the existing Bolivia-Brazil pipeline was suspended due to uncertainties surrounding Brazil's power sector and disagreement over gas prices.

Pacific LNG project

Bolivia continued to weigh plans for exporting LNG from a Pacific port in either Chile or Peru, while popular sentiment opposed any agreement with Chile, a traditional adversary of Bolivia. A consortium of companies called Pacific LNG plans to ship the LNG to a regasification plant in northern Mexico, where the gas would be piped into the US.

In July 2002, the Bolivian press reported that an agreement had been reached to export through Chile, but in August Bolivia officially postponed until year-end a decision on the gas pipeline route and liquefaction plant location, reported Rigzone.

Gas-to-liquids project

In October 2002 GTL Bolivia SA was planning a feasibility study for a 10,000 b/d gas-to-liquids (GTL) plant near Santa Cruz. Under an agreement with Rentech Inc. (Denver, CO), the proposed plant would use Rentech's patented GTL technology process primarily to make sulfur-free fuels.

GTL Bolivia, a new company whose principal shareholder is the Deane Group — a private US company that operates business ventures owned by Disque D. Deane — contracted Jacobs Engineering UK to conduct the 4-6 month study. Currently, Bolivia imports conventional, high-sulfur diesel fuels.

Rentech said its GTL process converts syngas into products using an iron-based catalyst technology. The primary source of syngas feedstock was expected to be natural gas wells that are not producing due to their remote location, or from the conversion of coal or low-value refinery bottoms.

Brazil

CAPITAL: BRASILIA
MONETARY UNIT: REAL
REFINING CAPACITY: 1,865,140 B/CD
OIL PRODUCTION: 1.488 MILLION B/D
OIL RESERVES: 8.3217 BILLION BBL
GAS RESERVES: 8.092 TCF

Brazil had weathered an energy crisis in 2001 and suffered persistently weak or negative economic growth during 2002. Early in the year, the government opened Brazil's market to refined product imports and linked domestic prices to the fluctuation of international oil prices through a price trigger mechanism.

However, in August the government reinstated oil price controls, ordering the National Petroleum Agency (ANP) to curb hikes in oil products prices. The government's biggest concern was with LPG, used in Brazil mainly for cooking. The government owns 33% of Petrobras, which produces or imports 96% of Brazil's fuels. Also, the legislature imposed an 18% tax on platforms and other offshore facilities that are built outside Brazil and brought into Rio de Janeiro state.

In June ANP reported that the oil and natural gas industry doubled its participation in the country's GDP during the first 4 years since the end of Petrobras's upstream monopoly (1997 to 2000).

New president

In October 2002 Brazil elected a left-wing president for the first time in its history. Luis Inacio Lula da Silva won in a landslide and took office in January 2003, breaking the grip of the conservative elite that had ruled South America's biggest economy. Analysts attributed Lula's leadership to his switching from a traditional socialist stance to a more market-friendly liberal philosophy.

In January 2003, Lula appointed as Petrobras president José Eduardo Dutra, a geologist said to have less experience on the business side of the oil industry. Dilma Rousseff, chosen as mines and energy minister, opposes privatizing state-owned electricity companies and plans to use the federal fuel tax as a cushion against crude oil price fluctuations.

During the campaign, Lula said he would cancel Petrobras's contracts with foreign companies to build floating production, storage, and offloading (FPSO) vessels. The FPSOs are destined for deepwater development projects in Brazil's prolific Campos basin.

However, Lula pledged to honor all other international agreements, notably those with foreign oil companies that hold exploration and production concessions and those providing products and services to the Brazilian petroleum industry. He also said he backed the terms of the IMF's $30 billion loan package to Brazil.

Issues of main concern for the Brazilian energy sector under a Lula government include its policies on price controls, simplification of the tax structure, investment incentives, upgrading the electricity sector, and Brazil's role as a key player on the volatile political and economic scene.

Of special concern is Petrobras's refining sector, which continues to lag its upstream operations. Lula's few statements involving Petrobras or the oil sector were all related to his wish that new investments in oil and gas generate jobs in Brazil.

Petrobras operations

Petrobras purchased assets in Argentina during mid-2002, including a controlling interest in Perez Companc and Petrolera Santa Fe (Devon Energy's Argentine subsidiary), and was considering purchasing the Argentine assets of Repsol YPF, the Spanish-Argentine giant. Petrobras planned to invest a total of $86 million in Argentina in 2002. The fall-off of oil and gas investments in troubled Argentina and Venezuela has spotlighted Petrobras's leadership in South American operations, said industry observers.

Early in 2002 Petrobras reported that it was producing record amounts of oil and expected new records as additional wells come onstream, especially in Marlim Sul (South) field, which could produce 150,000 b/d by December 2002, vs. 78,000 b/d at the start of the year.

In October Petrobras's P-34 FPSO was stabilized after suddenly sliding sideways in the water, listing as much as 32 degrees. The $200 million, 17,900-ton vessel, installed between Barracuda and Caratinga fields in the Campos basin, was handling 34,000 b/d of 25-degree gravity crude oil. A team was analyzing possible causes of the electrical failure that triggered the incident. Additionally, Petrobras faced fraud charges in relation to the company that built the P-37 FPSO as well as the stricken P-34 and the ill-fated P-36.

Exploration licensing

In mid-2002, diminished interest was shown by oil and gas companies in the ANP's fourth round of exploration licensing, netting only a fifth of the amount bid in the previous sale. Industry officials cited a shift in focus to onshore licenses, the paucity of significant discoveries on previously awarded acreage, and Brazil's burdensome fiscal regime. Companies that won fourth-round licenses will invest $1 billion in the next 9 years.

Once again Petrobras acquired the largest number of licenses in a Brazilian sale: eight blocks, including four in partnerships. However, one highlight of the round was the debut of a number of foreign companies, notably independents: Australia's BHP Billiton Ltd. (Melbourne), Oman's Partex Oil &Gas (Holdings) Corp., Devon Energy, and Newfield Exploration Co. (Houston). Cementing its participation in Brazil's E&D scene with a successful bid was a new Brazilian independent, Starfish Oil & Gas Co., which acquired 100% of Block BT-REC-7.

In November, ANP scheduled its fifth licensing round for June 2003 under a new bidding method designed to encourage participation by new small and medium-sized companies. Sedimentary basins will be divided into sectors and subdivided into cells, which can be offered individually or in groups. The method will enable companies to design exploration areas through their selection of cells.

Round 5 will offer 1,122 exploration cells, 824 offshore and 298 onshore, in 21 sectors of 9 sedimentary basins in 10 states. Cells will be offered in three sizes for locations onshore, in shallow water, and in deep water. For the first time, ANP also will permit companies to define their minimum exploratory programs, involving data research and drilling.

Campos discovery

In August 2002 Petrobras confirmed a major deepwater oil discovery at the edge of the northern Campos basin about 70 km (40 miles) off Brazil's Espírito Santo state — the first major hydrocarbon discovery there, and Petrobras's biggest discovery since 1996. Jubarte field lies in 1,300 m of water (4,300 ft) on Block BC-60. Appraisal drilling and production testing yielded a reserves estimate of 600 million bbl of 17-degree gravity crude.

The discovery well cut a 46-m thick (151-ft) pay zone with reservoir characteristics similar to those seen elsewhere in the prolific Campos basin. The confirmation well, 6-ESS-109D, was drilled in May 3 km (2 miles) from the discovery well. It cut a 120-m (390-ft) pay zone in the same formation as the discovery well but also encountered a 25-m (82-ft) oil-bearing reservoir in an older formation.

PETROBRAS DEEPWATER OIL STRIKE

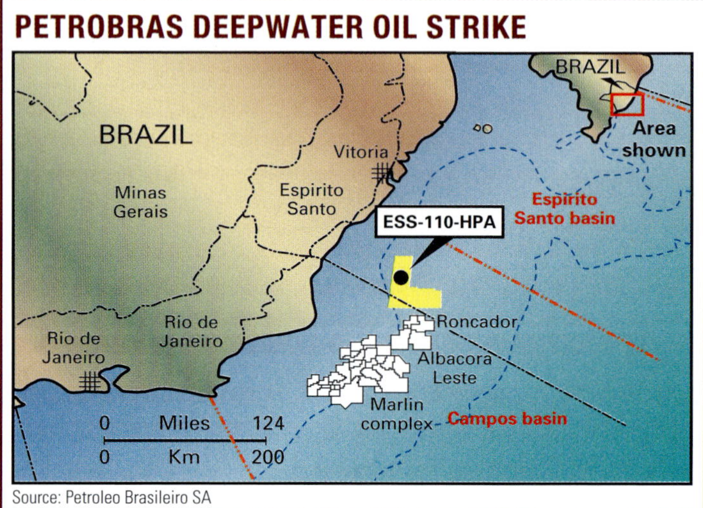

Source: Petroleo Brasileiro SA

An extended-reach appraisal well, 3-ESS-110-HPA, was drilled in July with a horizontal displacement of 1,000 m (3,300 ft). During initial tests, the well flowed 3,000 b/d of oil. Petrobras began a 6-month test of the extended-reach well, using the Seillean FPSO vessel, to appraise the reservoir's producibility. In addition, two delineation wells were scheduled.

In November, early production started from Jubarte. The oil was being stored on the Seillean FPSO. If development proves viable, Petrobras estimated that the field could produce as much as 120,000 b/d of oil within 3 years.

Near Jubarte on the same block, oil was discovered by the 1-ESS-116 wildcat, which was drilled 76 km (47 miles) offshore in 1,478 m of water (4,849 ft), said Petrobras in December. It encountered a 60-m (200-ft) thick, saturated 19-degree gravity oil-bearing formation. Area geological studies indicate postulated reserves of 300 million boe, bringing the block's total to 900 million boe.

Campos development

Petrobras is targeting offshore production of crude oil from the Campos basin averaging more than 1.8 million b/d before 2010, up from 2002's level of 1.26 million b/d. Petrobras also expects its combined oil and gas reserves to climb to 11.7 billion boe by 2005 from 9.7 billion boe in 2002. More than 75% of that would come from the deepwater and ultra-deepwater Campos.

Petrobras was operating 40 production units in 39 fields in the Campos basin — 14 fixed platforms, 16 floating drilling and production units (mainly converted semisubmersibles), and 10 FPSO units. Leading the initial surge in deepwater production in the Campos basin has been the development of Albacora and Marlim fields in 230-1,040 m of water (750-3,410 ft). These two giant fields were producing a combined 680,000 b/d of oil and 3.9 bcm/yr (395 MMcfd) of gas.

Another six giant fields were also under development in the deepwater Campos basin: Roncador, Albacora Leste (East), Marlim Sul, Marlim Leste, and the adjoining Barracuda-Caratinga complex. These fields contain a combined 4.2 billion boe of proved reserves and were under early production from pilot systems tied back to existing platforms in Marlim and Albacora or to floating production units (FPUs). Two FPUs were producing in Marlim Sul, and three FPSOs were under construction for installation in Barracuda, Caratinga, and Roncador.

Full development of Marlim Sul, Barracuda-Caratinga, Albacora Leste, and Roncador calls for installation of 11 FPUs and tie-in of 150 producing wells and 100 injection wells. All of this effort will require a capital investment of more than $10 billion, resulting in a net increment of new production totaling more than 1 million b/d.

Five FPSO conversions were slated for installation off Brazil. The Caratinga and the Barracuda conversions each will have capacity for processing 150,000 b/d and storing 2 million bbl of crude oil. The Stena Concordia and Stena Continent very large crude carriers were being used in the conversions. Water depths for the Barracuda-Caratinga complex are 670-1,200 m (2,200-3,900 ft).

In other Campos developments, 11 subsea guideline-less trees, rated to water depths of 2,000 m (6,500 ft), will be installed in Roncador and Albacora Leste fields. Petrobras and partner Repsol expect production in 2003 of 34,000 b/d of crude and 400,000 cu m/d (15 MMcfd) of gas, expandable to 180,000 b/d and 2.25 million cu m/d (83.8 MMcfd) by 2009.

Pipeline end equipment was to be delivered in 2002 for Barracuda and Caratinga fields, where Petrobras expects combined production to reach 247,000 b/d of oil and 3.4 million cu m/d (127 MMcfd) of gas by 2004-05.

Ultra-deep installation at Roncador

Petrobras in May 2002 set a world record in ultra-deep water for installation of a subsea natural gas lift manifold to a depth of 1,885 m (6,184 ft), marking the first time for such heavy equipment to be installed in waters deeper than 1,000 m (3,300 ft). The subsea equipment, installed in Roncador field, weighed 191 tonnes and measured 10 x 7.5 x 4 m (33 x 25 x 13 ft). The equipment was scheduled to be onstream by year-end 2002.

The manifold was designed to reduce gas lift costs by distributing gas coming from the production unit via a single gas line for injection into six wells. The installation was performed jointly by the Amethyst-1 (SS-47) semisubmersible drilling rig and the Asso 23 anchor-handling supply tug, as well as various supply boats.

Shell's Brazilian activities

Shell Brasil SA (Petroleo) said in December 2002 that it plans to invest $1 billion in oil and gas exploration and production in Brazil during the next 4 years, primarily in developing Block BC-10. Royal Dutch/Shell Group (Netherlands) purchased Enterprise Oil plc, including its Brazilian subsidiary, for which Shell paid $700 million and assumed its 80% ownership interest in Block BC-10.

The Sahara, one of the 15 largest tankers in the world with an oil storage capacity of 1.3 million bbl, was being converted into an FPSO and renamed the Fluminense for development of Bijupirá and Salema fields, where first oil is scheduled for mid-2003. The Fluminense, to be moored in 700 m (2,300 ft) of water, will have capacity to process 70,000 b/d of crude oil and 2 million cu m/d (75 MMscfd) of natural gas, plus oil storage of 1.3 million bbl.

Bijupirá and Salema, lying in water depths of 450-800 m (1,500-2,600 ft), hold reserves estimated at 170 million bbl of oil, and both fields have been extensively delineated. The fields' crude oil is relatively light (28-31 degree gravity), produced from sandstones at 2,740 m (9,000 ft). Gross reservoir thickness is 12-44 m (39-144 ft) with an average of 31 m (102 ft).

The subsea development will include 15 wells (9 production and 6 water injection) tied back via 3 manifolds near the FPSO, less than 2.4 km (1.5 mile) away. Gas will be exported using the existing pipeline from the Bijupirá area to the Petrobras P-15 installation. Oil will be stored for lightering by tankers for export.

Shell participates in exploration of 15 blocks throughout Brazil, including 7 discoveries that were being evaluated. One of these is a large oil reservoir 200 km (120 miles) southeast of Rio de Janeiro on Block BS-4 in the northern part of the Santos basin off Brazil. First projections indicate potential reserves of 300-500 million bbl of heavy crude.

Pipeline improvements under way

Petrobras said in October 2002 that it expects to invest more than $2.5 billion by 2005 in new pipelines. Under a program called PEGASO — a Portuguese acronym for Excellence in Environmental Management & Operational Safety Program — the company has already invested $800 million, of which $400 million was for improvement of company pipeline activities and $300 million for enhanced pipeline integrity. By June 2003, an additional $1.3 billion will be invested.

Under the integrity program, Petrobras has inspected more than 4,700 km (2,900 miles) of pipelines since the beginning of 2002 and eliminated 5,200 defects, rehabilitated more than 3,300 km (2,100 miles) of pipeline, and replaced 245 km (152 miles).

Petrobras also created an independent subsidiary, Transpetro, to build and operate pipelines for the Brazilian market and has been investing in automation. As of late 2002, 70% of the oil and gas pipeline network was remotely operated, and Transpetro intends to conclude this program by late 2003. The 20,000-km (12,000 miles) Brazilian pipeline transportation system should reach 25,000 km (16,000 miles) of oil, gas, and product lines by 2005.

Plans to extend the Brazil-Argentina gas pipeline were renewed in mid-2002. The extension would connect Uruguaiana, Brazil, to Porto Alegre on the Brazilian coast, where it could service a new power plant. The Mercosur pipeline currently supplies a 600-MW power plant in Uruguaiana.

Refinery upgrade

Petrobras planned to invest more than $290 million in Paulinia, Brazil's largest refinery, through 2005, mostly to boost diesel treatment and refining capacities and set up a coking unit. Daily refining capacity would increase to 60 million liters in 2005 from 54.2 million. Also, six new delayed coking units are scheduled for completion at various refineries by 2010, to convert bottom-of-the-barrel oil into higher value products.

Brazil imports one-fourth of its 30 billion liters/yr of diesel fuel consumption. In 2002 Petrobras incurred losses from these imports because the domestic market price could not be increased to account for sharp depreciation of the Brazilian real, which made diesel imports more expensive. Petrobras stopped importing naphtha because of rising costs and hopes to stop importing LPG products as well, said the National Fuel Distributors Association (Sindicom) in December 2002.

Products market

Petroleos de Venezuela SA (PdVSA) requested permission from Brazil's ANP to become a distributor of oil products in Brazil through its subsidiary Citgo Latin America, beginning with the northern and northeastern regions under the brand PdV do Brasil SA. The move into Brazil's lubricants and fuels market was confirmed in September, and the first service station was to be inaugurated before year-end 2002.

Chile

CAPITAL: SANTIAGO
MONETARY UNIT: PESO
REFINING CAPACITY: 204,849 B/CD
OIL PRODUCTION: 7,000 B/D
OIL RESERVES: 150 MILLION BBL
GAS RESERVES: 3.46 TCF

During 2002 Chile continued to compete strenuously to be involved in the $6 billion project to export Bolivian gas as LNG to Mexico and into the US. Either Chile or Peru will be selected as the site of a liquefaction plant and pipeline to be built by the Pacific LNG consortium. However, popular sentiment in Bolivia opposed any agreement with Chile, a traditional adversary, and in August 2002 Bolivia postponed its decision until year-end.

Chile is seeking to diversify its fuel mix by moving away from hydropower toward natural gas-fired generation, with eight new plants scheduled for completion by 2010. Labor unrest in Argentina triggered gas supply disruptions in March 2002, prompting Chile's Empresa Nacional de Petroleo (ENAP) to assess the security of its downstream operations.

Colombia

CAPITAL: BOGOTA
MONETARY UNIT: PESO
REFINING CAPACITY: 285,850 B/CD
OIL PRODUCTION: 583,000 B/D
OIL RESERVES: 1.84229 BILLION BBL
GAS RESERVES: 4.507 TCF

Colombia reported higher proved reserves estimates during 2002, reflecting discoveries that boosted oil reserves by 5.3% to 1.84 billion bbl, up from 1.75 billion bbl, and natural gas by 4.3% to 121 bcm from 115 bcm (to 4.5 tcf from 4.3 tcf), said *Oil & Gas Journal.*

In a move to encourage hydrocarbon exploration, outgoing Colombian President Andrés Pastrana signed a measure that significantly reduces the royalties oil and gas companies pay the state, and Congress approved the new law in June 2002.

The royalties on newly discovered oil fields producing fewer than 125,000 b/d (based on an average daily output) were cut to 8-20% from the previous flat rate of 20%. All of Colombia's biggest oil fields produce less than 125,000 b/d, except the Cusiana-Cupiagua field operated by BP plc (UK). Private companies producing 400,000-600,000 b/d would see royalties of 20-25%, based on average daily production.

Independent Alvaro Uribe became President in May 2002 and planned to crack down on rebel groups, which targeted politicians in numerous assassination attempts and kidnappings as well as civilian bombings.

In August 2002 US lawmakers writing an antiterrorism bill rejected a requirement that Occidental Petroleum Corp. (US) and Repsol YPF reimburse the US government for aid aimed at improving the security of the 777-km (483-mile) Caño Limon crude oil export pipeline in Colombia. Congress ultimately earmarked $6 million to the Colombian military for pipeline protection without requiring Oxy or Repsol to pledge funds.

Empresa Colombiana de Petroleos SA (Ecopetrol), Colombia's state oil company, green-lighted private plans in early 2002 to build a refinery at Sebastopol in Cimitarra. The new refinery would process crude from BP's Cusiana-Cupiagua fields and could be online by 2004.

E&D

Talisman Energy Inc. (Calgary) announced plans in early 2003 to drill one exploratory well on each of four blocks: Acevedo in the highly prospective Upper Magdalena Valley; the contiguous Altamizal and Huila Norte Blocks to the north; and the Tangara Block in the Llanos basin adjacent to the large Cusiana-Cupiagua discoveries.

Ecopetrol signed three new association contracts in August, covering about 500 sq km (200 sq mi) in the southern region of the country, for a total of seven contracts during 2002. With the Guayuyaco contract, Argosy Energy International (Colombia) intends to explore a 200 sq km (80 sq mi) area in the Putumayo basin. Argosy's Santana contract, which covers Toroyaco, Mary, Linda, and Miraflor fields, was producing an average of 3,200 b/d of crude oil.

Argosy plans to drill an exploratory well, Inchiyaco-1, on the former Mary East prospect. The well, to go to 2,500 m (8,100 ft), will test several formations that produce on the Mary structure and elsewhere in the Putumayo basin. Argosy estimated that new license terms will apply to 170-230 million bbl of potentially recoverable oil, 60-90 million bbl of which is represented by five currently defined prospects.

Guando area

Nexen's exploration program was yielding discoveries at Guando field, the company said in early 2002. Nexen planned to participate with Petrobras in the development of the field on Boqueron Block in central Colombia. Guando, located in the upper Magdalena Valley, had averaged 5,700 b/d, and more wells were being drilled. The field could contain as much as 200 million bbl of oil.

Gulf Sands Petroleum Ltd. (Houston) discovered two oil seeps on the Alborada Block, adjacent to the giant Guando oil discovery, said partner Gulf Shores Resources Ltd. (Vancouver) in October 2002. One seep is believed to be at the southern spill point of a footwall reservoir in Upper

Cretaceous Monserrate sand, and the other is near a surface expression of the Los Cauchos fault. Laboratory testing indicated the oil is from a Villeta (La Luna) source.

A 50-km (31-mile) seismic survey was completed on the 306 sq km (75,668 acre, or 118 sq mi) block in the Magdalena basin in Tolima state. Seismic data processing and interpretation were expected to take 2 months, and the first well will be spudded in 2003. Alborada is east of the Espinal Block, where Espinal field produces 12,000 b/d into the Upper Magdalena Valley pipeline system.

Other Magdalena prospects

Solana Petroleum Corp. (Calgary) and partner Petex de Colombia Ltda. agreed in October 2002 to buy an interest in the Magangue Association Contract in northwestern Colombia's Lower Magdalena basin near the Covenas oil export terminal. The contract includes Guepaje gas field and related production facilities.

Engineering studies showed 200-300 million cu m (7-10 bcf) of remaining proven reserves already developed; one well was producing 240,000 cu m/d (9 MMcfd). A further 400-500 million cu m (15-18 bcf) of proven undeveloped reserves could be produced, and two prospects adjoining Guepaje could hold an additional 4 bcm (150 bcf).

Seven Seas Petroleum Inc. (Houston) expected downward revision of its net proved reserves of 47.6 million bbl of oil in Guaduas field in the Magdalena basin and sold its interest in late 2002, including the 60-km (40-mile) Guaduas-La Dorada pipeline. The company retained rights to the Deep Dindal contract and was seeking financing or a partner to complete the Escuela 2 well there.

Llanos basin

An early production processing facility will be built on Recetor Block in the Casanare region about 160 km (100) miles northeast of Bogota, said a BP unit in October 2002. The plant will be designed to handle 30,000 b/d of oil and 2 million cu m/d (80 MMscfd) of natural gas and will generate its own power.

Oil from Recetor will be supplied to BP's Cupiagua central production, while gas will be reinjected for enhanced production. The Recetor reservoir is an extension of Cupiagua field in the neighboring Santiago de las Atalayas license. The Cusiana-Cupiagua complex is thought to contain about 110 bcm (4.1 tcf) of gas.

Cuba

CAPITAL: HAVANA
MONETARY UNIT: PESO
REFINING CAPACITY: 301,400 B/CD
OIL PRODUCTION: 40,000 B/D
OIL RESERVES: 750 MILLION BBL
GAS RESERVES: 2.5 TCF

Cuba's oil-import deal with Venezuela collapsed, leaving the island about 53,000 b/d short, which constitutes about one-third of Cuban demand.

Sherritt International Corp. (Toronto) continued to increase its gross operated production in Cuba, mainly from new wells in Yumuri, Canasi, and Seboruco fields. Sherritt holds an indirect interest in seven exploration and production-sharing contracts (PSCs) totaling 14,000 sq km (3.5 million acres, or 5,500 sq mi).

Canasi development

Pebercan (Montreal) in March 2002

reported that its Canasi No. 5 well, drilled to 3,705 m (12,155 ft) total depth and completed in late 2001, penetrated two successive hydrocarbon pools totaling 1,472 m (4,829 ft) of gross pay. On test, Canasi No. 5 flowed 3,500 b/d of oil.

Flow from the well brings total production from Block 7 to 13,000 b/d of oil, 6,200 b/d of which is net to Pebercan, which operates Block 7 on behalf of Cuban state oil monopoly Cubapetroleo. Pebercan is designated technical operator for the western half of the block, with Sherritt the technical operator for the eastern half.

In June production from Block 7 reached 15,000 b/d of oil with the latest well test, that of Canasi-6, which produced 3,500-4,500 b/d of oil in early tests, said Pebercan. Wellbore length is 3,745 m (12,290 ft), of which 1,572 m (5,157 ft) penetrated three hydrocarbon pools.

In August Pebercan spudded the Via Blanca West-1 exploratory test to evaluate a new structure on the western part of Block 7 along the north coast. The company was also production-testing Canasi-7, a 3,670-m (12,040-ft) directional well that penetrated three oil-productive lenses and flowed at rates of 4,500-6,000 b/d, depending on choke size. Pebercan also spudded Canasi-8 in August.

Ecuador

CAPITAL: QUITO
MONETARY UNIT: US DOLLAR
REFINING CAPACITY: 176,000 B/CD
OIL PRODUCTION: 398,000 B/D
OIL RESERVES: 4.6296 BILLION BBL
GAS RESERVES: 3.45 TCF

During 2002 new data received by *Oil & Gas Journal* resulted in a dramatic increase in Ecuador's estimated proved oil reserves, which more than doubled to 4.6296 billion bbl from 2.115 billion bbl. In November Ecuador elected a new president, Lucio Gutierrez, who vowed to defend the poor.

Ecuador offered four Pacific Coast blocks in the Gulf of Guayaquil region under its ninth international bidding round, said national oil company Petroecuador in November:

- Block 39, covering 3,850 sq km (1,490 sq mi), is wholly offshore in water depths of 10-2,000 m (30-6,500 ft)
- Block 40, covering 4,000 sq km (1,500 sq mi), is also offshore in 40-1,100 m of water (130-3,600 ft), with at least half the marine acreage in depths less than 200 m (650 ft), and it is adjacent to the prolific Amistad natural gas field
- Block 5 is primarily onshore in the Progreso sub-basin, with a small southwestern corner extending into the gulf
- Block 4 includes land on the southeastern tip of the mainland, a portion of the adjacent Puna Island, and shallow gulf waters of 10 m (30 ft) or less.

The participation contracts being offered provide a 4-year exploration period for oil and a 5-year period for natural gas, with possible extensions of 2 years each. Blocks in the Oriente oil-producing region could be offered in 2003.

Extensive production and pressure tests indicated that the first of two horizontal wells on the Oriente Block 17 appeared capable of producing oil at rates in excess of 10,000 b/d, reported Vintage Petroleum Inc. (Tulsa) in October. At year-end, Vintage sold its Ecuador unit to EnCana Corp. (Calgary). Previously, Vintage estimated its proved oil reserves there at 50.4 million bbl. Other independents such as Kerr-McGee Corp. (Oklahoma City) were also selling Ecuadorian assets during 2002.

The Palo Azul discovery on southern Block 18 in the Amazon region will be developed by a new group, reported OPEC News Agency in mid-2002. The Ecuador TLC SA group will spend $190 million and drill 35 wells during 4 years to establish initial production of 5,000-15,000 b/d of 28-29 degree gravity oil from an estimated 65.8 million bbl of recoverable oil.

OCP pipeline

Despite intense environmental opposition, construction on Latin America's highest profile pipeline project — Ecuador's Oleoducto Crudos Pesados (OCP) — was in full swing as 2002 began. The 502-km (312-mile) line was being built from producing areas near Lago Agrio to the marine terminal at Esmeraldas on the Pacific coast. Approved in mid-2001, the $1.2 billion line must be flowing in 2003 under conditions imposed by its permitting process.

Capacity will be 450,000 b/d.

The line will move crude oil of less than 24-degree gravity heated to 75 degrees C at four pump stations on the uphill side of the line. That portion will climb to more than 3,800 m (12,500 ft) above sea level. The existing Transecuadoran Pipeline System (known by its Spanish acronym SOTE) will carry lighter crude oils. OCP is being built parallel to most of SOTE.

Prompted by prospects of greater export capacity along the new line, Petroecuador was planning a major exploration effort in 2003 in preparation for completion of the pipeline. Many exploration and development plans have been held up because of lack of capacity on the existing 380,000-b/d SOTE line.

Private companies producing oil in the Oriente basin are also boosting exploration activities, with one estimating that $2.5 billion will be spent during 2002-2003 on seismic and drilling activity. Environmentalists, however, have fought the project vigorously since its inception. OCP undertook a significant 100-km (60-mile) route deviation near Quito from original plans solely to placate environmentalist opposition.

Guatemala

CAPITAL: GUATEMALA CITY
MONETARY UNIT: QUETZAL
REFINING CAPACITY: 16,000 B/CD
OIL PRODUCTION: 23,500 B/D
OIL RESERVES: 526 MILLION BBL
GAS RESERVES: 109 BCF

Guatemala is the only oil producing country in the Central American region, and all of its production is controlled by Perenco, a European exploration firm, which purchased Basic Resources International from Anadarko Petroleum Corp. (Houston) in 2001. Perenco's Guatemalan assets include all of the country's existing oil fields, a 275-mile crude oil pipeline, and a 2,000-b/d mini-refinery, as well as storage and loading facilities.

Perenco expected to boost production from its Guatemalan assets to 25,000 b/d.

Guatemala's 526 million bbl of proven oil reserves are located primarily in the country's northern jungles of the Peten basin and are most likely associated with those in Mexico's Tabasco formation.

Efforts were under way to explore and exploit potential reserves near Lake Izabal, Guatemala's largest lake, located in eastern Guatemala near the Gulf of Honduras. However, in May 2002, Guatemalan President Alfonso Portillo cancelled one of the contracts amidst strong pressure from environmental groups. The second contract still stands, yet faces similar opposition. The government reportedly was preparing more blocks across the country for exploration in the near future.

In early 2002, Duke Energy International's subsidiary Grupo Generador de Guatemala y Cia SCA begun construction of a $150 million, 165-MW power project in San Jose, Guatemala. The project was scheduled for completion by November.

Mexico

CAPITAL: MEXICO CITY
MONETARY UNIT: PESO
REFINING CAPACITY: 1,684,000 B/CD
OIL PRODUCTION: 3.180 MILLION B/D
OIL RESERVES: 12.622 BILLION BBL
GAS RESERVES: 8.776 TCF

Mexico was identified by international oil companies as one of the top 10 most favored countries for new exploration and production investment in 2002, according to a survey by Robertson Research International Ltd. (UK), marking the first time it's made the list. Mexico's Petroleos Mexicanos (Pemex) budgeted $14.7 billion in 2002 for investment in exploration infrastructure development, representing the largest annual increase in 20 years.

Mexico's real GDP was projected by the IMF to grow 4% in 2003. Through much of 2002 Mexico restrained its oil production in support of efforts by the Organization of Petroleum Exporting Countries (OPEC) to put a floor under crude prices.

In September 2002 Pemex lowered its estimate of proved reserves as of 2000; adjusted totals were lower by 6.755 billion bbl of oil, condensate, and natural gas liquids and by 2.334 billion boe of dry gas. The revised volumes, centering in northern Mexico's Chicontepec region, comply with the US Securities and Exchange Commission's definition of proved reserves, meaning that they will be developed in the short term, given existing technology and economic conditions, according to Fitch Ratings.

Gas supply outlook

To meet its natural gas needs, Mexico will rely increasingly on imports, despite efforts to boost domestic production, said Pemex in October. This outlook supports Pemex's campaign to promote controversial multiple service contracts (MSCs) with foreign companies, proposed to start in early 2003.

Mexico's gas demand will grow at 14%/yr through 2010, doubling from 2002's level, mostly to feed new power plants. Pemex was producing 45 bcm/yr (4.6 bcfd) and planning an output of 67 bcm/yr in 2006 and 88 bcm/yr in 2010 (6.8 and 9 bcfd, respectively), given contributions from MSCs.

Gas demand growth will also push expansion of the pipeline system and require construction of several cross-border pipelines by 2010, said Mexico's Energy Regulatory Commission. The national system will have to be reinforced with incremental compression, pipeline replacement, looping, and new pipelines in order to handle a total flow of 78-88 bcm/yr (8-9 bcfd) in 2010.

Pemex's strategic gas program, which is investing $8.1 billion in mostly nonassociated gas development during 2001-09, has achieved early successes including a discovery on the Lankahuasa prospect, which could contain 50-80 bcm (2-3 tcf) in shallow waters off Vera Cruz. Pemex is also venturing from its primary production areas in less than 100 m of water (300 ft) into deeper Gulf of Mexico waters of more than 2,000 m (6,600 ft) near the US border, where it is

MEXICO-CENTRAL AMERICA

Texas
Coahuila
US
Gulf of Mexico
Bay of Campeche
Gulf of Tehuantepec
Ulua
Buena Suerte
Monclova
Nuevo Laredo
Nuevo/Laredo
Laredo
Lampazos
Corpus Christi
La Presa
Arcos
Sta. Rosalia
Lojitas
Culebra
Canor
Carretas
McAllen
Reynosa
Mission-Trevino
Brasil
Fortuna
Brownsville
Matamoros
235
Cadereyta
Primavera
Cabeza
Ramirez
18 de Marzo
Monterrey
S. Bernardo
Benavides
Chapul
Huizache
Pobladores
San Fernando
Nuevo Leon
Zacatecas
Tamaulipas
Lerma
Ciudad Victoria
San Luis Potosi
195
Barcodon
Tamaulipas-Constituciones
Aguas
Ciudad Madero
Ebano Panuco Fields
Arenque
Tampico
Cabo Nuevo
Arrecife Medio
Tuxpan
Isla de Lobos
Guanajuato
Golden Lane Fields
Tiburon
Esturon
Marsopa
Queretaro
Hidalgo
Salamanca
Bagre
Escuala
320
Tula
Pachuca
Poza Rica
Pianche
Morelia
MEXICO
Tlax.
Veracruz
Mexico City
Jalapa
Michoacan
Toluca
D.F.
Tlaxcala
Remudauera
Cuernavaca
Puebla
Miralejos
Capite
M. Pancho
Mecayuman
Angastura
Mirador
Coapo
Novillero
Veinte
Mata Verde
Cacuite
Maculie
Tabasco
Morelos
Azcapotzelco
Cardenas
Minatitlan
200
Guerrero
Chilpancingo
Oaxaca
Acapulco
Tehuantepec
Juchitan
Salina Cruz
330
Tuxtla
Chiapas
Sinán
Citam
Bolontikú
Kix
Yum
May
Dos Bocas
Ciudad del Carmen
Fontera
Ciudad Pemex
Villahermosa
Merida
Yucatan
Campeche
Quintana Roo
Xan
6
Belmopan
BELIZE
Caribe
1 San Diego
W. Chinaja
Rubelsanto
Yalpemech
2
GUATEMALA
Guatemala
Escuintla
16
HONDURAS
21
San Salvado
Sonsanate
EL SALVADOR
100 0 100 200 Miles
100 0 100 200 300 Km
MEXICO-CENTRAL AMERICA

Straits of Florida
Canasi
Puerto Escondido
Boca de Jaruco
Seboruco
Varadero
122
Havana
Bacuranao
Motembo
2
76
Cabaiguan
Cienfuegos
Cristales
Jatibonico
CUBA
BAHAMA BANK
102
Santiago
20°
Yucatan
Cozumel
Quintana Roo
Chetumal
JAMAICA
34
Kingston
Caribbean
Sea
BELIZE
Belmopan
Gulf of
Honduras
ribe
San Diego
alpemech
Puerto Cortes
15°
HONDURAS
Tegucigalpa
22
San Salvador
La Union
NICARAGUA
20
Managua
Bluefields
Lake Nicaragua
10°
COSTA RICA
San Jose
15
Limon
Punta Cahuita
60
Panama City
100 0 100 200 Miles
100 0 100 200 300 KM
Golfito
David
PANAMA
La Palma
MEXICO-CENTRAL AMERICA
85°
80°

seeking technology-sharing agreements.

MSCs

MSCs, a new form of service contract for Mexico, will be structured as long-term (field life) agreements in which the contractor will be paid a unitized price for services and work performed. The success of efforts to open Mexico's Burgos basin to outside investment through MSCs will serve as a litmus test for Mexico's energy strategy, according to Cambridge Energy Research Associates (Cambridge, MA, US).

MSCs will allow outsiders to provide a much broader range of services than before, with payment in fees rather than profit shares. Pemex will retain the rights to reserves and profits. Essentially, the MSC treats an oil or gas development as a public works project, Pemex said.

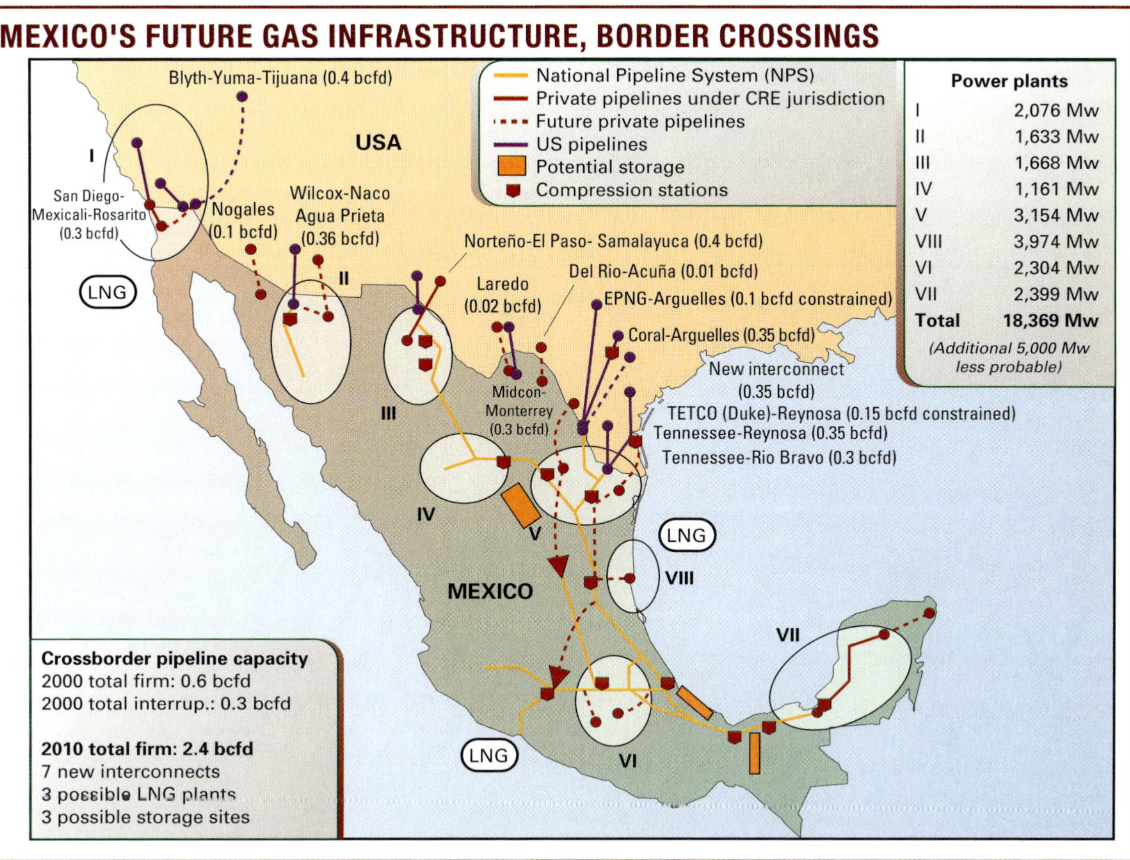

A tender for MSC bidding should start in February 2003. The aim is to double output from the Burgos basin by working with foreign companies in low-risk areas with certified reserves. Given Pemex's domestic gas supply cost of $45.65/cu m ($1.70/Mcf) and a projected 2006 gas import cost of $99.10/cu m ($3.69/Mcf), Pemex anticipates an MSC supply cost threshold at $66.06/cu m ($2.46/Mcf).

Opening up Burgos

For the first time in 50 years, Mexico was planning to invite private companies to bid on developing onshore natural gas fields starting in mid-2002 with a dozen blocks in the Burgos basin, which covers 50,000 sq km (19,000 sq mi).

Under a $270 million Pemex contract, at least 240 wells were being drilled during 2001-2003 in 11 Burgos gas fields by PD Mexicana, a joint venture of Precision Drilling Corp. (Calgary) and BJ Services Inc. (Houston). Pemex has conducted a detailed geological analysis of the Burgos basin's dominant Frio-Vicksburg play. In 2003-04, 100 Pemex wells in Burgos field will be drilled using rigs from Drillers Technology Corp. (Calgary).

Offshore E&D

Mexico continued during 2002 to make progress on Proyecto Cantarell, which could be the world's largest ongoing oil field development. The megaproject involves reservoir repressurization, redevelopment, and facilities refurbishment at the massive Cantarell fields complex in the Bay of Campeche. In early 2002 Pemex reported that it had achieved record oil production of 1.86 million b/d, reduced flaring of associated gas to only 10%, and completed the first successful year of nitrogen injection of 32 million cu m/d (1.2 bcfd).

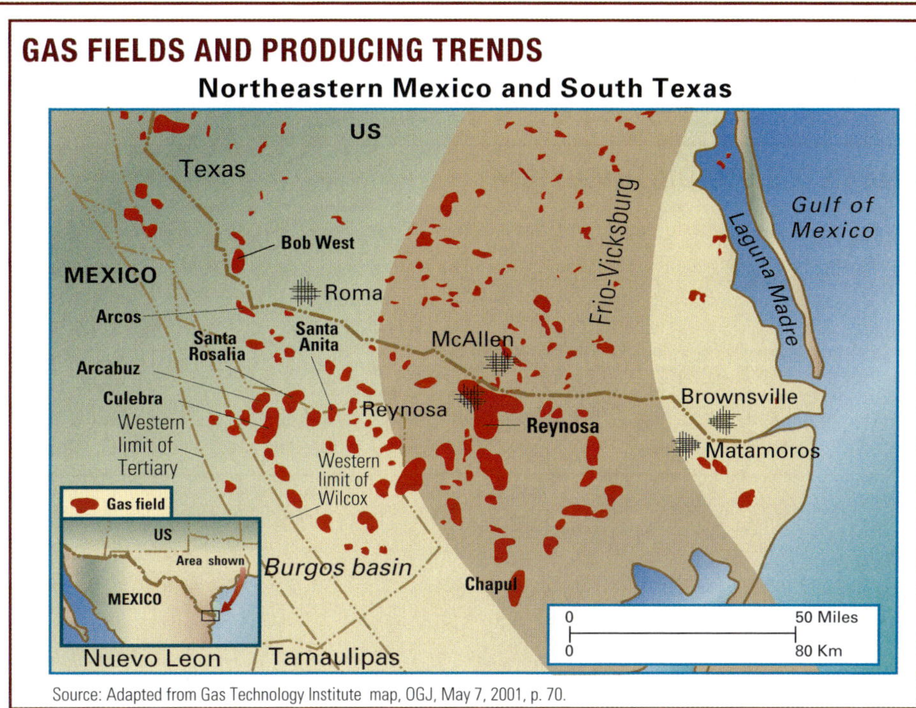

Source: Adapted from Gas Technology Institute map, OGJ, May 7, 2001, p. 70.

Remaining Cantarell reserves were pegged at 14 billion bbl in four fields; Akal alone, the biggest, is ranked as the world's sixth-largest oil field. Cantarell work during

2002 and beyond will entail drilling infill development wells, finishing debottlenecking efforts, eliminating gas flaring as gas compression capacity reaches 100%, and increasing the number of well workovers in Akal. Pemex also hopes to further delineate the Sihil reservoir, a fifth productive reservoir underlying Akal field. Sihil's preliminary reserves were estimated at 1.135 billion boe.

Rehabilitation of Cantarell's pipelines began in 2002 under a Pemex contract with Horizon Offshore, which will repair riser expansion curves, pig and test the repaired segments and related existing facilities, install topside spools and valves, perform line stabilizations, repair pipeline crossings, recover out-of-service lines, and trench the repaired segments and existing pipeline installations.

Mexico-US pipelines

A natural gas pipeline and dual propane-butane pipeline received US permits in July 2002. Both lines, being constructed by Tidelands Oil & Gas Corp. (US) subsidiary Reef International LLC, start in Eagle Pass, TX, and will travel to Piedras Negras, Mexico, where the gas pipe will interconnect with the Cia. Nacional de Gas SA de CV (Conagas) system. Several transportation and sales delivery agreements for both the liquids and the natural gas were in place.

In August Reef began directional drilling of the 12-inch gas pipeline and bilateral 6-inch liquids crossings, 1.5 m (5 ft) apart, under the Rio Grande River. The crossing was expected to be completed by year-end. Work on a terminal in Eagle Pass was also under way.

Tidelands was also planning another international pipeline crossing, a 30-inch natural gas trunkline at El Paso, TX, which will deliver natural gas to Juarez, Mexico. After permitting, Reef anticipated starting construction at year-end 2002.

Construction began in November on a 150-km (95-mile), 30-inch natural gas pipeline designed to carry initially 3.68 bcm/yr (375 MMcfd) of gas from Mier, TX, to Monterrey, one of Mexico's fastest growing industrial areas, said Kinder Morgan Energy Partners LP (US).

A Pemex unit subscribed for all of the capacity under a 15-year contract and will feed the gas into its transport system and to a 1,000-MW power plant complex. The pipeline

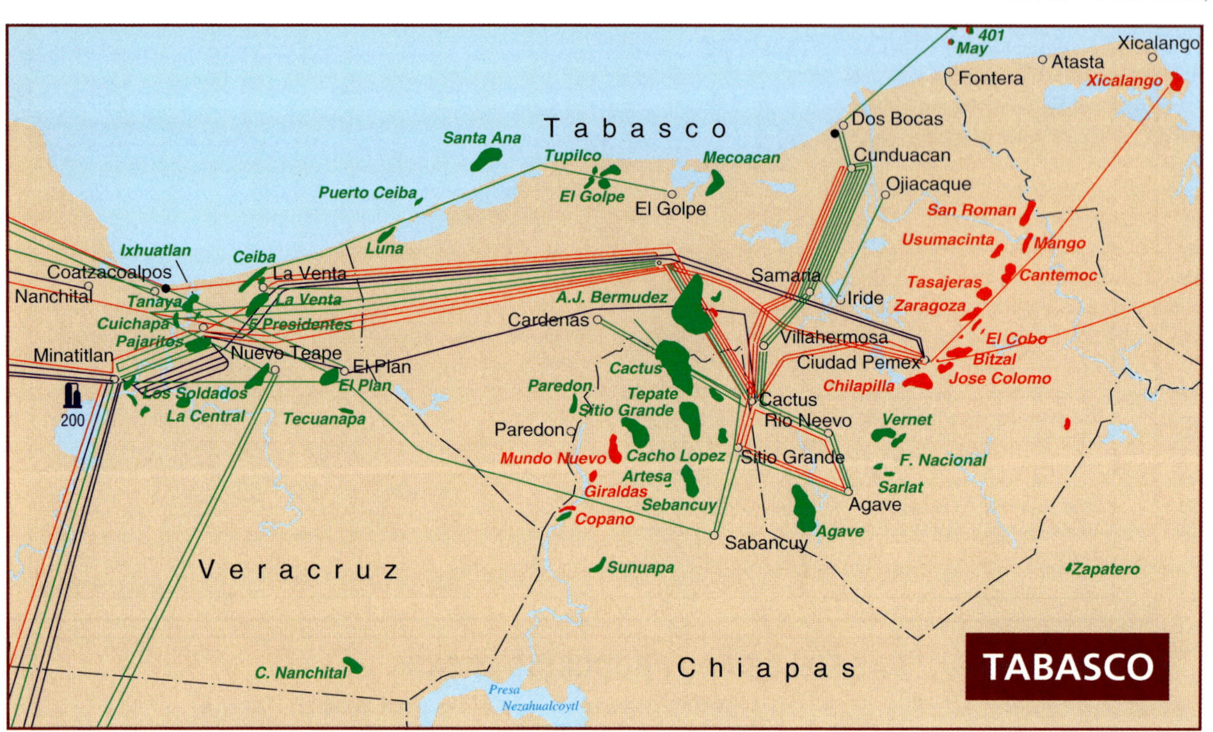

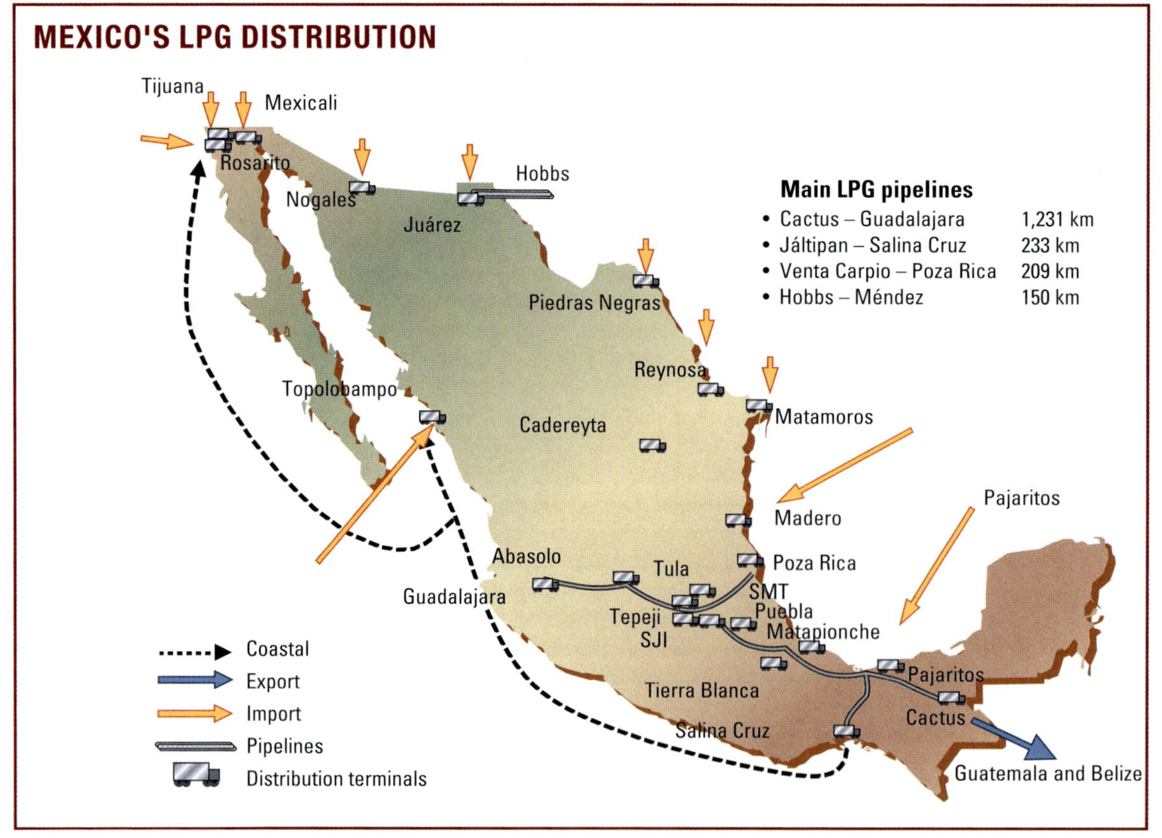

should be completed by mid-2003 and could be expandable or the flow reversed to export gas to the US.

A 3.08 bcm/yr (314 MMcfd) gas line was proposed from south Texas to Hidalgo County by an El Paso unit in April. Construction of the 15-km (9.3-mile), 30-inch lateral would feed power generators in Mexico.

LNG projects

Mexico is likely to import more LNG for its own use and to serve as a gateway to North America. Establishing at least one LNG terminal in Mexico is a goal of President Fox's administration, which ends in 2006, said Baker & Associates (Houston) in October. Although Mexico is reviewing some 18 proposals, four new LNG terminals over 20 years would be more realistic, said ANZ Investment Bank.

Of the new LNG receipt capacity proposed for North America, more than 75% is slated for Mexico, said a unit of CMS Energy Corp. Establishing an LNG infrastructure in Mexico will help advance the regional integration of gas markets in the Americas, but depends on multilateral trade agreements and long-term gas prices sufficient to support investment.

Marathon Oil Co. (Houston) and partners proposed plans in March 2002 for an LNG complex near Tijuana in Baja California, western Mexico. With a potential start-up in 2005, the facility would regasify up to 7.4 bcm/yr (750 MMcfd) of LNG, much of it from Asia-Pacific sources including Pertamina, and would also incorporate an LNG marine terminal, offloading terminal, pipeline infrastructure to transport the natural gas to local and export markets, and a 400-MW natural gas-fired power plant.

Other LNG projects proposed for Mexico include a west coast receiving terminal near Ensenada, to be developed by CMS Energy and Sempra Energy (both US), and a regasification terminal at Altamira, Tamaulipas state, on Mexico's eastern coast, to be developed by Shell Gas & Power. Shell was also planning a 13 bcm/yr (1.3 bcfd) LNG regasification terminal on the west coast in Costa Azul, Ensenada, and in April 2002 secured initial supplies of 7.5 million tonnes/year (tpy) from the Asia-Pacific region. The terminal, estimated to cost $500 million, was scheduled to come onstream in 2006.

Also to be built in Baja California is the La Rosita gas-fired combined cycle power plant. In May 2002 InterGen announced financing for the $748 million, 1,065-MW plant, which was to begin commercial operations by mid-2003. Natural gas fuel will be transported through a new 341-km (212-mile) pipeline from California and Arizona.

In addition to plans for LNG and electric power projects, LPG was already playing a major role in Mexico's economy during 2002, especially in urban areas with pollution concerns. Pemex's gas and petrochemicals subsidiary operates more than 1,750 km (1,100 miles) of LPG pipelines and numerous distribution terminals.

Peru

CAPITAL: LIMA

MONETARY UNIT: NEW SOL

REFINING CAPACITY: 190,950 B/CD

OIL PRODUCTION: 93,000 B/D

OIL RESERVES: 323.393 MILLION BBL

GAS RESERVES: 8.655 TCF

In October 2002 Peru ratified a 30% cut in royalties for oil and gas production from companies that make a discovery within the next 4 years. State oil company Petroleos del Peru SA (Perupetro) was authorized by President Alejandro Toledo to allow the reduction, which had been stalled since 2001. The government was also working to eliminate taxes on unsuccessful exploration.

Perupetro selected Robertson Research International to design and implement a strategy to promote crude oil exploration and exploitation. Peru's average production has fallen steadily for 20 years, from about 180,000 b/d to around 97,000 b/d. The Hydrocarbons Bureau in July estimated that oil and gas investments in 2002 would total $391 million, of which $280 million is earmarked for development and $48 million for exploration.

Earlier in the year, Energy and Mines Minister Jaime Quijandria survived a cabinet shake-up that arose from protests against privatization of energy companies, which remains unpopular. Quijandria helped write the government's master plan, including privatization, before President Toledo's election.

Peru continued to compete with Chile to be the site for Bolivia's LNG export facilities.

Camisea takes off

In May the government authorized appointment of an ombudsman for the $1.3 billion Camisea natural gas development, which includes a proposed LNG or GTL export project. Camisea is 500 km (300 miles) east of Lima, on the eastern slope of the Andes, in Peru's Ucayali basin.

A unit of Pluspetrol SA (Argentina) is operator of the Camisea fields' upstream consortium, which is developing 300 bcm (11 tcf) of gas and 600 million bbl of associated liquids from San Martin and Cashiriari fields.

The total project cost of the upstream and midstream portions is estimated at $1.3 billion, including field development, a gas pipeline and a liquids pipeline, and a gas distribution network in Lima and Callao.

According to Pluspetrol, Peru's petroleum future lies in natural gas, and its gas potential compares favorably with Bolivia's.

The country's large underexplored basins include the Ucayali and Madre de Dios, which are on trend with Bolivia's world-class gas reservoirs. Pluspetrol believes that with Camisea fulfilling domestic needs, additional discoveries could support Peru's growing role as a regional gas exporter.

In August the government held public hearings on the megaproject, due to persistent environmental opposition. The hearings were requested by the Inter-American Development Bank (IADB), which is developing financing for the project, to review the previously approved environmental impact assessments for each segment (upstream, transmission, and distribution/marketing).

The most environmentally sensitive areas of the project are the Las Malvinas gas fields region and the first third of the pipeline extending across the Andes through the jungle. Financing was expected to be obtained by early 2003 in order to begin producing and transporting natural gas and liquids to Lima by mid-2004.

In October 2002 Proinvestment, the government commission promoting privatization and concessions in Peru, offered a take-or-pay contract for Camisea gas that could require first-stage investors to install a 300-MW combined-cycle plant (a simple-cycle generator with a minimum 200 MW plus an additional 100 MW unit) using the gas within 3 years.

As of November, all seismic work had been completed, including 800 sq km (300 sq mi) of 3D, reported Pluspetrol, and gas compression, cryogenic gas processing, and liquids recovery facilities were on track to be completed in early 2003.

Camisea drilling

In mid-2002 Parker Drilling Co. (Houston) finished the first workover of wells drilled earlier in the San Martin field on Block 88. The timetable is to drill development wells to more than 2,000 m (6,600 ft) in the San Martin I structure by 2003, with production to be onstream by mid-2004. Pluspetrol said the wells would produce enough gas to supply Lima, as initial demand is expected to be low, so part of the gas will be reinjected.

As of late 2002, San Martin-1 was being tested, another San Martin well was due for completion by year-end, and the third development well on Block 88 was to be spudded around year-end. Two more San Martin wells are planned, while the San Martin III structure and Cashiriari field will be drilled as a function of gas demand and reinjection needs.

Pipelines and distribution

Transportadora de Gas del Peru (TGP) was planning to build and operate the gas and liquids pipelines to connect Camisea to the Peruvian coast. The Camisea project will have a guaranteed income for gas transportation and distribution through a contract covering the difference in the cost of transport until production reaches 3.7-4.4 bcm/yr (380-450 MMcfd). Pluspetrol was reportedly looking for a buyer for its share in TGP.

As of July 2002, 80 km (50 miles) of welded pipe had been laid on the route from Pisco on the south coast towards Ayacucho in the southern Andes. In November Pluspetrol said that infield pipeline construction would be 75% complete by year-end, and TGP said it had prepared 350 km of the total 720 km route (220 of 450 miles) for natural gas and liquids pipelines, buried almost 200 km of pipe (125 miles), and rebuilt 20 bridges.

In April the government postponed signing a contract with Belgium's Tractebel SA to distribute natural gas from the Camisea project in metropolitan Lima and Callao. Allegations appeared in the press that Tractebel had paid a commission for approval of its license. The reports were denied by Tractebel, which reported in November that its local unit had laid its first pipeline segment for entry into Lima.

LNG project

Results of a feasibility study for the Camisea natural gas export project, being conducted by Hunt Oil Co. (Dallas), were unveiled in November, and a decision was expected by spring 2003. Hunt was in discussions with potential purchasers of Peruvian LNG, and the consortium has screened several potential sites for the plant that meet design parameters. The LNG export project would target start-up by early 2007, with regasification facilities likely to be sited at Baja California or another site along Mexico's western coast.

The single-train LNG plant, to be based on Air Products Corp.'s liquefaction process, would have an initial inlet design capacity of 6.1 bcm/yr (620 MMcfd), but a more likely output level would be 4.5 million tpy, which would require input of 6.5-6.9 bcm/yr (670-700 MMcfd). Special attention is being paid to the design of two 110,000 cu m LNG storage tanks because of the intense seismic activity in the region. The gas, a mix of methane and 8% ethane, would not need scrubbers or liquids recovery and would be processed in facilities upstream at Las Malvinas.

Expansion of Camisea field development and pipeline infrastructure to support a coastal Peruvian LNG export project would require an additional $1 billion beyond the export plant costs, said Hunt in November, but Peru's plans should not depend on Bolivian gas because Peru's own infrastructure and gas market are already being created.

The greater Camisea area's 526 bcm of gas (19.6 tcf) can support the production levels required by both the LNG and pipeline projects. A projected reserves base of 113 bcm (4.2 tcf) needed to support a single-train, 4 million tpy LNG export project would leave at least 160 bcm (6 tcf) of gas still available in the greater Camisea area for future expansions.

In May Pluspetrol and partners were planning to build a $120 million natural gas liquids fractionating plant in Pisco on Peru's south coast. Construction was expected to begin by year-end 2002, with gas production scheduled to start up by mid-2004.

The fractionation plant would be built away from the Paracas Bay natural reserve, and only the terminal would be in the sea. Liquids will be piped to the plant for separation from the Camisea gas fields production center at Las Malvinas on the Urubamba River.

Auction aborted

In March 2002 Perupetro said it would auction concessions for Blocks 56, 57, and 58, which lie adjacent to the Camisea gas fields. Block 56 has estimated reserves of 80 bcm (3 tcf) of gas in Pagoreni and Mipaya fields, while the other blocks were to be explored.

However, most of the companies that had shown earlier interest backed out, so in October Perupetro decided to keep Block 56 until the export project materializes. The Camisea fields, with an estimated 240 bcm (9 tcf) of gas reserves, would be sufficient to fulfill immediate gas demands, Perupetro determined.

Other E&P

In addition to the high-priority Camisea, Pluspetrol was seeking to maintain maximum output from Block 8, which was producing 24,000 b/d, and Block 1AB, producing 40,000 b/d but declining at 20%/yr. Pluspetrol drilled a successful exploration well, Carmen Este X-1, which was yielding almost 4,000 b/d of oil, and two development wells on Block 1AB during 2002. Another four delineation-development wells are scheduled for the Carmen area in 2003. Pluspetrol has another 3 years remaining on the exploration concession covering Block 1AB, with an option to extend for a year and to drill another exploratory well.

Public hearings were scheduled for November 2002 on an environmental impact assessment of Occidental's proposed exploration program on Block 64 in Peru's Marañon basin. The block has been in force majeure since October 1996, when indigenous communities rejected oil companies in their areas.

During 2002 Perupetro signed one exploration contract, with Peruvian company Petrotech, to explore offshore block Z-6. Perupetro was awaiting final approval for a contract with Burlington Resources Inc. (US). Perupetro expected 10 wildcats to be drilled throughout the year, four in the jungle region and six on the northwestern coast. These included the Mashansha well on Block 35 and exploration wells on Blocks 87, 31-E, 13, 16, and 14. Additionally, contracts were being negotiated with Repsol (Block Z-4), with a Harken Energy Corp. unit (Z-5), and with Advantage Resources International Inc. (Block 70).

In September 2002 Syntroleum Peru Holdings Ltd. and BPZ Energy Inc. (Houston) were seeking partners in Block Z-1 offshore on Peru's far northern coast. The companies hope to exploit previous discoveries and declare a commercial find before the fourth year of the contract, in order to benefit from a 30% royalty discount.

Syntroleum Peru, a unit of Syntroleum Corp. (Tulsa), is also aiming to use its technology to install a GTL plant in Peru, provided it confirms sufficient natural gas reserves. Syntroleum and BPZ were conducting an environmental impact assessment on Block Z-1, including inspection of four 30-year-old platforms.

In March 2002 Petro-Tech Peruana SA was slated to sign a license contract with Perupetro for exploration and production of Block Z-6, also off Peru's northern coast. Block Z-6 is south of Z-2B, where Petro-Tech produces an average 12,800 b/d of crude oil and 1.6 million cu m/d (60 MMcfd) of associated gas.

Olympic Peru Inc., operator of Block 13 in the Sechura Desert, was planning to begin its first commercial sales of nonassociated gas in 2002. Previously Olympic had completed a 6-inch natural gas pipeline running about 50 km (30 miles) northwest from Olympic's fields to the gathering station system and Paita port. The Austral Group fisheries complex was to be the first customer for an initial 80,000 cu m/d (3 MMscfd) to be used in boilers converted from oil.

Olympic's reserves are certified at 5.234 bcm (194.90 bscf) of dry gas, contained in the Becara, La Casita, Loco, and Viru structures. Olympic completed two gas wells in February and March in the Mochica and Loco structures.

Trinidad & Tobago

CAPITAL: PORT OF SPAIN
MONETARY UNIT: DOLLAR
REFINING CAPACITY: 160,000 B/CD
OIL PRODUCTION: 127,000 B/D
OIL RESERVES: 716 MILLION BBL
GAS RESERVES: 23.45 TCF

Already the No. 1 supplier of LNG to North America, Trinidad and Tobago plans to more than double natural gas production to 37 bcm/yr (3.8 bcfd) by 2006. In 2003, the nation will become the world's fifth-largest LNG producer and the largest exporter to the US.

Trinidad and Tobago was identified by international oil companies as one of the top 10 most favored countries for new exploration and production investment in 2002, according to a survey by Robertson Research, marking the first time it's made the list. During 1995-2001, more than $6 billion was invested in the growth and diversification of Trinidad and Tobago's natural gas industry, the highest level of foreign direct investment per capita in the Americas.

During December, Trinidad and Tobago's security forces were on high alert following threats to American and British energy interests there from an Islamic fundamentalist group. Trinidad and Tobago is a key country particularly for BP and BG operations and a key supplier of natural gas to the US, particularly New England. Its Atlantic LNG Co. plant provides as much as 40% of the gas used for heating in the New England states during the peak winter periods.

Trinidad and Tobago was working with Venezuela to unitize several natural gas fields that straddle both countries' borders. An initial agreement included a major cross-border gas field, Loran, which lies in the Deltana Plataforma area with most of its reserves reportedly on the Venezuelan side. Venezuela wants to unitize Loran with Block 6d on the Trinidad side of the border.

PETROTRIN'S JV ACREAGE

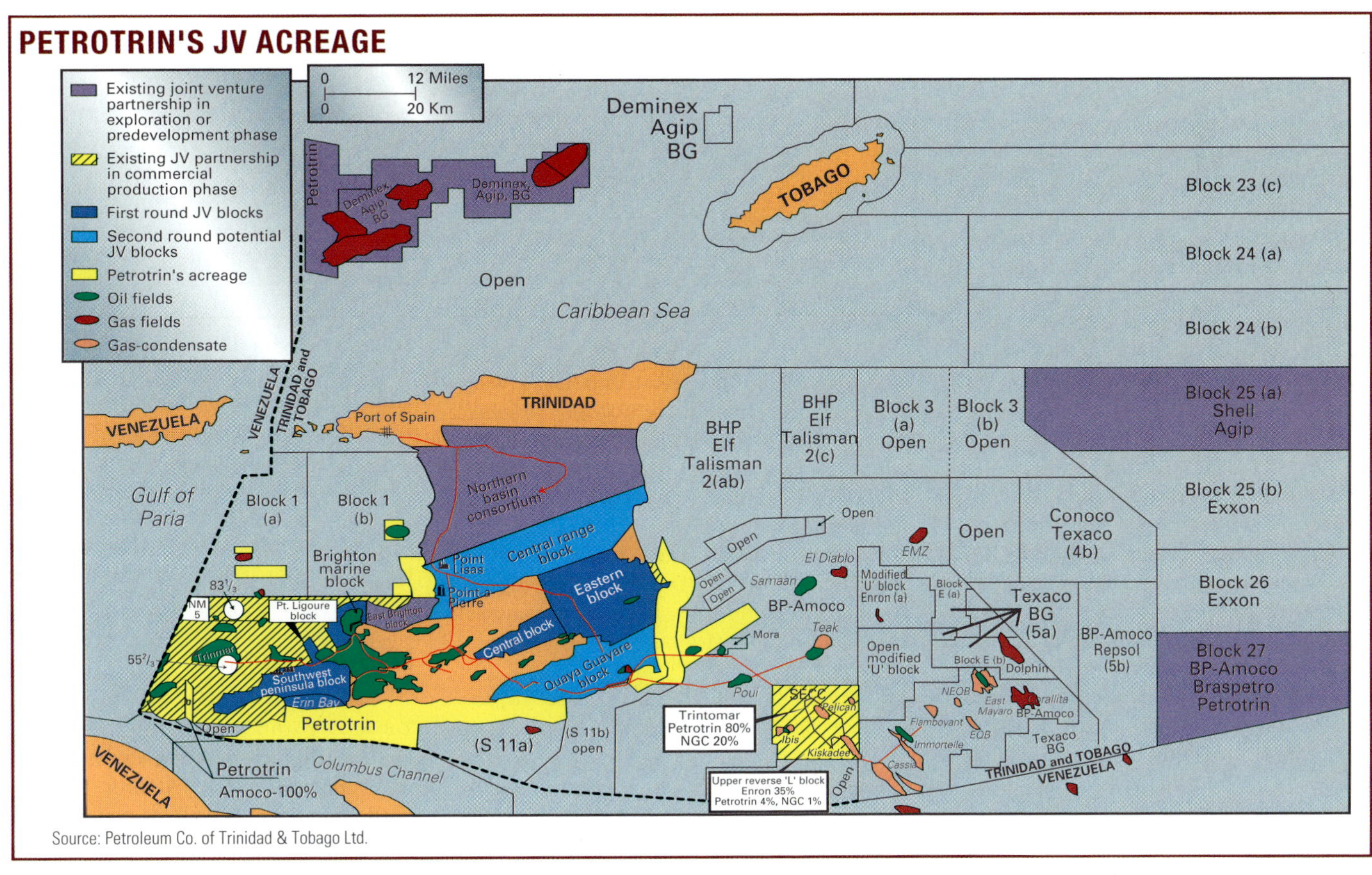

Source: Petroleum Co. of Trinidad & Tobago Ltd.

During 2002 the state's Petroleum Co. of Trinidad & Tobago Ltd. (Petrotrin) sought to increase foreign participation in Trinidad and Tobago's oil and gas operations. Through 2004 Petrotrin and its joint venture partners are expected to invest about $2 billion (TT) in properties, and Petrotrin will offer exploration acreage for bidding during 2003-04.

Exploration

In September 2002 BP Trinidad & Tobago (BPTT) announced the discovery of 27 bcm (1 tcf) of natural gas (180 million boe) in its Iron Horse field. The field was discovered in July, 39 km (24 miles) off the east coast, close to the Amherstia, Cassia, and Red Mango fields, in 70 m of water (230 ft). The well, BPTT's second major gas discovery for 2002 and its fourth in 3 years, raises the company's estimated gas reserves to 460 bcm (17 tcf).

BHP Billiton and partners in March 2002 gauged a hefty oil flow rate of 4,360 b/d in their Kairi-2 appraisal well on Block 2c in shallow waters off eastern Trinidad. The appraisal well, 2 km (1 mile) south of the Kairi-1 discovery, demonstrated greater hydrocarbon potential in Angostura field. Other discoveries on Block 2c include Angostura-1, 8 km (5 miles) southwest of Kairi-2, which yielded natural gas at a rate of 800,000 cu m/d (30 MMcfd) on test.

The Trinidad onshore Central Block was being operated by Vermilion Resources Ltd. (Calgary), which acquired interest in mid-2002. The block contains two successful Carapal Ridge and Corosan exploratory gas wells with proved reserves of 1.73 bcm (64.4 bcf) of gas and more than 1 million bbl of condensate.

The Carapal Ridge gas well was brought onstream in December for a 6-month test contract. The well was expected to flow at a gross rate of 540,000 cu m/d (20 MMscfd) of gas and 700 b/d of condensate. Following the test, the well will be shut in for as long as 3 months.

In September Trinidad Exploration & Development Ltd. was drilling two exploration wells, Rapso and Calypso, on the Southwest Peninsula block southwest of Point Fortin, which covers 45,000 sq km (17,000 sq mi) onshore and offshore.

Also in September two exploration wells were being planned by a joint venture of Vermilion and Aventura Energy Inc. (Calgary) to fulfill their work obligations. Saunders-1 on the southwestern Carapal Ridge feature was to spud by year-end 2002, with Barraca-1 slated for 2003. Three shallow oil wells along the southern edge of the Siparia syncline were also to be drilled. The joint venture has identified six plays.

Deepwater progress

Six wells have been drilled without commercial discoveries on four blocks in deep waters (750-1,500 m, or 2,500-5,000 ft) off Trinidad and Tobago, but Shell and BP are planning more work, they said in November.

BHP's oil discoveries on Block 2c have encouraged Shell's efforts on Block 25a, which lies 50 km (30 miles) to the east. Shell was drilling an exploratory well, Pepper Sauce, to 2,000 m (6,600 ft) below mud line in water depths of 1,050-1,100 m (3,440-3,610 ft). Under terms of its PSC, a second exploratory well was planned for December 2002.

In mid-2002 BPTT drilled its first deepwater well, Catfish 1 on Block 27, which turned out to be the most expensive well ever drilled in Trinidad and Tobago. Seismic reprocessing was under way to determine if this was a commercial discovery. The PSC obliges BPTT to drill at least two exploratory wells in the first phase.

Meanwhile, BPTT and 11 other companies are funding a 2D seismic survey in Trinidad and Tobago's ultra-deep waters. ExxonMobil Corp. (Irving, TX, US) was seeking to give up Blocks 25b and 26, where the PSC requires drilling 7 first-phase exploration wells.

LNG-RELATED GAS FIELD DEVELOPMENT OFF TRINIDAD

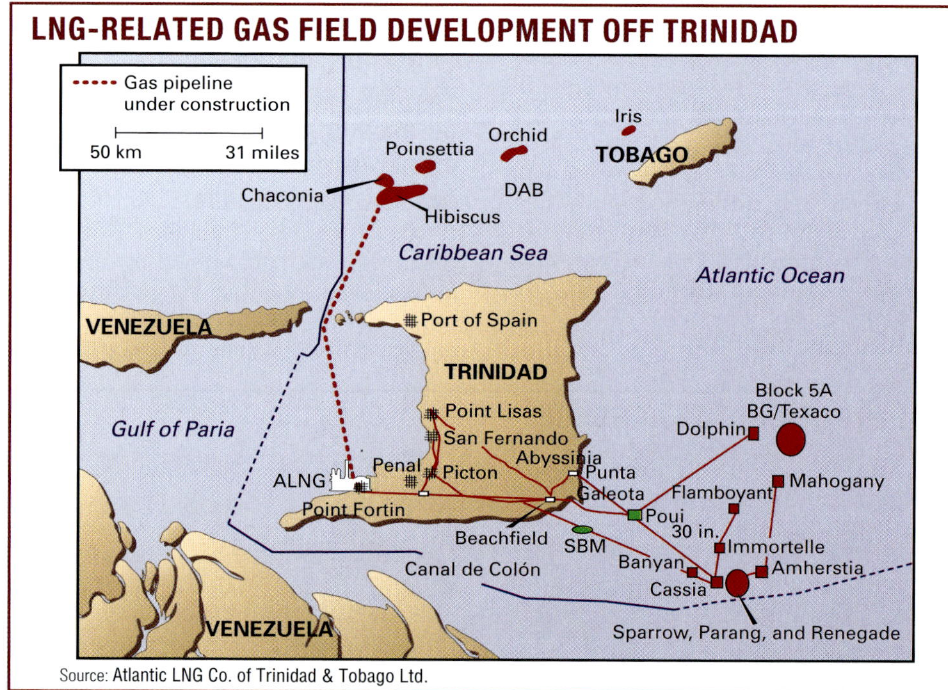

Source: Atlantic LNG Co. of Trinidad & Tobago Ltd.

LNG-related gas

Hibiscus field, in Trinidad's North Coast Marine Area (NCMA), produced first gas into Atlantic LNG's newly commissioned Train 2 at Point Fortin, reported BG in September. NCMA is 40 km (25 miles) offshore and also contains Poinsettia and Chaconia gas fields, with total proved and probable reserves of 64 bcm (2.4 tcf).

NCMA will supply 6.4 million cu m/d (240 MMscfd) to Train 2 for 20 years. Three wells will be drilled during the second phase of development and brought into production in 2003. In Phase 3, BG will drill six subsea wells in Poinsettia field, which will begin production in 2009.

BG and equal partner ChevronTexaco Corp. (San Francisco) also received government approval to develop Dolphin Deep and Starfish natural gas fields in the East Coast Marine Area, which will supply 2 million cu m/d (80 MMcfd) of gas as LNG to Southern LNG's Elba Island, GA, regasification terminal during 2005-23. Dolphin already delivers more than 7 million cu m/d (260 MMscfd) of natural gas for domestic use in Trinidad and Tobago.

BG and partners also signed a PSC in May for Block 3a, which covers 614 sq km (237 sq mi) in 30-90 m of water (100-300 ft), 40 km (25 miles) off the eastern coast and adjacent to Block 2c.

LNG disputes

The importance of Trinidad and Tobago's natural gas to the US market will grow with the two-train expansion under way during 2002 at Atlantic LNG's Port Fortin complex. Train 2 started up mid-year, and Train 3 was to begin running in 2003.

Natural gas from Dolphin Deep and Starfish fields will be processed at the third train. Construction of a fourth train, being discussed at year-end, would produce 5.2 million tpy, most of it for export to the US. Atlantic's $1.1 billion expansion will increase gross LNG production to 9.9 million tpy from 3.3 million tpy when Train 3 comes onstream.

BG was negotiating with the government to allow a fourth LNG train to begin operation in late 2005. In August the government had mandated that Train 4 start construction and that negotiations begin for supply contracts for Trains 5 and 6. However, in November, the government threatened to ban the expansion unless Atlantic LNG participants, including BP and BG, improve their offer.

The dispute reportedly centers on construction costs, part of which would be written off by the government. Ministry officials accused the LNG partners of significantly increasing the cost of the project, which would reduce government revenue. Although BG and its shareholders were concerned, BG expressed confidence that a compromise could be reached.

In a related dispute, Petrotrin refused to support the LNG expansion unless the company collects more than $60 million from Atlantic LNG. Petrotrin's management reportedly demanded that its subsidiary Trinmar Ltd.'s base be relocated if Train 4 is authorized. Trinmar's offshore production operations are serviced from a compound next to Atlantic LNG operations, and they share the same harbor channel.

If Train 4 is constructed, it would raise the country's LNG production to 14 million tpy. Trinidad's burgeoning gas reserves potential, including undiscovered prospective reserves, is estimated at an ultimate 3 bcm (100 tcf). But debate continued in 2002 on how best to transform this wealth into benefits for the nation's citizens. Trinidad and Tobago aims to achieve the status of a developed nation by 2020.

Regional pipeline

A subsea natural gas pipeline system could be built from Trinidad and Tobago northward through the Caribbean Sea and begin flowing gas by 2006, said government leaders in December at a meeting of Caricom (the 15-member Caribbean regional economic union).

A preliminary feasibility study was completed on the 20-inch pipeline, planned to extend up the island chain to Martinique and Guadeloupe, which would purchase nearly 70% of the gas. The line would also supply all the other islands, including Barbados, with spurs to Antigua, Barbuda, St. Kitts-Nevis, and Dominica. EOG Resources Inc. (Houston), is expected to supply the gas for the pipeline and has taken a lead role in the project.

Most of the gas is expected to be used by the islands' power companies, with total demand estimated at 1.5-2 bcm/yr (150-200 MMcfd). Demand is expected to increase with development of the islands' manufacturing sector and greater use of electricity due to growth in tourism and population. Trinidad and Tobago said the proposal could offer the islands a 30% reduction in energy costs, including electricity, and would guarantee price stability and predictability over the medium to long term.

Possibly, the gas pipeline could eventually extend to Miami via Puerto Rico, which would significantly complicate the project and increase costs, but also would provide an enormous potential market for Trinidad and Tobago's gas.

Refinery plans

The government approved in August plans to build a 224,000 b/cd export oil refinery and ancillary infrastructure, with targeted completion date of 2006. If constructed, the $2 billion refinery would be one of the largest in the world, with a 17.4 million bbl terminal and port facilities. However, previous plans for what would be the nation's third oil refinery have not succeeded.

Venezuela

CAPITAL: CARACAS
MONETARY UNIT: BOLIVAR
REFINING CAPACITY: 1,282,100 B/CD
OIL PRODUCTION: 2.415 MILLION B/D
OIL RESERVES: 77.8 BILLION BBL
GAS RESERVES: 148.0 TCF

Early in 2002 Venezuelan President Hugo Chávez briefly lost control of the country, but regained power after the coup failed. In May Alí Rodríguez Araque, president of PdVSA, reaffirmed support for OPEC's policies, but Venezuela later ramped up production above its OPEC quota.

A general strike began in early December 2002, which slashed Venezuela's oil production below 400,000 b/d from around 3 million b/d and removed about 50 million bbl of oil from world inventories that month, according to the US Energy Information Administration.

If Venezuela's production continues at year-end levels without being made up by other supplies, world oil markets would lose more than 70 million bbl per month, which would primarily have to be drawn down from stocks. Even if the strike is resolved quickly — which was anticipated by industry analysts in early 2003 — restoring Venezuela's full production could take several months.

PdVSA's fortunes

PdVSA, by far the country's largest employer and business, accounts for about half of government revenues and one-third of Venezuela's GDP. The company plays an important role in national politics and is run by presidential appointees.

However, budget cuts, inadequate foreign investment, and President Chávez's past policy of adherence to OPEC quotas have crimped PdVSA's long-term expansion plans. PdVSA's investment budget was down 28% in 2002.

At year-end, the strike had paralyzed PdVSA's refining and petrochemical operations, though 24 million cu m/d of gas (880 MMcfd) was being supplied to power generators. PdVSA declared force majeure on petroleum product exports, and seven of its top executives resigned (all except Rodríguez Araque).

The national guard was distributing gasoline imported from Brazil in exchange for future crude supplies. The administration was trying to use the military and unskilled workers to reactivate refineries, which reportedly resulted in damage to at least one unit. Many wells were reportedly unmanned, oil spills were reported, and 40 tankers were sitting idle outside Venezuela's ports.

In early 2003, the Chávez administration said it would split PdVSA into two operational centers, located in eastern and western Venezuela, to combat bureaucratic concentration at the Caracas headquarters and reduce costs.

Several hundred PdVSA employees and managers were fired for participating in the strike, and a transitional organization was formed which includes new managers for the eastern and western divisions and a different board of directors. Political opponents claimed the administration was trying to eliminate many of PdVSA's top managers, administrators, and technical experts who have been at the strike's forefront.

Deltana deals delayed

The Deltana Plataforma is Venezuela's first project for the extraction of nonassociated gas in the Delta Amacuro, 200 km (125 miles) from the coast near the maritime

Energy Investment Risks

Latin America's energy industry will require more than $225 billion in investments in the coming decade, but the region's political and economic environment suggests that external financing will be difficult to find and expensive to obtain. Concerns over sovereign-related issues could jeopardize Latin America's ability to raise the capital needed for all types of energy projects, said Fitch Ratings (New York) in November 2002. The $225 billion estimate includes upstream and downstream oil and gas projects, power projects, and associated infrastructure.

"Given the industry's close link with underlying sovereign risks, Latin America's energy sector is exposed to considerable volatility," said Fitch Ratings. "This has increased funding costs, undermined investor confidence, and raised concerns regarding the achieveme of long-term energy-related goals in the region."

With the exceptions of Chile and Mexico, Latin American cou tries' credit ratings are concentrated as speculative or non-investme grade category. As of late 2002, 8 of the 12 sovereign ratings assign by Fitch in Latin America have negative perspectives, including t major energy players of Venezuela, Brazil, Argentina, and Colomb Also, despite gradual privatization over the past decade, the st remains a key participant in Latin America's energy industry.

In cases where state-owned energy companies account for a s nificant portion of government revenues, company resources can diverted to satisfy governmental or political needs. Concerns abo interference deepen when the country's credit profile deteriorates

th Venezuela.

In times of extreme stress, sovereign governments can expropriate nsiderable value from private companies. Argentina is an example nere regulated energy companies are struggling to maintain their mmercial viability, noted Fitch.

Still, Latin America's energy fundamentals remain attractive. The gion accounts for up to 12% of the world's proved oil reserves and of its natural gas, assets that remain largely underdeveloped. More portant, Fitch said, pent-up demand is expected to increase Latin nerica's total energy consumption over the next 20 years at a comunded annual growth rate of 5.7% — the highest in the world after veloping Asia.

Besides this domestic market potential, Latin America is an important crude and products exporter because of its location close to North American markets. Yet, capital remains the key to tapping its significant potential, and investors require stable regulatory frameworks, real ROI opportunities, and in some cases the flexibility to be the majority shareholder in a venture. As Argentina is learning, investor confidence can take years to restore.

"The issue for Latin America today is one of credibility," explained Fitch Ratings. "Will regional governments take the necessary steps to make their volatile environments more attractive? If yes, the energy future is promising. If not, investments opportunities will exist, but efforts to secure the capital and sponsor participation will be difficult."

IPE

border with Trinidad and Tobago. The project involves an investment of $4 billion and is expected to generate up to $800 million annually in revenue beginning in 2007.

Natural gas reserves in the area, which lies on Venezuela's Atlantic continental shelf, are estimated at up to 1,000 bcm (38 tcf). In mid-2002 the first Deltana Plataforma test well began producing at a rate of 1,350 cu m/d (50 Mcfd).

The region's acreage is considered the linchpin of President Chávez's plans to stimulate investment in Venezuela's oil and gas sector. The government selected preferred bidders for the Deltana Plataforma area and was negotiating with several companies on tentative awards. However, the government ignored its December deadline to declare winning bidders, and no announcement had been made as of January 2003.

The government carved up the 27,000 sq km (10,000 sq mi) area into five blocks and received offers from three major international oil companies in late 2002:

- ChevronTexaco bid $19 million for Block 2
- TotalFinaElf SA (France) offered only $100,000 for Block 3
- Statoil ASA (Norway) bid $32 million for Block 4, in competition with TotalFinaElf's offer of only $5.15 million.

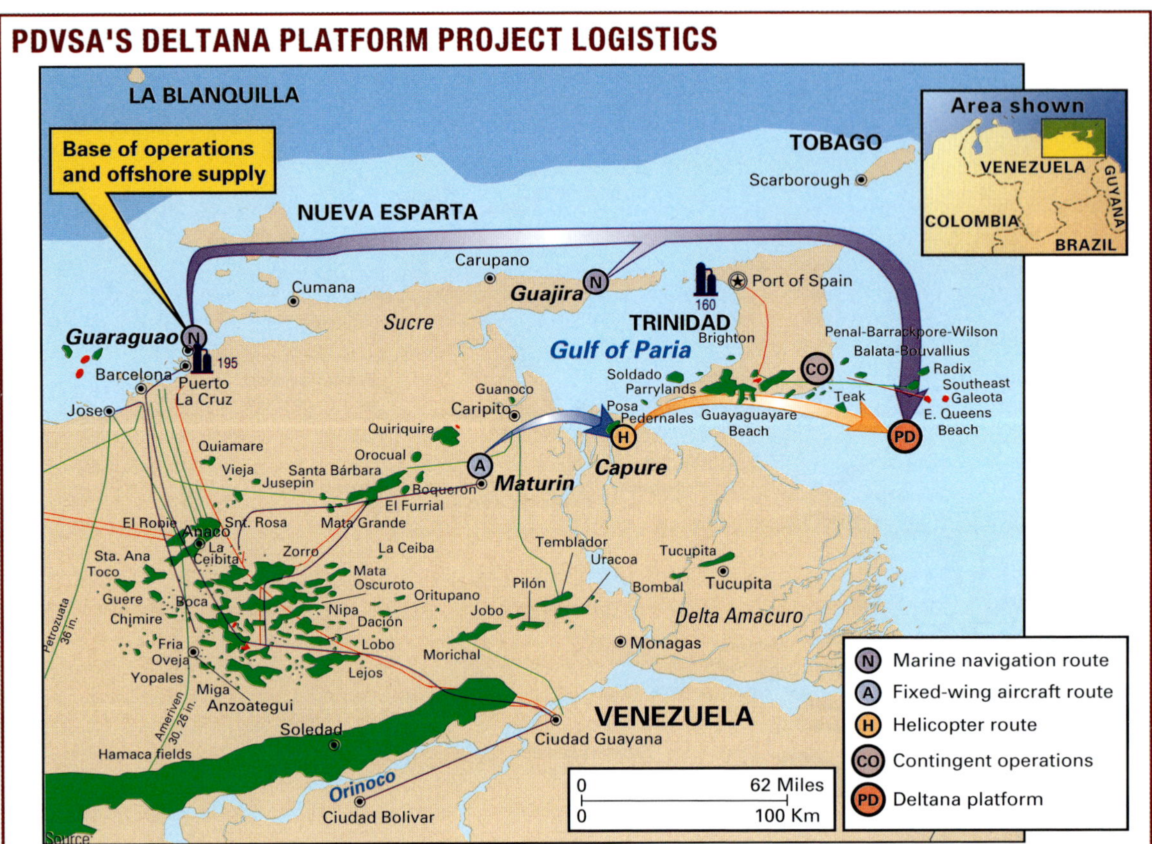

The companies also seek permits for the transport and distribution of gas extracted from the Deltana Plataforma. All the foreign majors will be required to take on PdVSA as a partner and can earn up to a 35% stake in the blocks.

BP was reportedly still negotiating for Block 1, which lies closest to Trinidad's operating gas fields. Earlier in 2002, BG had been selected as one of the preferred bidders for Block 2, but later decided not to make an offer for the gas. BG and ChevronTexaco are also partners in Trinidad's offshore Block 6d.

Mid-year, Venezuela was working with Trinidad & Tobago to unitize several of the natural gas fields that straddle both countries' borders. An initial agreement included a major cross-border gas field, Loran, which lies in the Deltana Plataforma area with most of its reserves reportedly on the Venezuelan side. Venezuela wants to unitize Loran with Block 6d on the Trinidad side of the border.

LNG revival

Venezuela decided in 2002 to try once again to develop an LNG export project by joining with Royal Dutch/Shell and Mitsubishi Corp. in a 4-4.7 million tpy venture on the Paria Peninsula, tentatively scheduled to come onstream by 2007.

The new Mariscal Sucre project is a scaled-down version of a previous LNG export scheme that foundered for lack of market and questionable economics. In early 2002, PdVSA unilaterally increased its share of Mariscal Sucre from 33% to 60% and began discussions that could lead to Qatar's inclusion in the project.

Venezuela continued to cling to the prospect of using reserves from Loran field for its own LNG export scheme, rather than quickly monetizing all of the field's reserves in Trinidad and Tobago's booming LNG export project. Despite the unitization of jointly owned gas fields, the two Caribbean neighbors could eventually compete on LNG exports to the Western Hemisphere and, to a lesser degree, Europe.

Orinoco Delta activity

PdVSA's $375 million, 2-year exploration program in the Orinoco Delta will use a mobile offshore unit to drill a dozen exploration and delineation wells in 30-300 m of water (150-1,500 ft). In early 2002 PdVSA set up logistical bases for the project in Guaraguao, Anzoátegui; Maturín, Monagas; Capure, Delta Amacuro; and Güiria, Sucre. If the project succeeds, it would be a first step toward the creation of a natural gas industrial center on Sucre's Paria Peninsula that would underpin the development of reserves in Venezuela's easternmost region.

The first well was spudded in February by the Pride Venezuela semisubmersible. Exploration is aimed at growth faults offshore, a different environment than that found around Pedernales oil field, the only producing area in the onshore portion of the delta.

The area to be explored is 1,000 sq km initially (400 sq mi), but PdVSA hopes to expand it to 2,250 sq km (870 sq mi). The nearest discoveries are Amherstia, Loran, Cocuina, and Sparrow off southeastern Trinidad. This area has seen discovery of

300 bcm (10 tcf) or more of gas in the past 5 years.

Uracoa, Corocoro development

Development of Uracoa oil field in eastern Venezuela was being accelerated in October 2002 by Harvest Natural Resources Inc. (Houston), following the signing of a contract to sell as much as 5.3 bcm (198 bcf) of natural gas for $27.66/cu m ($1.03/Mcf) through mid-2012 to PdVSA. The field, in south-central Monagas state, was producing 1 million cu m/d (40 MMcfd) of associated gas that was being reinjected for lack of a market.

Harvest was planning to spend $25 million in 2003 to build an 87-km (54-mile) gas pipeline, modify the Uracoa field processing plant, and provide other infrastructure. The company expects to invest an additional $21 million starting in 2004 for a gas pipeline, infrastructure, and drilling in Bombal field to sustain the gas production profile throughout the life of the gas sales contract.

Initial natural gas production will come from Uracoa field's 100 wells, which have been drilled through a gas cap. Production of the gas cap is also projected to provide incremental oil volumes. As a result of this expected increase, Houston Natural Resources agreed to sell 4.5 million bbl of oil to PdVSA at a price of $7/bbl during the gas contract's duration. Oil production initially will come from the 10-12 shut-in oil wells. Harvest plans to increase oil production by drilling infill wells.

Giant Corocoro oil field in the southwestern Gulf of Paria will be developed by ConocoPhillips (Houston), which said in December that it expects government approval in 2003. Oil production would begin in 2005 at a gross 50,000 b/d. During the first phase, reserves of 100 million bbl of oil net to ConocoPhillips's 50% interest would be developed. Second phase reserves would go online in 2007-08. Corocoro also contains associated gas.

Zuata Sweet production

The $4.2 billion Sincor project in March 2002 began producing and shipping "Zuata Sweet" syncrude, reported TotalFinaElf. Sincor operates the project, which produces 8.5-degree gravity crude, transporting it by a 200-km (125-mile) pipeline to Jose, and there upgrading it to a 32-degree gravity crude for export to TotalFinaElf's Port Arthur, TX, refinery.

Sincor is one of the four world-class integrated heavy oil projects in Venezuela's Orinoco oil belt. At peak production, the project is expected to yield 200,000 b/d of heavy crude to be upgraded into 180,000 b/d of syncrude.

Total output of Orinoco Belt extra-heavy crude ventures was estimated in early 2002 at 450,000 b/d of synthetic oil, which was expected to increase to 700,000 b/d by 2005.

PdVSA to retail in Brazil

PdVSA requested permission from Brazil's ANP to become a distributor of oil products in Brazil through its subsidiary Citgo Latin America. The move into Brazil's lubricants and fuels market was confirmed in September, and the first service station was to be inaugurated in late 2002.

Earlier in 2002 PdVSA announced plans to expand the 130,000-b/d El Palito facility and upgrade the 200,000 b/cd Puerto La Cruz refinery.

EUROPE

Albania

CAPITAL: TIRANA
MONETARY UNIT: LEK
REFINING CAPACITY: 26,300 B/CD
OIL PRODUCTION: 6,200 B/D
OIL RESERVES: 165 MILLION BBL
GAS RESERVES: 100 BCF

Drilling on Albania's South Tirana prospect was planned for 2003, said Lundin Petroleum AB (Stockholm). Preussag Energie GmbH (Lingen, Germany) took a farmout of a 33.3% interest in Blocks D and E in late 2002, subject to National Petroleum Agency approval. A Lundin subsidiary operates the blocks.

The South Tirana prospect, a large subthrust play for Ionian carbonates expected at 4,500 m (14,800 ft), has more than 35 sq km (13.5 sq mi) of closure. Lundin has acquired more than 300 km (190 ft) of 2D seismic on the blocks, which total 1,391 sq km (537 sq mi).

Albania was in the process of improving its electric power industry, with financial aid from the World Bank. The multi-year project was to cost $117 million, aimed at increasing the reliability and efficiency of the country's electrical power supply and interchanges with neighboring countries, as well as beginning the process of privatizing the power sector.

The Albanian Electroenergetic Corp. (KESH) and the government's Ministry of Public Economy and Privatization made progress during 2002 on meeting the World Bank's targets, after the project had been suspended the previous year. As of mid-2002, $10.2 million had been disbursed. An energy sector study was under way, and meters were being procured.

Austria

CAPITAL: VIENNA
MONETARY UNIT: EURO
REFINING CAPACITY: 208,600 B/CD
OIL PRODUCTION: 18,400 B/D
OIL RESERVES: 85.7 MILLION BBL
GAS RESERVES: 844 BCF

Austria's OMV AG could find itself overshadowed by Russia's giant oil firms and the major global oil companies, which are increasingly sharing their knowledge and expertise. To remain competitive, OMV was evaluating opportunities to expand market share into Poland and other Central and Eastern European countries, as well as making strategic acquisitions and divestitures.

A 3D seismic survey was under way in mid-2002 in northwestern Austria, where Deutsche Montan Technologie GmbH (Essen) was acquiring 530 sq km (205 sq mi) of data on the northern edge of the Austrian Alps east of Voecklabruck, between Salzburg and Linz, under a contract with Rohoel-Aufsuchungs AG (Vienna).

Refinery upgrades

To support future motor fuels specifications, a 30,000 cu m/hr (1.12 MMcf/hr) hydrogen production unit was being built at OMV's Schwechat refinery in Vienna, which has a crude charge capacity of 208,600 bbl/calendar day (b/cd). The $30 million project includes a reformer, high-temperature shift conversion, and high-recovery pressure swing adsorption (PSA) process. Expected completion is June 2003.

The refinery's clean diesel and gasoline units were also to be upgraded under another $15 million contract to modify two large columns and compressors, optimize a naphtha treater, and build a new reactor and dehexanizer. The revamping, contracted in September 2002 and scheduled for completion in March 2004, will enable the units to produce clean diesel with lower sulfur content and clean gasoline with lower aromatics content for use in automobiles.

Also at the Schwechat refinery, the stacked fluid catalytic cracking unit operated by UOP LLC (Des Plaines, IL, US) was revamped in mid-year to increase conversion and selectivity for propylene production. The project included Optimix feed distributors, vortex separation system riser disengaging, and a modern reactor stripper

EUROPE
0 50 100 150 Miles
0 100 200 Kilometers
Legend
Oil field
Oil sand
Gas field
Crude oil pipeline
Natural gas pipeline
Products pipeline
Pipeline planned or under construction
Refinery in operation
Refinery capacity in 1,000 b/d
600
Tanker terminal
Cities
Capital
International boundary
Water depth
0 to 200 m
200 m and deeper
IRELAND
N. IRELAND
ENGLAND
WALES
FRANCE
BELGIUM
NETHERLANDS
LUX.
GERMANY
DENMARK
SWITZERLAND
SPAIN
PORTUGAL
CORSICA
SARDINIA
MINORCA
MAJORCA
IBIZA
ALGERIA
TUNISIA
MOROCCO
North Sea
Irish Sea
Celtic Sea
Bristol Channel
Lyme Bay
Bay of Biscay
Mediterranean Sea
Gulf of Cadiz
Londonderry
Belfast
Sligo
Ballina
Galway
Dublin
Balough
Arklow
Foynes
Limerick
Tralee
Wexford
Waterford
Cork
Whitegate
Bantry
Waterford
Kinshale Head
Moffat
Carlisle
Newcastle
Darlington
Teesside
Middlesbrough
Barrow
Lancaster
Heysham
Morecambe
Formby
Bradford
Eastham
Manchester
Liverpool
Sheffield
Killingholme
Grimsby
Ellesmere Port
Theddlethorpe
Lincoln
Bacton
Aberystwyth
Birmingham
Milford Haven
Pembroke
Neath
Gloucester
Cambridge
Ipswich
Coryton
Oxford
London
Bath
Exeter
Southampton
Fawley
Portsmouth
Weymouth
Plymouth
Dunkerque
Antwerp
Brussels
Amsterdam
Rotterdam
Groningen
Emden
Leer
Bremen
Loningen
Salzbergen
Beckedorf
Herringhausen
Buckeburg
Salzgitter
Hamm
Dortmund
Gelsenkirchen
Duisburg
Essen
Dusseldorf
Cologne
Wesseling
Bonn
Verviers
Eisenach
Jena
Suhl
Frankfurt-am-Main
Fredericia
Kaergard
Kalundborg
Skaelskor
Kiel
Brunsbuttel
Heide
Cuxhaven
Wilhemshaven
Harburg
Hamburg
Grasbrook
Schwerin
Verden
Meerdt
Notre Dame de Gravenchon
Port Jerome
Le Havre
Cherbourg
Gonfreville L'Orcher
Petit Couronne
Gargenville
Maubeuge
Charlesville
Grandpuits
Luxembourg
Metz
Karlsruhe
Plestin Les Greves
Brest
Quimper
Lorient
Vannes
St. Malo
Laval
Paris
Coulommes-Vaucourtois
Vert-La-Gravelle
Velye-Germinon
Bussey-Lettree
Villeperdue
Dommartin-Lettre
Chaunoy
Marolles
Valence en Brie
Villemar
Coudemagnes
Grandville
Forcelles
Donnemarie
Aufferville
Flacy
St. Martin-de-Bossenay
Gisy-les-Nobles
Joigny
Chateaurenard
Donges
Nantes
Angers
Orleans
Blois
Briare
Bourbonne
Epinal
Trois Fontaines
Wallbach
Dijon
Besancon
Zurich
Zug
Hochst
Chur
Bern
Les Sables
Cholet
Poiters
Chauvigny
Vierzon
Dole
Roussiness
La Rochelle
Niort
Revigny
Macon
Bourg
Geneva
Martigny
Aosta
Lugano
Royan
Le Verdon
Pauillac
Limoges
Chazelles
Les Ancizes
Feyzin
Lyon
Vienne
Annemasse
Chamberry
Grenoble
Issoire
St. Etienne
Tulle
Bordeaux
Ambes
Bergerac
Lavergne
Cazaux
Mothes
Parentis
N. Mimizan
Lugos
Cabeil
Ychoux
Lucats
Lacq sup
Le Puy
Valence
Villefranche
Toulouse
Castres
Carcassonne
L'Isle
Gallician
Manosque
Fos-sur-Mer
Berre l'Etang
Marseille
Toulon
Nice
Monaco
Pamiers
Narbonne
Lourdes
Perpignan
La Junquera
Bayonne
Gaviota
Pont d'As
Lacq
St. Marcet
Proupiary
Rousse
St. Faust
Meilon
El Serrablo
Ferrol
La Coruna
Aviles
Santander
Vizcaya
Bilbao
Ayolvengo
Otomin
Logrono
Burgos
Vitoria
Huesca
Barbastro
Monzon
Balaguer
Tarrega
Gerona
Palamos
Zamora
Vallodolid
Zargoza
Fayon
Mataro
Barcelona
Tarragona
Chipirón NE
Angula
Salmonete
Tarraco
Dorado
Montanaza
Casablanca
Castellon
Amposta
Vinaroz
Castellon de la Plana
Vallencia
Madrid
Opporto
Cabo Ruivo
Lisbon
Setubal
Sines
Puertollano
Almendralejo
Cordoba
Huelva
Sevilla
Rota
Cadiz
Gulf of Cadiz
Granada
Malaga
Almeria
Cartegena
Murcia
Algeciras
Biella
Chivasso
Turin
Asti
Aquata
Savona
Busalla
Genoa
Brescia
Milan
Cremona
Rovigo
Padua
Mandello
Parma
Pistola
Leghorn
Bastia
Muro
Aleria
Zonza
Porto Torres
Maddalena
Alghero
Nuoro
Arbatax
Iglesias
Porto Scuso
Cagliari
Algiers
Bejaia
Skikda
Tunis
Oran
Arzew
Vohburg
Ingolstadt
Munich

design. The root cause of catalyst losses was identified, and post-revamp operation exceeded the expectations for propylene yield. Catalyst containment in the reactor, however, was not as good as expected, and OMV and UOP developed a joint risk-management plan to address the problem.

Belgium

CAPITAL: BRUSSELS
MONETARY UNIT: EURO
REFINING CAPACITY: 791,013 B/CD
OIL PRODUCTION: N/A
OIL RESERVES: N/A
GAS RESERVES: N/A

Belgium's Tractebel SA was active during 2002 in several projects worldwide. Tractebel owns 41.6% of the Zeebrugge LNG terminal in Belgium, which is also the terminus of the Interconnector pipeline from the UK. Zeebrugge, already Europe's largest natural gas trading hub, could emerge as an opportunity for arbitrage and trading among European, UK, and North American markets.

Also, a Tractebel unit is participating in Europe's first LNG export facility, Norway's Snøhvit project. Tractebel will supply product tanks and vessel loading systems, as well as design and build four cylindrical storage tanks and associated piping and loading systems connecting the tanks to export jetties. Two of the tanks, each with a capacity of 120,000 cu m, will store LNG; one will store up to 45,000 cu m of LPG; and the fourth will hold 470,000 bbl of condensate. Onsite construction began in 2002 and should be completed in late 2005.

Tractebel was also slated to distribute natural gas from Peru's Camisea project in metropolitan Lima and Callao. However, in April the government postponed signing the contract when allegations appeared in the press that Tractebel had paid a commission for approval of its license. The reports were denied by Tractebel.

Bulgaria

CAPITAL: SOFIA
MONETARY UNIT: LEV
REFINING CAPACITY: 115,240 B/CD
OIL PRODUCTION: 1,000 B/D
OIL RESERVES: 15 MILLION BBL
GAS RESERVES: 210 BCF

At a historic summit in Prague in late 2002, Bulgaria was invited to join the North Atlantic Treaty Organization (NATO), along with several other eastern European and former Soviet Union nations. Membership is expected in 2004. Bulgaria also was negotiating to join the European Union (EU) and could become a member around 2006.

Bulgaria was in the process of opening its natural gas market during 2002. Regulations to be adopted would result in opening 10% of the gas market by year-end. With World Bank funding, several companies will receive gas distribution licenses and begin constructing a distribution network. Bulgaria's State Energy Regulation Commission announced that the first public tenders for gasification of Bulgaria's urban centers would be organized in late 2002.

Also, the Energy Ministry decided to overhaul Bulgaria's monopoly gas distribution company Bulgargaz in 2002. Bulgargaz will preserve its role as a gas transportation company and a wholesale supplier, while retail distribution will be carried out by local and regional companies. Bulgargaz would be completely privatized and Bulgarian markets fully integrated with the rest of Europe by 2010.

Galata gas field in the Black Sea will be developed by Melrose Resources plc (London), which began awarding contracts for the $52 million project in late 2002. Melrose hopes to start production by early 2004. Under contract with a Melrose unit, Bulgargaz would purchase 379 million cu m/yr (14.1 bcf/yr) for 3 years. Galata, which lies 20 km (12 miles) offshore in 35 m of water (115 ft), has 1.3 billion cu m, or bcm (49 bcf) of proved reserves and 0.8 bcm (31 bcf) probable.

Melrose said it arranged $23 million in mezzanine financing through AIG Emerging Europe Infrastructure Fund LP and $34 million in senior debt through Black Sea Trade & Development Bank and International Finance Corp. Loan drawdown is subject to conditions and was expected to be available by year-end 2002.

Bulgaria would be the end point of a natural gas pipeline being considered by Turkmenistan in 2002. The line would pass through Iran and Turkey into Bulgaria, supplying about 23 bcm (856 bcf) in 2005 and 30 bcm (1.2 tcf) in 2010.

Bulgaria ended a longstanding dispute with Greece in late 2002 over plans to build the Trans-Balkan oil pipeline. The countries agreed to a three-way division of stakes in the pipeline company, along with third-party crude producer Russia. An initial capacity of 300,000 b/d is planned, rising to a maximum of 700,000 b/d.

Croatia

CAPITAL: ZAGREB
MONETARY UNIT: KUNA
REFINING CAPACITY: 260,337 B/CD
OIL PRODUCTION: 21,000 B/D
OIL RESERVES: 92.2 MILLION BBL
GAS RESERVES: 1.24 TCF

At year-end 2002, the deepwater terminal at Omisalj on Croatia's Adriatic Sea coast was to begin moving Russian crude oil into world export markets. Russian Urals blend flowed 3,198 km (1,987 miles) via the Druzhba and Adria pipeline systems west to the Croatian port, the final link being a newly reversed 178-km (110.5-mile), 36-inch line from the inland terminal at Sisak, Croatia.

At Omisalj, the oil is loaded into tankers up to 500,000 dwt for transit through the Adriatic, Ionian, and Mediterranean Seas to markets in the US and Asia, allowing Russia to bypass the Black Sea and its increasingly crowded Bosporus Straits.

The estimated cost of reversing the pipeline and related construction is $120 million, according to the Centre for Global Energy Studies (London). The plan is for the entire pipeline to have a capacity of 15 million tonnes/year (tpy), or about 300,000 b/d, of Russian crude oil for export, as well as about 24 million tpy (480,000 b/d) of oil imported from other sources for use in regional refineries.

The Omisalj oil terminal has two berths, with water depths alongside of 30 m (98 ft). The Centre said land is available to extend the port facilities by adding another 500,000-dwt berth plus a 150,000-dwt berth and by building crude oil storage tanks to increase capacity to 1.5 million cu m (79 MMcf) from the existing 680,000 cu m (24 MMcf). The use of very large and ultra-large tankers to carry Russian crude oil would make exports to the US and Asia-Pacific much more economic, said the Centre.

Another pipeline has been proposed to transport 660,000 b/d of Caspian oil from Constanta, Romania, to Trieste, Italy, but a route has not yet been selected. The southern route (South-East European Line, or SEEL), would pass through Croatia, using its Omisalj port as an intermediate transit point before linking with the existing Trans-Alpine Pipeline (TAP). In November 2002, Croatia, Romania, and Serbia agreed on the SEEL route, but financing remains a problem.

Adriatic gas development

In addition to transporting Russian oil,

Croatia will be involved in development of natural gas discoveries in the northern Adriatic Sea, which will include a 43-km (27-mile) subsea pipeline to Pola, Croatia, from Ivana field. Completion of the project will establish a link for the future transport of gas to Croatia from Italy.

The new $313 million project, being conducted by Croatia's state-owned INA Group and a unit of Eni SpA (Rome), involves development of 19 bcm (700 bcf) of gas reserves in the four fields near Ivana — Ika, Ida, Annamaria, and Marica. Only Annamaria field laps slightly into Italian waters. Development will entail drilling 18 directional wells, setting of nine platforms in 60 m of water (200 ft), and laying 120 km (75 miles) of subsea pipelines. Gross production should reach 1.9 million cu m/d (70 MMcfd) in 2005 vs. about 670,000 cu m/d (25 MMcfd) in late 2001.

In Croatia and other former Yugoslavian areas, oil processing and marketing will be conducted jointly by Tyumen Oil Co. (Russia) and Slovenia's Petrol. Tyumen will provide crude for processing and will manage processing of oil products. Petrol will provide marketing and sales services as well as manage logistics, marketing, and sales of oil products.

INA plans to support its ultra-low sulfur gasoline and diesel fuel projects with a new 20,000 tonne per day (tpd) sulfur recovery unit at its Rijeka refinery and a 15,000-tpd unit at its Sisak refinery. Both refineries will incorporate hydrocrackers and reformers as part of the upgrade packages. Earlier, INA restated the Rijeka plant's crude charge capacity, reducing it to 100,000 b/cd from the previous 132,000 b/cd.

Czech Republic

CAPITAL: PRAGUE
MONETARY UNIT: KORUNA
REFINING CAPACITY: 198,000 B/CD
OIL PRODUCTION: 5,000 B/D
OIL RESERVES: 15 MILLION BBL
GAS RESERVES: 140 BCF

The Czech Republic's growth forecasts for 2002 were cut to 3.3% because of continued low demand for Czech exports in the EU, where growth has remained slow. The International Monetary Fund (IMF) projected that the Czech Republic's real gross domestic product (GDP) would grow 3.2% in 2003. The country will probably become an EU member before 2006.

After 97% of the Czech national gas utility Transgas was sold in late 2001, privatization efforts continued in the refining sector. The government owns 63% of Unipetrol, holding company for Ceska Rafinerska AS, which owns and operates the country's two refineries. Four companies were competing for the government's share, with full privatization expected by year-end 2002.

In mid-2002 the residual fluid catalytic cracking unit at the 68,000-b/cd Kralupy refinery was onstream. UOP said the 27,740-b/d unit had met all guarantees for gasoline production yield and quality; light cycle oil, propylene, butylene, and main column bottoms quality; and catalyst consumption.

Denmark

CAPITAL: COPENHAGEN
MONETARY UNIT: KRONE
REFINING CAPACITY: 176,400 B/CD
OIL PRODUCTION: 365,000 B/D
OIL RESERVES: 1.347 BILLION BBL
GAS RESERVES: 2.975 TCF

Danish state oil firm Dansk Olie & Naturgas AS (DONG AS) acquired in July 2002 the assets of Statoil ASA (Norway) in the Danish North Sea, including Statoil's 40% interest in Siri and Stine fields and 18.8% interest in Lulita field. These shares were producing an average 7,000 b/d of oil.

Government authority to develop Nini and Cecilie North Sea oil fields was granted in July to DONG and other licensees. Nini and Cecilie will be tied in to the Siri platform, and DONG will operate all three fields. Production from the fields' reserves, estimated at 65 million bbl, was planned for mid-2003. Siri lies in 60 m of water (200 ft) on Block 5604/20 on the Ringkbing-Fyn high about 209 km (130 miles) west of Esbjerg, Denmark. The platforms will be built, installed, and connected to Siri by Saipem UK Ltd. and subcontractor Bladt Industries AS. Total costs were estimated at 2.5 billion kroner.

In the North Sea's Danish sector, recovery rates in existing fields were improving in chalk reservoirs, where enough gas is thought recoverable to allow exports, said

DRUZHBA-ADRIA PIPELINE TO OMISAJL PORT

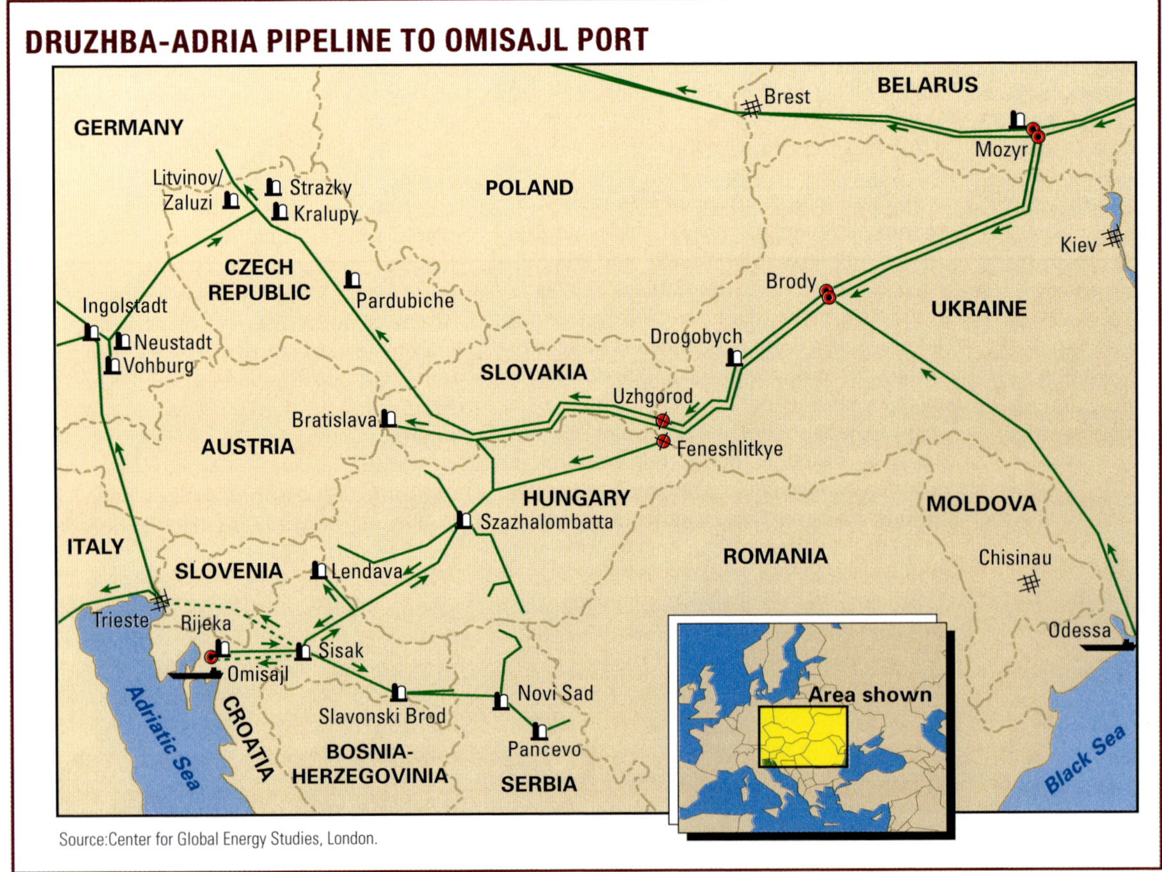

Source:Center for Global Energy Studies, London.

NORTHERN EUROPE
0 50 100 150 Miles
0 100 200 Kilometers
Midgard
Åsgard
Tyrihans
Draugen
Njord
Tjeldbergodden
Trondheim
Norwegian Sea
Åsgard Transport
FAEROE I.
Agat
NORWAY
Troll
200
Mongstad
Kollsnes
Sotra
Bergen
Sullom Voe
SHETLAND I.
Lerwick
Oslo
ORKNEY I.
Kirkwall
Flotta
Wick
Karsto
Kopervik
Stavanger
Sola
Slagentangen
110
Vall oy Tonsberg
Stornoway
Arendal
Nigg Bay
Moray Firth
Mandal
Goteborg
Inverness
St. Fergus
Skagen
Mallaig
Cruden Bay
Aberdeen
SCOTLAND
Dundee
Finnart
Ardyne Point
206
Grangemouth
Firth of Forth
DENMARK
Donegal Bay
Edinburgh
Glasgow
Ardrossan
12
Firth of Clyde
North Channel
70
Arhus
Fredericia
106
Kaergard
Corrib
Londonderry
North Sea
Moffat
Esbjerg
Kalundborg
Sligo
Belfast
Carlisle
Newcastle
Ballina
N. IRELAND
Port Clarence
100
Teesside
Darlington
Skaelskor
IRELAND
Middlesbrough
Barrow
Lancaster
Scarborough
80
Heysham
Morecambe
Bradford
Kiel
Galway Bay
Galway
Ballough
Irish Sea
Leeds
Heide
Dublin
Hull
Brunsbuttel
Formby
245
Manchester
Cuxhaven
Foynes
Eastham
192
230
225
Schwerin
Limerick
Liverpool
Grimsby
Wilhemshaven
Hamburg
Arklow
27
Tralee
Ellesmere Port
Theddlethorpe
Groningen
Harburg
Lincoln
Emden
Grasbrook
Wexford
Nottingham
78
97
Waterford
Aberystwyth
Bacton
Cork
Birmingham
Leicester
ENGLAND
71
Bantry
108
Whitegate
WALES
10
Amsterdam
Waterford
Coventry
Galley Head
Milford Haven
Helvick
Northampton
Salzbergen
Ardmore
Cambridge
Meerdf
Seven Heads
8
Kinsale Head
Llandarcy
Pembroke
NETHERLANDS
Roxforrde
210
Gloucester
6
Neath
Oxford
Ipswich
Rotterdam
Celtic Sea
London
180
182
76
399
390
149
246
GERMANY
Gelsenkirchen
Reading
Bath
Dortmund
307
Brugge
Antwerp
Duisburg
Essen
Exeter
Fawley
263
Dunkerque
Plymouth
Bournemouth
308
109
90
21
Dusseldorf
Eisenach
Southampton
Cologne
Portsmouth
Calais
184
Brussels
Liege
227
Wesseling
Weymouth
Lille
BELGIUM
Bonn
162
Suhl
Verviers
140
141
226
Notre Dame de Gravenchon
Maubeuge
Frankfurt-am-Main
Cherbourg
Port Jerome
LUX.
Le Havre
Grandpuits
Rouen
90
Luxembourg
Gonfreville L'Orcher
Gargenville
Vert-La-Gravelle
Plestin Les Greves
324
Petite Couronne
Coulommes-Vaucourtois
Reims
Velye-Germinon
Metz
Bussey-Lettree
285
Karlsruhe
Brest
St. Malo
Paris
Chauroy
Villeperdue
Dommartin-Lettre
Marolles
Soudron
Courdemagnes
Valence en Brie
Grandville
Trois Fontaines
Vohburg
Quimper
230
Rennes
Laval
Villemar
Donnemarie
Grandville
Forcelles
Ingolstadt
Lorient
Aufferville
Flacy
St. Martin-de-Bossenay
Vannes
Gisy-les-Nobles
Baden-Wurttemberg
Jolgny
Bourbonne
Reichstett-Vendenheim
78
106
Orleans
Donges
Chateaurenard
Zurich
Nantes
Blois
SWITZERLAND

EASTERN EUROPE
BORNHOLM
RUGEN
POLAND
BELARUS
UKRAINE
CZECH REP.
SLOVAKIA
MOLDOVA
AUSTRIA
HUNGARY
ROMANIA
SLOVENIA
CROATIA
BOSNIA &
HERZEGOVINA
SERBIA AND
MONTENEGRO
BULGARIA
ALBANIA
MACEDONIA
GREECE
TURKEY
ITALY
SICILY
MALTA
CRETE
Black
Sea
Adriatic Sea
Tyrrhenian
Sea
Ionian
Sea
Agean
Sea
Mediterranean
Sea
Gulf of
Venice
Gulf of Corinth
3 Lines
Copenhagen
Malmo
Kaunas
Vilnius
Kaliningrad
Krasnoborskoye
Ushakovskoye
Gdansk
Grodno
Baranovich
Unecha
Kursk
Kharko
Minsk
Davydovskoye
Vishanskoye
Mozyr
Pinsk
Kobrin
Brest
Ovruch
Kiev
Reinkenhagen
Kolberg
Rostock
Stettin
Schwedt
Frankfut-an-der-Oder
Plock
Warsaw
Berlin
Vestrup
Rudersdorf
Poznan
Frankfut-an-der-Oder
Roxforde
Magdeburg
Staakow
Grunberg
Lodz
Pulawy
Kovel
Lutsk
Rovno
Dubno
Cottbus
Guben
Radom
Sandomierce
Rozwadow
Kielce
Brody
Berdichev
Staro-Konstantinov
Leipzig
Jena
Zeitz
Dresden
Grobla
Partinia
Lubaczow
Jaroslaw
Ternopol
Vinnitsa
Kremenchug
Litvinov
Kralupy
Pardubice
Trzebina
Jasto
Rudki
Lvov
Dashava
Kradobnensk-Kalusz-
Waidhaus
Prague
Kolin
Czechowice
Gorlice
Smilov
Mikova
Olka
Stryy
Lugskoye
Starynj
Pribor
Zukow
Stanavo
Strazske
Dolina
Bitkov
Kosov
Brno
Trebio
Vacenovice
Breclav
Hodonin
Uzhgorod
Nadvornaya
Kosmachskoye
Chernortsy
Krasnoilsk
Salushsk
Znojmo
Lanzhof
Miskolo
Ugeny
Kishinev
Kherson
Odessa
Stefanov
Malacky
Wysok
Linz
Bratislava
Fedemes
Demjen
Mezokeresztes
Lasi
Munich
Vienna
Budapest
Hajduszoboszlo
Biharnagybajom
Sarmasel
Teleac
Zemes
Stanesti
Bacau
Victorovca
Baimaclia
Salzburg
Nadudvar
Zau-de-Cimpie
Miercurea
Cucaesti
Valeni
Dramanesti
Szanaszelos
Saros
Izmail
Pusztafoldvar
Battonya
Bogota
Singeorgiu
Graz
Lovaszi
Budafap
Nagylengyel
Tauni
Noul-Sasesc
Boldest
Moreni
Frasinu-Mislea
Bazna
Letrad
Ulles
Nagylanizsa
Szeged
Algyoe
Arad
Gura-Ocnitei
Paluzza
Maribor
Petisovci
Pecs
Knezevac
Mokrin
Timisoara
Midia
Lebada
Yelki Otok
Lepavina
Zagreb
Melve
Babocsa
Velebit
Srpska Crnja
Ploiesti
Navodari
Constanta
Ljubljana
Sandrovac
Gakovo
Bizovac
Ada
Besejci
Medja
Resita
Cimpina
Gura
Belluno
Dugo Selo
Zutica
Pepelana
Benicanci
Gospodjinci
Ticleni
Bilteni
Pitesti
Bucharest
Jezevo
Janja Cipa
Obod
Srbobran
Boka
Trieste
Ivanic Grad
Jamarica
Nova Gradiska
Pancevo
Nikolinki
Leonardi
Budisterni
Tjulenovo
Porto
Marghera
Koper
Rijeka
Sisak
Lipovljani
Bosanski Brod
Belgrade
Mramorak Selo
Turnu-
Severin
Giurgiu
Tolloukhin
Kavarna
Varna
Ferrara
Pula
Staro Orjachovo
Gigen
Bologna
Ravenna
Ivana
Gospic
Chiren
Dolni Dabnik
Gorni Dabnik
Sarajevo
Aleksinac
Devetaki
Vraca
Otmanli
Burgas
Florence
San Marino
Falconara
Ancona
Split
Nis
Pristina
Sofia
Stara Zagora
Plovdiv
Istanbul
Izmit
Dubrovnik
Shkoder
Skopje
Bellante
Cellino
Alexandroupolis
Bandirma
Bursa
Marta
Rospo
Allano
Chieuti
Lanciano
San Stefano
Valle Cupo
Cupello
Kavala
Prinou
S. Kavalla
Rome
S. Salvo
Vasto
Portocannone
Monte Stillo
Tirana
Ballaj
Divjaka
Thessaloniki
Ripi
Petroliard-Farnesina
Terracina
Candela
Giovinazzo
Bari
Seman
Marinze
Kucove
Bubullima
Burhaniye
Gaeta
Naples
Pietralunga
Grottole
Taranto
Brindisi
Panaja
Cakran
Gorisht
Vlore
Drashovice
Finiq
Larissa
Ferrandina
Cerro Falcone
Monte Alpi/Enoc
Tempa Rossa
Pomarico
Pisticci
Rotondella
Nova Siri
Izmir
Arta
Crotone
San Leonardo
Luna
C. Cimiti
Delfi
Aspropyrgos
Soke
Eleusis
Athens
Argos
Tripolis
Vibo Valentia
Rosarno
Lavinia
Bagnara
Messina
Palermo
Milazzo
Gaglioano
Rodhos
Mazara
del Vallo
Marinella
Bronte
Cisma
San Nicola
Catania
Ardizzone
Marsala
Porta Enpedocle
Commarata
Pone
Drillo
Augusta
Gela
Perla
Ragusa
Vega
Khania
Iraklion
Rethimnon
0
50
100
150 Miles
0
50
100
150
200 Kilometers

Douglas-Westwood Associates (Canterbury, UK) and Infield Systems (London). Statoil reported in May 2002 that Siri field's oil recovery was improved by mixing the associated produced gas with injection water at the wellhead and injecting a two-phase mixture.

The Danish government launched a licensing round off Greenland during 2002. In December an exploration and production license in the Davis Strait was awarded to EnCana Corp. (Calgary) and Greenland's state Nunaoil. The 3,985 sq km (1,539 sq mi) block is in water depths of 200-1,000 m (660-3,300 ft), 200 km (125 miles) west of Nuuk. Most of the license is in the southern Nuuk basin, but its eastern edge covers part of the Atammik structural complex.

No drilling has taken place on the license, but the block lies 50 km (30 miles) north of a well drilled earlier by Statoil which encountered reservoir sandstones. Geological Survey of Greenland and Denmark said four seismic surveys in Baffin Bay, the Davis Strait, and the Labrador Sea off western Greenland were completed during 2002.

Sulfur-free diesel

Diesel fuel with virtually no sulfur content was being produced in July 2002, 3 months ahead of schedule, by a new diesel oil unit at Denmark's Kalundborg refinery. The Statoil refinery's diesel sulfur content was cut by 80%, to 0.001% from 0.005%, surpassing the EU's 2005 standard. The unit will produce about 1 million tpy of sulfur-free diesel.

Finland

CAPITAL: HELSINKI
MONETARY UNIT: EURO
REFINING CAPACITY: 251,800 B/CD
OIL PRODUCTION: N/A
OIL RESERVES: N/A
GAS RESERVES: N/A

In September 2002 motorists in Finland began filling their cars with a 98-octane gasoline containing ethanol. The "bio-gasoline" became available in the Helsinki area and southeast Finland.

Fortum, which distributes the fuel from its Porvoo plant, was preparing for more extensive use of renewable raw materials to comply with anticipated EU directives. The bio-gasoline contains a maximum 5% ethanol, which is manufactured in Italy from surplus wine.

France

CAPITAL: PARIS
MONETARY UNIT: EURO
REFINING CAPACITY: 1,903,493 B/CD
OIL PRODUCTION: 26,200 B/D
OIL RESERVES: 148.473 MILLION BBL
GAS RESERVES: 506 BCF

France's economy during 2002 was tracking the eurozone as a whole, where growth was estimated at 1.4%. The IMF projected that France's real GDP would grow 3.2% in 2003. France's coastline was threatened in late 2002 by a fuel oil spill off Galicia, Spain.

Liberalization scheduled

In December 2002, French officials agreed to open the country's residential electric and natural gas markets to competition by mid-2007, under a compromise reached by EU energy ministers in November. The EU's original date for opening these markets was 2005. Following the announcement, France's parliament passed the EU directive on natural gas liberalization into local law in late December.

The new French law determines conditions for access to the network, transparency and regulation, transport and distribution, underground storage, and controls and sanctions. France's Regulatory Commission for Electricity will become the new Commission for Regulation of Energy, encompassing the gas market as well. Also in December, the commission said that more comparability is needed between Gaz de France (GdF) and third parties on natural gas tariffs to establish better access for competition.

The new energy commission, in charge of the gas liberalization effort, also called for broadening supply points into France by building new LNG terminals in southern France and by increasing France's interconnections with Spain. This competitive access to France's gas network applies to TotalFinaElf SA (France), which has its own natural gas network, as well as to GdF. The commission said that nine industrial users with 16 sites had switched from GdF as their gas supplier. This represents 25% of the volumes that were already open to free market competition under the first phase of liberalization.

LNG deal, new terminal

In July GdF, Europe's largest LNG buyer, approved plans to construct a second terminal at Fos-sur-Mer on the Mediterranean Sea in southeastern France. The 300-430 million euro project, set for completion by 2006, includes port, storage, and regasification facilities.

The facility will have a maximum natural gas sendout capacity of 8.25 bcm/yr (307 bcf/yr) and will be able to receive LNG carriers with cargoes of 160,000 cu m. The existing Fos I terminal, with a sendout capacity of 4.5 bcm/yr (168 bcf/yr), provides France with 11% of its total natural gas imports; another terminal at Montoir-de-Bretagne also imports LNG.

The new terminal will help supply a liberalized European gas market with gas from Egypt and other Mediterranean suppliers. In February 2002 BG Group plc (UK) and Edison International SpA (Italy) agreed to sell 3.6 million tpy of LNG to GdF over 20 years, starting in 2006-07. Gas source would be the West Delta Deep marine field off Egypt. GdF also took part in a joint venture with BG and Edison to build a natural gas liquefaction unit at Idku, near Alexandria, and holds a 5% stake in the Egyptian LNG joint venture. TotalFinaElf indicated interested in the Fos II terminal project.

Other developments

The offshore potential of the Bay of Biscay could be explored if Esso Rep succeeds in finding a partner to assess its Aquitaine Maritime and Cap Ferret Ocean permits. Esso Rep is the French exploration and production unit of ExxonMobil Corp. (Irving, TX, US). Although Esso Rep remained France's largest oil producer, its onshore production fell 11% in 2001.

GdF continued construction of the second, 300-km (186-mile) section of its Les Marches du Nord-Est pipeline, which was to begin operating in 2002. GdF said in late 2002 that it will acquire Preussag Energie's German assets, which raises GdF's gas production by nearly 50% to 4.6 bcm/yr (170 bcf/yr) and brings GdF nearer to its target of producing, in the short term, 15% of the natural gas it sells; so far, it has reached a 9% level.

TotalFinaElf's 114,000-b/cd Feyzin refinery in eastern France's Rhone Valley will be revamped during 2002-03. Plans are to increase its catalytic cracker capacity to 32,000 b/d from 30,000 b/d, revamp the hydrodesulfurization units, reduce flaring, install a new control room, and work on the petrochemical units.

The Lavéra complex on the French Riviera will become a "star site" for further integrated development of refining and petrochemicals, said Naphtachimie, a 50-50 joint venture of BP Chemicals SNC and TotalFinaElf's Atofina Petrochemicals Inc.

Investment of 250 million euros through 2011 will include modernization of the Lavéra steamcracker as a prelude to debottlenecking at the site. Production will be elevated to 1.1 million tpy from the current 720,000 tpy.

Germany

CAPITAL: BERLIN
MONETARY UNIT: EURO
REFINING CAPACITY: 2,267,100 B/CD
OIL PRODUCTION: 71,700 B/D
OIL RESERVES: 342.311 MILLION BBL
GAS RESERVES: 11.294 TCF

Although Germany's Social Democratic Party won September's elections, its victory was marginal, and the Green Party made significant gains. The IMF projected that Germany's real GDP would grow 2.0% in 2003. Liberalization of the European gas market was providing opportunities to build positions in Germany, but barriers still prevent free access to the German gas network.

Estimated reserves of Mittelplate oil field in the German sector of the North Sea were revised upward to more than 60 million tonnes from 35 million tonnes, said RWE DEA AG in November. The revision is based on a new 3D seismic survey as well as operational experience. Plans for the field include an increase in production capacity and a proposed pipeline to an onshore processing plant.

Controlling interest in a major German natural gas transporting and marketing company, GSV Gasversorgung Süddeutschland GmbH, was acquired in mid-2002 jointly by Germany's EnBW AG and Eni. GVS transports and markets about 8 bcm/yr of gas (300 bcf/yr) to about 750 locations and achieved annual revenues of 1.7 billion euros in 2001.

Preussag Energie's assets, including an estimated 25 billion cu m of natural gas (930 bcf) and 50 million bbl of oil in Germany's North West basin, will be acquired by GdF, the companies announced in late 2002. Preussag Energie's plateau production was about 1.4 bcm/yr of gas (52 bcf/yr) and 4 million bbl/year of oil.

The acquired assets also include an 11% stake in gas transportation and marketing company Erdgas Münster, which owns a 2,000-km (1,240-mile) network in the northwestern part of Germany; a 50% stake in the Reitbrook underground storage facility near Hamburg; a 33% stake in the Schmidhausen storage facility near Munich; and 100% of the Fronhofen storage facility near Stuttgart.

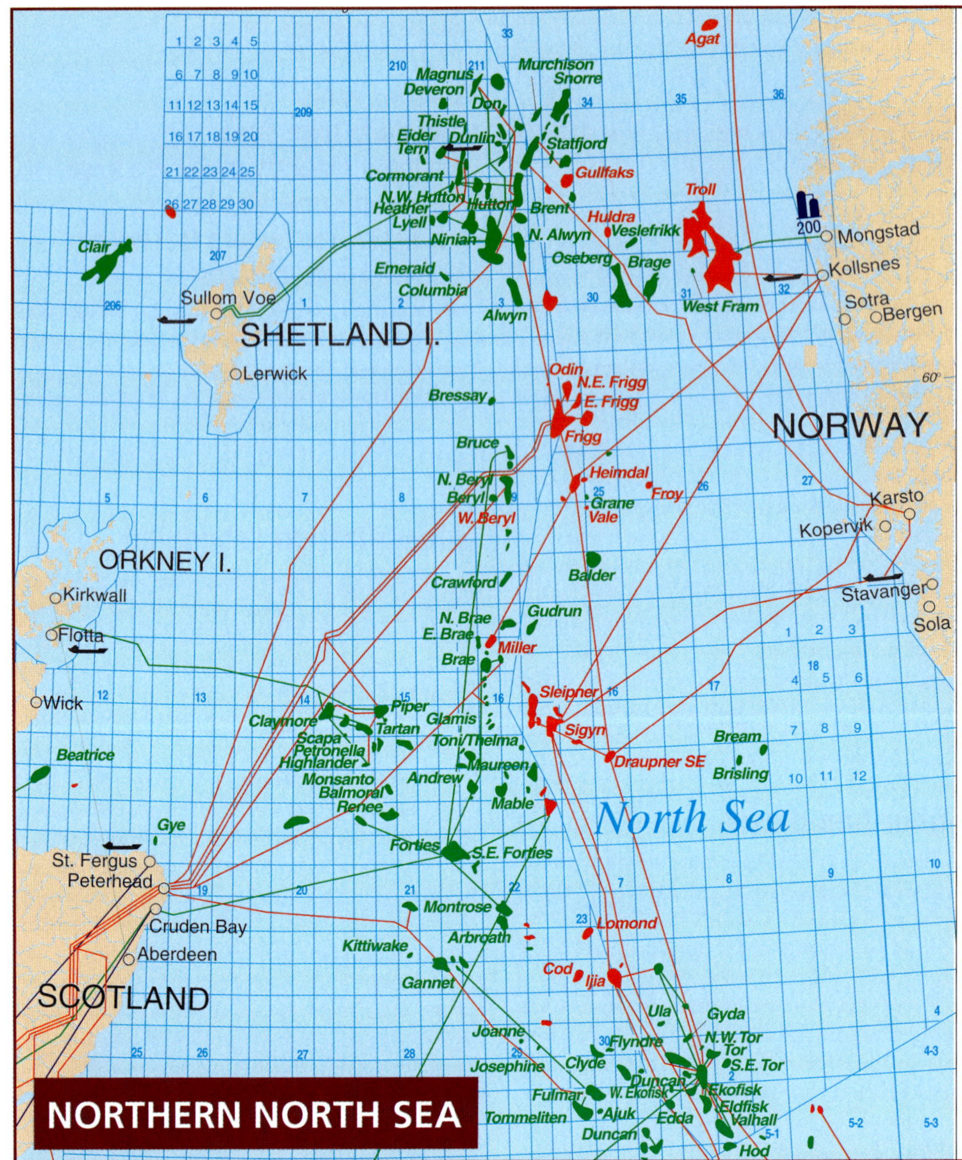

NORTHERN NORTH SEA

Germany continues to lead the world in wind power capacity, with 37% growth in 2002 and an estimated 12,000 MW installed.

Greece

CAPITAL: ATHENS
MONETARY UNIT: EURO
REFINING CAPACITY: 406,500 B/CD
OIL PRODUCTION: 3,200 B/D
OIL RESERVES: 9.0 MILLION BBL
GAS RESERVES: 18 BCF

Greece enjoyed strong GDP growth of nearly 3% during 2002, and the IMF projected that Greece's real GDP would grow another 3.2% in 2003.

As planned, the Greek government sold a chunk of its state-owned Hellenic Petroleum. The 23.17% share went to a joint venture of OAO Lukoil (Russia) and the Latsis Group (Greece) for $459 million. Latsis also owns Greece's third-largest refiner, Petrola. An employee strike protesting Hellenic's privatization was quashed in the courts.

The government also planned to sell of 35% of DEPA, the Greek Public Gas Co. Possible buyers include Russia's OAO Gazprom, Germany's Ruhrgas AG, and Algeria's Sonatrach.

Greece announced in May that it would hold its second round of oil exploration licensing by early 2003. The round will include unexplored areas in the Ionian Sea as well as offshore and onshore areas in northwestern and southwestern Greece. However, Aegean Sea exploration was not scheduled due to a continental-shelf boundary dispute with Turkey. Discussions were begun in March to resolve the dispute.

Also in March, Greece and Turkey agreed to extend the Iran-Turkey natural gas pipeline into Greece. The 280-km (175-mile) extension, 200 km in Turkey and 80 km in Greece (125 miles and 50 miles, respectively), was expected to be completed by 2005, connecting Ankara to Alexandroupolis at a cost of $300 million. Initial volumes will be 475 million cu m/yr (17.7 bcf/yr). Azerbaijan could begin supplying gas to Greece in 2006-07 through the extension.

Greece ended a longstanding dispute with Bulgaria in late 2002 over plans to build the Trans-Balkan oil pipeline. The countries agreed to a three-way division of stakes in the pipeline company, along with third-party crude producer Russia. An initial capacity of 300,000 b/d is planned, rising to a maximum of 700,000 b/d.

Hungary

CAPITAL: BUDAPEST
MONETARY UNIT: FORINT
REFINING CAPACITY: 161,000 B/CD
OIL PRODUCTION: 21,700 B/D
OIL RESERVES: 102.484 MILLION BBL
GAS RESERVES: 1.21 TCF

Hungary will probably become an EU member before 2006. The IMF projected that Hungary's real GDP will grow 4.0% in 2003.

The Hungarian Oil and Gas Co. (MOL Rt.) could find itself overshadowed by Russia's giant oil firms and the major global oil companies, which are increasingly sharing their knowledge and expertise. To remain competitive, MOL was evaluating opportunities to expand market share into Poland and other Central and Eastern European countries, as well as making strategic acquisitions and divestitures.

MOL agreed with Russia's Yukos in early 2002 to jointly develop western Siberia's Zapadno-Malobalysk field, which is near pipeline and other transportation infrastructure. The field, with estimated proven reserves of at least 20 million tonnes of crude oil, had been producing 10,000 b/d and was expected to peak at 55,000 b/d in 2005.

The dehydration process at the gas treating plant in southern Hungary's Szeged field became far more stable and reliable, MOL reported in October 2002. The Drizo glycol-enhancement process had been operating continuously at an outlet water content below design values for over a year.

Construction of an ethylene plant was planned for Tiszaujvaros in northeastern Hungary. The new plant, scheduled to begin operating in late 2004, is part of an extensive, 430 million euro petrochemical development project by Hungarian company Tiszai Vegyi Kombinat Rt. (TVK).

Using gas oil as the feedstock, the plant will increase ethylene production capacity at Tiszaujvaros to 610,000 tpy from 360,000 tpy, and the plant also will have the capability to process naphtha. TVK awarded Linde AG (Wiesbaden) a 160 million euro contract to provide its proprietary ethylene technology and cracking furnaces.

Ireland

CAPITAL: DUBLIN
MONETARY UNIT: EURO
REFINING CAPACITY: 71,250 B/CD
OIL PRODUCTION: 0
OIL RESERVES: 0
GAS RESERVES: 700 BCF

A gathering pipeline from the Greensand development in the Celtic Sea off Ireland will be installed by a unit of Technip-Coflexip Group (Paris) under a contract with Marathon International Petroleum Ireland Ltd., which was announced in December 2002. The line will tie back the Greensand well to the Kinsale Bravo platform, 7 km (4 miles) away. The gathering pipeline includes a 10-inch rigid flowline and riser system and a 200-m (660-ft) electrohydraulic control umbilical. First gas is slated for July 2003.

In August 2002, Enterprise Energy Ireland Ltd. with partners Agip Ireland and OMV drilled a deepwater well 125 km (78 miles) northwest of Donegal in 1,478 m of water (4,849 ft) in the Rockall basin. Drilling was suspended in October due to weather, but initial results were encouraging.

Pipelaying at Ireland's Corrib field, the country's second significant offshore natural gas discovery, was postponed in mid-2002 after operator Enterprise Energy Ireland's parent Enterprise Oil was acquired by Royal Dutch/Shell Group (Netherlands).

In early 2002 Ireland was planning to build the world's largest offshore wind farm, capable of generating 520 MW. Ireland already has some 125 MW of wind power.

Italy

CAPITAL: ROME
MONETARY UNIT: EURO
REFINING CAPACITY: 2,300,800 B/CD
OIL PRODUCTION: 87,000 B/D
OIL RESERVES: 621.7 MILLION BBL
GAS RESERVES: 8.0 TCF

Natural gas was discovered in the Sicily Channel off southwest Sicily in April 2002 by a consortium led by operator Eni and BG Group. The exploration well on the Panda exploration prospect was drilled 20 km (12 miles) offshore in 460 m of water (1,510 ft) and flowed a total 510,000 cu m/d of gas (19 MMcfd) from two intervals. Initial reserves estimates are 8-10 bcm (300-400 bcf).

The discovery, the first to be drilled by the consortium, lies on the G.R14.AG license. The group has rights to acreage totaling 2,400 sq km (930 sq mi) in this area. The natural gas could be earmarked for delivery into the Italian market.

More drilling was scheduled off Sicily and Calabria for Eni unit Agip. In August the Southern Cross semisubmersible drilling rig, owned by Atwood Oceanics Inc. (Houston), was upgraded to Italian requirements and inspected before beginning the job later in the year. The rig was to drill two wells, with a third well option over a 2-3 month period.

Field development

Gas discoveries in the northern Adriatic Sea were planned for development by Agip and Croatia's INA Group, with production to start by year-end 2004. Italian waters include part of Annamaria field, one of the four being developed. Completion of the project, which will include a 43-km (27-mile) subsea pipeline to Pola, Croatia, from Ivana field, will establish a link for the future transport of gas to Croatia from Italy.

Development of onshore Tempa Rossa oil field in the Basilicata region of southern Italy will begin in 2003, with production to start in 2006, said operator TotalFinaElf in September 2002. Production will be piped via the Val d'Agri pipeline to Agip Petroli's refinery in Taranto on Italy's coast for treatment and then exported. Proved reserves are estimated at 420 million boe, with associated gas accounting for about 7% of the total. Tempa Rossa is expected to reach a production plateau of 50,000 b/d of oil.

LNG deals

An LNG import terminal will be built in Brindisi Port on Italy's southeast coast by BG Group, which received approval in November 2002. BG expects to sanction the project by year-end 2003 and begin operating the 330-million euro terminal by year-end 2006. Phase I will be designed with throughput of 3 million tpy of LNG; Phase II will double this capacity to 6 million tpy.

The Brindisi terminal's location on the Mediterranean Sea is close to areas with high power generation demand in the Puglia region and is within 5 km (3 miles) of Snam Rete Gas's pipeline system. BG noted that Italy's energy demand is forecast to increase 25-30% by 2010 from 2002 levels. Italy is a net importer of natural gas and has one LNG receiving terminal, located at Panigaglia on the northwest coast. BG was finalizing the new terminal's design and negotiating for LNG supply and gas sales.

In July 2002, BG and Italy's Edison International agreed to purchase 4.8 bcm/yr of gas (180 bcf/yr) from GdF over 20 years, starting in 2006-07. Gas source would be the West Delta Deep marine field off Egypt. Italy currently imports Algerian LNG into its existing terminal, and Italy's electric company Enel SpA imports 3.5 bcm (130 bcf) from Nigeria via a swap agreement with GdF.

In August 2002 Edison received approval to build, with ExxonMobil, a 4 bcm/yr (150 bcf/yr) terminal near Rovigo on the northern Adriatic coast and signed contracts for the purchase of Qatari LNG. A preliminary agreement was also signed to purchase additional volumes of Nigerian LNG.

Under new Italian regulations, up to 20% of new LNG terminal capacity must be available to third parties. This will remain in effect until Italy's total regasification capacity reaches 25 bcm/yr (930 bcf/yr).

Refining, retail

The Sannazzaro refinery near Pavia will get a new absorbing system to treat flue gas from the fluid catalytic cracking unit (FCCU), said Agip Petroli in April. The Labsorb system will be the first FCCU flue gas treatment unit in Europe to incorporate a commercial-scale catalyst regeneration unit, said contractor Technip-Coflexip. Precommissioning will be completed by mid-2003.

Agip sold 195 of its Italian retail stations, with a total throughput of 270 million liters/yr, to TotalFinaElf in mid-2002, subject to government approval.

A 400-MW combined-cycle, natural gas-fired power plant will be built in Montenero di Bisaccia by Acea SpA, Rome's public electric utility, and Horizon Energy Development BV, a unit of National Fuel Gas Co. (Buffalo, NY, US). Start-up is slated for 2005.

Netherlands

CAPITAL: AMSTERDAM
MONETARY UNIT: EURO
REFINING CAPACITY: 1,206,842 B/CD
OIL PRODUCTION: 42,000 B/D
OIL RESERVES: 106 MILLION BBL
GAS RESERVES: 62.0 TCF

The Netherlands remained the EU's largest net exporter of natural gas and has been for years one of the top gas suppliers to western Europe. With production in decline, however, the country reduced output at Groningen — Western Europe's oldest and largest gas field — to maintain reserves for future use.

Production from the North Sea's Hanze field, 200 km (125 miles) north of Den Helder, increased in early 2002 to 31,500 b/d, representing almost two-thirds of Dutch oil production. Hanze, operated by Veba Oil & Gas Netherlands BV, is the first offshore chalk reservoir to be developed in the Netherlands. The field produces from two oil wells and includes two water injection wells.

Groningen redevelopment

To extend Groningen's life, operator Nederlandse Aardolie Maatschappij (NAM), a 50-50 joint venture of Shell and ExxonMobil, was in the midst of a $1 billion redevelopment program, including installing compression equipment to maintain reservoir pressure. The field is believed to have ultimate reserves of 2.7 trillion cu m, or tcm (100 tcf), with more than 1.5 tcm (55 tcf) already produced as of early 2002.

Groningen produces oil from 300 wells in 29 clusters. Total field production was down to 150 million cu m/day (5.6 bcfd), vs. domestic demand of 450 million cu m/day (16.8 bcfd), due to reservoir pressure decline to 2031 psi from its original 5033 psi.

NAM began installing compression equipment during 2002. Each well cluster is a standardized design with five treatment units. Initial work was to be completed by the end of 2004. Electric compressors were selected due to anticipated low loads and environmental requirements.

Downstream projects

A 100,000-tpy petrochemicals plant was brought onstream at Pernis by Shell Nederland Chemie in early 2002, producing propylene oxide glycol ethers, which are in demand as high-performance solvents to meet environmental legislation. The plant makes propylene glycol monomethyl ether, dipropylene glycol monomethyl ether, and propylene glycol monoethyl ether, some of which will be esterified to form acetates.

Work began in September 2002 on revamping a vacuum distillation unit and hydrodesulfurization units at Kuwait Petroleum Europoort BV's 75,500 b/cd refinery near Rotterdam. The vacuum unit will be increased to 28,000 b/d from 22,500 b/d, while three hydrodesulfurization units are being revamped at the same capacity to meet the EU's more stringent specifications on diesel fuel. Work by a subsidiary of Foster Wheeler Ltd. (Clinton, NJ, US) was slated for completion in late 2003.

Renewables

A 22.5-MW wind farm started operating in late 2002, reported BP plc (UK) and ChevronTexaco Corp. (San Francisco) in December. The $23 million wind project was constructed at the companies' jointly owned, 400,000 b/cd Nerefco refinery near Rotterdam.

The wind farm consists of nine Nordex turbines, each 120 m tall (394 ft) and with a generating capacity of 2.5 MW. The power will be sold into the Dutch national grid and will displace 20,000 tpy of greenhouse gas emissions. The project represents Europe's first full-scale wind farm on a brownfield refinery site.

The Dutch government has set a target to increase the amount of electricity generated from renewable sources to 5% by 2005. In 2002 the Netherlands had over 560 MW of wind power, accounting for roughly 1 terawatt-hour (TWh) of the total power market. This figure by 2010 is expected to approach 4 TWh, which would equal about 35-45% of the Netherlands' total renewable power generation.

Norway

CAPITAL: OSLO
MONETARY UNIT: KRONE
REFINING CAPACITY: 310,000 B/CD
OIL PRODUCTION: 3.150 MILLION B/D
OIL RESERVES: 10.265 BILLION BBL
GAS RESERVES: 77.3 TCF

Norway has been the world's third-largest oil exporter for several years, as well as the second-largest exporter of natural gas to western Europe. Norway's real GDP grew an estimated 2.3% during 2002, and the IMF projected 2003 growth at 1.9%.

Norwegian North Sea licensees agreed in May 2002 to coordinate ownership of their pipeline assets through a new state-owned company, Gassco AS, which will operate natural gas transportation from the Norwegian Continental Shelf.

In July Norway resolved its longstanding dispute with the EU over natural gas price competition. Long-term contracts with customers of Statoil and Norsk Hydro ASA were allowed to stand in return for an agreement to sell 14.2 bcm of gas (530 bcf) to new European customers over a 4-year period.

Norwegian and Iranian officials met in November to discuss expanding cooperation between the two countries' oil sectors, possibly including joint energy projects and greater collaboration between Norway and the Organization of Petroleum Exporting Countries (OPEC). Earlier, Iran's PetroPars Ltd. designated Statoil as operator of offshore development in Phases 6, 7, and 8 of the South Pars natural gas project.

Norway-UK cooperation

Norway agreed with the UK to increase cooperation, cut costs, and raise output, especially from aging fields. However, taxation rates remained unharmonized. Under a contract signed in mid-2002, Statoil will sell 5 bcm/yr of gas (186 bcf/yr) to British Gas Trading Ltd.'s Centrica plc unit for 10 years beginning in October 2005. The sale could drive construction of more gas pipelines from Norway to the UK. BP also solicited interest in new Norway-UK gas links, since the UK is expected to become a net gas importer in 2005.

Norway and the UK were also coordinating plans to decommission concrete substructures at Frigg, a giant gas field with associated condensate that straddles their two North Sea sectors. Given government approvals, Frigg could cease production as early as 2003, with decommissioning lasting another 8 years.

Norway agreed with Russia to open energy dialogue on shared areas of the Arctic Barents Sea.

Exploration

Even with high depletion rates and declining field size, the North Sea will continue to be a sizable producer of crude oil for the foreseeable future, said Simmons & Co. International (Aberdeen). As development shifts from large fields to small, the area is attracting more independent E&P companies, similar to how the US Gulf of Mexico has evolved.

This view was reinforced by Douglas-Westwood Associates (Canterbury, UK) and Infield Systems (London), which said that offshore Europe exhibits many of the characteristics shown earlier by the Gulf of Mexico — a great number of tiny, shallow-water prospects and a deep frontier. Considerable opportunities remain for players geared to the efficient production of small fields.

Results of 17th Licensing Round on NCS

Blocks or parts of blocks	Production license	Company	Share, %
6405/4, 7, and 10	281	Statoil ASA (operator)	30
		Royal Dutch/Shell Group	20
		BP PLC	20
		Phillips Petroleum Co.	10
		SDFI*	20
6504/6 and 6505/1, 2, 4, and 5	282	Royal Dutch/Shell Group (operator)	30
		Norsk Hydro	25
		RWE-DEA AG	25
		Statoil ASA	20
6605/5, 7, 8, and 9 and 6606/7	283	Norsk Hydro (operator)	30
		ChevronTexaco Corp.	25
		Conoco Inc.	25
		SDFI	20
6607/11 and 12	284	Phillips Petroleum Co. (operator)	40
		ChevronTexaco Corp.	30
		DONG	30
6607/5	285	Agip SPA (operator)	70
		Gaz de France	30
6609/5 and 6	286	Norsk Hydro (operator)	60
		Agip SPA	40

*Norway's State Direct Financial Interest.

Source: Norway Ministry of Petroleum and Energy

Licensing round

Norway's government in June 2002 reported the results of its 17th licensing round. Several blocks are in challenging deepwater areas.

Norsk Hydro enhanced its position in areas including the Ormen Lange project, Fles Nord, and the Solsikke prospect to be drilled during 2002. The Fles Nord blocks are in water depths of 900 m (3,000 ft) in the center of the Voring basin, about 150 km (90 miles) northwest of Åsgard field.

Statoil will operate on the Grip Ridge in 700-900 m of water (2,300-3,000 ft) and gained interests in PL 281, where the first well will be spudded in 2002, and in PL 282, in 1000 m of water (3,200 ft).

Staer discovery

In April 2002 Statoil found oil on the Staer structure close to Norne field and plans to assess its commerciality. The find could contribute to a unitized development of several reservoirs in the area. Staer lies about 3 km (2 miles) northeast of the Norne production ship, and its oil quality is similar to that of Norne.

Exploration well 6608/10-8 on production license 128 was drilled vertically from the Stena Don rig to 2,660 m (8,730 ft). A sidetrack, drilled 600 m out (2,000 ft) to delineate the areal extent of the field, had a total measured depth of 2,600 m (8,530 ft) below the seabed.

Tampen discoveries

Condensate and light oil were discovered in May 2002 in the Tampen area's License 152 on the Dole and Ole prospects near Gullfaks and Statfjord fields. Because of similarities with producing fields, the wells were not production-tested. The Dole well (33/12-8S) went to 3,350 m (10,990 ft) below mean sea level and Ole (33/12-8A) to 3,369 m (11,053 ft).

Statoil began to evaluate tie-back options for the new finds, which could begin producing in 2004. Statoil became sole operator of the Tampen complex at year-end 2002 when it assumed Snorre, Visund, Vigdis, and Tordis fields from Norsk Hydro.

Statoil expected to begin delivering Visund gas to a new 26 million cu m/day (970 MMcfd) processing plant at Kollsnes in 2005. Visund's reserves are an estimated 55 bcm (2 tcf). A proposed 34-km (21-mile) pipeline would connect Visund to the 8.5 bcm/yr (316 bcf/yr) Kvitebjørn pipeline, which is being planned to carry gas from

Norne Area Fields

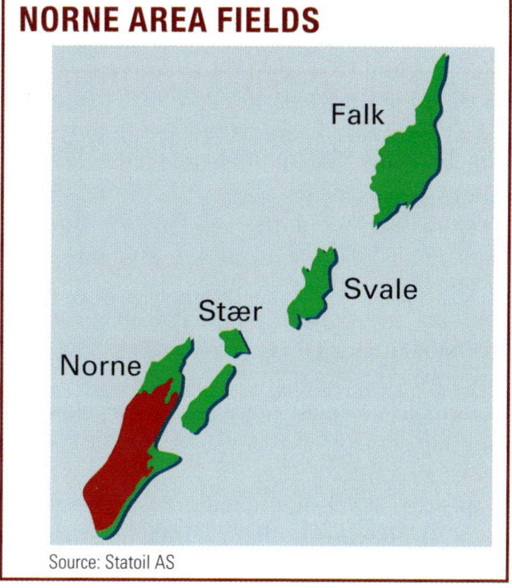

Source: Statoil AS

several fields. The new Kollsnes plant will recover propane, butane, and naphtha for transport via the Vestprosess pipeline to Mongstad for fractionation.

More oil and gas was proved in the Tyrihans South discovery, said Statoil in December, and oil was found in its Dolly prospect in the Tampen area. The Tyrihans South appraisal well, drilled to 4,000 m (13,000 ft), was being logged and could be tied in to Åsgard for development as a satellite. The 34/10-47S wildcat on Dolly prospect 5 km (3 miles) north of Gullfaks satellites Gullveig and Rimfaks encountered hydrocarbons and was being evaluated. Statoil also spudded an exploration well in November on Blåmeis prospect, east of Norne field in an area where Statoil has found oil.

Other exploration

Seismic data covering 22,000 sq km (850 sq mi) over deepwater portions of the More and Voring basins off mid-Norway was being processed in September 2002. Fugro Survey AS acquired the data for the Norwegian Deepwater Programme Seabed Project. The survey aimed at assessing slope stability and potential seabed drilling hazards.

Two Norwegian sector players exchanged interests in exploration licenses in early 2002, with Statoil moving into the F Prospect of the Barents Sea and Tampen area in the North Sea, while ExxonMobil received interest in exploration territory near Grane field.

Halten Bank development

Kristin

Development of Kristin gas and condensate field continued in 2002. A floating production platform will be tied back to 12 subsea wells in four complexes via six flowlines totaling just under 40 km of pipe (25 miles). Kristin field, which will deliver 35 bcm (1,300 tcf) of gas until 2016, is expected to come onstream in fall 2005. Its output of condensate and natural gas liquids is estimated at 220 million bbl and 8.5 million tonnes, respectively

Construction of the platform by Aker Maritime will begin in 2003. The subsea production system contract, awarded in February to Kværner Oilfield Products, calls for template delivery in mid-2003. Steel pipe for the flowlines was ordered in September through Marubeni-Itochu Steel Inc. (Japan) from Kawasaki Steel Corp. for delivery in fall 2003.

In early 2003, Statoil announced several more contracts for the 17 billion kroner ($1.9 billion) Kristin development, including installation of a 30-km (19-mile) dual gas export pipeline and a 23-km (14-mile) oil export line. The work will be carried out largely in spring and summer 2004, with the remainder of the tie-ins to the platform due in 2005 when this installation is scheduled to arrive on-site.

Mikkel

Southern Norway's Kårstø gas treatment complex was to be expanded under a Statoil contract signed in May 2002 with Fabricom Contracting NV. The expansion will enable the gas treatment plant to receive and treat gas from Mikkel field, which would begin producing in fall 2003. Kårstø will be ready to receive Mikkel gas by October 2003.

The plant has been receiving rich gas through the Statpipe and Åsgard transport pipelines, which are operated by Gassco.The condensate is carried through a separate pipeline from the Sleipner area. Following processing at Kårstø, LPG and condensate are exported by tanker, while sales gas is exported to continental Europe through the Europipe II and Statpipe trunklines.

The Alpha North satellite in Sleipner West field was set for drilling of three wells in 2003, which would be tied back to Sleipner A platform. Production of gas and condensate was scheduled to begin in late 2004. Production started in December 2002 from two Sigyn gas and condensate satellites. Their output is processed on Sleipner A. Sigyn holds an estimated 5.6 billion cu m of gas (209 bcf), an equal volume of condensate, and 20.5 million bbl of gas liquids.

Ormen Lange development

Ormen Lange, on deepwater Block 6305/5-1, is the largest undeveloped natural gas field on the Norwegian Continental Shelf, with estimated gas reserves of 375 bcm of dry gas (14 tcf) and 138 million bbl of condensate. In December its licensees agreed to develop the field using subsea production installations linked to a new onshore natural gas processing plant at Nyhamna, near Aukra, in Møre county, Norway. Ormen Lange is 100 km (60 miles) offshore in 800-1,000 m of water (2,600-3,300 ft).

Development would include construction of the largest pipeline on the Norwegian Continental Shelf, Norsk Hydro said. Gas will be exported to continental Europe via existing pipelines from Statoil's Sleipner R riser platform in Sleipner East field. First gas from Ormen Lange is planned for late 2007. The field, expected to produce for 30-40 years, will peak at a minimum 20 bcm/yr (745 bcf/yr), about 20% of total anticipated Norwegian gas output in 2010.

Troll production

Although Troll's production has been relatively flat, the field achieved a daily record of 440,000 bbl in May 2002. In Troll's Olje field, lying in water depths of 315-340 m (1,033-1,115 ft) 100 km (60 miles) northwest of Bergen, multilateral well completions increased the total wellbore drainage area from existing subsea template structures, avoiding the addition of more templates.

Continued development of Troll Olje will require 15 additional multilateral wells, which will include two three-branched multilateral completions. Norsk Hydro expected the multilateral well program to total 28 wells by 2003 and contribute 88 million bbl of additional oil reserves. Development of additional oil zones within Troll should recover another 47 million bbl.

In other work at Troll, emissions of carbon dioxide and nitrogen oxides will be reduced by powering the A platform's natural gas compressors with electricity, rather than using turbine-generated power. ABB AS (Zurich) was to develop and install the power system, including new compressor drives, submarine cables, and a transformer and rectifier station at the platform's associated 120 million cu m/d (4.5 bcfd) gas treatment plant at Kollsnes near Bergen.

Other development

Natural gas flowlines and other equipment were ordered in October to connect Jotun and Balder fields with Ringhorne field, which lies about 260 km (160 miles) west of Haugesund in 124-129 m of water (407-423 ft). The 81-km (50-mile) system — including rigid flowlines, along with flexible risers and jumpers — will be installed by summer 2003 by Technip-Coflexip's Norwegian unit. Its plant in France will manufacture the piping.

More than 100 days of rig time was saved by increasing the rate of penetration at Jotun field using rotary steerable, directional-drilling systems. The operations experienced no hole problems or wellbore instability during the 15-well Jotun drilling campaign. The rotary-steerable systems provided greater accuracy and efficiency in extended-reach drilling applications because drillstring sliding in the wellbore was not required to maintain directional control.

For the first time in the world, multiple fiber-optic sensors, combined with surface controlled, flow-control components, were installed downhole, said supplier Weatherford International Ltd. in September. The downhole sensor arrays in Oseberg East field, which included pressure and temperature gauges and a distributed temperature sensing line, provide data for day-to-day production management and enable strategic reservoir management during the field's life.

Snøhvit LNG project

In March 2002 Norway's government approved Statoil's plans to develop the $5 billion Snøhvit LNG project in the Barents Sea, which will be the largest subsea LNG project in the world, as well as the most northerly, and Europe's first LNG export facility. Gas will be produced from subsea installations in water depths of 250-345 m (820-1,132 ft).

Construction of the processing facilities began during 2002. Snøhvit is the first major project on the Norwegian continental shelf without fixed or floating surface facilities, said Statoil. Natural gas will be piped to the coast, liquefied, and shipped on four new, 140,000-cu m LNG carriers.

Development costs are estimated at 40 billion kroner ($5.3 billion), excluding the carriers, estimated to cost 5.8 billion kroner ($770 million). Statoil will be part owner of three of the vessels, all being built in Japan using a spherical tank design.

Statoil expects to produce about 5.75 bcm/yr of LNG, along with 747,000 tpy of condensate and 247,000 tpy of LPG. A total of 70 LNG cargoes will be shipped annually from the terminal near Hammerfest, carrying 2.4 bcm/yr to the US and 1.6 bcm/yr to Spain under existing long-term contracts. Production and gas deliveries would start in 2006 and last until 2030.

The project includes a subsea development tied back by pipeline to a receiving terminal and a single-train liquefaction plant on the small island of Melkøya at the entrance of the shipping channel into Hammerfest. Statoil was considering using electric compressors to minimize emissions, since Snøhvit field lies in an ecologically sensitive area north of the Arctic Circle.

In addition to Snøhvit field, substantial gas reserves have been found in nearby Albatross and Askeladd fields in Norway's Hammerfest basin. The fields extend across seven production licenses, which were unitized and will not expire until 2035.

The core recoverable resource is about 420 bcm of natural gas (15.6 tcf) and 170 million bbl of condensate in the fields, estimated Wood Mackenzie Ltd. (Edinburgh). Statoil estimates the three fields' recoverable reserves at 193 bcm of gas (7.187 tcf) and 113 million bbl of condensate.

Immediately upon government approval of the LNG project, Statoil began to qualify technology for the record-long umbilical

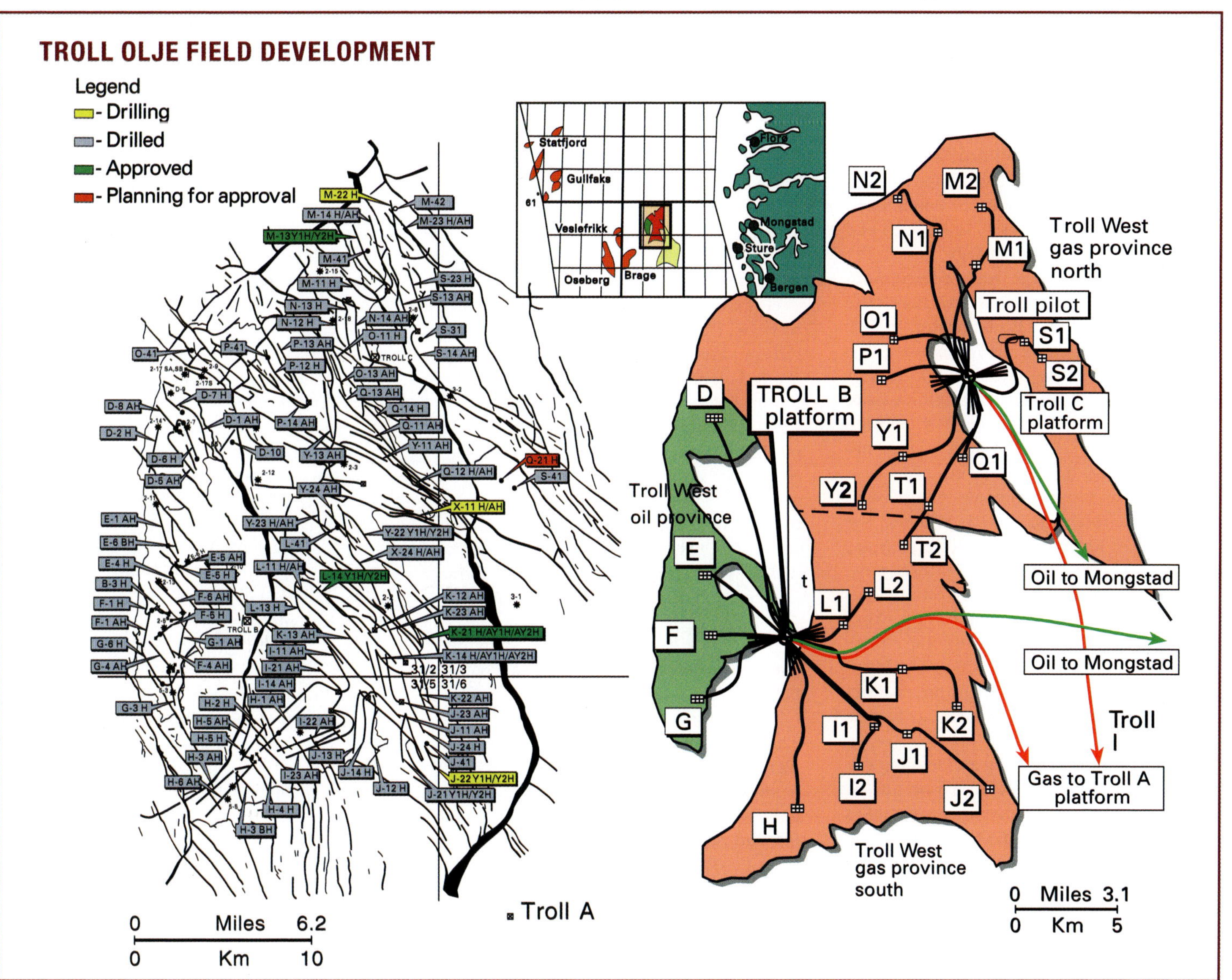

TROLL OLJE FIELD DEVELOPMENT

that will connect Snøhvit to land facilities. Statoil and the subsea system suppliers were testing existing technology for the 160-km (100-mile) bundle of control lines and cables, with verification to continue until year-end 2002. All the wells planned for Snøhvit — 21 for gas and condensate and one for carbon dioxide injection — will be remotely controlled from land through the umbilical.

With a diameter of roughly 11 cm (4.3 inches), the umbilical will contain high-voltage power lines, fiber-optic cables, and hydraulic piping. The fiber-optic cables will transmit control signals to subsea valves and will return information from sensors mounted in the wells. Snøhvit will involve the world's longest distance between the control station on land and the first subsea installation.

Plans call for 18 horizontal wells in the fields — five in Snøhvit, five in Askeladd, and eight in Albatross — and a 160-km (100-mile), 27-inch multiphase pipeline to deliver condensate and natural gas from the fields to the plant. Pipelaying is scheduled for mid-2005.

Phased development, using the one-train facility, would have condensate-rich Snøhvit field production onstream first, followed by Askeladd 8 years later and Albatross 14 years after initial production. Output could reach 4.5-5.6 bcm/yr of gas (435-542 MMcfd) and about 20,000 b/d of condensate.

In July 2002 a barge was ordered from Izar Carenas (Spain) to support the Melkøya's floating liquefaction plant. Scheduled for completion in August 2003, the barge will house about 24,000 tonnes of processing equipment and will measure 154 m long, 54 m wide, and 9 m high (respectively, 505, 177, and 30 ft), with a net weight of 8,500 tonnes of steel.

Germany's Linde will engineer and build the liquefaction plant, which will use the Statoil-Linde technology for liquefying gas at —163 degrees C. A low-NOx turbopower generation plant fueled with Snøhvit gas will be integrated in the liquefaction plant. Heat recovery will help boost the plant's energy efficiency to 70%.

Belgium's Tractebel will supply product tanks and vessel loading systems, as well as design and build four cylindrical storage tanks and associated piping and loading systems connecting the tanks to export jetties.

Two of the tanks, each with a capacity of 120,000 cu m, will store LNG; one will store up to 45,000 cu m of LPG; and the fourth will hold 470,000 bbl of condensate. On-site construction began in 2002 and is scheduled to be completed in late 2005. In October TotalFinaElf hired Mitsubishi Heavy Industries Ltd. to build a 145,000 cu m carrier for Snøhvit LNG.

Poland

CAPITAL: WARSAW
MONETARY UNIT: ZLOTY
REFINING CAPACITY: 350,000 B/CD
OIL PRODUCTION: 16,500 B/D
OIL RESERVES: 96.375 MILLION BBL
GAS RESERVES: 5.829 TCF

Poland will likely become an EU member in 2004. The IMF projected that Poland's real GDP will grow 3.0% in 2003. Due to flat demand for natural gas, some of Poland's gas purchase agreements with Russia, Denmark, and Norway reportedly could be amended or canceled.

The state's Polish Oil and Gas Corp. (PGNiG) planned to launch 200 new drilling sites in 2002 at a cost of Zl 700-800 million, spend Zl 600 million in domestic oil and gas exploration, and begin liquidating 1500 old drilling sites.

Exploration of three areas was being planned by FX Energy Inc. (Salt Lake City): the Fences area, covering 1,200 sq km (300,000 acres, or 500 sq mi) in western Poland's Permian basin; the Pomerania area, also in the Permian basin, covering 8,900 sq km (2.2 million acres, or 3,400 sq mi); and the Wilga area, covering 1,000 sq km (250,000 acres, or 390 sq mi) in central Poland.

PGNiG and Russia's Gazprom planned to boost the capacity of the Yamal-Western Europe natural gas pipeline by building a second section and a system-connecting link in Poland via Belarus. The "Contract of the Century," which envisions the shipment of 250 bcm (9.3 tcf) of Russian gas to Poland by 2020, will possibly be extended to 2029. There would be 12.5 bcm (465 bcf) delivered in 2010 just for Poland's consumption. The Yamal-Europe system ultimately will have a capacity of 60 bcm/yr (2.2 tcf/yr) and cost at least $30 billion.

The 300-km, 7.6 bcm/yr (186-mile, 283 bcf/yr) Baltic Pipe Line, destined to carry natural gas imports from Denmark, was to begin construction in mid-2002.

Revamping began on a Polski Koncern Naftowy Orlen SA (PKN Orlen) ethylene plant in Plock, about 100 km (60 miles) from Warsaw. The work will increase ethylene capacity to 660,000 tpy from 360,000 tpy and propylene capacity to 315,000 tpy from 130,000 tpy.

ABB will apply new proprietary technology to the ethylene cracker, which is Poland's largest. Expansion will be complete by 2005. PKN Orlen is Poland's largest fuel manufacturer and distributor and operator of the largest retail station network.

Portugal

CAPITAL: LISBON
MONETARY UNIT: EURO
REFINING CAPACITY: 304,172 B/CD
OIL PRODUCTION: 0
OIL RESERVES: 0
GAS RESERVES: 0

The Portuguese economy continued to slow in 2002, and the IMF projected that Portugal's real GDP will grow only 1.5% in 2003. The state's share of conglomerate Galpenergia continued to dwindle as portions were sold off. Galp includes the state oil company Petroleos de Portugal (Petrogal), natural gas distributor Transgas, and Gas de Portugal. However, an offer to sell an additional 20% of Galp was postponed due to weak international stock markets.

To diversify its energy supply, Portugal was building its first LNG regasification terminal in Sines, 90 km (56 miles) south of Lisbon, which will go online in 2003 as a gas-processing facility. In the meantime, Transgas was taking delivery of 350 million cu m of Nigerian LNG through Spain's Huelva terminal.

In September Mohave Oil & Gas Corp. (Houston) began a six-well shallow drilling program onshore in the Lusitanian basin. The wells were to be drilled to 300 m (345 ft) in the Torres Vedras area 40 km (25 miles) north of Lisbon. Mohave plans to drill directionally to optimize fracture intersections and to produce the wells on the pump. The company also plans to acquire 450 km (280 miles) of new onshore seismic data in the basin by year-end.

Mohave reported in December on its exploration activity over the past 5 years, including a previous four-well campaign in the Lusitanian basin in west central Portugal and offshore. Prospects for hydrocarbon development and production in the basin are excellent, and Portugal's main gas trunkline traverses it.

The government held a licensing round for deepwater areas along the Atlantic coast in 2002 with a bid deadline in December. Portugal offers the best royalty and tax incentives in the EU, said Mohave.

Romania

CAPITAL: BUCHAREST
MONETARY UNIT: LEU
REFINING CAPACITY: 501,182 B/CD
OIL PRODUCTION: 118,000 B/D
OIL RESERVES: 955.62 MILLION BBL
GAS RESERVES: 3.556 TCF

At a historic summit in Prague in late 2002, Romania was invited to join NATO, along with several other eastern European and former Soviet Union nations. Romania also was negotiating to join the EU and could become a member around 2006.

New and reprocessed data were being evaluated in April 2002 by Tullow Oil plc (UK), which completed the acquisition of 557 km (346 miles) of 2D seismic and acquired 1,400 sq km (540 sq mi) of surface geochemical data. Tullow operates two licenses in Romania, the EPI-3 (Brates) and EPI-8 (Valeni de Munte). An infill seismic program was conducted in EPI-3 in mid-2002, and an exploration well was scheduled for late 2002 in EPI-8.

Development of Romania's coalbed methane resources was being considered by Galaxy Energy Corp. (Denver), which said in November it would acquire the stock of Pannonian International Ltd. (US), which will remain under direction of Thomas Fails, a former Shell senior geologist.

Galaxy plans to focus initial efforts on development of Pannonian's 30-year concession on 87.16 sq km (21,538 acres, or 33.65 sq mi) in the Jiu Valley. The concession is in an area mined for decades that contains as many as 18 coal seams 300-1,000 m deep (985-3,280 ft). The target seam thickness averages 22 m (72 ft). Pannonian plans to drill two coalbed methane wells in the Jiu Valley by spring 2003 at a site 3.6 km (1.5 miles) from a 20-inch gas pipeline.

In March Romania opened a 200-km (125-mile) gas pipeline linking the country's borders with Bulgaria and Ukraine and completing southeastern Europe's natural gas transit corridor from Russia. With the Isaccea-Negru Voda pipeline operational, Romania can transport up to 26.5 bcm/yr of gas (988 bcf/yr), up from previous capacity of 9.5 bcm/yr (353 bcf). In addition, by 2004, Romania will increase its underground gas storage capacity to 4.3 bcm from 1.4 bcm (to 159 bcf from 53 bcf).

A pipeline has been proposed to transport 660,000 b/d of Caspian oil from Constanta, Romania, to Trieste, Italy, but a route has not yet been selected. The southern route (South-East European Line, or SEEL), would pass through Serbia and Croatia before linking with the existing Trans-Alpine Pipeline (TAP). In November 2002, Romania, Croatia, and Serbia agreed on the SEEL route, but financing remains a problem.

Upgrading of the Petrotel refinery was started in April 2002 by Lukoil, which owns 51%, to bring products into compliance with EU requirements and Euro-3 standards by 2004. Lukoil also planned to invest $30 million to increase its network of filling stations in Romania from 30 to 150 by 2004.

Slovakia

CAPITAL: BRATISLAVA
MONETARY UNIT: KORUNA
REFINING CAPACITY: 115,000 B/CD
OIL PRODUCTION: 1,000 B/D
OIL RESERVES: 9.0 MILLION BBL
GAS RESERVES: 530 BCF

In early 2002 Slovakia's state pipeline company Transpetrol sold 49% of its shares to Russia's Yukos, with an option to purchase the remaining 51%, and 49% of Slovakia's state natural gas monopoly, Slovensky Plynarensky Priemysel (SPP), was sold in March to a consortium of GdF, Gazprom, and Ruhrgas.

SPP is integrated across the natural gas chain and owns a pipeline linking Russian gas producers with European gas distribution companies. SPP was planning to invest 1.643 billion korunas during 2002 for new gas mains to connect additional households.

Liberalization of the country's gas market began in July 2002 for large customers (consuming more than 25 million cu m/yr, or 882 MMcf/yr) and will extend to smaller customers in 2003 (15 million cu m/yr, or 530 MMcf/yr) and 2008 (5 million cu m/yr, or 177 MMcf/yr).

Spain

CAPITAL: MADRID
MONETARY UNIT: EURO
REFINING CAPACITY: 1,321,500 B/CD
OIL PRODUCTION: 6,500 B/D
OIL RESERVES: 157.626 MILLION BBL
GAS RESERVES: 94 BCF

Although Spain's rapid economic expansion was slowing, its real GDP growth in 2002 exceeded the eurozone's rate, and the IMF projected growth of 2.7% in 2003.

During 2002 new data received by *Oil & Gas Journal* resulted in dramatic increases in Spain's estimated proved reserves, with oil rising by a factor of 7.5, from 21 million bbl to 157.6 million bbl, and natural gas by a factor of 5.2, to 2.5 bcm from 483 million cu m (to 94 bcf from 18 bcf).

Oil spill

In November 2002, the Prestige single-hull oil tanker split in two and sank in 3,500 m of water (11,500 ft) in the Atlantic Ocean 270 km (170 miles) off the Galicia coast of northwest Spain, in an area known as the "Coast of Death" for its history of shipwrecks.

Most of its cargo of 73,000 tonnes of fuel oil remained aboard the sunken vessel. An estimated 5,000 tonnes of oil spilled from the ship when one of its tanks was punctured earlier in November during the storm that precipitated the accident.

About 4,000 tonnes of the oil had already spread over the Galician coastline between Cape Finisterre and Seixo Branco, a major fishing zone. Spanish authorities banned commercial fishing and shell gathering in the area as a result of the spill.

Some specialists claimed the heavy fuel oil could remain on the ocean bottom for some time before spreading toward the coastlines of Spain, Portugal, and France. However, according to the World Wildlife Fund, Galicia could face perhaps the worst tanker oil spill in history, twice the magnitude of the Exxon Valdez spill off Alaska in 1989. Several EU countries responded to Spain's call for equipment and vessels to handle the spill.

The cargo of fuel oil was reportedly bound for Singapore from Riga, Latvia, for Crown Resources AG, a consortium member of the Alpha Group based in Gibraltar. Neither Latvia nor Gibraltar is a member of the international body in charge of controlling merchant ships, officials said. The single-hull tanker was built in 1976 in Japan and is one of the oldest tankers in service. It had already been sanctioned for its safety risk both in Rotterdam and in New York in 1999.

The EU had earlier banned single-hull tankers, but that ban does not take full effect until 2015. In December, the European Council endorsed a host of proposals by its transport and environment ministers aimed at preventing oil spills, but the measures have intensified criticism by shipowners and environmentalists alike.

LNG plans

Spain was poised to become Europe's largest LNG importer as natural gas consumption grew 8% during 2001 to 17.3 bcm (644 bcf). Spain imports gas via pipeline from Algeria and as LNG into three receiving terminals owned by Gas Natural SDG SA, which were being expanded during 2002. LNG imports are expected to reach 26 bcm in 2005 and 31 bcm by 2010. Spain was also the world's second-largest spot-market LNG importer.

As Spain's gas pipelines and terminals were being opened to other companies, electric utilities, independent power producers, and oil companies were aggressively developing markets and lining up gas supplies, sometimes by developing their own LNG projects. Plans call for the construction of four new Spanish terminals.

Exploration

Repsol YPF SA (Madrid, Buenos Aires) received permits in early 2002 to an area 32-80 km (20-50 miles) off the Canary Islands, just west of Lanzarote and Fuerteventura islands. Repsol was acquiring and processing 3,000 sq km (1,160 sq mi) of 3D seismic and would begin drilling in 2004.

Plans for exploring nine blocks were submitted to local Fuerteventura authorities in April. The blocks, Canarias-1 through Canarias-9, cover 6,160 sq km (2,378 sq mi) in water depths of 1,000-1,500 m (3,300-4,900 ft), within territory administered by Spain.

SPANISH IMPORT TERMINALS

Terminal	Owners	2001	2002	2003	2004	2006
		Capacity, billion cu m*				
Barcelona	Gas Natural	9.5	10.7	10.7	14.7	24.7
Huelva	Gas Natural	3.5	3.5	7.0	7	11
Cartagena	Gas Natural	2.3	4.6	5.8	5.8	8.1
Bilbao	BP, Repsol, Iberdrola, EVE	—	—	3.1	6.2	6.2
Ferrol	Enedesa, Union Fenosa, Sonatrach, Galician govt., Sp. Cos.	—	—	—	3.1	3.1
Sagunto	Union Fenosa, Iberdrola + BP & Gas Natural	—	—	—	5.9	5.9
Castellon	Iberdrola,	—	—	—	—	3
Total		**15.3**	**18.8**	**26.6**	**42.7**	**62**

*Projected.

Source: Cedigaz[5]

Transport, downstream

Work was under way to increase the capacity of the Pedro Duran Farell natural gas pipeline to 10.4 bcm from 7.6 bcm (to 388.5 bcf from 282.5 bcf) by adding a compressor station. Completion was scheduled for 2003. The 1,400-km (870-mile) pipeline travels under the Strait of Gibraltar, connecting Spain to Algeria's gas. A pipeline connection with France, also to be completed in 2003, was being built through Irun in the Basque Country.

The 120,000 b/cd refinery at Castellon de la Plana was being upgraded in mid-2002 by Foster Wheeler under contract to BP. The $100 million project, including a new diesel hydrotreater, a hydrogen plant, and a naphtha hydrotreater, will enable production of ultra-low sulfur fuels. The upgrade will be finished in 2004.

TotalFinaElf's entire gasoline distribution system and network of retail stations in Spain were sold in mid-2002 to Galp and Agip, subject to government approval. The deal includes 186 stations with a total throughput of 620 million liters/yr, as well as a 100 million liter storage site on the Mediterranean Spanish coast.

Sweden

CAPITAL: STOCKHOLM
MONETARY UNIT: KRONA
REFINING CAPACITY: 423,500 B/CD
OIL PRODUCTION: 0
OIL RESERVES: 0
GAS RESERVES: 0

The IMF projected that Sweden's real GDP will grow 2.5% in 2003.

Swedish oil and gas exploration company Lundin Petroleum AB appointed president Ian Lundin to also serve as chairman, succeeding Adolf Lundin. In September 2002 Lundin acquired 95.3% of Coparex International SA (France), an exploration and production company with 55 million bbl of proved and probable reserves.

Sweden's Preem Petroleum AB sold its 79 service stations in Poland to Statoil.

Switzerland

CAPITAL: BERN
MONETARY UNIT: FRANC
REFINING CAPACITY: 132,000 B/CD
OIL PRODUCTION: 0
OIL RESERVES: 0
GAS RESERVES: 0

The IMF projected that Switzerland's real GDP will grow 1.9% in 2003.

Switzerland's UBS Warburg AG investment bank was selected in early 2002 as the successful bidder for Enron Corp.'s North American wholesale electricity and natural gas trading business. Before collapsing, Enron was the world's largest energy merchant, accounting for about 25% of all trades in a market it helped pioneer.

Turkey

CAPITAL: ANKARA
MONETARY UNIT: LIRA
REFINING CAPACITY: 719,275 B/CD
OIL PRODUCTION: 47,000 B/D
OIL RESERVES: 300.0 MILLION BBL
GAS RESERVES: 300 BCF

Following Turkey's 2001 financial crisis, an IMF assistance package of $17 billion hinged on privatizing state-owned industries. Corrupt energy officials were convicted and sentenced in July. Turkey's real GDP growth was estimated at 2.6% in 2002 and projected at 5.0% in 2003.

A petroleum market reform bill was being considered, which would liberalize pricing, integrate industry functions, and privatize refining and fuel retailing. In July the government announced plans to sell its 25.8% share of Petrol Ofisis (Poas) and privatize most of the energy sector during 2003.

In November 2002, Turkey initiated a policy banning supertanker traffic through the Bosporus and Dardanelles Straits between the Black Sea and the Aegean Sea. The policy also severely restricts the passage of smaller tankers carrying dangerous cargo, including chemicals, LPG, LNG, and explosives. For years, Turkey has expressed concerns about the increasing tanker traffic passing through the two straits and in the Sea of Marmara between them.

The Caspian News Agency reported that Turkey prevented supertankers from Russia and Kazakhstan from exporting oil through the straits during October and November, citing its new regulations limiting tankers to 200 m (660 ft) in length to pass through the narrow, 30-km long (19-mile) Bosporus and restricting tanker length to 250 m (820 ft) for passage through the Dardanelles.

Turkey's new regulations also prohibit loaded oil tankers from passing through the straits at night and in other instances when visibility is obscured. Regulations apply to vessels carrying cargo classified by the international agreements as dangerous, including petroleum, its derivatives, and petroleum products.

E&D

Exploratory drilling could start in mid-2003 in a lightly explored basin on the southwestern Black Sea shelf. Toreador Resources Corp. (Dallas) identified several gas prospects from 1,275 km (792 miles) of 2D seismic data on four permits that hug the shore off Eregli, a gas pipeline terminus 193 km (120 miles) east of Istanbul. Toreador also plans to complete the Barbaros-1 discovery in the Thrace basin 100 km (60 miles) west of Istanbul, where logs indicate an initial rate of 54,000 cu m/d (2 MMcfd).

Barbaros is 50 km (30 miles) southwest of Gocerler gas field, which began delivering gas in early 2002 to a power station in the Misinli Industrial Area. Gocerler was producing 190,000 cu m/d of gas (7 MMcfd) and 7.5 bbl/MMcf of condensate in August.

Amity Oil Ltd. (Sydney) and Turkiye Petrolleri Anonim Ortakligi acquired 3D seismic data along 24 km (15 miles) of the Gocerler structural trend in mid-2002 to define exploratory structures and optimize development locations. Amity also planned to start 2D seismic data collection on some blocks in the Thrace basin and spud the East Hasancik-1 exploratory well on an oil prospect 5.6 km (3.5 miles) west-southwest of Zeynel oil field in south-central Turkey.

Oil pipeline

Construction will begin in 2003 on the 1 million b/d oil pipeline from Azerbaijan's Caspian Sea port of Baku through Tbilisi, Georgia, to Ceyhan, Turkey. The 1,760-km (1,090-mile), 42-inch line, to be completed by 2005, will be operated for 40 years by Turkey's Botaintergral Petroleum Pipeline Corp. The Baku-Tbilisi-Ceyhan (BTC) pipeline will carry Caspian region crude oil to Turkey's Mediterranean Sea coast for export.

The pipeline will include seven pumping stations and three metering systems and will pass over mountains in Turkey up to 2,500 m high (8,000 ft). Turkey's state pipeline company BOTAS will build the 1,080-km (670-mile) section within its borders and serve as the turnkey contractor for the Turkish section.

Although the Azerbaijan state oil company reduced its stake in the group sponsoring the pipeline, leader BP announced in June 2002 that ownership was complete, and the partners in the $2.9 billion project formed the BTC Pipeline Co. to construct, own, and operate the system.

Control and safety systems for the pipeline will be supplied by ABB under a $34 million contract announced in December 2002. Greece's Consolidated Contractors International Co. will handle pipelaying in Azerbaijan, while a joint venture of France's Spie-Capag SA and Petrofac Ltd. (London) will handle pipelaying in Georgia and related facilities work in both Azerbaijan and Georgia.

Gas pipelines

The $3.3 billion Blue Stream pipeline from Russia was set to supply 16 bcm/yr (596 bcf/yr) to Turkey by 2003. Construction of the 483-km (300-mile) Turkish onshore section was complete, while the 375-km (222-mile) Russian section including compressor stations and underground storage was scheduled for completion in September 2002.

Work was also nearly complete on the 378-km (235-mile) seabed section under the Black Sea, which was being laid by the Saipem 7000 in waters up to 2,150 m (7,054 ft) — the world's deepest pipelaying. Russia was to begin delivering 1.9 bcm (70.6 bcf) to Turkey in late 2002, and the line would reach peak capacity of 15.2 bcm/yr (565 bcf/yr) by 2009.

Construction of the 1,014-km (630-mile) Baku-Erzurum line from Azerbaijan to Turkey was scheduled to begin in late 2002 and become operational by year-end 2004 with an initial capacity of 20.9 bcm/yr (777 bcf/yr), rising eventually to 28.5 bcm/yr (1.06 tcf/yr). Turkey had agreed to purchase 1.9 bcm/yr (70 bcf/yr) from Azerbaijan starting in 2005 and increasing to 4.75 bcm (177 bcf) in 2008 and around 6.2 bcm/yr (230 bcf/yr) in 2008-20.

Also, the proposed 1,690-km (1,050-mile) Trans-Caspian pipeline would transport gas from eastern Turkmenistan to Turkey. Given that Turkey has agreed to take about 15 bcm/yr (1.5 bcfd) of gas from Turkmenistan, partners in the project remained committed despite delays. Turkey also has deals to purchase gas from Iran and Greece. Some contracts could penalize Turkey if it fails to take full volumes.

In July 2002 Turkish State Petroleum Co. (TPAO) was planning to build Turkey's first natural gas storage facility 80 km (50 miles) west of Istanbul on the Marmara Sea coast.

United Kingdom

CAPITAL: LONDON
MONETARY UNIT: POUND
REFINING CAPACITY: 1,788,500 B/CD
OIL PRODUCTION: 2.250 MILLION B/D
OIL RESERVES: 4.715 BILLION BBL
GAS RESERVES: 24.6 TCF

In a surprise move in April 2002, UK tax officials hit offshore producers with an immediate new 10% supplementary tax on North Sea oil and gas profits, while hinting at possible abolishment of the 12.5% overriding royalty for older fields. To encourage continued investment, 100% of most first-year North Sea upstream capital expenditures will be allowable for both the corporate tax and the supplementary tax for the year incurred, compared with 25% currently.

Then in November, the government kept its promise to abolish royalty on the UK North Sea's 30 oldest oil and gas fields, as of January 1, 2003. Like all British companies, oil and gas producers also pay corporate tax at the rate of 30% of company profits. The supplementary tax imposed in April had lifted the corporate tax rate for oil and gas producers to 40%. Wood Mackenzie said the cost to the government of royalty abolition was comparatively low for the benefit of restoring investor confidence in the North Sea fiscal system.

With royalty abolition, fields approved for development before March 31, 1982, will avoid the 12.5% royalty on the gross value of oil and gas, adjusted for costs of transportation, treatment, and initial storage. Royalty-paying fields include such venerable producers as Brent, Forties, Buchan, Claymore, Frigg, Heather, and Viking. Fields receiving development approval before March 16, 1993, pay petroleum revenue tax (PRT) at the rate of 50% of profit, field by field. Some fields pay both royalty and PRT.

The North Sea's Forties oil field was sold to Apache Corp. (Houston) in early 2003 by BP, which had held a 96.14% stake. Production from the field was 48,000 boe/d, said BP, and the transaction should take effect in mid-2003. The Forties pipeline was not included in the sale.

The acquisition signals a trend in development of the UK's North Sea industry, said Wood Mackenzie. Since UK oil production peaked in 1999, major players could increasingly divest their non-core assets and consolidate holdings in key areas, particularly around infrastructure hubs.

UK-Norway cooperation

The UK agreed with Norway to increase cooperation, cut costs, and raise output, especially from aging fields. However, taxation rates remained unharmonized. Under a contract signed in mid-2002, Norway's Statoil will sell 5 bcm/yr (186 bcf/yr) of gas to British Gas Trading Ltd.'s Centrica plc unit for 10 years beginning in October 2005. The sale could drive construction of more gas pipelines from Norway to the UK. BP also solicited interest in new Norway-

UK gas links, since the UK is expected to become a net gas importer in 2005.

A natural gas pipeline from Norway's Heimdal area to the UK's Bacton terminal was proposed as an open-access contract carrier. Marathon Oil Co. (Houston) held an open season during 2002 and said the 675-km (420-mile), 8.8 bcm/yr (900 MMcfd) Symphony line could begin operating in 2005. The pipeline would also pass through the UK Brae complex and near the UK Miller and Britannia complexes.

In November Marathon announced positive feedback from the open season, leading to a second phase of discussions with interested parties to determine how best to supply UK gas demand. Marathon received expressions of interest totaling 17 bcm/yr (1.65 bcfd) of term transportation capacity from prospective gas shippers on the pipeline.

The UK and Norway were also coordinating plans to decommission concrete substructures at Frigg, a giant gas field with associated condensate that straddles their two North Sea sectors. Given government approvals, Frigg could cease production as early as 2003, with decommissioning lasting another 8 years.

Offshore prospects

Even with high depletion rates and declining field size, the North Sea will continue to be a sizable producer of crude oil for the foreseeable future, said Simmons & Co. International (Aberdeen). As development shifts from large fields to small, the area is attracting more independent E&P companies, similar to how the US Gulf of Mexico has evolved.

This view was reinforced by Douglas-Westwood Associates (Canterbury, UK) and Infield Systems (London), which said that offshore Europe exhibits many of the characteristics shown earlier by the Gulf of Mexico — a great number of tiny, shallow-water prospects and a deep frontier. Considerable opportunities remain for players geared to the efficient production of small fields.

In March 2002 producers endorsed a UK plan designed to encourage development of the country's remaining North Sea reserves, estimated at up to 35 billion boe. A growing amount of that production is expected to come from independents, which typically are more willing to accept tighter margins.

The new rules set stricter investment deadlines on existing licenses and will expand information available to companies interested in acquiring a concession. The UK had 250 fallow discoveries and 200 unused licenses.

Blake Flank development

Blake Flank oil field in the UK North Sea will be developed by BG Group and partners, which received approval for the project in late 2002, only a year after drilling the final appraisal well. First oil production from Blake Flank, which is an extension of BG-operated Blake field in the same production license, is scheduled for late 2003.

Facilities will include two production wells and a single water injector tied into the existing Blake production manifold. Previous work includes the appraisal well, testing of two zones, and a sidetrack program. Due to the increased oil production rate of the combined development, the economic life of Blake field is expected to be extended by 2 years.

Blake Flank lies 103 km (64 miles) from Aberdeen in the Outer Moray Firth, directly northeast of, and immediately adjacent to, Blake oil field (originally known as Blake Channel), which started production in mid-2001. Blake field has estimated gross field reserves of 70 million bbl of oil and produced an average of 40,000 b/d during 2001.

Caister Murdoch development

Conoco (UK) Ltd. and partners successfully tested in August their first development well in the North Sea's Caister Murdoch System III, which comprises five natural gas fields. The six-well development scheme involves tying in the cluster of discoveries — Hawksley, McAdam, Murdoch K, Boulton H, and Watt — through subsea wellheads to the existing Murdoch platform.

The well, drilled horizontally through the reservoir in 18 m of water (60 ft) in Hawksley field, found more than 300 m (1,000 ft) of net pay in the Carboniferous Westphalian formation and encountered an additional reservoir zone that could affect field reserves.

Production from Murdoch K began in December, attaining 5.48 million cu m/d (204 MMscfd) on test, which was more than expected. Hawksley field came onstream in September with a sustained production rate of 4.6 million cu m/d (170 MMscfd). Another McAdam development well should begin producing in early 2003.

The five Caister Murdoch System fields, on Blocks 44/22a and 44/23a, hold an estimated 500 bcm of gas (19 tcf). In mid-2003, compression capacity will be doubled with installation of a new module.

Other E&P

In July partners in the Buzzard oil discovery in the UK North Sea completed the last of three appraisal wells and will begin development, said PanCanadian Energy Corp. (Calgary). The discovery, which lies in license areas P986/P928, contains more than 400 million bbl of reserves.

Earlier test wells confirmed the nature and extent of the field, while a third well established a significant reservoir section in license area P928. First production was expected in 2005-06. A 305 sq km (118 sq mi), proprietary 3D seismic survey of Buzzard was performed by WesternGeco in mid-2002.

Development of Helvellyn field on Block 47/10 was approved in August, with production expected to commence in 2003. ATP Oil & Gas (UK) Ltd. (Guildford) said a development well would be tied back to the Amethyst field via 16 km (10 miles) of subsea pipeline.

A gas accumulation was found on the South West Seymour prospect east of Armada field, said BG in November. The 22/5b-A12 discovery well encountered a 46-m (150-ft) hydrocarbon column and another oil column. The area will be drilled as part of BG's Armada development.

Tullich and Maclure fields started production in mid-2002 using existing infrastructure in Quadrant 9, reported Kerr-McGee Corp. (Oklahoma City). Both fields were developed as subsea tiebacks to the Gryphon A floating production, storage, and offloading (FPSO) vessel.

Tullich field, on Block 9/23a 5 km (3 miles) southeast of Gryphon in 110 m of water (370 ft), contains reserves of 40 million boe and was producing at a rate of 7,500 b/d from two wells in August. Production was expected to peak at 15,000 b/d from four wells in early 2003. Maclure field, on Block 9/19, began producing 12,000 b/d from one well in July.

Also in July Halley oil field on Block 30/12b began production at 15,000 b/d, said Talisman Energy Inc. (Calgary). An increase to a plateau of 20,000 b/d was planned. Halley holds 11 million boe of oil and gas reserves and has an anticipated field life of 5 years. It was developed with an extended-reach well from Shell's Fulmar platform, where the oil and gas is processed and exported. A water injection well was being drilled to provide pressure support to the Halley reservoir.

Earlier in the year Talisman also began producing 15,000 b/d from a single horizontal well in Hannay field on Block 20/5c. Hannay contains an estimated 10 million bbl and has a life expectancy of 8 years. Talisman also signed contracts in early 2002 for development of Kildrummy (formerly Lucy) field, which is being tied back 6.9 km

(4.3 miles) west to the Piper Bravo platform, also on Block 15/17.

Otter field began production in October and was expected to reach a plateau of 30,000 b/d by year-end 2002, said TotalFinaElf. Otter lies on Block 210/15a in 183 m of water (600 ft) about 150 km (90 miles) northeast of the Shetland Islands. Otter was being developed as a satellite of the Eider platform 21 km away (13 miles). Three producing wells and two water injectors were planned.

The first permanent pipeline system in the West of Shetlands area was commissioned in August as part of a miscible-gas enhanced oil recovery (EOR) project for Magnus field. The system carries gas from the Foinaven, Schiehallion, and Loyal fields to the Sullom Voe terminal in Scotland, where it is enriched with natural gas liquids. Another pipeline transports the enriched gas to the Magnus platform for injection. BP expects the $500 million EOR plan to increase Magnus reserves by 50 million bbl of oil and extend the field's life beyond 2015.

The high-temperature, high-pressure Jade field on Block 30/2c began production at 1.6 million cu m/day of natural gas (60 MMcfd) and 4,500 b/d of oil, said Phillips Petroleum Co. UK Ltd. in early 2002. Phillips and its partners expected Jade to peak at 5.4 million cu m/d of gas (200 MMscfd) and 16,000 b/d of oil in late 2002.

Plans were to drill the remaining development wells and further test the field's oil and gas potential with an exploration tail on one of the development wells. Jade was connected to Phillips's Judy platform 19 km south (12 miles) via a 16-inch multiphase pipe-in-pipe pipeline.

Composite materials were used to repair corrosion damage to a high-pressure, high-temperature carbon-steel gas compression line on the AH001 platform in Ivanhoe field. A welded repair was not an option because the plant was live, and complex geometries made a clamp impractical.

An offshore depth record was set for electromagnetic measurement-while-drilling telemetry, reported Shell UK Exploration & Production. The tool, deployed by Precision Drilling Corp. (Calgary), transmitted downhole-measured data during drilling operations from measured depth of more than 4,420 m (14,500 ft), or total vertical depth of 1,951 m (6,400 ft), in a southern North Sea well.

Block 42/27 in 55 m of water (180 ft) was awarded to a joint venture led by RWE Dea UK Ltd., as part of the UK's 20th round of offshore licensing in early 2002. Initial plans called for conducting a 3D seismic survey during the license's first 2 years.

In July the West Compton-1 well, which is the first to examine prospects updip from the giant Wytch Farm field in the Wessex basin, was spudded in License 048 by Bow Valley Energy Ltd. (Calgary).

Prospects near Saltfleetby gas-condensate field were being delineated by Roc Oil Ltd. (Sydney), which gathered 400 sq km (154 sq mi) of 3D seismic data in early 2002. If results are positive, Roc could begin a multiwell drilling program by 2003.

A coalbed methane pilot project will be developed jointly by StrataGas PLC (Nottingham) and the Halliburton (Houston) Energy Services Group. The four-well project, at Swallowcroft near Stoke-on-Trent, 220 km (137 miles) northwest of London, will prove the commerciality of Staffordshire field.

LNG prospects

UK imported LNG from Algeria from 1964 to 1994, when the Canvey Island terminal was dismantled. As North Sea production declines, however, the UK is again looking at LNG. In June Qatar Petroleum and ExxonMobil agreed to supply LNG to the UK, starting with shipments in 2006-07 for 25 years, with the gas coming from Qatar's North Field.

Under the agreement, two LNG trains were being developed at the Ras Laffan Industrial City LNG complex in Qatar. ExxonMobil was investigating numerous UK sites for the LNG import facility, including a 2.55 bcm/yr facility on the Isle of Grain in the Thames River to be developed with Lattice Group plc (owner of Transco, the UK gas network). Another site under consideration was Mildford Haven in Wales.

Cogeneration

In early 2002 a 730-MW cogeneration plant was announced by Conoco Global Power (UK) Ltd., to be built next to its 186,000 b/cd Humber refinery. Construction was to begin immediately and finish by 2004.

The cogeneration facility will supply steam and power to the refinery, steam to a neighboring refinery, and electricity to the grid. The plant will also make better use of certain refining by-products, reduce surplus gas flaring, and conserve water by using wastewater for cooling.

IPE

MIDDLE EAST

Bahrain

CAPITAL: MANAMA
MONETARY UNIT: DINAR
REFINING CAPACITY: 248,900 B/CD
OIL PRODUCTION: 174,000 B/D
OIL RESERVES: 124.56 MILLION BBL
GAS RESERVES: 3.25 TCF

Bahrain's monarch approved constitutional amendments that called for electing a parliament in October 2002, the first legislature in the country since 1975, and Bahrain became the second Arab state on the Persian Gulf currently to have a directly elected legislature (the other is Kuwait). Among the gulf Arab states, any move toward democracy is significant. Qatar, which has been rapidly modernizing, was considering a similar move.

In June the merger was completed between upstream Bahrain National Oil Co. (Banoco) and the Bahrain Petroleum Co. (Bapco), which took the name Bahrain Petroleum Company BSC and will perform all industry functions at home and abroad.

High-level discussions continued through late 2002 on the proposed delivery of natural gas from Qatar's North field to Bahrain, via either a spur from a proposed Qatar-Kuwait pipeline or a separate, independent line.

Natural gas liquids-related opportunities will be explored by Bahrain National Gas Co. (Banagas) and Dynegy Global Liquids Inc. in a joint venture formed in early 2002 and headquartered in Bahrain.

ChevronTexaco Corp. (San Francisco) was drilling an exploratory test in the gulf on Block 5 off Bahrain's east coast, said partner EnCana Corp. (Calgary) in November. ChevronTexaco acquired a 3D seismic survey on the 391 sq km (151 sq mi) block, where the company has an initial 3-year exploration period.

In early 2002, ChevronTexaco relinquished blocks 1, 2, and 3 off Bahrain. Exploratory drilling was scheduled for late 2002 in blocks awarded previously to Malaysia's Petronas Sdn. Bhd. and ChevronTexaco.

Bahrain worked with Saudi Arabia on a 50% production increase by 2004 at their jointly owned Abu-Safah field in the Persian Gulf, which was producing 140,000 b/d.

Iran

CAPITAL: TEHRAN
MONETARY UNIT: RIAL
REFINING CAPACITY: 1,474,000 B/CD
OIL PRODUCTION: 3,450,000 B/D
OIL RESERVES: 89.7 BILLION BBL
GAS RESERVES: 812.3 TCF

Iran's economy grew by 4.8% in real terms in 2001-02, though the country's oil output and gross domestic product (GDP) per person remained below historic levels. Iran holds nearly 10% of world oil reserves and more natural gas than any country but Russia, said *The Economist*, yet Iran's oil production of 3.4 million b/d was below its claimed capacity of 4.1 million b/d. The oil industry provides 10-20% of GDP, 40-50% of government revenue, and 80% of export earnings.

The US vowed not to soften its stance on trade with Iran and also decried Iran's position on sharing out the Caspian Sea area with its neighbors. The longstanding boundary dispute between Iran and Azerbaijan looked likely to continue beyond 2002.

Iranian and Norwegian officials met in November 2002 to discuss expanding cooperation between the two countries' oil sectors, possibly including joint energy projects and greater collaboration between Norway and the Organization of Petroleum Exporting Countries (OPEC).

Iran was granted observer status in December in the Energy Charter Conference (ECC), the governing body of an intergovernmental organization devoted to promoting East-West energy cooperation among European and Asian states. Member states already include all the countries of the former Soviet Union.

Iran is the only littoral state of the Caspian Sea that is not a member of the Energy Charter process. The charter is aimed at promoting market-oriented, non-discriminatory principles as the common basis for international energy relations.

Oil field development

Masjid-I-Suleiman oil field will be redeveloped by Sheer Energy Cyprus Ltd. under a buyback contract signed in March 2002 with Iran's government. In buyback deals, the contractor funds all investments and receives a share of production from National Iranian Oil Co. (NIOC), then transfers operation of the field to NIOC at the end of the contract. Sheer is a unit of Sheer Energy Inc. (Calgary). Repayment will be made from oil sales over 3 years.

Masjid-I-Suleiman, the first field found in the Middle East, originally held an estimated 6 billion bbl, of which 1.2 billion has been pumped. The producing horizon has 60-90 m (200-300 ft) of net oil pay at an average depth of 1,000 m (3,300 ft).

Sheer will invest $88 million (US) over 4 years for a reservoir simulation study, recompletion of four to six wells, drilling of two vertical wells and eight horizontal wells, and construction of processing and water reinjection facilities. The project would add 20,000 b/d to the current 4,500 b/d being produced by NIOC. In May 2002 the US planned to review Sheer's project to see whether it violates US sanctions.

Iran's offshore Foroozan and Esfandiar oil fields, which straddle the Saudi border, will also be developed under a buyback contract, signed in May between PetroIran and Iran's Oil Ministry. PetroIran will increase production at the two fields from 40,000 b/d to 109,000 b/d within 3 years.

Giant Azadegan oil field could be developed under a buyback contract with Japan that was scheduled to be signed by March 2003. Japan had loaned Iran $3 billion in exchange for the top negotiating spot for the Azadegan project. The second installment of $1 billion was disbursed in April 2002.

The 1,500 sq km (580 sq mi) field is in Khuzestan Province 80 km (50 miles) west of Ahwaz near Dash-e Azadegan. Production is expected to average 120,000 b/d for 7 years starting a year after contract signing and 350,000 b/d in a second phase.

Shell's development of two offshore oil fields came partially online in early 2002 and was expected to increase output to 150,000 b/d by 2003. Nowrooz was due to come online by year-end 2002 at 60,000 b/d, while Soroush will produce 195,000 b/d by 2004. The fields lie about 80 km (50 miles) west of Kharg Island and contain estimated recoverable reserves of around 1 billion bbl.

Rehabilitation of giant Parsi (Paris) oil field in the Zagros foothills of southwestern Iran was being studied in mid-2002 by Hydrosearch Group (Woking, UK) and Kanaz Engineering (Iran).

South Pars gas development

Iran's South Pars natural gas field, which lies in 70 m of water (230 ft) in the Persian Gulf, began production in March 2002 under Phases 2 and 3 of a $2 billion development program being operated by TotalFinaElf SA (France). Production from South Pars field could reach a total of 25 billion boe at full development over 20 years. Iran is speeding up the process for utilizing South Pars gas with a massive investment in the treatment complex, one of the largest gas construction projects in history, involving expenditures of $15-20 billion over a 5-7 year period.

Production from Phases 2 and 3 is expected to plateau at about 20 billion cu m (bcm)/yr of gas (2 bcfd) and 80,000 b/d of condensate from 20 wells tied into two unmanned platforms. Two 32-inch, 105-km (65-mile) pipelines carry gas and condensate to the Assaluyeh onshore facility, where the first gas processing train was commissioned and another three to come onstream before year-end. The Assaluyeh complex will also include export compressors, condensate stabilization and storage units, and sulfur recovery units.

Gas produced from South Pars will be used in Iran, with condensate to be exported from a buoy. However, a liquefaction plant at Assaluyeh, being considered by NIOC, BP plc (UK), and Reliance Industries Ltd. (India), could also use South Pars gas for LNG exports to Asian and European markets.

Previously, Phase 1 development was performed by Iran's state-owned PetroPars Ltd. Phases 4 and 5, to be handled by a unit of Eni SpA (Italy) and PetroPars, will come onstream in 2004-05 and are expected to have an eventual output of 80,000 b/d of condensate, 1 million tonnes/year (tpy) of LPG, and more than 1 million tpy of ethane. Gas processing facilities in Phases 9 and 10 will be built by a unit of LG Group (South Korea) and partners. South Pars could undergo as many as 12 development phases, which could include the LNG project. However, Assaluyeh has grown so fast that there is no longer room for an LNG terminal there, and new locations must be considered.

In November Statoil was tapped to operate offshore development in Phases 6, 7, and 8, as well as taking ownership interest and committing capital of $300 million during 2003-07. Production of condensate and LPG would begin in late 2004 and should reach 36 bcm/yr (3.7 bcfd). Most of the output (29 bcm/yr, or 3 bcfd) will be exported to other Iranian oil fields for injection as pressure support; the remainder will be sold.

Oil trades

To evade restrictions in dealing with Iran, several players send Caspian oil to Iranian refineries and receive Iranian oil from Persian Gulf terminals. Iran had already begun retooling its oil infrastructure to facilitate such swaps, including a 390-km (240-mile) pipeline from Iran's Caspian port of Neka to its upgraded Rey refinery near Tehran.

Iran expected to transport 175,000 b/d of Caspian crude by year-end 2002 and possibly up to 300,000 b/d by late 2003. Iran also was planning to boost capacity at its northern refineries to accommodate this oil.

Pipelines

Iran and Turkey officially inaugurated a much-delayed natural gas pipeline link in January 2002, which was to export a total of 225 bcm (8.4 tcf) of Iranian gas to Turkey over 25 years, beginning with 3 bcm (110 bcf) through year-end and rising to 10 bcm/yr (370 bcf/yr) by 2007. In case Turkish gas demand fails to keep pace with this supply, Iran agreed with Greece in March 2002 to extend the pipeline to Alexandroupolis, where the gas could flow into Europe.

Iran was also considering building a gas pipeline to India via Pakistan, as well as another line that would carry up to 1.5 bcm/d (56 bcfd) to Armenia. Iran and Armenia had agreed earlier to construct the line, of which 100 km (60 miles) is in Iranian territory.

During 2002 Iran imported about 6.5 bcm (242 bcf) of natural gas from Turkmenistan.

Iraq

CAPITAL: BAGHDAD

MONETARY UNIT: DINAR

REFINING CAPACITY: 417,500 B/CD

OIL PRODUCTION: 2,030,000 B/D

OIL RESERVES: 112.5 BILLION BBL

GAS RESERVES: 109.8 TCF

As 2002 ended, United Nations (UN) inspectors were searching Iraq for weapons of mass destruction and were due to report results in early 2003. The prospect of war overshadowed the oil market, said UBS Warburg LLC (New York).

If Iraqi exports are interrupted, replacement supplies from US and International Energy Agency (IEA) strategic stocks, along with higher OPEC supply, could replace Iraqi barrels and mitigate upward pressure on prices, UBS analysts said. However, OPEC members alone cannot replace supply losses from both Iraq and Venezuela, where widespread strikes in late 2002 had crippled oil exports.

Throughout most of 2002, Iraq's crude exports fluctuated on a downward trend, due to its own month-long embargo in April as well as pressure from the US and other countries to clamp down on Iraq's illegal surcharges. In September, however, Iraqi exports rose sharply to over 2 million b/d as President Saddam Hussein sought to win international support.

Also, Iraq reportedly was smuggling some 200,000-400,000 b/d to Syria, Turkey, Jordan, and Iran. Much of the illegal volume was piped from Iraq's northern Kirkuk fields to Syria's Mediterranean port of Baniyas. The Iraq-Turkey pipeline into Ceyhan was also operational at a reduced capacity of 900,000 b/d, Iraq's largest working pipeline. A parallel 500,000 b/d line reportedly was inoperable, and work to repair it was well behind schedule.

UN actions

In May 2002, the UN Security Council agreed to a "smart sanctions" plan designed to increase civilian goods to the Iraqi people by streamlining import controls. The protocol was predicated on the assumption that Iraq allows the UN to resume weapons inspections.

The smart sanctions retained the UN escrow account for Iraqi oil revenue and restrictions on items of potential military and military-related use. A complete lifting of the decade-plus economic sanctions would occur only when Iraq dismantles its weapons stockpile. However, the updated sanctions had no provisions to deter oil smuggling and illicit trade or to reintroduce weapons inspectors, according to an analysis by the US General Accounting Office.

In December the UN renewed the Iraq oil-for-aid program for 6 months. Because of US concerns, the council also agreed to review a restricted goods list to ensure that President Hussein does not use funds from oil sales to build up a military war chest.

Flurry of deals

During 2002 Iraq was poised to develop new oil fields and rework old ones. Over 30 deals were in place and ready to be implemented, according to *The Economist*. Deutsche Bank estimated that these contracts totalled nearly 50 billion bbl of reserves, 4 million b/d of potential production, and investment potential of over $20 billion.

Russian companies were planning to drill 70-100 wells in North Rumaila field and were rumored to hold oil concessions worth up to $90 billion, said *The Economist*. Iraq and Russia agreed on economic cooperation in energy and related sectors, a deal that could be worth as much as $40 billion.

OAO Lukoil (Russia) had held a majority stake in West Qurna, a huge field with 11-15 billion bbl. Lukoil, Zarubezhneft, and Machinoimport were to provide most of the $4 billion investment in West Qurna under a production sharing agreement (PSA).

However, late in the year Iraq canceled some of the planned work with Lukoil, and Russian officials met with Iraq in early 2003 to mend their commercial relationship. This resulted in new oil field development deals including a contract to develop Block 4.

In addition to Russian companies, French, Italian, and Spanish companies may have signed contracts to bring Iraq's oil to market. One enormous field slated for development is Majnoon, with reserves of 12-30 billion bbl. The field was producing 50,000 b/d in mid-2002. Iraq's oil minister projected output of 100,000 b/d.

Other foreign companies planning to develop Iraqi fields are based in China, Indonesia, and Malaysia, as well as Japan, Algeria, Turkey, Viet Nam, and other countries.

Iraq was also planning to repair its Mina al-Bakr and Khor al-Amaya terminals.

Although Iraq requires companies to start work immediately on its contracts, UN sanctions have prevented any efforts to do so, despite Iraqi threats. Russia's oil industry has pressured its government to secure promises from America to honor Russian contracts under a post-war regime.

Pipeline to Jordan

Iraq and Jordan plan to build an oil pipeline to transport Iraqi crude oil to Jordan starting in 2003, and Iraq will supply Jordan with all its crude oil supplies that year (4 million tonnes plus 1 million tonnes of oil products), reported OPEC News Agency in November. Jordan's oil purchases from Iraq are exempt from UN economic sanctions.

Work would start in early 2003 on the first section of the pipeline, which will extend from the Iraqi border to Jordan's 90,400 bbl/calendar day (b/cd) refinery at Zarka, northeast of Amman. Also in 2003 Iraq plans to begin building its section.

Israel

CAPITAL: JERUSALEM

MONETARY UNIT: SHEKEL

REFINING CAPACITY: 220,000 B/CD

OIL PRODUCTION: 100 B/D

OIL RESERVES: 3.81 MILLION BBL

GAS RESERVES: 1.375 BCF

The Israeli economy was expected to remain flat or even contract during 2002, due in part to heightened security.

A natural gas pipeline will be built from Israel's offshore Mari-B field to Ashdod, where it will begin supplying gas to a power station by year-end 2003. Noble Energy Inc. (Houston) obtained the pipeline license in May 2002 and reported that field development was on schedule. Mari-B, in 240 m of water (790 ft), 25 km (16 miles) from Ashqelon, is part of a hot E&D play in the eastern Mediterranean. Mari field, along with Noble's Noa gas field, contain more than 32 bcm (1.2 tcf) of reserves.

Noble has a contract to supply 17 bcm of gas (630 bcf) to Israel Electric Co. over 11 years. The gas line will handle up to 5.9 bcm/yr (600 MMcfd), anticipating rapid growth in Israeli demand. Construction of decks, production facilities, and the jacket was under way, with installation scheduled for early 2003 and completion of development and pipeline facilities by late 2003.

A significant Triassic reef trend in northern Israel could hold substantial gas reserves and several hundred million bbl of liquids, according to Zion Oil & Gas Inc. (Dallas), which holds permits in the Ma'anit License, which covers 117 sq km (28,800 acres, or 45 sq mi) and the Joseph area of 555 sq km (137,250 acres, or 214 sq mi). Seismic work has identified and delineated two specific reef prospects, and Zion planned to begin drilling in late 2002.

Offshore drilling was scheduled by Oil Fields Ltd. (Pakistan) for mid-2002, using a semisubmersible contracted with Atwood Oceanics Inc. (Houston).

Kuwait

CAPITAL: KUWAIT CITY

MONETARY UNIT: KUWAITI DINAR

REFINING CAPACITY: 889,200 B/CD

OIL PRODUCTION: 1,600,000 B/D

OIL RESERVES: 94.0 BILLION BBL

GAS RESERVES: 52.2 TCF

Kuwait remains one of the few oil-producing countries with significant excess production capacity. With oil accounting

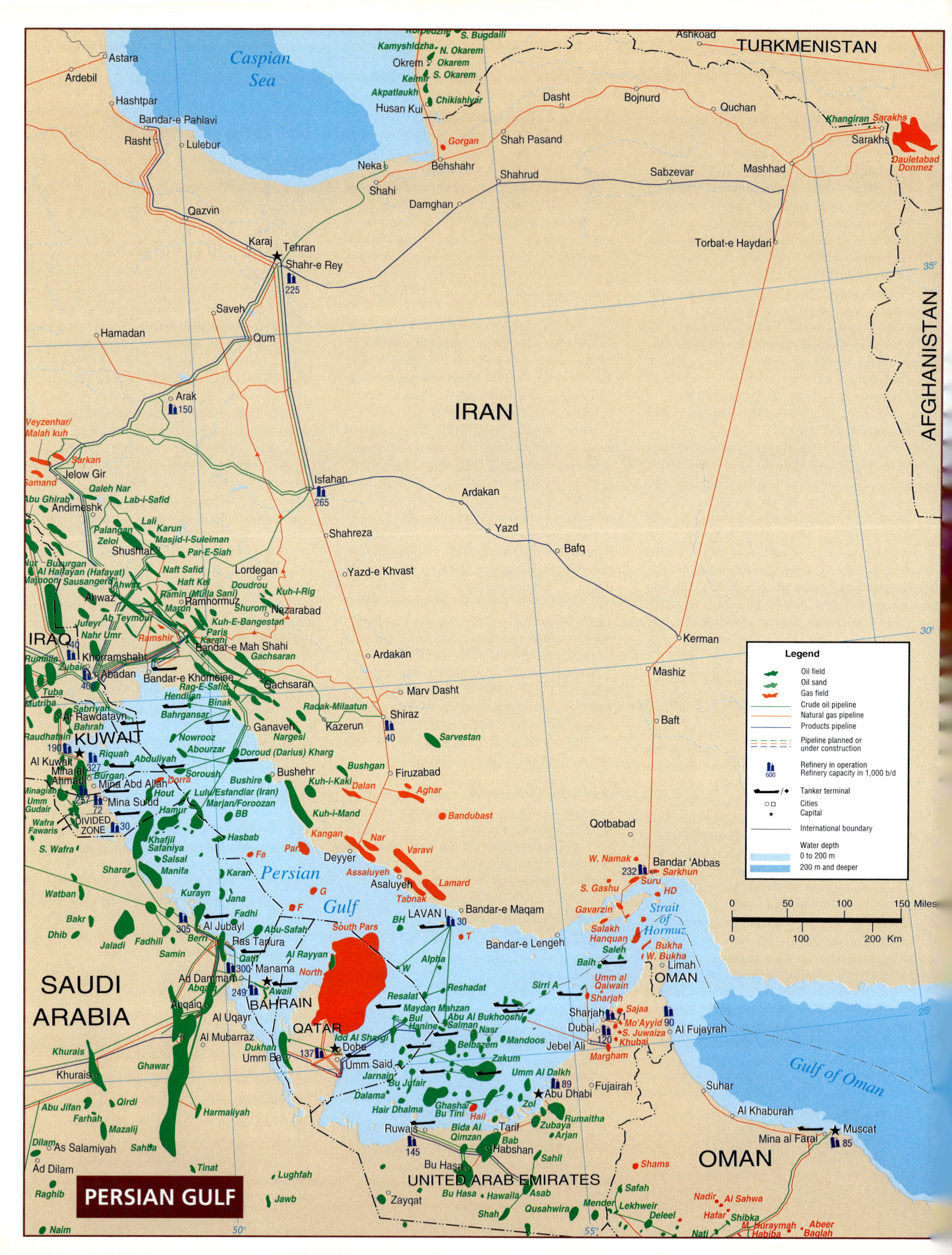

PERSIAN GULF
Caspian Sea
TURKMENISTAN
IRAN
AFGHANISTAN
IRAQ
KUWAIT
SAUDI ARABIA
BAHRAIN
QATAR
UNITED ARAB EMIRATES
OMAN
Persian Gulf
Strait of Hormuz
Gulf of Oman
Legend
Oil field
Oil sand
Gas field
Crude oil pipeline
Natural gas pipeline
Products pipeline
Pipeline planned or under construction
Refinery in operation
Refinery capacity in 1,000 b/d
Tanker terminal
Cities
Capital
International boundary
Water depth
0 to 200 m
200 m and deeper
0 50 100 150 Miles
0 100 200 Km
35°
30°
25°
50°
55°
Astara
Ardebil
Hashtpar
Bandar-e Pahlavi
Rasht
Lulebur
Qazvin
Karaj
Tehran
Shahr-e Rey
225
Saveh
Hamadan
Qum
Arak
150
Isfahan
265
Ardakan
Yazd
Bafq
Shahreza
Yazd-e Khvast
Lordegan
Nazarabad
Ramhormuz
Ahwaz
Shushtar
Andimeshk
Jelow Gir
Khorramshahr
Abadan
Bandar-e Khomeine
Bandar-e Mah Shahi
Gachsaran
Marv Dasht
Shiraz
40
Kazerun
Ganaveh
Bushehr
Firuzabad
Deyyer
Asaluyeh
Kerman
Mashiz
Baft
Qotbabad
Bandar 'Abbas
232
Bandar-e Maqam
Bandar-e Lengeh
LAVAN
30
Limah
Ashkoad
Dasht
Bojnurd
Quchan
Sarakhs
Mashhad
Sabzevar
Shahrud
Damghan
Shah Pasand
Behshahr
Neka
Shahi
Torbat-e Haydari
Okrem
Husan Kui
Al Kuwait
Mina al Ahmadi
Mina Abd Allah
Mina Su'ud
DIVIDED ZONE
Al Jubayl
Ras Tanura
Ad Dammam
Manama
Al Uqayr
Al Mubarraz
Umm Bab
Doha
Umm Said
Abqaiq
Khurais
As Salamiyah
Ad Dilam
Ruwais
Tarif
Habshan
Bu Hasa
Zayqat
Abu Dhabi
Jebel Ali
Dubai
Sharjah
Fujairah
Al Fujayrah
Suhar
Al Khaburah
Mina al Fahal
Muscat
Kamyshldzha
S. Bugdaili
N. Okarem
Okarem
Keimir
S. Okarem
Akpatlaukh
Chikishlyar
Gorgan
Khangiran
Dauletabad Donmez
Veyzenhar/ Malah kuh
Sarkan
Samand
Abu Ghirab
Qaleh Nar
Lab-i-Safid
Lali
Karun
Masjid-i-Suleiman
Palangan
Zeloi
Par-E-Siah
Naft Safid
Haft Kel
Ramin (Mulla Sani)
Marun
Kuh-I-Rig
Doudrou
Shurom
Kuh-E-Bangestan
Ab Teymour
Jufeyr
Nahr Umr
Ramshir
Paris
Karanj
Buzurgan
Al Halfayan (Hafayat)
Majnoon
Sausanger
Ahwaz
Rumaila
Zubair
Rag-E-Safid
Hendijan
Binak
Gachsaran
Radak-Milaatun
Sarvestan
Tuba
Mutriba
Sabriyah
Ar Rawdatayn
Bahrah
Raudhatain
Bahregansar
Nowrooz
Abourzar
Nargesi
Doroud (Darius) Kharg
Riquah
Abduliyah
Soroush
Bushire
Kuh-i-Kaki
Bushgan
Dalan
Aghar
Bandubast
Dorra
Burgan
Minagish
Umm Gudair
Hout
Lulu/Esfandiar (Iran)
Marjan/Foroozan
BB
Kuh-i-Mand
Hamur
Wafra
Fawaris
Khafji
Safaniya
Hasbab
Kangan
Nar
Varavi
Pars
Fa
Assaluyeh
Tabnak
Lamard
S. Wafra
Salsal
Manifa
Sharar
Karan
G
F
South Pars
Kurayn
Jana
Fadhi
Abu-Safah
Berri
Watban
Bakr
Dhib
Jaladi
Fadhili
Samin
Qatif
Abqaiq
Awali
Al Rayyan
North
BH
T
W
Alpha
Reshadat
Resalat
Maydan Mahzan
Bul Hanine
Abu Al Bukhoosh
Salman
Nasr
Belbazem
Mandoos
Zakum
Umm Al Dalkh
Sirri A
Sirri
Idd Al Shargi
Dukhan
Jarnain
Bu Jufair
Dalama
Hair Dhalma
Ghasha
Bu Tini
Hail
Zol
Rumaitha
Zubaya
Arjan
Bida Al Qimzan
Bab
Sahil
Bu Hasa
Hawaila
Asab
Shah
Qusahwira
Mender
Lekhweir
Safah
Deleel
Shams
Nadir
Al Sahwa
Hafar
Shibka
M. Huraymah
Habiba
Abeer
Baqlah
Nati
W. Namak
Sarkhun
Suru
HD
S. Gashu
Gavarzin
Salakh
Hanquan
Saleh
Baih
Bukha
W. Bukha
Umm al Qaiwain
Sharjah
Sajaa
Mo'Ayyid
S. Juwaiza
Khubai
Margham
Khurais
Ghawar
Qirdi
Abu Jifan
Farhah
Mazalij
Harmaliyah
Dilam
Sahba
Tinat
Lughfah
Jawb
Raghib
Naim
190
327
257
72
30
40
460
305
300
249
137
145
89
120
71
90
85

for 95% of exports, Kuwait's budget depends almost entirely on oil prices. Throughout most of 2002, Kuwait was producing well below capacity, but was still planning to expand production and increase foreign investment in certain fields under operating service agreements (OSAs), which allow Kuwait to retain full ownership and control.

A series of oil industry accidents during 2002 prompted the Oil Minister to resign and led to renewed efforts to repair damage inflicted by the 1990-91 Iraqi invasion. Analysts also blamed insufficient technical expertise and maintenance.

An explosion and fire at an oil-gathering center near the northern Raudhatain field in January 2002 killed 4 workers and injured 17, followed by an ammonia plant explosion in April injuring 6 and a substantial oil spill from a broken pipeline near the al-Ahmadi refinery in June.

The oil-gathering facility explosion also damaged a power plant and a gas booster facility, which led Kuwait to halt LPG exports to Asia in early 2002. Production of 600,000 b/d from the northern oil fields was expected to be restored before mid-year. Meanwhile, Kuwait was exporting oil from its 14 million bbl of storage.

The Kuwaiti-Saudi Neutral Zone oil fields will be operated by Kuwait Gulf Oil Co., established in March 2002 as a subsidiary of Kuwait Petroleum Corp. The new company will receive up to $1 billion of development financing from Japan's Arabian Oil Co., as well as training and technical assistance, beginning in 2003, when Arabian Oil surrenders its drilling rights in Khafji and Hout fields. The Japanese firm will receive rights to 100,000 b/d of Khafji crude, with an option for 70,000 b/d more.

Kuwait was planning to purchase natural gas from Qatar's North field through a proposed 7.8-13.7 bcm/yr, 590-km (0.8-1.4 bcfd, 370-mile) subsea pipeline that would take 3 years to build. However, following a protocol agreement in early 2002, a firm deal had not yet been signed. Kuwait also agreed to import gas from Iran, but the likelihood of both projects going forward was uncertain.

Production was restored and reserves increased from new and existing wells by stimulating the reservoirs with a polymer-free viscoelastic diverting acid, according to the results of 18 field trials in northern Kuwait's Mauddud formation.

During 2002 Kuwait National Petroleum Co. reported that its refinery in Mina Al-Ahmadi had increased capacity to 442,700 b/cd from 326,800 b/cd.

KUWAITI FIELDS

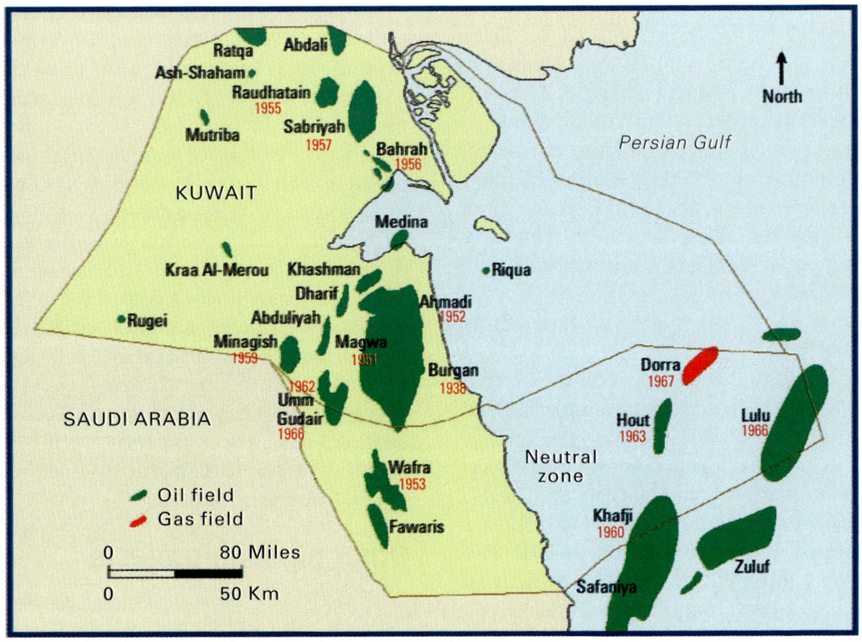

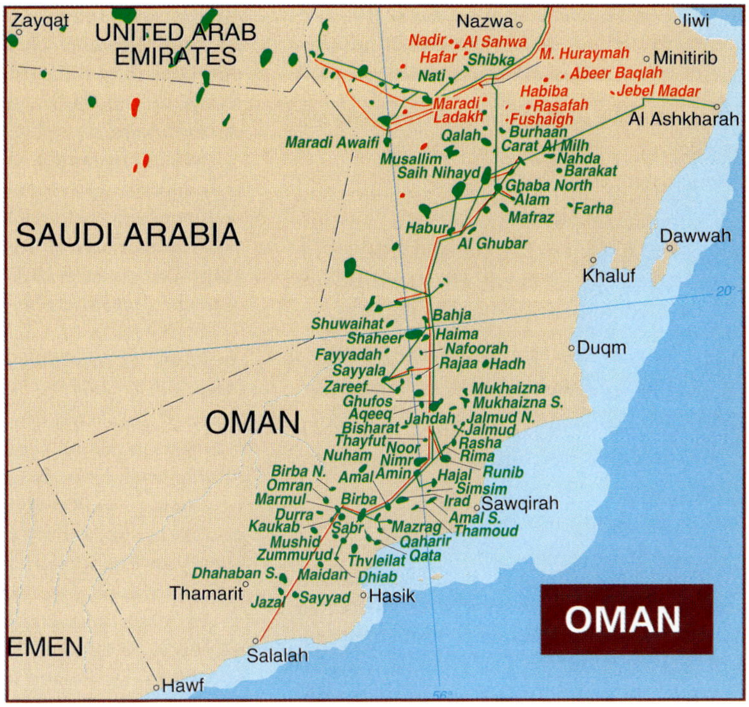

Oman

CAPITAL: MUSCAT

MONETARY UNIT: OMANI RIYAL

REFINING CAPACITY: 85,000 B/CD

OIL PRODUCTION: 895,000 B/D

OIL RESERVES: 5.506 BILLION BBL

GAS RESERVES: 29.28 TCF

LNG production in Oman nearly tripled in 2002 as its new plant ramped up. A third train was approved for the 6.6 million tpy plant in May, and sales agreements were signed with Union Fenosa SA (Spain) — its third supply contract with Oman — and with Gaz de France.

The third train, to be located next to the existing Qalhat facilities, will add production capacity of 3.3 million tpy and is expected to begin production in early 2006, increasing total output to about 10 million tpy. The LNG facility is operated by state-owned Oman LNG Co. A joint venture of Chiyoda Corp. (Yokohama) and Foster Wheeler Ltd. (Clinton, NJ, US) is developing the third train under terms of a contract awarded by Royal Dutch/Shell Group (Netherlands).

As of December 2002, Oman LNG had completely sold out its current LNG capacity to long-term purchasers and spot buyers. Existing contracts included:

- Union Fenosa, 1 bcm of LNG in 2004 and an additional 800 million cu m in 2005, to be transported by Oman LNG's Lakshmi carrier

- Gaz de France, 9 spot cargos of 138,000 cu m each for delivery to the Montoir-de-Bretagne terminal through year-end 2002
- TotalFinaElf, 600,000 tonnes in 2003 for delivery to European and American markets
- Tractebel SA (Belgium), medium-term supplies of 600,000 tonnes for sale into North America, Europe, and Asia, to be shipped by the newly chartered Hoegh Galleon and Excalibur LNG carriers
- Osaka Gas Co. (Japan), 700,000 tpy over a 25-year period (November 2000 contract)
- Korea Gas Corp., 4.1 million tpy over a 25-year period (April 2000 contract)
- Tokyo Electric, 2 two spot cargos.

Oman also continued to expand its natural gas pipeline network in 2002, with two lines completed in August, connecting central gas deposits to the coastal cities of Suhar in the north and Salalah in the south.

Exploration

A major gas field was found in early 2002 about 50 km (30 miles) south of Adam in the Al Dakhiliyah region, possibly the largest discovery in the last 6 years, said Petroleum Development Oman (PDO). Kauther-1 cut a thick interval of gas-bearing sandstone at 4,200 m (1,300 ft). PDO will run tests and conduct a seismic survey over nearby acreage.

Drilling began in October 2002 on the Tibat prospect in the Persian Gulf off the Musandam Peninsula, 12 km (7 miles) south of the Bukha field platform, said Heritage Oil Corp. (Calgary) and partner Novus Petroleum Ltd. (Sydney). The Tibat prospect, in 50 m of water (160 ft) on Block 8, targets the same producing zones as in Bukha field, which produces gas with high condensate yields.

In early 2003 Heritage reported that the Tibat-1 well was drilled to 3,097 m (10,161 ft), which was 250 m (820 ft) deeper than originally projected due to a considerably larger reservoir section, and it was logged successfully. The Mauddud and Thamama Cretaceous reservoirs are in a fault-bounded structure originally thought be at 2,550 m and 2,710 m, respectively (8,366 and 8,891 ft). Good shows were encountered while drilling the Thamama target reservoir, and its potential was supported by the logging data, so a testing program of the section is being prepared.

The Tibat-1 well apparently encountered a reservoir section that is two or possibly three times larger than originally mapped — a total of some 350-400 m (1,150-1,300 ft), said Heritage. However, the test program has to confirm the reservoirs' productivity before Heritage decides on field development.

Novus operates Bukha field, Oman's only offshore production, where a play fairway runs south into Oman. With the addition of the Ras al Khaimah block in the United Arab Emirates (UAE), Novus holds acreage along the play's entire length.

Exploration of 1,162 sq km (449 sq mi) on Block 44 near Safah field, about 300 km (185 miles) west of Masqat (Muscat), was licensed to Thailand's PTT Exploration & Production plc in August 2002, marking its Middle East exploration debut. Safah was producing 30,000-40,000 b/d. The Thai company's subsidiary planned a 2D seismic survey by year-end and $9 million in exploration spending over 3 years.

In Safah field, geosteering of a 1,436-m (4,712-ft), 6.125-inch slimhole lateral well succeeded, achieving a world first, claimed Occidental of Oman Inc. and Schlumberger Oilfield Services in June 2002. Steerable systems increase penetration rates and lateral length. Two more wells were planned.

Southeastern Oman's 11,500 sq km (4,400 sq mi) Block 34 will be explored by TotalFinaElf under a March 2002 agreement, which calls for geological and seismic work for the first 2 years. Block 51 in the Sharqiyah region was awarded to Hunt Oil (US) in May 2002, and a 50% stake in Block 5 was acquired by China National Petroleum Corp.

In February Mitsui & Co. Ltd. acquired a 35% interest in the producing Suneinah concession, which covers several oil fields.

Petrochemicals

A world scale ammonia urea fertilizer complex will be developed by a joint venture of Engro Chemical Pakistan Ltd. and Oman Oil Co. under an agreement signed in November 2002. Plans call for relocating and updating an existing ammonia plant and constructing a new, 650,000 tpy urea plant. The fast track project is targeting completion of detailed technical and economic feasibility studies by mid-2003.

The world's largest grassroots fertilizer complex will be built at Sur, 150 km (90 miles) south of Masqat, consisting of two 1,750 tonne/day (tpd) ammonia plants, two 2,350 tpd urea plants, and two granulation units. Construction by Technip-Coflexip Group (Paris) and Snamprogetti SpA (Italy) is slated for completion in mid-2005, under contract to Oman-India Fertilizer Co. The complex will be fed by Omani gas, with all of the urea output exported to India.

Qatar

CAPITAL: DOHA
MONETARY UNIT: RIYAL
REFINING CAPACITY: 200,000 B/CD
OIL PRODUCTION: 640,000 B/D
OIL RESERVES: 15.207 BILLION BBL
GAS RESERVES: 508.54 TCF

Qatar's economy continued to expand in 2002, with GDP growth estimated at 3.8% in real terms. The country also recorded a large budget surplus for 2001-02.

Dolphin gas development

The $3.5 billion Dolphin project, one of the Middle East petroleum industry's most ambitious undertakings, centers on developing natural gas and condensate in Qatar's offshore supergiant North field and exporting the gas to UAE under a long-term, phased expansion program.

In mid-2002 state-owned UAE Offsets Group (UOG) selected Occidental Petroleum Corp. (US) to assume a 24.5% interest in the Dolphin project, replacing bankrupt Enron Corp. as the third partner in Dolphin Energy Ltd., the project's leader. TotalFinaElf also holds a stake.

Dolphin Energy hopes to build, own, and operate a $1.5 billion export pipeline to transport 20 bcm/yr (2 bcfd) of dry natural gas to markets in the UAE for at least 25 years. The 420-km (260-mile), 48-inch pipeline will connect Qatar's Ras Laffan terminal to Tawilah in Abu Dhabi. The pipeline system's final capacity is designed for 31 bcm/yr (3.2 bcfd). Dolphin will supply natural gas initially to Abu Dhabi and Dubai, and eventually to the UAE's northern emirates.

In mid-2002 North field development drilling was underway, and engineering studies for the upstream and pipeline segments were in progress. Construction of production and processing facilities and the pipeline will begin in 2003, with production to start by year-end 2005.

Upstream engineering was being conducted by a joint venture of Sofresid, a subsidiary of Bouygues Offshore SA (France), and Foster Wheeler, under a contract signed in February 2002. The work includes two offshore production platforms, two subsea pipelines from the platforms to Ras Laffan, onshore gas processing and compression facilities in Ras Laffan, and two gas receiving and metering stations.

Other pipeline projects

High-level discussions continued through late 2002 on the proposed delivery

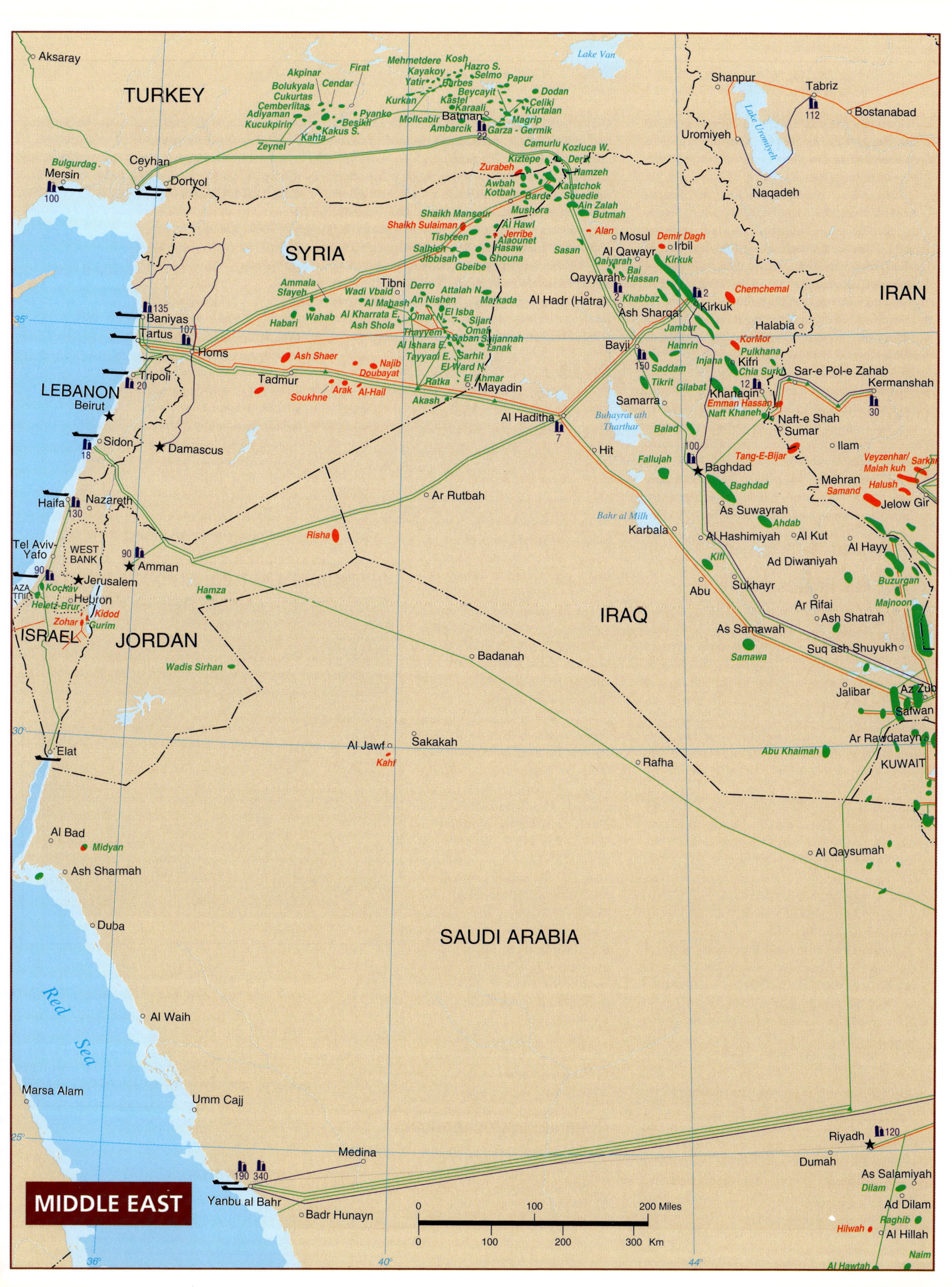

MIDDLE EAST
TURKEY
SYRIA
LEBANON
ISRAEL
JORDAN
IRAQ
IRAN
KUWAIT
SAUDI ARABIA
WEST BANK
Red Sea
Lake Van
Lake Uromiyeh
Buhayrat ath Tharthar
Bahr al Milh
Aksaray
Mersin
Ceyhan
Dortyol
Bulgurdag
Akpinar
Firat
Bolukyala
Cendar
Cukurtas
Cemberlitas
Adiyaman
Kucukpirin
Zeynel
Kahta
Kakus S.
Besikli
Pyanko
Mehmetdere
Kosh
Hazro S.
Selmo
Kayakoy
Yatir
Barbes
Papur
Beycayit
Kurkan
Kastel
Karaali
Dodan
Celiki
Kurtalan
Molicabir
Batman
Ambarcik
Magrip
Garza - Germik
Camurlu
Kozluca W.
Kiztepe
Deriki
Zurabeh
Hamzeh
Awbah
Kotbah
Karatchok
Barde
Souedie
Mushora
Ain Zalah
Butmah
Shaikh Mansour
Shaikh Sulaiman
Al Hawl
Jerribe
Alaounet
Tishreen
Salhieh
Jibbisah
Hasaw
Shouna
Gbeibe
Sasan
Alan
Mosul
Demir Dagh
Irbil
Al Qawayr
Qaiyarah
Kirkuk
Bai Hassan
Qayyarah
Khabbaz
Chemchemal
Ammala
Sfayeh
Tibni
Derro
Wadi Vbaid
Al Mahash
An Nishen
Attalah N.
Markada
Al Hadr (Hatra)
Ash Sharqat
Kirkuk
Wahab
Habari
Al Kharrata E.
Omar N.
Al Isba
Ash Shola
Sijan
Omar
Thayyem
Saban
Sajannah
Tanak
Al Ishara E.
Tayyani E.
Sarhit
El Ward N.
El Ahmar
Ratka
Akash
Mayadin
Baniyas
Tartus
Homs
Ash Shaer
Najib
Doubayat
Tadmur
Arak
Al-Hail
Soukhne
Tripoli
Beirut
Sidon
Damascus
Haifa
Nazareth
Tel Aviv-Yafo
Jerusalem
Amman
Kochav
Hebron
Heletz-Brur
Kidod
Zohar
Gurim
Hamza
Wadis Sirhan
Elat
Halabia
Jambur
KorMor
Pulkhana
Bayji
Hamrin
Injana
Kifri
Chia Surkh
Sar-e Pol-e Zahab
Kermanshah
Saddam
Tikrit
Gilabat
Khanaqin
Emman Hassan
Naft Khaneh
Samarra
Naft-e Shah
Sumar
Al Haditha
Balad
Tang-E-Bijar
Ilam
Hit
Fallujah
Baghdad
Veyzenhar/ Malah kuh
Sarkan
Mehran
Samand
Halush
Jelow Gir
Ar Rutbah
As Suwayrah
Ahdab
Risha
Karbala
Al Hashimiyah
Al Kut
Al Hayy
Ad Diwaniyah
Kifl
Buzurgan
Abu
Sukhayr
Ar Rifai
Majnoon
Ash Shatrah
As Samawah
Samawa
Suq ash Shuyukh
Badanah
Jalibar
Az Zubayr
Safwan
Ar Rawdatayn
Abu Khaimah
Al Jawf
Sakakah
Kahf
Rafha
Al Qaysumah
Al Bad
Midyan
Ash Sharmah
Duba
Al Waih
Marsa Alam
Umm Cajj
Medina
Yanbu al Bahr
Badr Hunayn
Riyadh
Dumah
As Salamiyah
Dilam
Ad Dilam
Raghib
Hilwah
Al Hillah
Naim
Al Hawtah
Shanpur
Tabriz
Bostanabad
Uromiyeh
Naqadeh
0 100 200 Miles
0 100 200 300 Km

of natural gas from Qatar's North field to Bahrain, via either a spur from a proposed Qatar-Kuwait pipeline or a separate, independent line.

Qatar also was planning to sell North field natural gas to Kuwait through a proposed 7.8-13.7 bcm/yr, 590-km (0.8-1.4 bcfd, 370-mile) subsea pipeline that would take 3 years to build. However, following a protocol agreement in early 2002, a firm deal had not yet been signed.

An offshore gas pipeline through Iranian waters to Pakistan was also in the works, as studies were completed in early 2002 by Crescent Petroleum. The proposed $3 billion, 16 bcm/yr (1.6 bcfd) pipeline would extend from Ras Laffan through Hamriyah, then Dibba on the Gulf of Oman, terminating at Gadani, Pakistan, said OPEC News Agency. The section between Hamriyah and Dibba would be onshore. However, the project has been under discussion for more than 10 years. Gas for the pipeline could be supplied by TotalFinaElf via its stake in the Ras Laffan LNG project in Qatar.

LNG, gas liquids projects

Qatar was the world's fourth-largest LNG exporter in 2001, with 11.5% market share, and Qatari output increased substantially during 2002. Qatar was the largest spot-market exporter in 2001, selling 2.8 bcm of LNG.

Qatar's Ras Laffan complex contains two existing LNG trains, which together produce 6.6 million tpy of LNG. The third and fourth trains were being built by Snamprogetti under contracts signed in 2001 and 2002. Each will produce 4.7 million tpy of LNG, making them the two largest trains currently being built. The third train will be completed in 2004 and the fourth in 2005. Rasgas's plant expansion is in line with Qatar's revised 2010 LNG production target of 40 million tpy, up from 30 million tpy.

Qatar Petroleum and ExxonMobil Corp. (Irving, TX, US) agreed in mid-2002 to supply LNG to the UK, starting with shipments in 2006-07 for 25 years, using North field gas. Additional tankers also have been ordered for increased delivery of LNG. Qatar Shipping Co. will acquire interests in several more LNG vessels in early 2003 and other ships in 2004, reported OPEC News Agency in November.

Conversion of North field gas to liquid products was being considered by Ivanhoe Energy Inc. (Vancouver), which launched a study in March 2002 to investigate Japanese markets for gas liquids. Ivanhoe has licensed gas-to-liquids (GTL) technology from Syntroleum Corp. (Tulsa). Ivanhoe was evaluating plants with capacities of 185,000 b/d of GTL fuels and 155,000 boe/d of natural gas liquids (NGL) products.

A GTL plant was also being developed at Ras Laffan by Qatar Petroleum Co. and Sasol Synfuels International (Pty) Ltd. (South Africa). The plant will be commissioned in 2005. Its slurry-phase Fischer-Tropsch reactor will be supplied by an alliance between Sasol Technology Pty. Ltd. (South Africa) and Japanese engineering firms Ishikawajima-Harima Heavy Industries Co. Ltd. and Nissho Iwai Corp.

Oil E&D

Block 10, which covers 3,170 sq km (784,000 acres, or 1,230 sq mi) in shallow water 100 km (60 miles) northeast of Doha, will be explored by Talisman Energy Inc. (Calgary) under a PSA announced in November, subject to government ratification. The block is between Al Khalij oil field to the east and to the west, Al Shaheen oil field and North gas field.

Talisman believes the block holds potential for discoveries in the 100-300 million bbl range. During the first exploration term of 5 years, Talisman will reprocess existing 2D seismic data, acquire new 2D and 3D seismic data, and drill three wildcat wells.

Al Rayyan oil field and other areas of Blocks 12 and 13 will be explored and operated by Anadarko Petroleum Corp. (Houston), which purchased BP's interest in mid-2002. Anadarko planned to boost its 2002 investment from $80 million to $102 million and increase production by installing a permanent production platform and drilling additional horizontal development wells in Al Rayyan field on Block 12. When the expansion project is completed in early 2003, gross production is expected to increase from 12,000 b/d to 35,000 b/d.

Qatar's offshore al-Khalij field increased production to 60,000 b/d in early 2002 when TotalFinaElf's capacity expansion was completed.

Refining, petrochemicals

Upgrading and expansion of the Umm Said refinery was completed in early 2002, increasing capacity to 137,000 b/cd from 57,500 b/cd.

The Q-Chem petrochemicals plant was also finished in 2002, and an expansion project was being planned.

Saudi Arabia

CAPITAL: RIYADH

MONETARY UNIT: RIYAL

REFINING CAPACITY: 1,745,000 B/CD

OIL PRODUCTION: 7,380,000 B/D

OIL RESERVES: 259.3 BILLION BBL

GAS RESERVES: 224.2 TCF

Repeatedly during 2002, Saudi Arabia insisted it would unilaterally make up any shortfall in global oil supply in the event of a US-led attack on Iraq. However, Saudi Arabia and other OPEC members would be unable to replace the loss of Iraq's exports plus output from Venezuela, which was virtually shut down in late 2002 by widespread strikes.

Saudi Arabia exports oil through terminals at Ras Tanura and Juaymah, with a combined loading capacity in excess of 8 million b/d, said J.P. Morgan Securities Inc. (London) in September. In addition, Saudi oil from Abqaiq and Ghawar fields moves through the 1,200-km (750-mile) Petroline pipeline to Yanbu on the Red Sea and north through the Sumed pipeline to the Egyptian port of Sidi Kerir. The loss of Ras Tanura terminal would effectively eliminate use of Saudi Arabia's spare production capacity, said J.P. Morgan.

KEY MIDDLE EAST OIL EXPORT FACILITIES

Source: J.P. Morgan Securities Inc.

Gas ventures

International oil companies were awaiting word at year-end 2002 on Saudi Arabia's

OPEC Timing Crucial to Balancing World Supplies

A general strike began in Venezuela in early December 2002, which slashed the country's oil production below 400,000 b/d from around 3 million b/d and removed about 50 million bbl of oil from world inventories that month, according to the US EIA. Venezuela is one of 11 OPEC members, along with Algeria, Indonesia, Iran, Iraq, Kuwait, Libya, Nigeria, Qatar, Saudi Arabia, and UAE.

Oil prices moved above $28/bbl in mid-December — outside of OPEC's $22-$28/bbl price band, based on a basket of 7 crudes — and reached 2-year highs late in the year. (The Nymex benchmark is usually about $2/bbl higher than OPEC's basket price.) However, during November and December, Saudi Arabia had increased its exports to the US. The arrival of this oil cushioned the loss of short-haul Venezuelan supplies and prevented further shortfalls and price increases, said EIA.

OPEC production and reserves

In mid-December, OPEC decided on a production ceiling of 23 million b/d for the OPEC 10 countries (excluding Iraq), effective January 1, 2003. However, rising oil prices and a drawdown in inventories, along with the continued threat of war in Iraq, prompted OPEC to increase oil output by 1.5 million b/d in January, bringing total quotas for 2003 to 24.5 million b/d. Timing of extra production is crucial to avoid OPEC shipments coinciding with restoration of Venezuelan supplies. OPEC promised to reconvene if too much oil came to market.

Saudi Arabia and other OPEC countries repeatedly reassured world oil markets during 2002 that OPEC has plenty of spare capacity to make up for interruption of Iraqi exports if war erupts. However, simultaneous loss of Iraqi and Venezuelan supplies would require tapping the US's SPR and other countries' strategic stocks. Iraqi oil-for-food exports officially averaged 1.27 million b/d during 2002, but were up to 1.5-1.7 million b/d late in the year (not counting illegal exports estimated at 200,000-300,000 b/d). Major non-OPEC producers were reportedly unable to boost output immediately.

EIA estimated OPEC's total production, including Iraqi exports, condensate, and other liquids, at 29.239 million b/d in fourth quarter 2002 and 28.3-30.3 million b/d in first quarter 2003. Saudi Arabia is the only country that can increase its sustainable production capacity quickly and substantially. The Saudis' maximum output, including surplus capacity of 2-2.5 million b/d, is estimated at 10 million b/d within 30 days and 10.5 million b/d within 90 days.

The share of world oil reserves controlled by OPEC members diminished by more than 10 percentage points in 2002 — not because their reserves decreased, but because Canada's jumped by a factor of 37, from less than 5 billion bbl to 180 billion bbl (reflecting the inclusion of oil sands). As a result, proved reserves attributed to OPEC amounted to 67.5% of all crude and condensate, down from the 2001 estimate of 79.4%, said *Oil & Gas Journal*. OPEC reserves are estimated at 819 billion bbl of the world's total of 1,213 billion bbl. OPEC also controls about 45% of global natural gas reserves, with 66.89 trillion cu m of a total 147.7 trillion cu m (2,491 of 5,501 tcf).

Inventories and demand

If Venezuela's production continues at year-end levels without being made up by other supplies, world oil markets would lose more than 70 million bbl per month, which would primarily have to be drawn down from stocks. Unless lost Venezuelan supplies are at least partially made up from other sources, oil inventories held by countries in the Organization for Economic Cooperation and Development (OECD) would spend most of 2003 near the bottom of their 5-year minimum/maximum range, said EIA.

Even if the strike is resolved quickly — which was anticipated by industry analysts in early 2003 — restoring Venezuela's full production would take several months, resulting in new 5-year lows in OECD inventories by spring 2003. With strike resolution plus a 1-month OPEC offset of 1-1.5 million b/d, OECD inventories could still be close to their 5-year minimum.

Although supply concerns about Venezuela and Iraq were dominating oil markets in late 2002, demand for OPEC oil will be driven by world oil demand, projected to grow 1.3 million b/d in 2003 and another 1.4 million b/d in 2004. EIA expects much of this new demand to be met by non-OPEC production, particularly from the Caspian Sea region.

IPE

proposed $25 billion program to develop and utilize its natural gas reserves. Seven foreign companies, most US-based, had signed preliminary agreements to develop three "core" ventures designed to direct new gas exploration and production into downstream Saudi petrochemical, power, and water projects.

ExxonMobil was tapped to oversee the first venture in South Ghawar field and the second at the Red Sea, both with several partners. Royal Dutch/Shell was slated to run the third core venture at Shaybah field in the Empty Quarter, also with partners.

Of the three prospects, the third venture remains the most promising, with a final deal expected in March 2003, said *Oil & Gas Journal* in January. The South Ghawar venture is also hopeful, with finalization anticipated by mid-2003, but prospects for the Red Sea project are cloudy, and it could be reopened for bid.

Development

In March 2002 Saudi Arabian Oil Co. (Saudi Aramco) awarded a major turnkey contract for its $1.2 billion Qatif development, located in the east near Dhahran. Technip-Coflexip will expand the Berri gas processing facility to handle increased output of sour gas from Qatif fields.

The work will include construction of a low-pressure gas sweetening unit, which will increase capacity by 6.7 million cu m/d to 23.4 million cu m/d (by 250 MMscfd to 871 MMscfd), and two new sulfur recovery units, which will increase capacity by 1,330 tpd to a total of 3,313 tpd. The Berri expansion began in 2002 and was scheduled for completion in late 2005.

Saudi Aramco intends to produce 500,000 b/d of Arabian light crude from Qatif field and 300,000 b/d of Arabian medium crude from Abu Safah field and process a respective 8.9 and 1.1 million cu m/d (330 and 40 MMscfd) of associated gas from these fields. At Khurais field, plans are to install four gas-oil separators at 200,000 b/d each.

Saudi Arabia was also in the midst of an aggressive nonassociated gas development program, which encompasses drilling new wells, developing known reserves in Ghawar and new fields, constructing new facilities, and upgrading and expanding existing facilities.

The 43 million cu m/d (1.6 bscfd) Hawiyah gas treatment plant was built as part of this program and was inaugurated in October 2002. Hawiyah, the first Saudi facility to process only nonassociated gas, receives sweet gas from the Jauf reservoir and sour Khuff gas. To maintain gas production from the Khuff carbonate reservoirs, Aramco initiated an acid fracturing campaign and has performed over 20 stimulations.

Haradh will become the second nonassociated gas processing facility when it is completed in mid-2003. Haradh will process 40 million cu m/d (1.5 bscfd) of gas from the south Ghawar area..

Oil production

Saudi Arabia worked with Bahrain on a 50% production increase by 2004 at their jointly owned Abu Safah oil field in the Persian Gulf, which was producing 140,000 b/d. Abu Safah and Zuluf offshore producing oil fields will be surveyed by Petroleum Geo-Services ASA (Oslo) to gather more than 1,000 sq km (400 sq mi) of data for Saudi Aramco. High-resolution two-component and four-component seafloor surveys will characterize the producing reservoirs and identify potential deeper targets.

The Kuwaiti-Saudi Neutral Zone oil fields will be operated by Kuwait Gulf Oil Co., established in March 2002 as a subsidiary of Kuwait Petroleum Corp.

An underground oil storage facility in Jeddah was formally inaugurated in August 2002. The facility has a capacity of 945,000 bbl of crude oil and refined products at five sites and is linked to refineries.

Petrochemicals, refining

Saudi Basic Industries Corp. (SABIC) continued work at its Al Jubayl petrochemicals complex, awarding a contract in June 2002 to Linde AG (Germany) to build an air separation unit that will increase air products capacity to 8600 tpd. Earlier, a $1 billion addition to the complex was announced, expected to start producing ethylbenzene, styrene, and other products in 2006.

In May, Gulf Farabi Petrochemicals Co. Ltd. named a Foster Wheeler subsidiary as project manager to oversee development of a $250 million petrochemical plant in Al Jubayl. The plant is slated to produce 120,000 tpy of n-paraffin and 70,000 tpy of linear alkylbenzene (LAB). About half of the n-paraffin will be dedicated to LAB production while the rest will be exported to Asia. The adjacent Aramco/Shell refinery will supply the kerosine feedstock. Commercial production will begin in late 2004.

At Aramco's Riyadh refinery, a sulfur recovery plant will be added by Technip-Coflexip. Work includes two new, 70-tpd recovery units, sulfur tankage, a new sour-water stripper, and a new amine-treating unit.

Syria

CAPITAL: DAMASCUS
MONETARY UNIT: SYRIAN POUND
REFINING CAPACITY: 239,865 B/CD
OIL PRODUCTION: 490,000 B/D
OIL RESERVES: 2.5 BILLION BBL
GAS RESERVES: 8.5 TCF

Syria's real GDP growth rate, estimated at 1.8% in 2002, was far below what the country would need to make economic progress, given its rapidly rising population.

Syria reportedly was receiving at least 100,000 b/d of crude oil smuggled via pipeline to the Mediterranean port of Baniyas from Iraq's northern Kirkuk fields, in violation of UN sanctions. Syria was considering building a new pipeline from Iraq that would be brought under UN sanctions.

Syria was expected to award five exploration areas to international oil companies during 2002 (Blocks II, IV, X, XII, and XIX). In September Syria's Oil Minister announced that a major gas field would be awarded by year-end, attracting an anticipated $800 million of investments.

In November Petrofac Resources International Ltd. (Sharjah, UAE) sold its bidding rights in three blocks to Stratic Energy Corp. Petrofac said it would continue to look for investment opportunities in Syria.

Earlier the assets of Veba Oil & Gas Netherlands BV were purchased by Petro-Canada (Calgary), including Veba's stake in a joint venture with Syrian Petroleum Co. and Shell, the Al Furat Petroleum Co., which produces most of Syria's oil.

United Arab Emirates

CAPITAL: ABU DHABI
MONETARY UNIT: DIRHAM
REFINING CAPACITY: 514,250 B/CD
OIL PRODUCTION: 1.9845 MILLION B/D
OIL RESERVES: 97.8 BILLION BBL
GAS RESERVES: 212.1 TCF

Although Abu Dhabi remained an exporter of natural gas and LNG during 2002, discoveries have not kept up with rising demand, reported FACTS Inc. (Honolulu), and Abu Dhabi needs gas over the short and medium terms. Some 500,000 tpy of LNG capacity, originally committed to the Dabhol project in India, remained available, and an additional 200,000-300,000 tpy could become available through debottlenecking. Pipeline exports of 4.9 bcm/yr (500 MMcfd) were flowing to neighboring Dubai and were expected to rise to 7.8 bcm/yr (800 MMcfd) around year-end.

To meet growing demand, Abu Dhabi will receive 10 bcm/yr (1 bcfd) from Qatar's Dolphin project, which will go to the city's power sector and other areas, and Dubai will receive 7.8 bcm/yr (800 MMcfd) of Dolphin gas. Conversion of industrial facilities from oil products to gas was continuing throughout UAE, which could require another 2.9-4.9 bcm/yr (300-500 MMcfd) of gas from Dolphin's Phase II.

In addition to Dolphin gas, UAE continued work on its $1 billion onshore natural gas development program. During the third phase under way in 2002, two processing trains are being added to the four trains constructed during the project's second phase. Under a May 2002 contract, the two trains will be designed and built by Bechtel, with completion scheduled for 2003.

Novus Petroleum obtained 100% interest in the 600 sq km (230 sq mi) onshore Ras al Khaimah block in July. Novus operates Oman's Bukha field, where a play fairway runs south into Oman. With the addition of the UAE block and several blocks in Oman, Novus holds acreage along the play's entire length.

A major gas injection project was underway in 2002 at Zakum oil field off Abu Dhabi. Gas from UAE's giant Umm Shaif field will be moved to the Zakum West Complex for injection. Abu Dhabi Marine Operations Co. (Adma-Opco) awarded the project to Halliburton KBR, which will install a 5.4 million cu m/d compressor (200 MMcfd) on the gas injection platform. The project completion date is 2004.

The Ruwais petrochemical plant began shipments in early 2002, producing its full capacities of 600,000 tpy of ethylene and 450,000 tpy of propylene (after running at 70% propylene capacity). The plant is operated by the Borouge joint venture of Abu Dhabi National Oil Co. (Adnoc) and Borealis. Borouge was exporting one-third of Ruwais output to the Middle East, including ethylene via tanker to Dubai's Jebel Ali.

Production of condensate and NGLs from Abu Dhabi plants will be increased by Foster Wheeler under a June 2002 contract with Abu Dhabi Gas Industries Ltd. The project will include two new gas plants at Habshan and Asab fields (OGD-III and AGD-II), addition of an NGL fractionation train at an existing plant at Ruwais, and an upgrade of existing NGL storage at the Takreer refinery at Ruwais. Following liquids removal, dry gas will be reinjected.

The Ruwais oil refinery was being expanded to 500,000 b/cd and equipped with units for producing unleaded gasoline and low-sulfur gas oil, under a contract between operator Abu Dhabi Oil Refining Co. and Technip-Coflexip. New equipment includes heavy and light naphtha hydrotreating units, a hydrogen sulfide removal unit and sulfur recovery units, a reformer, a gas oil hydrotreater, and an isomerization unit. The work was due for completion in mid-2005.

Yemen

CAPITAL: SANAA
MONETARY UNIT: RIAL
REFINING CAPACITY: 130,000 B/CD
OIL PRODUCTION: 350,000 B/D
OIL RESERVES: 4.0 BILLION BBL
GAS RESERVES: 16.9 TCF

Yemen entered the world's news spotlight in early October when an explosion and fire struck a 300,000-dwt, double-hull, very large crude carrier, killing 1, injuring 12, and leaving an 8,000-tonne oil slick. The 2-year old French supertanker was carrying 400,000 bbl from Saudi Arabia to Malaysia and was about to load another 1.5 million bbl at the Mina Al-Dabah terminal near Al Mukalla, Yemen, when the incident occurred.

Within days, French, US, and Yemeni investigators determined that the explosion's cause was a terrorist attack by a small boat seen speeding toward the tanker. These initial results were confirmed by France's Foreign Affairs ministry, and the investigation was continuing. Reportedly, responsibility for the attack was claimed by a Yemeni militant group, Islamic Army of Aden-Abyan. Analysts noted the potential impact on global oil markets if the industry becomes a terrorist target.

E&P

New oil pay was found in Tasour field by DNO ASA (Oslo) and TransGlobe Energy Corp. (Calgary) in October. Tasour 7 well, drilled on the 570 sq km (220 sq mi) Block 32 in the Masila basin, extended the main field pay south and is expected to increase the field's reserves. Additional appraisal wells were being considered, and 2 of 11 seismically defined prospects were ready for drilling.

An exploration well, Osaylan 1, was spudded on the Alif/Lam prospect by Vintage Petroleum Inc. (Tulsa) in October. Although it encountered hydrocarbons, results were disappointing. The well is on Block S-1 south of Halewah oil field. The rig moved on to drill the An Nagyah 2 exploratory well on a 3D seismic defined structure.

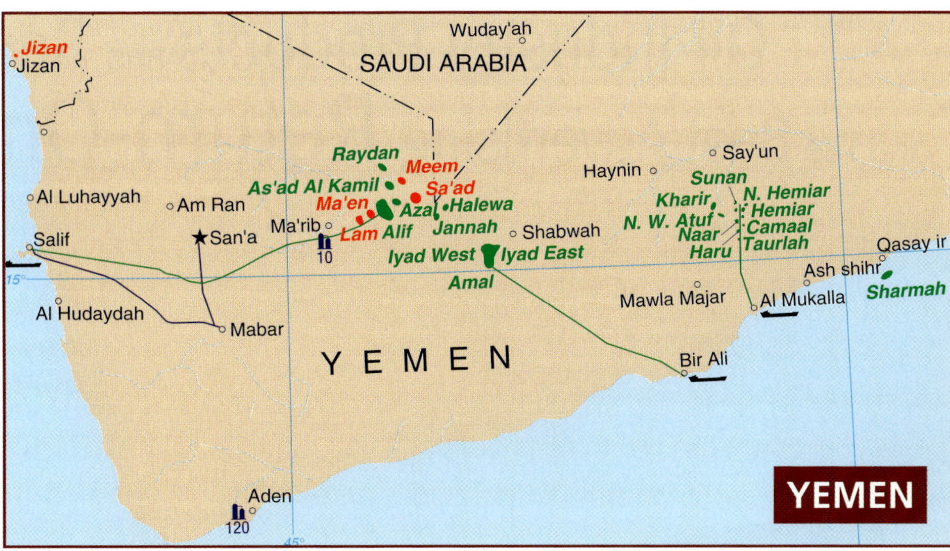

Exploration of the eastern al-Mahrah province was started by Rosneft (Russia) and Avirex (UAE), which planned to drill nine wells in 2003.

Dove Energy (UK) began producing 13,500 b/d in early 2002 from two wells in Hadramawt province.

IPE

AFRICA

Algeria

CAPITAL: ALGIERS
MONETARY UNIT: DINAR
REFINING CAPACITY: 450,000 B/CD
OIL PRODUCTION: 850,000 B/D
OIL RESERVES: 9.2 BILLION BBL
GAS RESERVES: 159.7 TCF

Algeria's state oil and gas company Sonatrach was being reorganized to separate the state's role from Sonatrach's role as a commercial enterprise. In late 2002 Algerian officials were debating a proposal to strip Sonatrach of its monopoly market powers, put it on a nearly equal competitive footing with foreign energy companies, and establish a new agency, ALNAFT, to award upstream contracts, approve development plans, and collect state royalties

This would relax Sonatrach's obligations, such as building pipelines, and free the company to compete for prospects with higher returns such as exploration and development. Also, Sonatrach plans to increase its non-Algerian assets to 30-40% from 10%.

Algeria has ambitious plans to boost natural gas exports, via pipeline and as LNG, to 83 billon cu m (bcm)/yr (8.5 bcfd) by 2005 from about 61 bcm/yr (6.2 bcfd) in 2002. Companies were bidding on an integrated project to develop reserves, lay a pipeline, and expand liquefaction facilities at Gassi Touil gas field 150 km (93 miles) southeast of Algeria's biggest oil field, Hassi Messaoud.

Algeria also plans to increase its crude oil production capacity to 1.5 million b/d by 2005. In late 2002, Algeria's production quota was increased to 735,000 b/d from 693,000 b/d at the meeting of the Organization of Petroleum Exporting Countries (OPEC), and in early 2003 OPEC increased total quotas. Sonatrach indicated that Algeria will continue to push for a higher production quota to match its existing capacity of 1.1 million b/d.

Five offshore terminals for loading crude and condensate will be installed by a unit of FMC Technologies Inc. (Houston) under a Sonatrach contract announced in December 2002. FMC Sofec Floating Systems will install two crude loading quays off the Mediterranean port of Arzew, two others in Skikda, and one in Bejaia. Construction begins in 2003 and is scheduled for completion within 24 months.

The oil terminals project would double Algeria's exports of liquid hydrocarbons to 220 million tonnes per year (tpy). Sonatrach said the new units would be able to receive large crude tankers with a capacity of 300,000 tonnes regardless of weather conditions. That would enable Sonatrach to supply distant markets in the US and, eventually, Asia. In 2002 Algeria was also producing more than 5 million tpy of LPG and exporting over 3.5 million tpy. By 2005, its LPG exports could exceed 8.5 million tpy, said Purvin & Gertz Inc.

Exploration

An appraisal well was spudded on Ledjmet block 405b, said First Calgary Petroleums Ltd. in November. The MLE-1 discovery well had production tested 1.15 million cu m/d of gas (43 MMcfd) and 1,745 b/d of condensate from three zones. The MLE-2 appraisal well, 2.5 km (1.5 miles) updip from MLE-1, reached a depth of 3,100 m (10,170 ft), and 7-inch production casing was being set. The well will be drilled to 4,450 m (14,600 ft) to evaluate all potential pay zones, including the Triassic TAGI formation, the Carboniferous F1 sandstone, and the Devonian. An extensive production testing program of all prospective zones will follow.

The north Reggane basin will be explored under a production-sharing agreement (PSA) by Repsol YPF SA (Madrid, Buenos Aires) and partners. The licensed area, covering 12,217 sq km (31,642 sq mi) on Blocks 351c and 352c, contains probable and potential reserves of 81 bcm (3.0 tcf) of gas. The initial 3-year work program includes acquiring seismic data and drilling one appraisal well and one exploration well.

The Sbaa basin in southwestern Algeria will be explored under a permit granted in July to Gaz de France (GdF). The Touat license, covering 15,392 sq km (5,943 sq

WEST AFRICA
Legend
Oil field
Oil sand
Gas field
Crude oil pipeline
Natural gas pipeline
Products pipeline
Pipeline planned or under construction
Refinery in operation
Refinery capacity in 1,000 b/d
Tanker terminal
Cities
Capital
International boundary
Water depth
0 to 200 m
200 m and deeper
0 300 600 Mi.
0 200 400 600 Km.
Atlantic Ocean
Gulf of Guinea
Bight of Biafra
Mediterranean Sea
Ionian Sea
MOROCCO
ALGERIA
TUNISIA
LIBYA
WESTERN SAHARA
MAURITANIA
MALI
NIGER
CHAD
SENEGAL
GAMBIA
GUINEA-BISSAU
GUINEA
SIERRA LEONE
LIBERIA
IVORY COAST
BURKINA FASO
GHANA
TOGO
BENIN
NIGERIA
CAMEROON
CENTRAL AFRICAN REPUBLIC
EQUATORIAL GUINEA
GABON
CONGO (FORMER ZAIRE)
CABINDA
ANGOLA
NAMIBIA
BOTSWANA
REPUBLIC OF SOUTH AFRICA
SICILY
GREECE
CRETE
Rabat
Mohammedia
Casablanca
Kechoula
Essaouira
Djebel-Jeer
Safi
Sidi Rhalem
Sidi-Kacem
Las Palmas
Sidi Ifni
Tarfaya
Oran
Arzew
Algiers
Skikda
Bejaia
Bizerte
Tunis
Athens
El Golea
Hassi Messaoud
Tripoli
Sider
Tobruk
Bel Rhazi
Touat
Ilatou
SBAA
Azzene
Reggane
In-Salah
In Amenas
Amasralad
Le Camp
Murzuk
Farrud
Umm Farud
Amal
Nafoora
Augila
Defa
Messla
Sarir
2 Lines
3 Lines
Nouakchott
Bamako
Niamey
Ouagadougou
Bissau
Conakry
Freetown
Monrovia
Abidjan
Accra
Lome
Porto Novo
Lagos
Kano
Kaduna
Makurdi
N'Djamena
Lake Chad
Yaounde
Bangui
Espoir
Panthere
Bellier
Foxtrot
Saltpond
Ceiba
Sao Tome
Libreville
Cape Lopez
Port Gentil
Tschengue
Rembo-Kotto
Asswe
Sette-Cama
Gamba-Ivinga
Mossaka
Brazzaville
Kinshasa
Pointe Noire
Nkossa
Emeraude
GCO
Essungo
Mavanga
Sulele
Guntala
Maleva North
Luanda
Benfica
Tobias
Lobito
Windhoek
Walvis Bay
Gaborone
Luderitz
Johannesburg
Kudu
Ibhubesi
Cape Town
Cape of Good Hope
Mossel Bay
Sable
F-A-
Oribi
60
300
34
125
31
60
88
30
115
8
220
27
110
10
15
65
45
118
150
60
42
17
21
15
39
88
105
600
30°
25°
20°
15°
10°
5°
0°
5°
10°
15°
20°
25°

EAST AFRICA
Mediterranean Sea
Red Sea
Gulf of Aden
Indian Ocean
Mozambique Channel
SICILY
GREECE
CRETE
TURKEY
CYPRUS
SYRIA
LEBANON
ISRAEL
JORDAN
IRAQ
IRAN
PAKISTAN
SAUDI ARABIA
OMAN
YEMEN REPUBLIC
TUNISIA
LIBYA
EGYPT
NIGER
CHAD
SUDAN
ERITREA
DJIBOUTI
ETHIOPIA
SOMALIA
NIGERIA
CAMEROON
CENTRAL AFRICAN REPUBLIC
CONGO (FORMER ZAIRE)
UGANDA
KENYA
GABON
CONGO
RWANDA
BURUNDI
TANZANIA
CABINDA
ANGOLA
ZAMBIA
MALAWI
MOZAMBIQUE
ZIMBABWE
NAMIBIA
BOTSWANA
SWAZILAND
LESOTHO
REPUBLIC OF SOUTH AFRICA
MADAGASCAR
Tunis
Tripoli
Cairo
Khartoum
Asmera
Addis Ababa
Djibouti
Mogadishu
Nairobi
Kampala
Kigali
Bujumbura
Dar es Salaam
Lilongwe
Lusaka
Harare
Maputo
Mbabane
Maseru
Pretoria
Windhoek
Gaborone
Antananarivo
Luanda
Kinshasa
Brazzaville
Libreville
Bangui
N'Djamena
Riyadh
Sanaa
Baghdad
Damascus
Amman
Jerusalem
Beirut
Nicosia
Athens
Kuwait
Manama
Doha
Abu Dhabi
Port Sudan
Mombasa
Durban
Cape Town
Mossel Bay
Walvis Bay
Luderitz
Kano
Makurdi
Lobito
Ndola
Beira
Mutare
Kilwa
Zanzibar
Toamasina
Cabinda
Alexandria
Suez
Asyût
Tobruk
Sider
Bizerte
In Amenas
Kassala
Kosti
Muglad
Melut
Bentiu
Aden
Ma'bar
Assab
Jizan
Mossaka
Thohoyandou
Johannesburg
Sasolburg
Umtata
Hassi Messaoud
Le Camp
Murzuk
Farrud
Umm Farud
Amal
Nafoora
Augila
Defa
Messla
Sarir
Alamein
Abu Qir
Abu Gharadig
Mango
Ramadan
Ras Gharib
Geisum
Midyan
Ghawar
Hilwah
Sheraf
Tabaldi
Heglig
Abu Gabra
Munga
Unity
Adar
Calub
Tschengue
Rembo-Kotto
Asswe
Sette Came
Gamba-Ivinga
Nkossa
Emeraude
GCO
Essungo
Mavanga
Sulele
Guntala
Maleva North
Luanda
Benfica
Tobias
Songo-Songo
Pande
Kudu
Ibhubesi
Sable
Oribi
F-A
Cape of Good Hope
Lake Chad
Lake Albert
Lake Victoria
Lake Tana
Lake Tanganyika
Lake Malawi
Lake Kariba
Ionian Sea
Aegean Sea
Persian Gulf
0 300 600 Miles
0 200 400 600 Km

mi) on Blocks 352a and 353, holds potential gas reserves estimated at 60-120 bcm (2.2-4.5 tcf). The area lies 200 km (125 miles) northwest of the Ahnet basin and an equal distance southwest of the In Salah gas region.

In central Algeria's Timimoun basin, TotalFinaElf SA (France) received an exploration permit in July covering 13,250 sq km (5,116 sq mi) on Blocks 325a-329. Permit partners plan to conduct 2D and 3D seismic surveys and drill three wells.

Field development

Southern Algeria's In Amenas natural gas fields will be developed by BP plc (UK) and Sonatrach, which contracted with JGC-KBR in November to build gas processing facilities, product pipelines, and basic infrastructure. First gas is expected in 2005.

The project, which includes four fields, will first develop Tiguentourine gas field 40 km (25 miles) southwest of In Amenas. Pipeline-grade gas, LPG, and condensate products will be piped 110 km (68 miles) for delivery to Sonatrach's pipeline grid at Ohanet, a gas and condensate field expected to begin production in 2004.

Oil recovery from giant Zarzaitine field in east-central Algeria will be enhanced by Sinopec Corp. (China) under a 20-year agreement. Zarzaitine is in the Illizi basin near the Libyan border. The field's original reserves were nearly 1 billion bbl of oil and almost 81 bcm of gas (3 tcf). Sinopec will increase the recovery factor to 50% from 40% by drilling 41 wells, replacing seven production centers, and installing a water treatment unit.

Rhourde Oued Djemaa oil field and five nearby oil fields in southeastern Algeria's Berkine basin will be developed jointly by Saipem SpA (Italy) and Bouygues Group (France), said Sonatrach in early 2002. With reserves of 300 million bbl, the six fields are to begin producing 80,000 b/d in 2004. The project includes installation of a central treatment unit, a gathering network, and oil storage tanks. Additionally, the companies will drill 20 production wells.

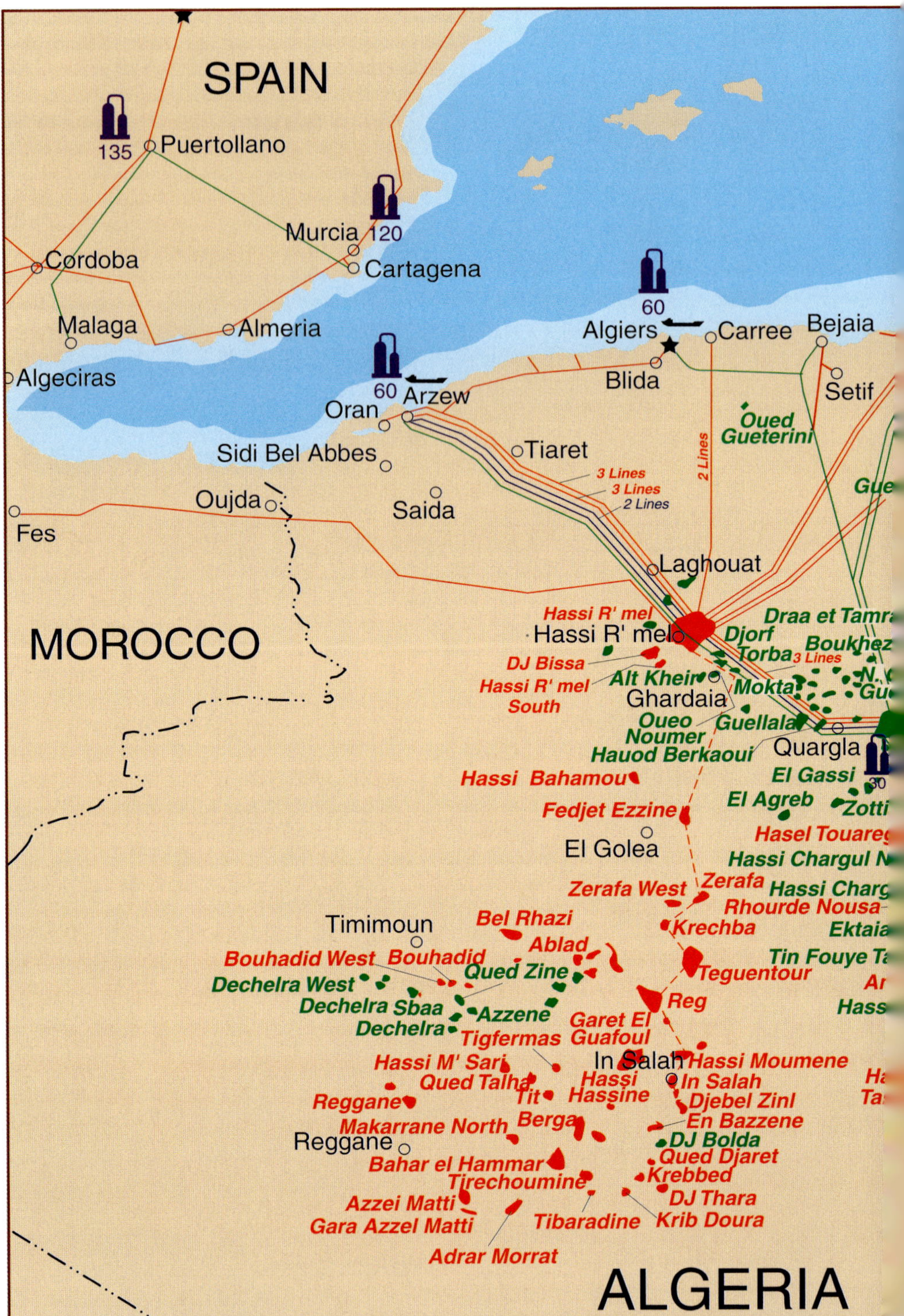

Production

Algeria's second-largest oil field, Ourhoud, started production of 75,000 b/d, said Sonatrach in late 2002, according to OPEC News Agency. Ourhoud lies in the Ghadames basin in northeastern Algeria, about 125 km (78 miles) west of the southern tip of Tunisia and about 250 km (155 miles) east-southeast of Hassi Messaoud. Ourhoud reportedly came onstream 1 month ahead of schedule. Sonatrach expects Ourhoud production to peak at 230,000 b/d by March 2003, boosting Algeria's total capacity above 1.3 million b/d.

At Hassi Berkine oil field's central processing facility, a fourth production train started up in April, said Anadarko Petroleum Corp. (Houston), marking completion of the project. The 75,000 b/d train ties in production from five satellite fields on Block 404 and will eventually increase output from the four-train facility to 285,000 b/d of oil.

Earlier, Anadarko began oil production at Hassi Berkine nearly 2 months ahead of schedule. A gathering system, injection lines, crude oil storage, and export facilities had been installed as part of the facility expansion. Hassi Berkine is on Blocks 403 and 404.

Also in the Berkine basin, BHP Billiton Ltd. (Melbourne) was working on an integrated oil development project that would begin production in 2003 and peak at 80,000 b/d.

Pipelines, LNG

Sonatrach was planning a feasibility

study of a second natural gas pipeline to Italy, and a similar study was under way on a second gas line to Spain. In a longer term project, Sonatrach also was working with Nigerian National Petroleum Co. to study the 4,000-km (2,500-mile) trans-Sahara pipeline, which would export Nigerian gas through Algeria's port of Beni Saf to Europe via Spain.

Algeria was the world's second-largest exporter of LNG during 2001, with a 16.5% market share, and also the second-largest spot-market exporter, selling 2.4 bcm.

An LNG master sales agreement was signed between Sonatrach and the Shell Gas & Power unit of Royal Dutch/Shell Group (Netherlands), the companies announced in September. The contract will provide a framework for the terms and conditions of LNG sales on an individual cargo basis. The two firms had also agreed to identify and develop upstream and downstream projects in Algeria and elsewhere.

Angola

CAPITAL: LUANDA
MONETARY UNIT: KWANZA
REFINING CAPACITY: 39,000 B/CD
OIL PRODUCTION: 900,000 B/D
OIL RESERVES: 5.412 BILLION BBL
GAS RESERVES: 1.62 TCF

Angola was positioned to become Africa's second-largest oil producer, as international petroleum companies planned to spend $20 billion on deepwater developments through 2006, said Centre for Global Energy Studies (London) in March 2002. Although Angola's oil production slipped in 2001, recent discoveries such as Girassol began boosting output in early 2002.

Four prolific blocks (1, 14, 15, and 18) could almost double the country's production by 2006. The Centre projected that Angola will reach an annual average level of 1.85 million b/d in 2007 and remain in the 1.9-2.0 million b/d range through 2010.

The government, to mitigate a potential boom-bust cycle, offered no new exploration licenses during 2002, but did ensure producers that no restrictions would be placed on existing offshore developments. According to the US Energy Information Administration, Angola's gross domestic product (GDP) was growing at an 11.4% annual rate as of mid-2002, and oil production could reach 1 million b/d in 2003.

Block 0 (Zero) production, including an estimated average 42,000 b/d from Nembe Norte platform during 2002, could be maintained or possibly raised to 600,000 b/d with further development and enhanced recovery. Operator Cabinda Gulf Oil Co. Ltd., the Angola affiliate of ChevronTexaco Corp. (San Francisco), plans to invest nearly $4 billion in field development over the next few years.

Exploration

The ninth significant discovery on Block 14, Negage, was announced in December 2002 by Cabinda Gulf Oil. The 4,040 sq km (1,560 sq mile) deepwater Block 14 lies off Angola's enclave of Cabinda. Negage was drilled in 1,444 m (4,738 ft) of water, 47 km (29 miles) southwest of the Gabela-1 discovery, an exploration well completed in August on Block 14 by Eni SpA (Rome).

Gabela-1, drilled in more than 320 m of water (1,000 ft), flowed more than 1,000 b/d during tests. The Negage well encountered a hydrocarbon column greater than 98 m (320 ft), producing more than 8,630 b/d of 33-degree gravity oil on test.

Oil was discovered on ultra-deepwater Block 31, 400 km (248 miles) northwest of Luanda, reported BP Exploration (Angola) Ltd. in September. The block lies in 1,500-2,500 m of water (4,900-8,200 ft) and covers 5,349 sq km (2,065 sq mi). Plutão-1, BP's second exploration well on the block, was drilled by a semisubmersible in 2,020 m (6,628 ft) of water and tested at a maximum rate of 5,357 b/d.

Plutão marks the first ultra-deepwater discovery following numerous deepwater finds off Angola. On the adjacent ultra-deepwater Block 32, the first well on the Gindungo prospect was to be drilled in late 2002.

ANGOLA'S ULTRA-DEEPWATER BLOCKS

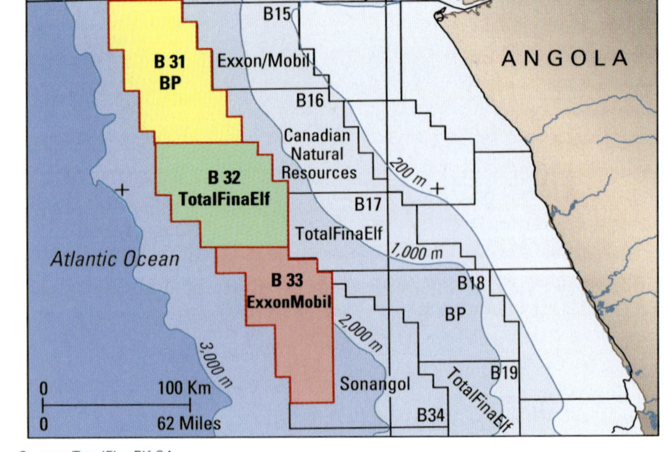

Source: TotalFinaElf SA

ANGOLA BLOCK 17

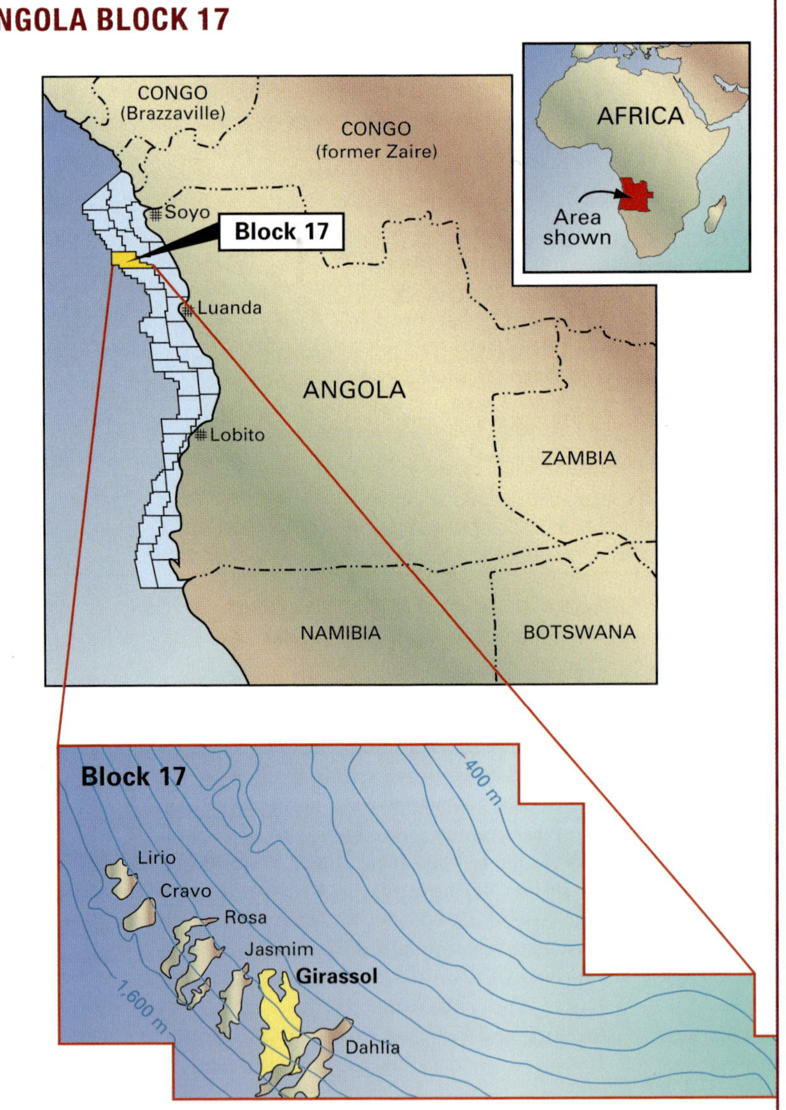

Block 16, covering 4,937 sq km (1,219,831 acres, or 1,906 sq mi) in 300-1,500 m of water (1,000-5,000 ft), will be explored by a subsidiary of Canadian Natural Resources Ltd. (Calgary) under a September PSA with Angola's state oil company Sonangol.

Two exploratory wells each on Blocks 24 and 19 were planned for 2002 by Ocean Energy, Inc. (Houston), which also planned to start shooting 3D seismic and drill exploratory wells in 2003 on Block 10, a 4,900 sq km (1.2 million acres, or 1,900 sq mi) concession. Block 10 lies in water as deep as 460 m (1,500 ft) in Benguela subbasin of Kwanza basin.

LPG FPSO planned

The world's first new-build LPG floating production, storage, and offloading (FPSO) vessel is slated for operation in Sanha gas-condensate field on Block 0C, said Cabinda Gulf Oil, which is leasing the unit from Single Buoy Moorings Inc. The FPSO should be completed in mid-2004 and begin shipping product in 2005.

With an LPG storage capacity of 135,000 cu m (4.77 MMcf) and a daily processing capacity of 6,000 cu m (210 Mcf), the Sanha LPG FPSO will be the largest LPG hull ever built and the first floating production facility to combine all LPG processing and export functions onboard the same unit. The LPG production plants will include gas separators, gas refrigerators, and boil-off gas reliquefaction units.

Mixed LPG, received from two platforms in Block 0, will be fractionated onboard to separate butane and propane. Each product will be chilled for storage in atmospheric-pressure tanks and periodically transferred to LPG export tankers. ABS Corp. will provide classification services. The Sanha project will greatly reduce the amount of gas flaring on Block 0.

Sanha's development will also include five offshore platforms with two linking bridges, modification of three existing platforms, and installation of 100 km (60 miles) of subsea pipelines (4-30 inch diameters) in 100 m (330 ft) of water. Cabinda Gulf contracted this work in June.

Block 17 plans

Girassol became the first deepwater field to produce on Block 17, one of several discoveries on the block that TotalFinaElf hopes to develop over the next decade. Girassol's production in early 2002 had already reached 200,000 b/d from eight wells. TotalFinaElf estimated that the project will cost $2.8 billion after drilling and connecting the rest of the planned 36 wells. Girassol, along with Dalia, Rosa, Lirio, Tulipa, Orquidea, Cravo, Jasmim, and Perpetua, contain an estimated 3.5 billion bbl of recoverable oil.

Jasmim is next in line for development, with production scheduled for late 2003. Subsea completed wells will be connected to the Girassol FPSO, the world's largest when installed in 2001. Five producing wells and five injectors are planned for Jasmim. Rosa, Lirio, and Cravo could also be connected to Girassol, with the longest tie-in at 30-40 km (20-25 miles).

Dalia is a larger field than Girassol but contains heavier oil. Its development plan, awaiting approval, would use a 240,000 b/d FPSO, with an anticipated production plateau of 225,000 b/d. Other specifications include a 400,000 b/d total liquid processing rate, 410,000 bw/d injection rate, 8 million cu m/d gas handling rate (300 MMcfd), and 2 million bbl oil storage capacity.

Dalia's base-case development scheme involves 67 subsea wells (34 producers, 30 water injectors, and 3 gas injectors). Production is expected by year-end 2005. Tulipa, Orquidea, and Perpetua could be tied into the Dalia FPSO.

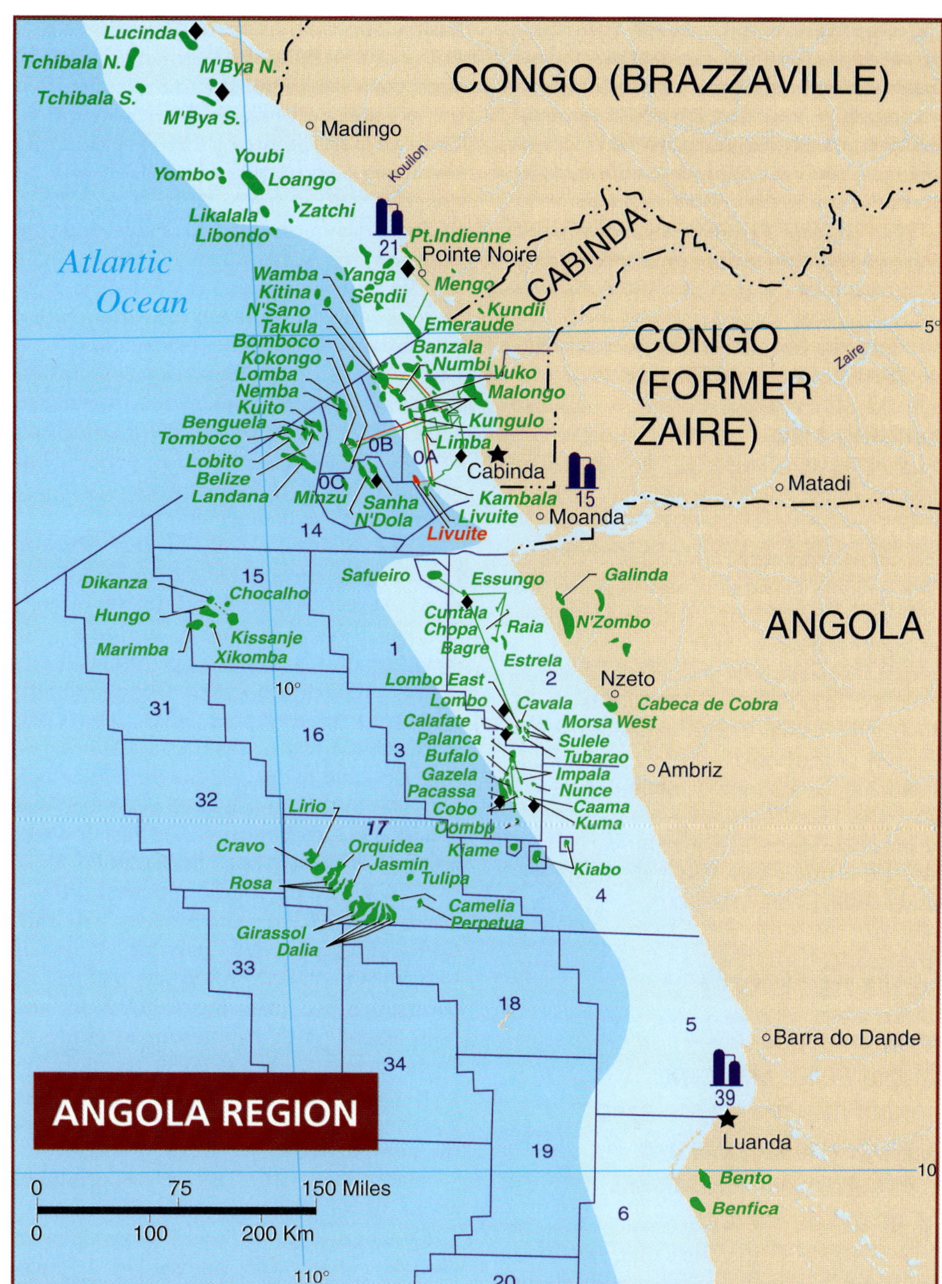

Block 15 development

The surface wellhead platform for the Xikomba project was being built during 2002. The platform, a new generation called the extended tension leg platform (TLP), will be installed in 1,177 m (3,863 ft) of water and will be the first floating dry-tree unit to operate off West Africa. ExxonMobil Corp. (Irving, TX, US) is Block 15 operator.

The Xikomba project, which targets Xikomba and Marimba fields, will contribute an additional 31,000 b/d of oil output to the Angolan total when it comes onstream in 2003, with production rising to a peak level of 80,000 b/d in 2004, said the Centre for Global Energy Studies.

Angola's oil production should average more than 1 million b/d in 2004 as the Kizomba A development on the same block begins producing from Hungo and Chocalho fields in 2004, yielding an additional 250,000 b/d at peak production.

A second FPSO vessel was ordered for the Kizomba B project, said ExxonMobil in early 2003. The Kizomba B development targets Dikanza and Kissanje fields. The $760 million FPSO will be built by Hyundai Heavy Industries Co. Ltd. (South Korea), with installation scheduled for mid-2005.

The FPSO consists of a hull designed to store 2.2 million bbl of oil, a topside with a production capacity of 250,000 b/d, refining facilities, and living quarters for 100 crew members. Hyundai said the FPSO will weigh 81,000 tons and measure 285 m long, 63 m wide, and 32 m high (935, 207, and 105 ft, respectively), similar in size and capacity to sister Kizomba A's FPSO.

The Kizomba A and B project, considered West Africa's largest deepwater development, is 370 km (230 miles) off Angola in water depths of 1,000-1,300 m (3,300-4,200 ft). First oil from Kizomba A is anticipated by the end of 2004, while production from Kizomba B is not expected to begin before the end of 2005. A third phase of the field's development, Kizomba C, is expected to follow 12-18 months after Kizomba B.

Refining, LNG

In October 2002 Sonangol said that work on developing a 200,000 bbl/calendar day (b/cd) refinery at Lobito on Angola's coast was proceeding smoothly. After a year's delay, construction was expected to begin by year-end 2003, with the refinery coming onstream by 2007.

An LNG facility is planned for Luanda adjacent to Angola's only existing refinery. The facility, which would process natural gas from several offshore blocks, will consist initially of a single 4 million tpy LNG train, with room for later expansion. The construction start date for the $2 billion project was pushed back to 2004, with service expected to begin in 2007.

Cameroon

CAPITAL: YAOUNDE
MONETARY UNIT: CFA FRANC
REFINING CAPACITY: 42,000 B/CD
OIL PRODUCTION: 69,000 B/D
OIL RESERVES: 400 MILLION BBL
GAS RESERVES: 3.9 TCF

During 2002 Cameroon's oil production continued to decline, averaging about 71,000 b/d in mid-year. Wood Mackenzie Ltd. (Edinburgh) predicted output will fall to 50,000-60,000 b/d by 2005. However, the government hoped for development of two largely unexplored basins, Logone Birni and Douala. Cameroon also received bids on developing the Sanaga Sud gas field, with awards expected by year-end 2002.

Cameroon and Nigeria both claim ownership of the Bakassi Peninsula, which could hold oil reserves. The countries had agreed to abide by the International Court of Justice's decision, but when it gave the territory to Cameroon, Nigeria rejected the ruling.

The World Bank decided in September to continue its endorsement of the 1,070-km (663-mile) pipeline from Chad's Doba basin oil fields to Cameroon's port of Kribi. Both countries allowed the bank to take an aggressive role in tracking oil revenue from the pipeline in order to deter corruption. The two host countries will have minority ownership in the pipeline, which will be financed with a minimum of $186 million in loans from the World Bank and the European Investment Bank.

The pipeline represents the largest infrastructure project of its kind in Africa, with a consortium of subsidiaries from ExxonMobil, ChevronTexaco, and Petronas making investments. The project is expected to generate some $500 million in revenue for Cameroon over its 25-30 year production period. The revenues, pegged to future oil prices, will represent 3% of the country's budget.

An exploratory well in the offshore Douala basin successfully flowed 3,000 b/d of 34-degree gravity oil and 48,000 cu m/d of natural gas (1.8 MMcfd) during a drill-stem test of a Lower Tertiary reservoir pay interval, reported operator ConocoPhillips (Houston) in December 2002. The Coco Marine No. 1 test marks the first successful flow of liquid hydrocarbons from a Tertiary reservoir in the Douala basin. The well was spudded in October in 23 m (75 ft) of water and was drilled to 2,627 m (8,620 ft).

The well's exploration permit PH 77 covers 1,093 sq km (2,830 sq mi). ConocoPhillips and partner Petronas Carigali Sdn. Bhd., the Malaysian national oil company, were analyzing well results and will work with Cameroon's National Hydrocarbon Corp. (SNH) to develop forward plans for evaluating the Coco Marine discovery and other identified leads within the permit.

Shell's Pecten Cameroon Co. produces about 30,000 b/d total fluid offshore from about 84 gas-lifted wells and two wells with electric submersible pumps on 11 platforms. Control and data acquisition at the offshore facilities was upgraded by integrating new software and installing wellhead control valves.

The Ngosse block in the Rio del Rey basin near Nigerian waters will be explored by Addax Petroleum Cameroon Ltd. and Tullow Cameroon Ltd. under a concession announced in December 2002. The 474 sq km (183 sq mi), shallow water block consists of the former MLHP-9 and PH-48a blocks. The companies plan to confirm the commerciality of the Narendi, Odiong, and Oongue discoveries and evaluate other prospects.

The partners will acquire 200 sq km (77 sq mi) of 3D seismic data in early 2003, and two commitment wells are set for 2004. Shallow deltaic Miocene targets and deeper Eocene turbidite plays are present, said Tullow. Miocene fields have been exploited in Cameroon since the 1970s, and modern seismic techniques have highlighted the possibility of larger and deeper structures similar to those discovered to the south off Equatorial Guinea.

Chad

CAPITAL: N'DJAMENA
MONETARY UNIT: CFA FRANC
REFINING CAPACITY: N/A
OIL PRODUCTION: N/A
OIL RESERVES: N/A
GAS RESERVES: N/A

The World Bank decided in September to continue its endorsement of the 1,070-km (663-mile) pipeline from Chad's Doba basin oil fields to Cameroon's port of Kribi. Both countries allowed the bank to take an aggressive role in tracking oil revenue from the pipeline in order to deter corruption.

The project is expected to generate $3 billion in revenue for Chad over its 25-30 year production period. The revenues, pegged to future oil prices, will represent about half of the national budget of Chad, one of the five poorest countries in the world.

The two host countries will have minority ownership in the pipeline, which will be financed with a minimum of $186 million in loans from the World Bank and the European Investment Bank. The pipeline represents the largest infrastructure project of its kind in Africa, with a consortium of subsidiaries from ExxonMobil, ChevronTexaco, and Petronas making investments.

The Doba basin's three fields (Bolobo, Kome, and Miandoun) will produce 900 million to 1 billion bbl of low-sulfur oil over the project life. Peak production is projected at 225,000-250,000 b/d from a total of 300 wells, which began drilling in late 2001. First oil is due in 2003.

Congo (Brazzaville)

CAPITAL: BRAZZAVILLE
MONETARY UNIT: CFA FRANC
REFINING CAPACITY: 21,000 B/CD
OIL PRODUCTION: 250,000 B/D
OIL RESERVES: 1.50591 BILLION BBL
GAS RESERVES: 3.2 TCF

Congo's political parties jousted over the government's management of the country's oil resources, and its state oil company, Société Nationale des Pétroles du Congo (SNPC) expressed disappointment that some exploratory wells had come up dry.

Production from Kouakouala oil field onshore near the border with Cabinda should begin to climb with completion of an export pipeline, said Heritage Oil Corp. (London) in April. The field rate averaged 1,238 b/d during 2001. The light oil is trucked to the Coraf refinery at Pointe-Noire. A seismic survey and more drilling are planned on the Kouilou license for 2002.

Gross field output at Kouakouala is expected to rise to more than 3,000 b/d once production starts from the Kouakouala 202 well in December 2002, said Heritage. The well, drilled to 1,710 m (5,610 ft), had a gross vertical hydrocarbon column of 51 m (167 ft) from Vandji sandstone of probable Lower Cretaceous age. The Kouakouala 202 well tested 40-degree gravity oil at a stabilized flow rate of 1,561 b/d.

Reportedly, ChevronTexaco and TotalFinaElf were considering joint exploration along the border between Angola and the Republic of Congo (Brazzaville). The two nations have agreed to explore the area jointly to avoid conflict over reserves rights.

Congo (Former Zaire)

CAPITAL: KINSHASA

MONETARY UNIT: ZAIRES

REFINING CAPACITY: 15,000 B/CD

OIL PRODUCTION: 23,000 B/D

OIL RESERVES: 187 MILLION BBL

GAS RESERVES: 35 BCF

Along Congo's eastern border with Uganda, exploration has focused on the Albert graben, where Uganda's Block 3 was licensed to Heritage oil subsidiary Heritage Oil & Gas Corp. (London). The Ugandan government has made significant efforts to stabilize its western border with Congo, although security remains a matter of concern.

Signing of the Lusaka Accord, a treaty among factions fighting in eastern Congo, along with the Congo's change of presidency in 2001, has stimulated cooperation between the two countries on border security, improving stability. As a result, Heritage has been able to undertake exploration activities in the Albert graben area.

Egypt

CAPITAL: CAIRO

MONETARY UNIT: POUND

REFINING CAPACITY: 726,250 B/CD

OIL PRODUCTION: 750,000 B/D

OIL RESERVES: 3.7 BILLION BBL

GAS RESERVES: 58.5 TCF

During 2002 new data received by *Oil & Gas Journal* resulted in substantial increases in Egypt's estimated proved reserves, with natural gas rising by 66%, to 1,571 bcm from 945 bcm (to 58.5 tcf from 35.18 tcf) and oil rising by 25%, to 3.70 billion bbl from 2.948 billion.

Gas in place at Ha'py field was reevaluated at 54.9 bcm (2.045 tcf), compared with the original estimate of 26.6 bcm (989 bcf), reported BP Egypt Gas Business Unit (Cairo) and Baker Atlas (Houston) in October 2002. Ha'py is in the Pliocene producing trend in the frontier development area on the Nile Delta's outer shelf.

The Nile Delta and Mediterranean off Egypt are emerging as a giant natural gas province, situated ideally for regional markets. The deepwater trend along Egypt's north coast could hold largely untapped potential, said BP Egypt, despite the area's already explosive growth in discovered gas reserves. More than 100,000 sq km (38,600 sq mi) was being explored off the Nile Delta, which accounts for nearly all of Egypt's gas discoveries.

Apache's E&P

Apache Corp. (Houston) budgeted $145 million for exploration in Egypt during 2002 and has nearly 10,000 sq km (3,861 sq mi) of 3D seismic, most of it recent. The company has averaged about 40 wildcats annually for 10 years.

Deepwater gas

Four deepwater gas discoveries on the West Mediterranean (Block I) concession were announced in 2002, moving Apache toward its goal of establishing at least 80 bcm (3 tcf) of gross natural gas reserves on the block's deepwater portion, enough to support development. The concession encompasses 31,200 sq km (12,050 sq mi) within the gas-prone Nile Delta.

Abu Sir-1X, located 68 km (42 miles) off Egypt in 992 m of water (3,255 ft), tested at 467,000 cu m/d (17.4 MMcfd) under non-ideal conditions. The well's absolute open flow potential could be 2.43 million cu m/d (90.4 MMcfd), Apache said in May.

About 16 km (10 miles) southwest of Abu Sir, Apache found gas with its Al Bahig-1X well in water depth of 1,070 m (3,510 ft). Wireline logs and pressure data indicated reservoir quality as good as or better than that encountered in Abu Sir, so the discovery was not tested, said Apache in July.

The third deepwater gas discovery in 2002 was El Max-1X, in 945 m of water (3,100 ft) 9 km (5.5 miles) south of Abu Sir. The El Max potential pay interval is similar to Abu Sir and Al Bahig, but with a larger interpreted gas column, Apache reported in November.

Apache's fourth consecutive natural gas discovery was announced in December, also in the block's deepwater portion. El King-1X, drilled 46 km (28.5 miles) offshore in water depth of 720 m (2,361 ft), tested at 832,000 cu m/d (31 MMcfd) of gas and 757 b/d of condensate. Because the test rate was limited by surface test equipment, an absolute open flow rate of 10 million cu m/d (372 MMcfd) was calculated based on the wellhead pressures and the gas flow rate. Apache plans to drill a fifth well to evaluate correlative Miocene intervals at depths approaching 3,400 m (11,000 ft).

The El King-1X well found not only significant gas reserves in both the primary objective Miocene age formation and the secondary Pliocene objective, but also the first Miocene deepwater oil in the Nile Delta, said Apache.

Also, the discovery well tested rich gas in the Miocene's middle Abu Madi formation. A test conducted in a 2.4-m (8-ft) interval in the Abu Madi at 2,359-2,361 m (7,738-7,746 ft) flowed at a maximum of 2,630 b/d of 32-degree gravity oil and 32,000 cu m/d of gas (1.2 MMcfd). A second test, in an 11.6-m (38-ft) interval at 2,336-2,348 m (7,664-7,702 ft), flowed 832,000 cu m/d (31 MMcfd) and 757 b/d of condensate.

In late 2002 RWE Dea AG (Hamburg) increased its share of the West Mediterranean (Block I) concession to 35% by purchasing interest from BP Egypt, subject to government approvals.

Onshore E&P

Apache also explored onshore during 2002, resulting in several discoveries. On the western desert's Khalda concession, which covers 9,300 sq km (2.3 million acres, or 3,600 sq mi), Ozoris-1X flowed at 2,504 b/d early in the year. The discovery is 10 km (6 miles) from the nearest well and 14 km (9 miles) from the nearest field.

In June Apache announced its second Khalda oil discovery, Selkit-1X, which flowed 5,103 b/d from high-quality Kharita sands pay. Selkit represents an entirely new play on the concession, and Apache hopes that it might extend beneath Renpet field, 6 km (4 miles) west of the discovery. Selkit also set a record turnaround time of 2 months between seismic acquisition and development of a completely new prospect.

Two more Khalda discoveries were completed successfully, said Apache in July. Tut 51 flowed 3,200 b/d of oil and 26,900 cu m/d of gas (7.7 MMcfd). Tut 52, which tested at 784,000 cu m/d (29.2 MMcfd) and 781 b/d of condensate, was drilled to extend the limits of the Khatatba reservoir on the southern flank of Tut field. A 6.8-km (4.2-mile) pipeline was being built to connect Tut 52 to the Salam gas plant.

In December gas and condensate were discovered on Apache's Ras El Hekma concession 29 km (18 miles) northeast of the prolific Khalda area. The Emerald-1X well, which logged a total of 66.4 m (218 ft) of net pay in multiple Alem El Bueib reservoirs, flowed 454,000 cu m/d of natural gas (16.9 MMcfd) and 4,285 b/d of condensate. The well tested a structure of 5 sq km (1,200 acres, or 2 sq mi) and was drilled to 3,792 m (12,440 ft).

Apache expected Emerald-1X to be on production by year-end 2002, with gas and condensate going to the gas plant in Tarek field about 8.5 km (5 miles) away. A new 3-D seismic interpretation indicates that the discovery has significant development and delineation potential, which Apache plans to test with a second well, Emerald-2X.

Apache began producing oil and gas in April from its Ras Kanayes lease at rates of 2,130 b/d of crude and condensate and 478,000 cu m/d of gas (17.8 MMcfd). Ras Kanayes, which covers 314 sq km (77,690 acres, or 121 sq mi), is also part of Apache's Khalda concession operations.

On another western desert concession, the 5,700 sq km (1.4 million acre, or 2,200 sq mi) East Bahariya, Apache drilled the Southeast Karama-1X discovery well, which flowed on test as a rate of 1,140 b/d. The well is in the Abu Gharadig basin, about 1.2 miles southeast of Karama field.

Production from Karama and Southeast Karama fields was averaging 1,800 b/d of oil, including the discovery. Six additional prospects have been identified in the area.

WEST DELTA DEEP CONCESSION

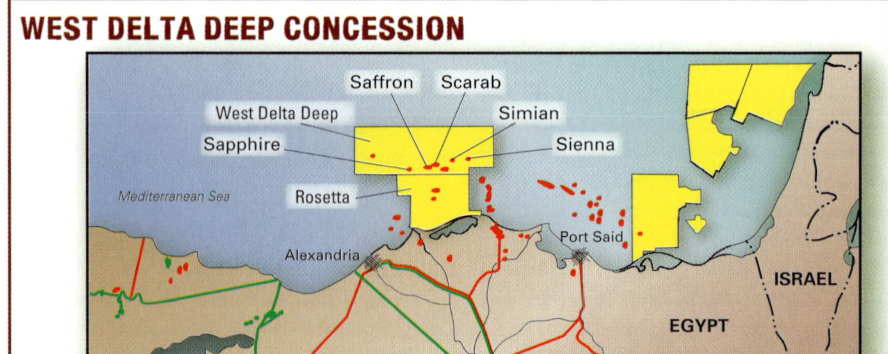

West Delta Deep development

Egypt's deal to sell 3.6 million tpy of LNG to GdF, signed in early 2002, spurred plans for an LNG export plant at Idku, east of Alexandria. BG Group plc (UK) is leading the project through its BG Egypt unit, which is partner in the Egyptian LNG joint venture with Egyptian General Petroleum Corp., Egyptian Natural Gas Holding Co., Edison International SpA (Italy), and GdF. Plans were to begin producing LNG from the first train in mid-2005.

The offshore West Delta Deep concession will provide gas to the project and holds enough reserves to supply a second train. Scarab and Saffron fields are the first deepwater gas development off the Nile Delta and Egypt's largest gas field development. Scarab and Saffron's gas production, expected to be about 5.9 bcm/yr (600 MMscfd) for the first 4 years, will go to the local Egyptian market, while the other fields could feed the LNG project.

West Delta Deep is being explored and developed by Burullus Gas Co., an international joint venture. BG and partners drilled four exploration and appraisal wells in Egypt during 2002 and conducted a 1,200 sq km (463 sq mi) 3D seismic survey over the eastern portion of West Delta Deep.

Eight subsea wells were being completed at Scarab-Saffron in water up to 622 m (2,040 ft) deep, with first gas scheduled for early 2003. Plans call for four more wells. Development was challenging due to large variations expected in the wells' production rate, from 8% to 100% of system capacity, which could cause high transient liquid flow rates.

Also, the design had to allow for future tie-ins from additional fields. Burullus plans to produce Simian and Sienna initially with six wells, increasing to 18, in water depths up to 1,100 m (3,600 ft), with start-up expected in mid-2005. Sapphire will be developed third and start up in 2006.

SCARAB/SAFFRON DEVELOPMENT

As a result, the Scarab-Saffron development incorporates a long, 90-km (56-mile) tie-back using large-diameter flowlines, with ODs ranging up to 36 inches to accommodate future flow rates as low as 4 million cu m/d (150 MMscfd).

Other gas projects

Another Egyptian LNG project was also in the works during 2002, being propelled by Union Fenosa (Spain) through its SEGAS unit, which is building an LNG complex at Damietta port with a capacity of nearly 5 million tpy, the largest ever for a single train. The plant, being constructed by a Halliburton (Houston) unit, should begin operating in late 2004. Spain's power plants will consume most of the gas.

Also, Ivanhoe Energy Inc. (Vancouver) is planning a 90,000 b/d gas-to-liquids (GTL) plant, probably to be sited on Egypt's Mediterranean coast.

A natural gas liquids extraction plant was being built at Port Said by United Gas Derivatives Co., an international consortium. The plant's molecular sieve gas dehydration package, to be delivered in mid-2003, will treat 29.5 million cu m/d of natural gas (1,100 MMscfd) and will be the largest molecular sieve dehydration unit ever built in Egypt. The plant will produce about 330,000 tpy of LPG, 280,000 tpy of propane, and 1 million bbl/yr of condensate, as well as ethane for petrochemicals manufacture.

An acid gas treatment plant will be built for Alexandria Mineral Oil Co. by Technip-Coflexip Group (Paris) under a turnkey contract announced in December 2002. The Alexandria facility, to start up in August 2003, will produce 12 tonnes/day (tpd) of sulfur and will be the first plant built to use Thiopaq process technology on an industrial scale. The technology converts acid gas from the desulfurization processes into elementary sulfur by a biological process; the sulfur produced is water-soluble and therefore has considerable benefits for agricultural applications.

Equatorial Guinea

CAPITAL: MALABO

MONETARY UNIT: CFA FRANC

REFINING CAPACITY: 0

OIL PRODUCTION: 135,000 B/D

OIL RESERVES: 12 MILLION BBL

GAS RESERVES: 1.3 TCF

Oil production resumed in early 2002 from Ceiba field on Block G at a rate of more than 50,000 b/d, said Amerada Hess Corp. (New York). The field's new FPSO, Sendje Ceiba, has onboard processing capacity of 160,000 b/d of liquids and a water injection capacity of 135,000 b/d.

Building on Ceiba production, Amerada Hess reported a string of oil discoveries in Block G off Equatorial Guinea during 2002, following its late 2001 Oveng-1 and Oveng-2 discovery and appraisal wells on Block G and its Ebano discovery in the Rio Muni basin on Blocks F and G.

Akom was discovered in February in 443.8 m of water (1,456 ft). The G-7 well found 49.4 m (162 ft) of net oil pay. Then in March, the G-8 well in 64 m of water (210 ft) discovered 47.9 m (157 ft) of net oil pay, called the Elon discovery, in the Rio Muni basin. Elon is 24 km (15 miles) northeast of Ceiba field and 10 km (6 miles) southeast of Akom. The G-8 well established an extension of the play fairway into the shallower water portion of the Rio Muni basin, said Amerada Hess.

A significant oil discovery dubbed Abang was announced in June. The G-10 exploration well in 98.5 m of water (323 ft) encountered 51.8 m (170 ft) of net pay, 47.9 m of oil and 4.0 m of gas (157 ft and 13 ft, respectively). The discovery is 4 km (2.5 miles) northwest of Elon.

The Elon appraisal well, drilled in water depth of 50.3 m (165 ft), encountered 96.3 m (316 ft) of net oil pay in a single, continuous column, Amerada Hess reported in September. The successful appraisal increases the areal extent and size of Elon field and will be incorporated into the development plan that Amerada Hess will submit to the Equatorial Guinea government for approval. In addition to Elon, that submission includes development plans for Okume, Oveng, Ebano, Akom, and Abang fields in the northern part of Block G.

In November Amerada's G-13 wildcat on Block G encountered 77 m (251 ft) of net oil pay over a 294-m (963-ft) interval. The well was drilled to 4,187 m (13,737 ft) in 1,001 m of water (3,284 ft) in the southern part of the block 16 km (10 miles) south of Ceiba field. Wireline sampling recovered 34-37 degree gravity oil and indicated good reservoir characteristics. After technical review and evaluation, Amerada Hess plans an appraisal well in 2003 that also will explore deeper objectives.

Other exploration

Seismic acquisition began in mid-2002 on Blocks I and J near Bioko Island, and interpretation of a new 3D seismic survey of Block H was under way. Block H lies on trend with Amerada Hess's discoveries on Blocks F and G.

Corisco Bay Block N in the Rio Muni basin will be explored by a Petronas affiliate under a PSA signed in early 2002, joining block partner Ocean Energy. Block N covers 2,740 sq km (678,000 acres, or 1,060 sq mi) in as much as 200 m of water (660 ft).

Zafiro development

Subsea pipelines and systems were ordered for installation in Zafiro field off Bioko Island as part of the Southern Expansion development project, said an ExxonMobil subsidiary in May. First production from the project is due in late 2003. The Southern Expansion area of Zafiro field lies on Block B, about 65 km (40 miles) northwest of Malabo in water depths of 300-

FIELDS OFF EQUATORIAL GUINEA

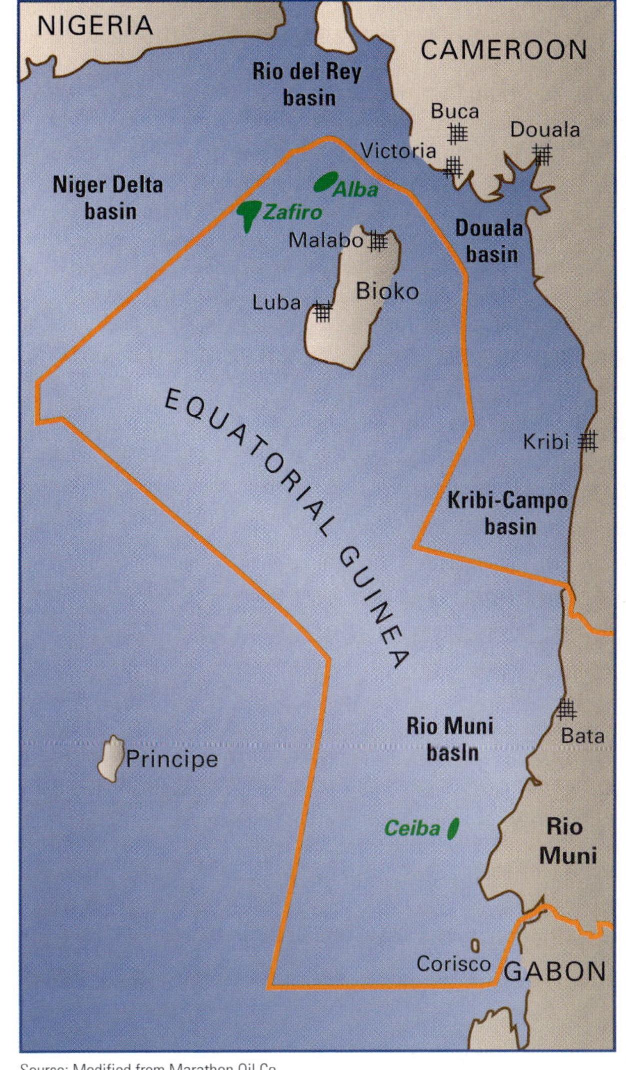

Source: Modified from Marathon Oil Co.

850 m (1,400-2,800 ft). The project will recover about 150 million bbl of oil, in addition to existing production of 150,000 b/d.

New development will consist of subsea wells tied back to an FPSO vessel capable of processing 110,000 b/d of oil and having about 2 million bbl of storage capacity. A total of 19 subsea wells are planned, and drilling is expected to continue until 2004. Subsea lines totaling 48 km (30 miles) will be installed by Technip-Coflexip. Subsea systems including 19 trees and 5 production manifolds will be installed in 430-800 m of water (1,411-2,625 ft) by an FMC Technologies unit.

Gabon

CAPITAL: LIBREVILLE

MONETARY UNIT: CFA FRANC

REFINING CAPACITY: 17,300 B/CD

OIL PRODUCTION: 294,000 B/D

OIL RESERVES: 2.499 BILLION BBL

GAS RESERVES: 1.2 TCF

Production began from Etame field off Gabon following completion of the third development well in August, said Vaalco Energy Inc. (Houston). The ET-1VA well, vertically drilled, flowed 4,130 b/d; previously, ET-3H and 4H wells, drilled horizontally, tested 7,600 b/d and 9,800 b/d of oil, respectively.

Etame is estimated to hold more than 150 million bbl of original oil in place, of which 30 million bbl is recoverable. The field was producing 15,000 b/d in late 2002. Oil from Etame is being produced using the Petróleo Nautipa FPSO, which is anchored in 76 m of water (250 ft) and has a 1.1 million bbl storage capacity. Vaalco was evaluating a number of prospective leads with particular attention to North and South Tchibala fields, also on the Etame concession.

Vaalco announced the first sale of Etame oil in November, when 639,000 bbl was lifted; some 200,000 bbl remains on the FPSO. Vaalco also said a 355 sq km (137 sq mi) 3D seismic survey began in the Etame Marin Block.

Exploration

Oil was discovered in 21 m of water (69 ft) by Perenco plc (London) in early 2002. On test, the GSA-1 discovery well, named Ompoyi, flowed 6,300 b/d from a 110-m (33.5-ft) gross hydrocarbon column. No water-hydrocarbon contact could be identified.

The discovery is 8 km (5 miles) from Gombe field, where Perenco has production and export facilities, allowing rapid development. Well data and 3D seismic indicate 85 million bbl of original oil in place. Perenco planned up to three appraisal wells to confirm the column height and the accumulation's northerly extent.

An appraisal well was drilled in mid-year on the Olowi Block by Pioneer Natural Resources Co. (Dallas). The Olowi Marin-2 well is 4 km (2.5 miles) southeast of the Olowi Marin-1 discovery well. The same Lower Gamba reservoir found in the discovery well flowed at sustained rates of just over 2,000 b/d with no water.

Pioneer planned to drill the Gnadi Marin-1 well to test the same reservoir and move forward with development in 2003. The Olowi Block, in water less than 40 m (130 ft) deep, lies about 5 km (3 miles) southeast of giant Gamba oil field, along an oil-productive trend that connects Gamba with smaller oil fields.

Shell discovered an oil field on the Ozigo concession that is expected to produce 20,000 b/d when fully developed.

Ivory Coast

CAPITAL: ABIDJAN

MONETARY UNIT: CFA FRANC

REFINING CAPACITY: 65,200 B/CD

OIL PRODUCTION: 5,000 B/D

OIL RESERVES: 100 MILLION BBL

GAS RESERVES: 1.05 TCF

In late 2002 the Ivory Coast was in a state of political upheaval, following a failed military coup in September. The country's real GDP grew an estimated 3.5% in 2002 after shrinking the previous year.

Espoir field started oil production from one well in early 2002 and began gas sales in August, said Canadian Natural Resources. The gas production was going via pipeline to onshore power generators in the capital of Abidjan. The field's new FPSO vessel, Espoir Ivoirien, has crude oil production capacity of 40,000 b/d and water injection capacity of 60,000 b/d. During the first half of 2002, Espoir averaged 7,400 b/d and was on target to achieve 30,000 boe/d. Production of 537,000 cu m/d (20 MMcfd) of associated gas was expected by year-end.

Espoir field's reserves are estimated at 93 million bbl of oil and 4.8 bcm of gas (180 bcf). The field had been abandoned due to high production costs but was redeveloped at an estimated cost of $265 million. Espoir field life is anticipated to be 20-25 years. Peak production could reach 28,000 b/d of oil and 940,000 cu m/d of gas (35 MMcfd).

A second phase of development might include installing another wellhead tower and drilling in the western lobe of the reservoir. Also, the first oil well contains separate satellite prospects that could be tied in to the FSPO.

The Emien exploration well was being drilled to evaluate a previously untested part of the reservoir and could result in addition of 10-15% to reserves, said partner Tullow Oil plc (London). A well on the Acajou prospect was planned for 2003.

The deepwater CI-400 Block, in waters up to 2,000 m (6,500 ft), will be explored by Canadian Natural Resources and partners under a production sharing contract (PSC) signed in June.

A second successful well was completed in early 2002 on Block CI-40, which contains Canadian Natural's deepwater Baobab discovery. The well tested at a rate of 10,000 b/d. The block is 8 km (5 miles) south of Espoir field and contains about 150 million bbl of recoverable oil reserves. The first well, which flowed 6,700 b/d of 22-23 degree gravity oil, is expected to come onstream by 2005.

Libya

CAPITAL: TRIPOLI
MONETARY UNIT: DINAR
REFINING CAPACITY: 343,400 B/CD
OIL PRODUCTION: 1.3 MILLION B/D
OIL RESERVES: 29.5 BILLION BBL
GAS RESERVES: 46.4 TCF

For the third year in a row, Libya topped the list of countries and regions most favored by international oil and gas companies for new exploration and production investment in 2002, said Robertson Research International Ltd. (UK). Although Libya has been slow to award blocks, its unexploited hydrocarbon potential continues to motivate interest.

In early 2002 Libya's National Oil Corp. appointed a new chairman to work on attracting foreign investment. US sanctions in place since 1996 continued to deter American companies, although in March Marathon Oil Corp. (Houston) was allowed to discuss projects with Libyan officials. Libya reportedly agreed with China to offer Chinese companies a wider role in the Libyan oil sector.

Western gas development

Libya has outstanding potential to increase its exports of natural gas to Europe. Libya's National Oil is working on the Western Libyan Gas Project with Eni and other participants to develop and export 7-8 bcm/yr (280 bcf/yr). The gas would be transported via Gela, Sicily, to Italy and France beginning in 2004 through a 595-km (370-mile), 32-inch pipeline called Greenstream, which was being laid in 1,150 m (3,770 ft) of water. Italian and French companies have committed to take gas volumes for power generation and other uses.

Huge Libyan gas reserves will be developed for the project in offshore Block NC-41 in the Gulf of Gabès and onshore in the Wafa gas and oil field on the Algerian border. Both areas are expected to begin producing gas by mid-2004, along with condensate estimated at 70,000 boe/d.

Oil and gas infrastructure near Mellitah on Libya's Mediterranean Coast and in the Wafa Desert will be constructed by a consortium of European and Asian companies. Facilities at Mellitah will include a gas treatment plant producing 98,000 b/d of liquid hydrocarbons and 10 bcm/yr (370 bcf/yr) of gas. Also, an offshore gas platform will be built northwest of Tripoli by Eni affiliate Saipem under a contract signed in June.

Other pipelines being considered include a dual link with Egypt, which would carry Libyan oil to Alexandria and move Egyptian gas into Libya for export to Europe. Libya also has plans to build a second gas pipeline to Spain via Morocco and has agreed to export 1.9 bcm/yr (70 bcf/yr) of gas to Tunisia starting in 2003.

Field development

An oil field on Block NC-186, the site of several discoveries in 2001, was being developed by Repsol and partners, with production to start in early 2004. "A" field, which holds an estimated 140 million bbl, lies in the remote Murzuq basin of the southwestern Sahara Desert, 720 km (450 miles) from the coast.

Development will include a 31-km (19-mile) pipeline to carry the oil to facilities at Repsol's giant El-Sharara field, where it can be transported via pipeline to Zavia, a Mediterranean port. Peak field production is expected to reach 40,000 b/d. Of eight exploration wells drilled in the area, five tested at rates of 1,600-2,300 b/d, demonstrating a reserves potential in excess of 300 million bbl, said Repsol.

Other E&P

Oil discoveries continued in the Murzuq basin, where Repsol and partners in April made their first discovery on Block NC-190 in the Hawaz formation.

Seismic surveys were being conducted in September on two onshore blocks by Oil & Natural Gas Corp. (India) in Libya's prolific Sirte basin (2,100 sq km, or 800 sq mi) and Ghadames basin (6,600 sq km, or 2,500 sq mi). Exploratory wells could be drilled in 2003.

The 900,000-bbl Farwah FPSO vessel was launched on schedule in October to operate on Block C137 in Aquitaine field.

Production from Elephant field in the Murzuq basin was delayed again and rescheduled for 2003. Elephant should reach 150,000 b/d within a year or two of startup.

Refining, gas treatment

Libya's 120,000-b/cd Az Zawiya refinery will be upgraded and virtually reconstructed by LG Petrochemicals (South Korea) under a contract signed in May.

A grass-roots natural gas treatment plant will be built at Mellitah for Agip Gas by Snamprogetti SpA (Italy) and consortium partners under a turnkey contract announced in November. Work is scheduled for completion in 2.4 years. The 6.66 bcm/yr (248 bcf/yr) plant will treat raw gas and liquid hydrocarbons from Libyan offshore fields and produce gas suitable for transportation in the Italian gas network.

ONE VIEW OF BLOCK STATUS OFF NORTHWEST AFRICA

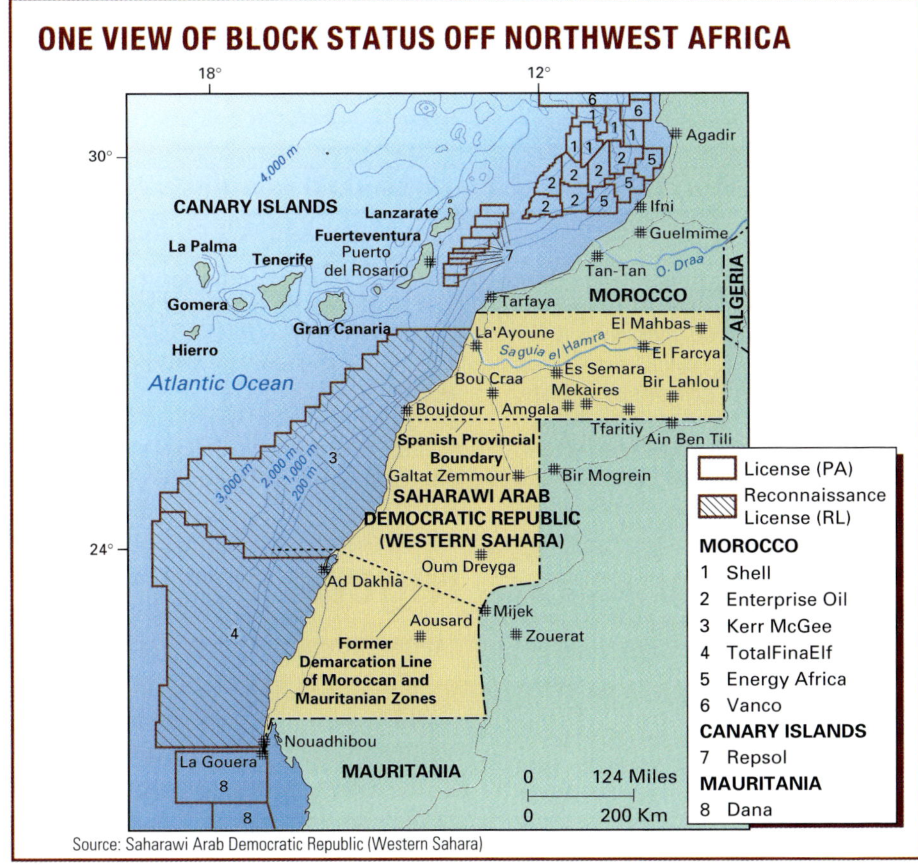

Source: Saharawi Arab Democratic Republic (Western Sahara)

Morocco

CAPITAL: RABAT

MONETARY UNIT: DIRHAM

REFINING CAPACITY: 154,901 B/CD

OIL PRODUCTION: 3,000 B/D

OIL RESERVES: 1.6 MILLION BBL

GAS RESERVES: 43 BCF

An exploration group was gearing up to acquire a 3,000 sq km (1,200 sq mi) 3D seismic survey in 2003 that will represent the first exploration in the Fustercasas-Hesperides basin, which is prospective for gas and oil, said RWE in November. The basin is part of Spain's Canary Islands near Morocco's southern coast.

Nine blocks granted to Repsol in early 2002 cover 6,160 sq km (2,380 sq mi) in water depths of 700-1,500 m (2,300-5,000 ft) east of Lanzarote and Fuerteventura islands, 100 km (60 miles) off Tarfaya. Besides the 3D survey, the group is committed to drilling two wells starting in 2004.

A comprehensive geological and geophysical study in the Atlantic Ocean off Western Sahara will be conducted by Fusion Oil & Gas plc (London), which has an agreement with the government of the northwest African state of Western Sahara. The 210,000 sq km area (81,000 sq mi) extends from Mauritania to Morocco and the Canary Islands. Western Sahara authorities refer to the area as Saharawi Arab Democratic Republic.

Morocco's onshore Missour basin and offshore Rabat Safi Haute Mer area will be explored by a Petronas unit under contracts signed in April 2002 with Morocco's national oil company, Office National de Recherches et d'Explorations Pétrolières (ONAREP). Petronas will perform field work and geological and geophysical studies that will lead to reconnaissance licenses.

The largest license holder in offshore Morocco is Vanco Energy Co. (Houston), which was the first company to shoot 3D seismic in the area and plans to be the first to start deepwater drilling there. Vanco has 20,000 sq km (5 million acres, or 8,000 sq mi) in the Safi Haute Mer permit and 14,000 sq km (3.4 million acres, or 5,300 sq mi) in the Ras Tafelney permit. Vanco said the geology of these licenses is similar to that of the Chinguetti oil discovery nearby in Mauritania.

Nigeria

CAPITAL: ABUJA

MONETARY UNIT: NAIRA

REFINING CAPACITY: 438,750 B/CD

OIL PRODUCTION: 1.93 MILLION B/D

OIL RESERVES: 24 BILLION BBL

GAS RESERVES: 124 TCF

The Nigerian government said in early 2002 that it will not privatize Nigerian Petroleum Development Co., the exploration and production unit of Nigerian National Petroleum.

Nigeria and Cameroon both claim ownership of the Bakassi Peninsula, which could hold oil reserves. The countries had agreed to abide by the International Court of Justice's decision, but when it gave the territory to Cameroon, Nigeria rejected the ruling.

Nigerian lawmakers were developing proposals in late 2002 to establish a national oil spill contingency plan and to implement new penalties aimed at discouraging oil spills, said OPEC News Agency. The government is expected to implement regulations outlining a compensation process in case of environmental damage. Pollution in Nigeria's oil-producing areas has prompted protests by Niger Delta citizens.

E&P

Oil was discovered off Port Harcourt on oil prospecting license (OPL) 222, said TotalFinaElf in June. Usan-1, 100 km (60 miles) offshore in 747 m of water (2,450 ft) near the Equatorial Guinea border, flowed 5,000 b/d from one interval. An appraisal well was planned before year-end.

A significant deepwater oil discovery was announced in early 2002 by Shell's Nigerian subsidiary. The Bolia-1X well on OPL 219, drilled in 1,100 m of water (3,600 ft), flowed 6,000 b/d on test.

Deepwater Erha oil and gas field will receive an FPSO vessel to be built by Bouygues Offshore for an ExxonMobil unit under a turnkey contract announced in October. The vessel will operate in 1,200 m of water (3,900 ft) on Block 209, about 160 km (100) miles southeast of Lagos. The FPSO is scheduled to arrive at Erha field in mid-2005 for production to start up by the end of that year. Erha field could contain 1 billion bbl of reserves.

The FPSO will have a hull 285 m long, 63 m wide, and 32 m high (935, 207, and 105 ft, respectively) and will contain 24,000 tonnes of production modules and living quarters. It will have a storage capacity of 2.2 million bbl of oil and an initial production capacity of 165,000 b/d of oil.

The offshore Yoho development project will include an early production system to produce first oil almost 2 years ahead of full field start-up in 2004, said an ExxonMobil subsidiary in September. The Falcon FPSO will be used temporarily to begin production of about 90,000 b/d of oil in late 2002. Yoho field, on shallow-water Oil Mining Lease 104, has reserves of 400 million bbl of oil.

Yoho facilities will develop discoveries in both Yoho and Awawa reservoirs in waters 60-90 m deep (200-300 ft). Peak oil production is targeted at 150,000 b/d. Produced gas will be reinjected.

A dispute was resolved in September over operations on Oil Mining Lease 109, which contains the producing Ejulebe oil field 16 km (10 miles) offshore in about 15 m of water (50 ft). Atlas Petroleum of Nigeria and TransAtlantic Petroleum Corp. (Calgary) split the lease into an area of 61 sq km (15,000 acres, or 23 sq mi) containing Ejulebe and a joint exploration and development area of 800 sq km (200,000 acres, or 300 sq mi).

The companies are performing AVO analysis of the 3D seismic data from the smaller area to help evaluate whether infill drilling would be economic in the field, which is producing 7,000 b/d and starting to decline.

Bonga, Aparo development

Bonga is Nigeria's first deepwater oil and gas field to be developed and is at least twice as large as other discoveries in Nigeria. Bonga, on OPL 212, covers about 60 sq km (23 sq mi) in water more than 1,000 m deep (3,200 ft). Bonga will produce up to 220,000 boe/d starting in 2003. To develop Bonga, Shell will utilize an FPSO with 2 million bbl of oil storage capacity and 4.6 million cu m/d of gas capacity (170 MMcfd), which will form part of the gas feed to the Bonny LNG terminal.

In addition, subsea gas pipelines will link Bonga's riser platform to the Forcados-Yokri area via 87.9 km (54.6 miles) of 24-inch line and to South Forcados via 18.0 km (11.2 miles) of 16-inch line. The pipelines are part of Shell's Nigerian offshore gas gathering system, which began construction in 2002.

Aparo and Bonga discoveries could be developed jointly by Nigerian affiliates of ChevronTexaco and Shell. An Aparo appraisal well indicates the two discoveries might share a common structure, ChevronTexaco said in September. Aparo-2, drilled in mid-2002 in 1,250 m (4,100 ft) of water on OPL 213, encountered a substantial amount of net oil sand. The appraisal well lies 5.5 km (3.4 miles) north of the Aparo-1 oil discovery and 1.2 km (0.75 miles) south of Bonga field's appraisal well OML 118 SW-2, which was drilled earlier in 2002 about 120 km (75 miles) off the Niger Delta's southwestern coastline.

Gas pipeline, monetization

Work on the 600-km (370-mile) West African pipeline will start in 2003, said the Economic Community of West African States in early 2002. The natural gas pipeline, with capacity of 3.9 bcm/yr (400 MMcfd), will link Lagos, Nigeria, to Takoradi, Ghana, with supply points at Cotonou, Benin; Lome, Togo; and Tema, Ghana. Operation is scheduled for year-end 2004.

Gas flaring at TotalFinaElf's oil production center in Nigeria will halt before 2003, in line with Nigeria's gas monetization policy, which seeks to derive maximum revenue from its gas resources. Offshore gas flaring also will be terminated before 2006.

Downstream activities

A GTL plant is being developed in the Escravos Delta region of southern Nigeria by a joint venture of Sasol Ltd. (South Africa) and ChevronTexaco. A slurry-phase Fischer-Tropsch reactor for the plant was ordered in 2002. The plant will incorporate Sasol's low-temperature synthesis technology to convert methane-rich natural gas into high-quality, low-emissions fuel suitable for use in diesel engines.

Nigeria hopes to license about five refineries as part of its strategy to liberalize its downstream sector. Two private firms previously were licensed to build refineries, but those projects have yet to get under way.

South Africa

CAPITAL: CAPE TOWN
MONETARY UNIT: RAND
REFINING CAPACITY: 489,547 B/CD
OIL PRODUCTION: 21,500 B/D
OIL RESERVES: 15.68 MILLION BBL
GAS RESERVES: 1 BCF

South Africa's highly developed synthetic fuels industry was being converted from coal to natural gas feedstock, under a project officially launched in May by President Mbeki and by President Chissano of Mozambique, which will supply the gas.

Sasol, South Africa's coal-to-oil company, will convert its existing synthetic-gas pipeline network to natural gas, and a new 865-km (536-mile) pipeline from Mozambique's Pande and Temane fields will carry natural gas to Sasol's 150,000 b/d facilities at Secunda starting in 2004. Upstream gas gathering and processing facilities were being built in Mozambique.

President Mbeki in October launched a new state company, Petroleum Oil and Gas Corp. of South Africa (PetroSA), to enable South Africa to tap West Africa's oil and gas reserves. PetroSA includes the former Mossgas Pty. Ltd., the state company that produces liquid petroleum products from natural gas and condensate supplied by offshore fields in Mossel Bay, and the former Soekor Exploration & Production.

South Africa's Cape provinces could receive gas from Kudu field off Namibia under a government agreement with Shell South Africa. The Cape's western, eastern, and northern provinces extend from Namibia's Atlantic coast border with South Africa, including Cape Town and Port Elizabeth, as far east as Lesotho and almost to Durban, with markets tending to lie along the coast. Kudu will start producing in 2005.

Orange basin E&D

A new oil play in the Orange basin will be explored by a unit of BHP Billiton, which acquired a 90% stake in Block 3B/4B off the western coast in 2002. Partners in the block identified a large play concept named Pinotage, which covers 272 sq km (105 sq mi) in 600 m of water (2,000 ft). The structure was confirmed by studies of seismic and geological data. Block 3B/4B covers 28,839 sq km (11,135 sq mi) in water depths of 300-2,500 m (1,000-8,000 ft). A 2D seismic survey covering 2,750 km (1,710 miles) was completed over the block and was being marketed.

Also in the offshore Orange basin on Block 2, Ibhubesi field holds 43 bcm (1.6 tcf) of proved and probable reserves. Gas sales deals were in the works, said Forest Oil Corp. and Anschutz International (both in Denver) in March, and eventually as much as 403 bcm of gas (15 tcf) could be recoverable on Block 2, where they have identified 176 prospects.

ORANGE BASIN E&D SCENE

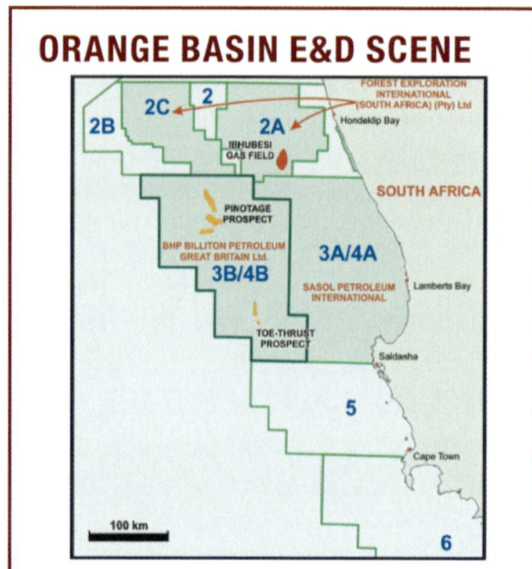

Source: Global Energy Holdings LLC

Development of the field, to begin in 2003, will involve a mini-TLP with compression and liquids stripping facilities. A pipeline will be built to shore at Saldanha and then to an electric power peaking plant in Cape Town. Gas production is to start in 2004.

Bredasdorp basin development

Sable oil field production will start in early 2003, said Pioneer Natural Resources in March, making Pioneer the first outside company to produce oil off South Africa. Sable holds 25 million bbl of estimated recoverable oil reserves and will initially yield a gross flow rate of up to 40,000 b/d of light, sweet crude from two reservoirs. Sable is in 100 m of water (320 ft), 95 km (60 miles) off the southern coast on Block 9 on the Bredasdorp basin. Pioneer's Gabon Viper prospect, which remains undrilled, is on the same block.

Sable is being developed with six subsea wells tied back to an FPSO vessel that can process 60,000 b/d of oil, reinject 2 million cu m/d (80 MMcfd) of gas, and recover natural gas liquids. Associated gas will be reinjected to improve liquids recovery but could be produced later as part of a larger gas development project.

Oribi and Oryx fields, 18 km (11 miles) east of Sable, have been producing for 3 years via an FPSO. Also nearby on Block 9 is Pioneer's Boomslang oil and gas discovery, which was being appraised. The discovery confirmed a prospective oil and gas trend on the Bredasdorp basin's southern flank that is similar to the gas-producing trend on its northern flank and validates several analogous prospects, Pioneer said.

Pioneer also has natural gas discoveries in the vicinity of Sable and foresees South Africa's power generation driving a future gas market, with initial volumes used as feedstock to the Mossel Bay GTL plant. On Block 14 off Port Elizabeth Pioneer drilled one unsuccessful well and plans more exploration.

Downstream activities

A semicommercial GTL plant with a nominal capacity of 1,000 b/d being built at

WHERE PIONEER IS WORKING OFF SOUTH AFRICA

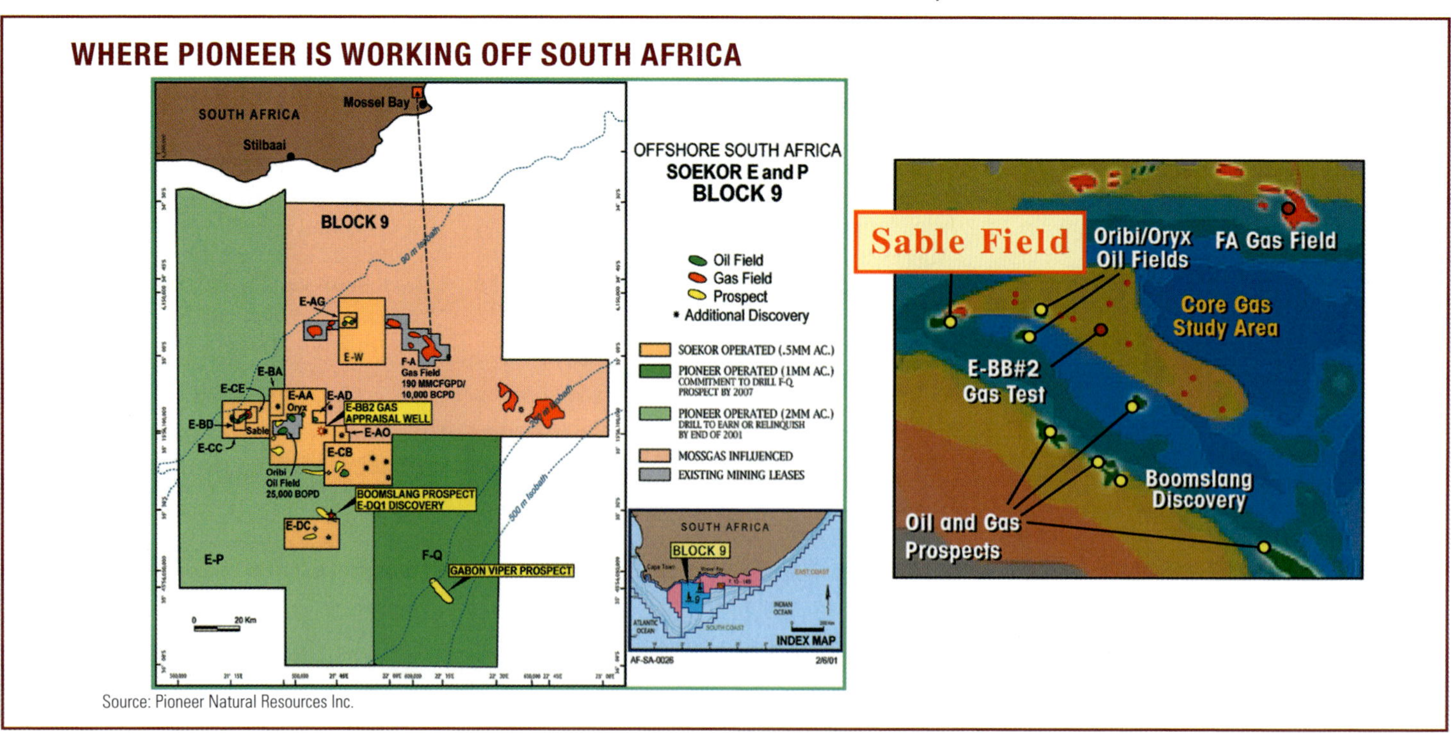

Source: Pioneer Natural Resources Inc.

BLOCKS AND GNPOC FIELDS IN SOUTHERN SUDAN

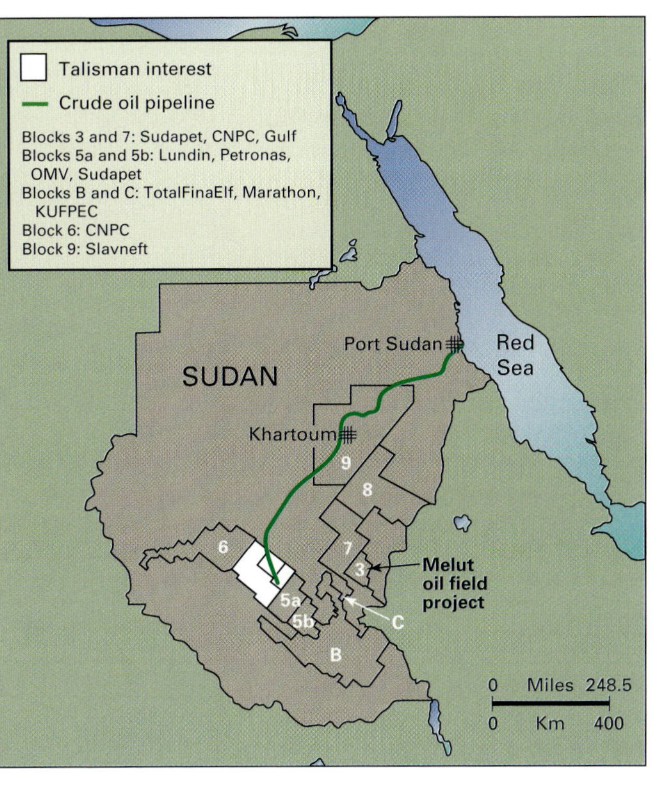

Source: After Talisman Energy Inc.

Mossel Bay will start up by year-end 2003. The plant is based on low-temperature Fischer-Tropsch technology. Mossel Bay is also the site of a major GTL complex that produces liquid fuels from natural gas supplied by offshore fields.

The Natref (Sasolburg) refinery was being expanded during 2002 to increase its capacity by 17,000 b/cd and enable the plant to produce low-sulfur diesel fuel.

A second sulfur granulating plant was commissioned in early 2002 at Sasol's 87,547 b/cd Secunda refinery.

Sudan

CAPITAL: KHARTOUM
MONETARY UNIT: POUND
REFINING CAPACITY: 121,700 B/CD
OIL PRODUCTION: 210,000 B/D
OIL RESERVES: 563 MILLION BBL
GAS RESERVES: 3 TCF

Southern Sudan fields in the Muglad basin averaged 242,400 b/d in the second quarter of 2002, up 3% on the first quarter and 14% higher than the same period in 2001, said Talisman Energy Inc. (Calgary) in August. Talisman is a member of the Sudan consortium, Greater Nile Petroleum Operating Co. (GNPOC), which manages the fields' production.

In late 2002, Talisman was in the process of selling its 25% interest in the Greater Nile Oil Project to ONGC Videsh Ltd. (India). Although the $758 million sale was delayed until the end of January 2003, Talisman said it would be completed under the same terms previously announced. Talisman had experienced heavy public criticism for its work in Sudan, as well as complaints from shareholders weary of monitoring events in that country.

Talisman's 2002 capital budget included start-up of Munga field on Block 1A in mid-year at 30,000 b/d, which is constrained by pipeline capacity. Plans also included water injection at Unity field, which was increased to 38,000 b/d of water in mid-2002, and the drilling of 38 wells.

Talisman disclosed that in addition to discoveries on the area operated by GNPOC, covering 20,000 sq km (4.9 million acres, or 7,700 sq mi), other operators have revealed discoveries on blocks 3, 7, 5a, 6, and 9. When developed, these discoveries could provide significant third-party revenue to GNPOC, which operates Sudan's only pipeline, a 1,530-km (950-mile) line to Port Sudan. GNPOC has committed to add pumping capacity that will increase pipeline capacity to 300,000 b/d by late 2003.

Talisman was evaluating potential on Block 4, where it had previously reported a large oil discovery at Shelungo. In spring Talisman drilled one development well and nine exploration wells, three of which were dry. A successful exploration well at Diffra West tested 6,000 b/d of oil from two zones, increasing expectations for additional discoveries on Block 4.

Talisman called the well significant for being the first demonstration of a working hydrocarbon system on the west flank of the graben. An appraisal well at Diffra West located the oil-water contact, proving the structure's commerciality. Talisman was drilling a well at Hamam on an adjacent fault block and planned to drill Diffra-1 on a separate fault block in 2002.

Tunisia

CAPITAL: TUNIS
MONETARY UNIT: DINAR
REFINING CAPACITY: 34,000 B/CD
OIL PRODUCTION: 71,000 B/D
OIL RESERVES: 307.56 MILLION BBL
GAS RESERVES: 2.75 TCF

Isis oil and gas field started production in early 2002, said a group led by Coparex Netherlands BV in July. The first two wells flowed a combined 11,500 b/d, and a third well began producing in January. Isis lies in 70 m (230 ft) of water and is Tunisia's farthest field from land, at 145 km (90 miles) offshore in the Gulf of Gabès. The concession's eastern edge is close to Libyan waters. Planned gas lift will involve injection of 80,000 cu m/d (3 MMscfd) at 3,000 psi. Oil flows to an FPSO vessel with facilities that can handle 30,000 b/d.

An FPSO vessel will also be used to develop a discovery in the Gulf of Hammamat, said Eni's Tunisian unit in September. The Baraka South East-1 discovery, 30 km (19 miles) offshore in 90 m of water (300 ft), flowed crude oil at 4,600 b/d, the country's highest rate for a vertical well.

The Miskar natural gas field in the Gulf of Gabès will receive a new compression package to begin operating in 2004, said BG Group's Tunisian unit. The upgrade will allow the field to maintain peak production for an additional 5 years and increase total gas recovery. BG has a long-term contract to supply more than 2 bcm/yr of gas (205 MMscfd) to Tunisia's state electric and gas company.

Exploration of the Borj El Khadra permit in the Ghadames basin in southern Tunisia will include Pioneer Natural Resources, which bought into the project in mid-2002. Pioneer also operates the Bazma, Jorf, and El Hamra permits to the north, which cover 11,000 sq km (2.7 million acres, or 4,200 sq mi) in the TAGI sand play of the Ghadames basin. The Borj El Khadra permit borders Oued Zar field, where 10 wells produce oil and gas as close as 1 km (0.6 mile) from the boundary.

A 27-MW independent power plant was to start up at mid-year, fed by gas from two marginal fields developed by Centurion Energy International Inc. (Calgary). The plant, near the port of Zarzis in south-central Tunisia, was 70% complete in April.

The Zarzis plant, adjacent to the El Biban field production facility, will consume more than 160,000 cu m/d (6 MMcfd) of gas now being flared from Centurion-operated Ezzaouia and El Biban fields. Each field will supply about 80,000 cu m/d (3 MMcfd) initially. Production is declining from Ezzaouia and increasing from El Biban.

IPE

FORMER SOVIET UNION

Armenia

CAPITAL: YEREVAN
MONETARY UNIT: DRAM
REFINING CAPACITY: 0
OIL PRODUCTION: 0
OIL RESERVES: 0
GAS RESERVES: 0

Armenia will be supplied with 2 billion cu m (bcm)/yr (74 bcf/yr) of natural gas from Turkmenistan when a 141-km (88-mile) pipeline is built from Iran into Armenia. Iran is already connected with Turkmenistan via a 150-km, 8 bcm/yr (90-mile, 300 bcf/yr) gas pipeline to Kurt Kui from western Turkmenistan's Korpedje field. Armenia and Iran had agreed earlier to construct the line, of which 100 km (60 miles) is in Iranian territory. Armenia also imports Russian gas.

Although Armenia produces no oil or gas of its own, Armenian American Exploration Co., which operates the country's only license, said that the right conditions exist for hydrocarbon accumulations. The question remains whether commercial deposits can be found.

The company believes that the Garni-Shorakhpur area is the most prospective part of the Near Yerevan basin, which has had oil shows. Two oil prospects, Shorakhpur and Nubarashen, each contain potential recoverable volumes of 20 million bbl. Together the fields could produce 11,000 b/d over 10 years, which is about equal to Armenia's petroleum product consumption.

The Oktemberyan basin also has been of interest due to gas shows in the area. Prospects have been identified that could each hold at least 0.3-1.2 bcm (10-40 bcf).

Azerbaijan

CAPITAL: BAKU
MONETARY UNIT: MANAT
REFINING CAPACITY: 441,808 B/CD
OIL PRODUCTION: 300,000 B/D
OIL RESERVES: 7.0 BILLION BBL
GAS RESERVES: 30.0 TCF

Azerbaijan's estimated proved reserves increased nearly six-fold in 2002, to 7 billion bbl from 1.178 billion, according to *Oil & Gas Journal,* which called the new estimate conservative. The country's natural gas reserves multiplied by a factor of almost seven, to 805 bcm from 118 bcm (to 30 tcf from 4.4 tcf), also a conservative figure.

Azerbaijan and the other four nations with Caspian Sea coastlines — Iran, Kazakhstan, Russia, and Turkmenistan — failed to agree on surface water boundaries or seabed mineral rights at a summit in April, prompting individual countries to pursue bilateral agreements.

Azerbaijan continued discussions with Iran over their long-standing border dispute, which includes the Araz-Alov-Sharg prospects claimed by both countries.

Pipelines

The Baku-Tbilisi-Ceyhan Pipeline Co. (BTC) was created in August, allowing project partners led by BP plc (UK) to begin building the line, which will carry 1 million b/d of crude oil from Sangachel on Azerbaijan's Caspian coast through Georgia to a new marine terminal at Ceyhan on Turkey's southern Mediterranean coast. The 1,760-km (1,090-mile), 42-inch pipeline, estimated to cost $2.9 billion, will provide a critical piece in the puzzle of how to export oil and gas from the landlocked Caspian region.

The project is on schedule to begin construction in 2003 and receive first oil in early 2005 from Azeri-Chirag-Gunashli (ACG) giant fields off Azerbaijan in the Caspian Sea. The fields were already producing an average 120,000 b/d of oil in early 2002, due to increase to 130,000 b/d by year-end.

WESTERN FSU
0 100 200 300 Miles
0 200 400 Km
Barents Sea
Kara Sea
Eniseisk Gulf
NOVAYA ZEMLYA
Admiralteyskaya
Shtockmanovskoye
Tulomskaya
Severo-Kildinskoye
Murmanskoye
Kolguyev Island
Kolguyev
Peschanoozerskoye
Kurensovskaya
Pomorskoye
Severo-Gulyaevskoya
Prirazlomnoye
Korovinskoye
Kumzhinskoye
Vaneiyisskoye
Layavozhskoye
Naryan-Mar
Vozeyskoye
Leningradskoye
Rusanovskoye
Zapadno-Sharapovskoye
Severo-Tambel
Utrenyeye
Yuzhno-Tambel
Gydanskoye
Messoyakh
Solenin
Dudinka
Norilsk
Bovanenkovskoye
Solyet
Antipiutin
Yamburg
Zapolyarnoye
Russkoye
Yuzhno-Russkoye
Urengoi
Chernichnoye
Verkhne-Chasel
Kharampur
Novy Port
Novo-Port
Vorkuta
Salekhard
Medvezhye
Nadym
Muravlenkovskoye
Sevro-Kazym
Sredne-Likhmin
Usinsk
Pechora
Pechorogorodskoye
Yuzhno-Syninskoye
Pechorokozhvinskoye
Aranetskoye
Kola Peninsula
White Sea
Arkhangel
Kamenskoye
Ukhta
Pashninskoye
Vuktylskoye
Zapadno-Tebukskoye
Punga
Gornoye
Sote
Verkhne-Kondin
Krasnoleninsk
Shaim
Surgut
Salym
Nizhnevartovsk
Samotlor
Al-Yaun
Tallakov
Igolskoye
Urnen
FORMER SOVIET UNION
Kurinskoye
Serov
Nizhnyaya Tura
Lemyuskoye
Zapadno-Izkosgorinskoye
Kush-Kodzhskoye
Pachginskoye
Rassokhinskoye
Unvin
FINLAND
Helsinki
Primorsk
Vyborg
Petrozavosk
St. Petersburg (Leningrad)
Gulf of Finland
Tallin
ESTONIA
Tartu
Kirishi
Cherepovets
Pskov
Novgorod
Valdai
Ventspils
Riga
LATVIA
LITHUANIA
Kaunas
Vilnius
Grodno
Minsk
BELARUS
Mozyr
Brest
Polotsk
Torzhok
Moscow
Kaluga
Ryazan
Smolensk
Yaroslavl
Vladimir
Nizhniy Novgorod
RUSSIA
Kirov
Okhansk
Perm
Kopdinip
Sverdlovsk
Tyumen
Kurgan
Omsk
Chelyabinsk
Atbasar
Karaganda
Izhevsk
Neftekamsk
Nizhnekamsk
Kazan
Ufa
Krasnovsolskiy
Ishimbay
Magnitogorsk
Salavat
Ul'yanovsk
Samara (Kuibyshev)
Syzran
Orenburg
Orsk
KAZAKHSTAN
Karachaganak
Orel
Yelets
Michurinsk
Saratov
Kursk
Voronezh
Unecha
Alexandrou Gai
Ostrogoshsk
Kharkov
Kiev
Brody
Lvov
Dolina
UKRAINE
Uzhgorod
Vinnitsa
Kremenchug
Lisichansk
Lugansk
Volgograd
Zaporozhye
Rostov
Kishinev
MOLDOVA
Odessa
Kherson
ROMANIA
Bucharest
Constanta
BULGARIA
Sofia
Burgas
Istanbul
Black Sea
Novorossiysk
Simferopol
Tuapse
Sochi
Krasnodar
Tikhoretsk
Stavropol
Astrakhan
Atyrau
Kulsary
Tengiz
Beineu
Aral Sea
Kungrad
UZBEKISTAN
Aktau (Shevchenko)
Caspian Sea
TURKMENISTAN
Grozny
Batumi
GEORGIA
Tbilisi
ARMENIA
Yerevan
AZERBAIJAN
Evlakh
Baku
Nakhichevan
Astara
Tabriz
Rasht
Neka
Tehran
Rey
Bandar Shah
Ashgabat (Ashkhabad)
IRAN
Samsun
Ankara
TURKEY
Aegean Sea
Blue Stream
2 Lines
3 Lines
4 Lines
6 Lines

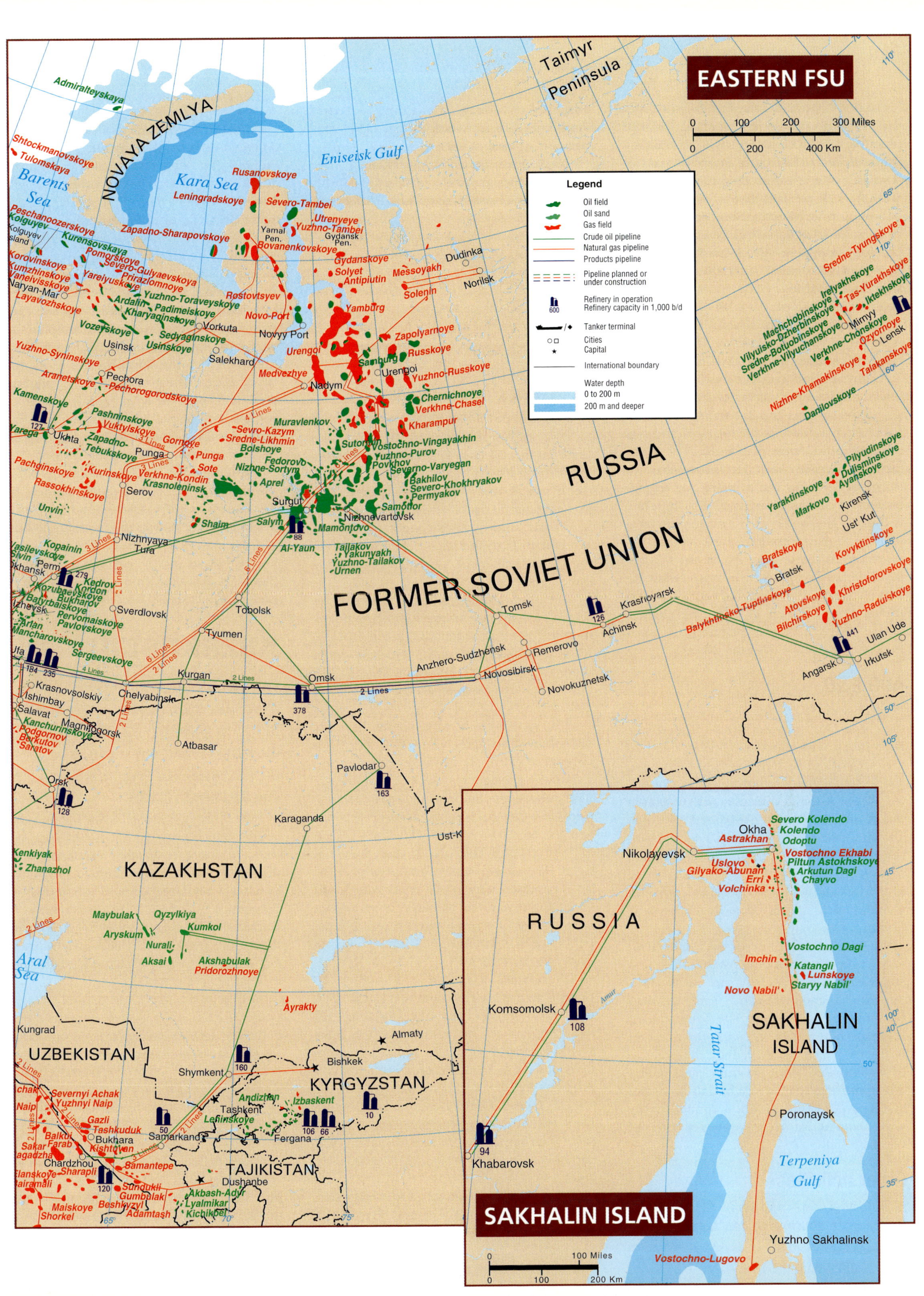
EASTERN FSU
0 100 200 300 Miles
0 200 400 Km
Legend
Oil field
Oil sand
Gas field
Crude oil pipeline
Natural gas pipeline
Products pipeline
Pipeline planned or under construction
Refinery in operation
Refinery capacity in 1,000 b/d
Tanker terminal
Cities
Capital
International boundary
Water depth
0 to 200 m
200 m and deeper
RUSSIA
FORMER SOVIET UNION
KAZAKHSTAN
UZBEKISTAN
KYRGYZSTAN
TAJIKISTAN
NOVAYA ZEMLYA
Kara Sea
Barents Sea
Eniseisk Gulf
Taimyr Peninsula
Aral Sea
Yamal Pen.
Gydansk Pen.
Vorkuta
Salekhard
Nadym
Urengoi
Surgut
Nizhnevartovsk
Tomsk
Krasnoyarsk
Achinsk
Novosibirsk
Novokuznetsk
Angarsk
Irkutsk
Ulan Ude
Omsk
Tyumen
Tobolsk
Chelyabinsk
Sverdlovsk
Perm
Ufa
Orsk
Pavlodar
Karaganda
Shymkent
Tashkent
Samarkand
Bukhara
Chardzhou
Dushanbe
Bishkek
Almaty
Fergana
Kungrad
Dudinka
Norilsk
Mirnyy
Lensk
Kirensk
Ust' Kut
Bratsk
SAKHALIN ISLAND
RUSSIA
Okha
Nikolayevsk
Komsomolsk
Khabarovsk
Poronaysk
Yuzhno Sakhalinsk
Tatar Strait
Terpeniya Gulf
SAKHALIN ISLAND
0 100 Miles
0 100 200 Km

Under contracts awarded in mid-2002, the 444-km (276-mile) section of the pipeline within Azerbaijan will be built by Consolidated Contractors International Co. (Greece); the 250-km (155-mile) Georgia section will be laid by a joint venture between Spie Capag SA (France) and Petrofac (US, UK); and Turkey's state-owned pipeline company BOTAS will build the 1,080-km (670-mile) Turkish section, as announced earlier. About 70% of project financing will be private loans.

Control and safety systems for the BTC pipeline will be supplied by ABB (Zurich) under a contract announced in December 2002. In addition to operational control, the integrated systems will provide emergency shutdown options and fire and gas monitoring of the pipeline, marine terminal, and two offshore production platforms. The work includes remote-controlled subsystems to isolate sections of the pipeline for regular inspection and maintenance. The pipeline will be linked over its entire length by a fiber optic telecommunications system.

Maintenance on the Shirvanovka-Qazax pipeline was completed in April, allowing Azerbaijan to increase gas imports from Russia to 4.8-5.7 bcm (177-212 bcf) if necessary. Azerbaijan was planning to boost Russian imports 13% during 2002.

ACG E&D

The second phase of ACG's full field development project, sanctioned in September, will complete development of the eastern and western parts of Azeri field. These sections, according to project operator BP, are estimated to contain 1.6 billion bbl of oil. Production is planned for 2006. The first phase covered Azeri's central portion, which should begin producing about 1 million b/d of oil by early 2005. Phase three of the ACG project will develop deepwater Gunashli field, expected to reach production in 2008.

A 3D 4C seismic survey over Azeri field was completed which involved the first successful imaging of the subsurface beneath a mud volcano, said Caspian Geophysical in October. The study sought to obtain better data on continuity of target horizons, which would reduce the number of wells drilled, improve their positioning, cut nonproductive time while drilling, and attain earlier field start-up. A similar survey was planned for the deepwater part of Gunashli field.

Other activities

Kursangi and Karabagli oil fields 100 km (60 miles) southwest of Baku will be explored by units of China National Petroleum Corp., which joined other field partners in early 2002.

Modernization of Azerbaijan's two refineries progressed with a feasibility study contract awarded to ABB Lummus in early 2002.

Georgia

CAPITAL: TBILISI

MONETARY UNIT: LARI

REFINING CAPACITY: 106,436 B/CD

OIL PRODUCTION: 2,000 B/D

OIL RESERVES: 35 MILLION BBL

GAS RESERVES: 300 BCF

Due to unpaid debts, Georgia was required to prepay for natural gas shipments from Russia beginning in early 2002. Georgia was also planning to import 200-300 million cu m (7-10 bcf) of gas from Azerbaijan during 2002, under a previous agreement that includes rehabilitation of the 37-km (23-mile) pipeline connecting the countries.

The agreement allows Georgia to receive 5% of gas volumes shipped beginning in 2004 via the Baku-Erzurum pipeline from Azerbaijan to Turkey. Georgia could purchase up to 500 million cu m/yr (18-19 bcf/yr) for 20 years at a cost of $55 per 1000 cu m.

The government of Georgia in early 2003 signed an agreement with Northrop Grumman Corp. (Los Angeles) to develop an aerial surveillance system to monitor the BTC export pipeline and its adjacent area. Georgia would receive radar configurations similar to those used by the US in Afghanistan, said Georgian International Oil Corp. The Georgia government is plagued by separatist elements in Abhkazia, South Ossetia, and Ajaria, as well as terrorists from neighboring Chechnya where a war of independence has been waged against Russia since the early 1990s.

US officials allocated $11 million to Georgia to form a 400-member special military unit to protect the BTC pipeline, said the company. The rapid reaction unit would consist of Georgian military personnel trained by US instructors under a US program for development of anti-terrorist forces. One 500-member unit has already been trained under this program, while a second group, the Sachkhere mountain rifle battalion, is to commence training in early 2003.

Oil and gas exploration and production rights to Block XI(G) (Tbilisi) and Block XI(H) (Rustavi) in the Mtsketa, Tetritskaro, and Gardabani regions of eastern Georgia were awarded in November to a unit of CanArgo Energy Corp. (London) by the Georgian State Agency for Regulation of Oil and Gas Resources. CanArgo will receive a license for both blocks once a production sharing agreement (PSA) has been negotiated.

The blocks, which cover 486 sq km (188 sq mi), neighbor CanArgo's existing acreage, where oil and gas shows and limited production have been obtained. Potential exists in the Eocene and Cretaceous formations. CanArgo plans further exploration and appraisal activities in line with a continuing program in its nearby acreage. The Norio MK72 exploration well had previously targeted a large oil prospect on Norio block near Samgori field, but it was suspended until CanArgo finds a partner to help complete the well.

Other blocks in Georgia that were up for bidding in 2002 included Block IIC, 2,260 sq km (873 sq mi), and Block IID, 3,480 sq km (1,344 sq mi), both in the Black Sea; Block VIII (Achara-Trialety), 4,760 sq km (1,838 sq mi), in eastern Georgia; and Block XA, 2,680 sq km (1,035 sq mi), in the eastern Caucasus.

Kazakhstan

CAPITAL: ASTANA

MONETARY UNIT: TENGE

REFINING CAPACITY: 427,093 B/CD

OIL PRODUCTION: 800,000 B/D

OIL RESERVES: 9.0 BILLION BBL

GAS RESERVES: 65 TCF

Kazakhstan's gross domestic product (GDP), which grew 13.2% in 2001, was projected to increase another 7% during 2002, driven by foreign investment mainly in the country's booming oil and gas industries. Estimated proved oil reserves for Kazakhstan increased by 66% in 2002, to 9 billion bbl from 5.417 billion, according to *Oil & Gas Journal*, which called the new estimate conservative.

A new national oil and natural gas company, KazMunaiGaz, was established in February 2002 by merging state oil company Kazakhoil and the national oil and gas transportation firm TransNefteGaz. KazMunaiGaz also took control of the country's pipeline assets, previously owned by KazTransGaz.

In June KazMunaiGaz created a joint venture called KazRosGaz with Russia's Gazprom, which will allow Kazakhstan to pipe its natural gas through Russia's system for the first time. KazRosGaz could transport 3.4 bcm (125 bcf) of Kazakh gas via

Russia, increasing to 47.5 bcm (1.77 tcf).

Kazakhstan and the other four nations with Caspian Sea coastlines — Azerbaijan, Iran, Russia, and Turkmenistan — failed to agree on surface water boundaries or seabed mineral rights at a summit in April, prompting individual countries to pursue bilateral agreements.

In May Kazakhstan and Russia agreed on a joint boundary dividing their interests in the northern Caspian Sea, which should encourage more foreign investment in the region. Petroleum reserves will be developed in previously disputed areas including Kurmangazi, Tsentralniye, and Khvalinskoye fields, and a joint state-controlled company will build ships, barges, and offshore drilling rigs to service foreign companies operating in the Caspian.

Tengiz development halted

Further development of giant Tengiz oil field was suspended in November by joint venture TengizChevroil because the partners had not agreed on a funding plan. Parker Drilling Co. (Houston) was instructed to suspend drilling upon completion of wells. The suspension applied to expansion of the second-generation production and sour gas injection projects. TengizChevroil partners include ChevronTexaco Corp. (San Francisco), ExxonMobil Corp. (Irving, TX, US), and Kazakh state interests. Reportedly, Kazakhstan officials threatened to impose up to $70 million in environmental fines against ChevronTexaco for storing sulfur near the company's Tengiz oil field, according to the Sulfur Institute.

Tengiz field had been averaging 260,000 b/d of oil and could yield 700,000 b/d by 2010. Its production is expected to nearly double by 2006. The sour gas injection plant would have reduced the sulfur content of oil shipped from Tengiz. TengizChevroil had planned a new 2,400 tonne per day (tpd) Claus and tail gas unit to manage sulfur from the field's sour natural gas. Expected completion was 2005.

Kashagan to be developed

The Kashagan supergiant oil discovery in the northeastern Caspian Sea was officially declared commercial in June, opening the way for development plans. A 2-year appraisal program undertaken by partners in the North Caspian PSA estimated Kashagan's producible reserves at 7-9 billion bbl. First oil from the field is expected by 2005.

Kashagan, which lies 70 km (40 miles) southeast of Atyrau, could also hold 38 billion bbl of probable reserves, said Agip Kazakhstan North Caspian Operating Co. (Agip KCO). However, the estimates were considered to be conservative, said Kazakh officials and energy analysts.

Integrated drilling services for Kashagan's development will be provided by Halliburton Energy Services Group (Dallas) under a 2-year contract extension awarded in June by Agip KCO. Agip also hired a 3,000-hp drilling rig to work in Kashagan, said Pride International Inc.

(Houston) in July. Under a 1-year contract plus two 1-year options, Rig 319 will be modified and enhanced for deployment to an artificial island in the Kazakh sector of the Caspian Sea. Operations were expected to begin in spring 2003. Three sulfur-recovery trains of 1,900 tpd each are also planned for Kashagan field, to be complete in 2006.

Partners in Kashagan continued to explore other structures in the North Caspian PSA, which covers 5,600 sq km (2,200 sq mi). The Kalamkas-1 well yielded an initial flow rate of 2,300 b/d, said Agip KCO in October. Plans call for the companies to explore the Kashagan Southwest, Aktote, and Kairan structures in 2003.

Other exploration

The Liman license, which covers 6,500 sq km (1.6 million acres, or 2,500 sq mi) on the Caspian Sea's north shore near Atyrau, will be explored by Aurado Exploration Ltd. (Toronto), which took a 90% interest that was approved by Kazakh authorities in late 2002. Aurado identified 59 drilling prospects from seismic and well data, 15 of which could hold a combined 706 million bbl recoverable at 500-3,000 m (1,600-10,000 ft). Up to 21 locations at East Tegen, each on 0.3 sq km (80 acres, or 0.125 sq mi), could yield 30 million bbl. A road, railway, and pipeline cross the block.

The first well, East Tegen-1, found a small, shallow gas discovery, said Aurado in December 2002. A water well drilled as part of the East Tegen-1 program flowed 48,000 cu m/d of gas (1.8 MMcfd) from 126 m (413 ft) and was capped pending sales to a nearby town. East Tegen-1 encountered gas at less than 200 m (650 ft) but was not tested. A larger rig was moved in to deepen the well to its primary objective, an oil zone in Albian sands at 600 m (2,000 ft).

The first of six deep play exploration wells on license 260 D1 provided encouragement, said Hurricane Hydrocarbons Ltd. (Calgary) in February. North Nurali 1 was drilled to 2,265 m (7,431 ft) and produced 47.6-degree gravity, low-sulfur crude from deep sands during an open hole test.

Karachaganak production

Liquid hydrocarbon production from Karachaganak gas condensate field will eventually be increased to 7 million tonnes/year (tpy) by constructing additional production and export capacities, said joint venture Karachaganak Integrated Organization (KIO) in June. The field holds an estimated 2.25 billion bbl of recoverable oil and gas condensate reserves. During 2002 production should reach 5 million tonnes of condensate and 4.7 bcm (175 bcf) of gas.

Construction through 2003 will include a gas processing plant, a 240-MW power plant, and three high-pressure compressors. Sour gas reinjection was to start in October 2002. Later, KIO will build a 650-km (400-mile) export pipeline from Karachaganak to Atyrau, which will ultimately connect with the Caspian Pipeline Consortium system.

New oil pipeline connections

Western Kazakhstan's large fields can begin producing or increase their production, since transportation agreements were signed in October between KazMunaiGaz and Hurricane Hydrocarbons. The fields most affected are in the South Torgai and eastern North Caspian basins, north and northeast of the Caspian and Aral seas. Hurricane is the second-largest non-Kazakh oil producer, accounting for 15% of the country's average 950,000 b/d production through August 2002.

Under the agreements, Hurricane is building a pipeline that will reduce export shipment distance by 1,300 km (800 miles) and cut transportation costs by $2.50/bbl. Hurricane has interests in 10 fields containing a combined 512 million bbl of proved and probable reserves, as well as productive exploration acreage. The fields are Kumkol South, South Kumkol (a separate field), Kumkol North, East Kumkol, Maybulak, Aryskum, Qyzylkiya, Akshabulak, Aksai, and Nurali.

A second, 443-km (275-mile) pipeline is being laid by KazTransOil from Kenkiyak oil field, 200 km (125 miles) south of Aktiubinsk, to Atyrau, Kazakhstan's port and oil pipeline hub on the northeastern Caspian Sea. Start-up is set for early 2003. Meanwhile, KazTransOil also completed and pressure tested a 16-km (10-mile), 19.5-inch, 120,000-b/d pipeline connecting giant Alibekmola field to the Kenkiyak pipeline terminal. In addition to Alibekmola, the line will later serve giant Kozhasai field; together the fields have estimated reserves of 500 million bbl of crude oil and 18-19 bcm of recoverable natural gas (690 bcf).

According to the October agreements, Hurricane will also support and assist KazMunaiGaz in its future plans to build the 700-km (435-mile) Kumkol-Aralsk-Kenkiyak (KAK) oil pipeline. Together, the KAK and Kenkiyak-Atyrau pipelines would fully connect the South Torgai basin oil fields to Atyrau. The KAK pipeline might be built in phases, with the first section from Kumkol to Aralsk.

From Atyrau, three trans-shipment pipelines would be available: to Samara, Russia; to the Black Sea via the Caspian

NEW PIPELINES FOR KAZAKH OIL

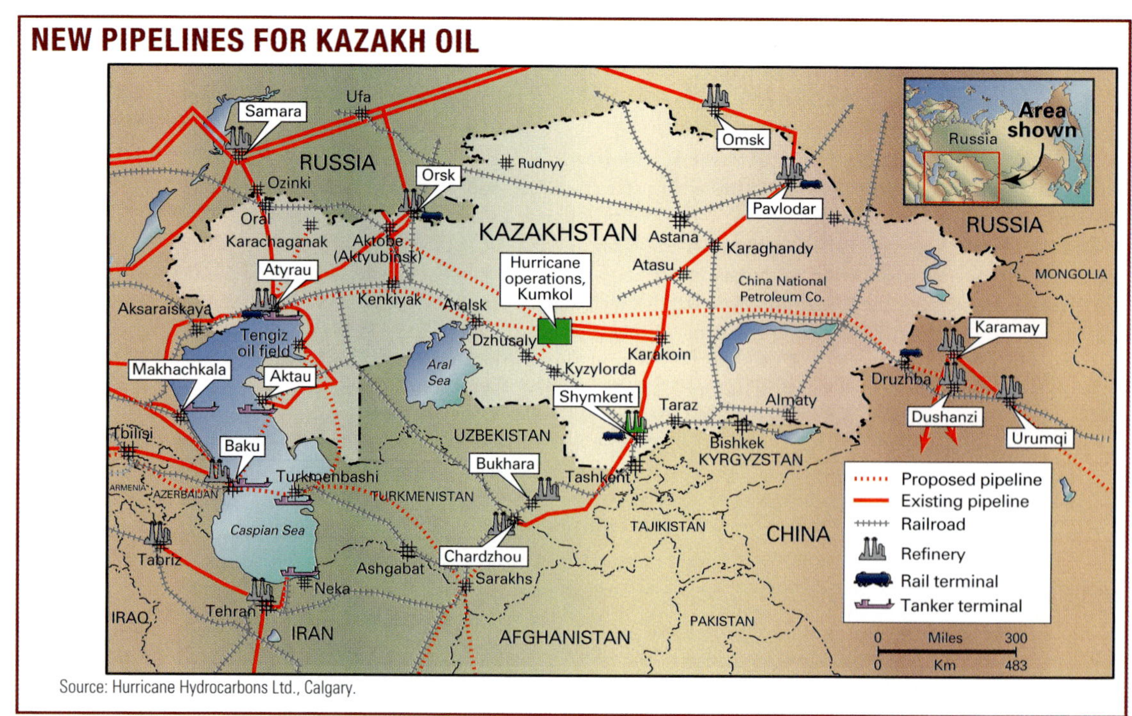

Source: Hurricane Hydrocarbons Ltd., Calgary.

Pipeline Consortium line; or to the eastern Caspian port of Aktau, Kazakhstan. Alternatively, the KAK pipeline could be reversed in the future to ship western Kazakh crude oil to southern or eastern markets, adding valuable strategic options for future exports.

The agreements confirm KazMunaiGaz' support for Hurricane's construction of the QAM pipeline, named for Qyzylkiya, Aryskum, and Maybulak fields in the South Torgai basin. The 16-inch QAM pipeline, under construction with an initial design capacity of 100,000 b/d, should begin full operation in late 2003. The line will facilitate early development of Hurricane's known oil fields and stimulate exploration in the South Torgai basin. Capacity could later be hiked to 140,000 b/d.

The QAM pipeline consists of two sections: 1) a 105-km (65-mile) section from the Aryskum field pump station southwest to Dzhusaly will connect the three fields to a new rail terminal at Dzhusaly on the Syr Darya River for further rail shipments to western destinations; and 2) a 69-km (43-mile) section will connect Hurricane's producing Kumkol fields to Aryskum.

Alibekmola-Kozhasai fields

At Alibekmola field, production reached 3,500 b/d in mid-2002, when a fourth appraisal well was spudded. Following a fifth well by year-end and more drilling over several years, Alibekmola's output should reach 55,000 b/d, said Nelson Resources Ltd. (Toronto) in October. The field holds an estimated 159 million bbl of proved, 80 million bbl of probable, and 168 million bbl of possible reserves, according to a September report by outside engineers.

The proved and probable figures are based on 69 projected producing wells on spacing of 0.55 sq km (135 acres, or 0.21 sq mi), each averaging 990 b/d initially. Crude is currently exported by rail and pipeline to Europe. With completion of pipeline service to Atyrau, Nelson anticipates almost halving the $12.60/bbl differential between Brent and Alibekmola crudes.

Development of Kozhasai, with reserves estimated at 224 million bbl, is to follow 1 year behind that of Alibekmola. Nelson anticipates acquisition of 230 sq km (89 sq mi) of 3D seismic and workover of two Soviet wells by year-end 2002. The first new well would be drilled in late 2003, followed by a new reserves audit. Alibekmola and Kozhasai also contain 18-19 bcm (690 bcf) of recoverable natural gas.

Kazakhstan has expressed support for the BTC Azerbaijan-Turkey oil pipeline, but has not officially pledged to use the line, preferring to keep its export options open. Kazakhstan agreed with Russia in June to export at least 350,000 b/d via the Russian pipeline system. Also, Kazakhstan had begun oil swaps with Iran and was discussing a pipeline connection.

Tanker traffic restricted

Turkey's concerns about the increasing tanker traffic through its Bosporus and Dardanelles Straits between the Black Sea and the Aegean Sea led to a policy banning supertanker traffic through the straits and severely restricting the passage of smaller tankers carrying cargo such as LPG and LNG.

Beginning in October, Turkey prevented supertankers from Kazakhstan and Russia from exporting oil through the straits, citing new regulations limiting tankers to 200 m in length (660 ft) to pass through the narrow, 30-km long (19-mile) Bosporus and restricting tanker length to 250 m (820 ft) for passage through the Dardanelles. The regulations also prohibit loaded oil tankers from passing through the straits at night and in other instances when visibility is obscured.

Kyrgyzstan

CAPITAL: BISHKEK
MONETARY UNIT: SOM
REFINING CAPACITY: 10,000 B/CD
OIL PRODUCTION: 2,000 B/D
OIL RESERVES: 40 MILLION BBL
GAS RESERVES: 200 BCF

Kyrgyzstan's state oil and natural gas company, Kyrgyzneftegaz, was planning to begin drilling exploration wells in the Dzhalalabad region in 2002, investing $30 million of its own funds.

Kyrgyzneftegaz was also entering into partnerships with several foreign energy companies. During 2002, Kyrgyz and Chinese operators were repairing more than 100 idle oil wells in Kyrgyzstan. Also, a Kyrgyz-Austrian joint venture that performed exploration work in 2001 was planning to increase its 2002 exploration budget to $30 million.

A Kyrgyz-Kazakh joint venture, Bigmao Oil, was building a mini-refinery in Kyrgyzstan with a fuel oil production capacity of 400 bbl/calendar day (b/cd). It was to begin operating by year-end 2002.

Russia

CAPITAL: MOSCOW
MONETARY UNIT: RUBLE
REFINING CAPACITY: 5,435,480 B/CD
OIL PRODUCTION: 7.385 MILLION B/D
OIL RESERVES: 60.0 BILLION BBL
GAS RESERVES: 1,680 TCF

Russia is the world's largest natural gas exporter, second-largest oil exporter, and third-largest energy consumer. Estimated proved oil reserves for Russia increased by more than 23% in 2002, to 60 billion bbl from 48.573 billion, according to *Oil & Gas Journal*, which called the new estimate conservative.

Russia's GDP was estimated to grow 4.1% in 2002 and projected at 4.9% for 2003. To decouple its economic growth from oil and gas exports, Russia was seeking to liberalize its energy sector and encourage foreign investment. Russia's largest oil companies are now focused more on profits and operate more freely as participants in the international oil industry.

However, the business climate remained unpredictable. Large international companies were still seeking a stable PSA framework before making huge investments, especially in frontier areas. At a government-industry summit in October, US representatives called for a transparent and fair tax regime and definitive mineral rights laws in Russia, in addition to PSAs. Russian officials said again they hope to have legislation in place soon. Some small independent foreign companies such as Teton Petroleum Co. (Steamboat Springs, CO) reported successful, low-risk operations in Russia.

Lukoil shares sold

In response to a rising tide of competition from several emerging Russian oil companies, Russia offered for sale in late 2002 a 5.9% stake, or 50 million shares, in OAO Lukoil. The share issue garnered close to $800 million — nearly $87 million more than if the shares had been sold in mid-2002 as originally planned. This capital will help Russia meet its huge foreign debt payments, which are expected to peak in 2003 at $17.9 billion ($10 billion in principal and the rest in interest).

Lukoil is facing more competition from other Russian oil companies coming up the ranks, such as OAO Yukos, which are overtaking Lukoil in terms of production and efficiency, said Petroleum Finance Co. (Washington, DC). Bloomberg LP reported

Our world for

VAM® ACE

VAM® FJL

VAM® PRO

VAM® MUST

VAM® TOP HT

VAM® RISER

VAM® HW ST

VAM® TOP HC

VAM® SLIJ-II

V

that if Yukos achieves its plans to boost oil production 19% in 2003, it could overtake Lukoil as Russia's biggest oil producer. Yukos intends to produce 83 million tonnes (1.66 million b/d) of oil in 2003, compared with an average 1.4 million b/d in 2002, while Lukoil says it expects to increase production 5.1% to 82 million tonnes of oil that year. Lukoil's net profits could tumble to $1.5-$1.7 billion in 2003, versus $2 billion in 2002 and $2.11 billion in 2001. However, the company's long-term corporate credit rating was upgraded in December, giving it a stable outlook.

Western firms dealing with Lukoil have voiced concerns over its management tactics and lack of internal transparency. Aware of those concerns, Lukoil in mid-2002 named two Western executives to its board of directors. Other Russian oil companies are also trying to become more like Western oil companies by boosting production and exports, hiring Western managers or appointing Western board members, and contracting with Western oil service companies, said Petroleum Finance. The big question is whether Western oil companies will get PSAs in Russia and gain access not only to onshore Russian PSAs but also to pipelines, which would allow greater control over all their Russian investments.

Petroleum Finance cited current forecasts that have Russia producing more than 8 million b/d of oil in 2003, with 3 million b/d available for export. Russia's goal is to reach 9 million b/d of production by 2005, which will strain the capacity of the country's oil export pipeline infrastructure. During 2002, despite output cuts by the Organization of Petroleum Exporting Countries (OPEC), Russia's oil production surged to 74.752 million tonnes, or 13%, according to Interfax, a Russian news agency. Reaching the 9 million b/d goal would require financing for major exploration and infrastructural projects, primarily from foreign investment.

US ties

Russian President Putin agreed with US President Bush in May to encourage US investment in Russia's oil and gas sector as part of a new energy partnership, which supports improvements to Russia's export infrastructure. The first direct shipment of Russian crude oil to the US, about 2 million bbl of Urals blend, arrived in July.

At another meeting in late 2002, Russia confirmed plans to consider building an Arctic port to export oil to the US. The proposed 1 million b/d facility and pipeline, to be completed in 2005 or later, would be built at Murmansk port in northwestern Russia on the Barents Sea, in water deep enough to allow supertanker traffic. The export route across the Atlantic would be considerably shorter than from the Middle East. Yukos has begun trial shipments of large tanker volumes of crude from its terminal at Vitino through Murmansk and could soon begin test-loading of very large crude carriers at the same location.

An agreement to study the plan's feasibility was signed by Lukoil, Yukos, Siberian Oil Co. (Sibneft), and Tyumen Oil Co., which together produce more than half of Russia's oil output. Yukos' rapidly increasing production is forcing it to seek new markets, the company said earlier. However, obstacles to the Murmansk project include financing, ownership questions, pipeline route, and inadequate Russian pipeline legislation.

The proposal would include a 1,500-km (900-mile) link between Murmansk and Russia's existing pipeline network. US investment in a Murmansk terminal would reduce Russian producers' cost of exporting crude to US markets. Russia already produces more oil than it can export, due to the lack of pipelines and ports and stagnant domestic demand, so its producers are scrambling to build infrastructure to reach new markets. At the same time, the US is seeking to diversify its oil supplies. Alfa Bank (Moscow) estimates that by mid-2003, Russia will be exporting 1.2 million b/d more than it did in 2001 and early 2002.

Russia's lack of oil terminals capable of loading tanker consignments of 2 million bbl or more has long been viewed as the main obstacle to large-scale exports to the US market. That obstacle could be overcome through development of Murmansk, the only ice-free port in northern Russia capable of servicing supertankers year-round. It would reduce the cost of Russian oil exports, which now start at Novorossiysk on the Black Sea and pass through the Bosporus Straits and the Mediterranean Sea before heading across the Atlantic Ocean.

Lukoil and ConocoPhillips (Houston) compiled a feasibility report on the so-called Northern Transportation System, which would include land and subsea pipelines, an ice-free stationary marine terminal at Varandei oil field, and a fleet of icebreaking tankers, as well as the oil export terminal at Murmansk. Lukoil officials said the project would provide the shortest sea route to northwest Europe and to the US while resolving the shortage of export capacity for oil produced in the Timan-Pechora region and the Barents Sea. At the May 2002 US-Russia meeting, Lukoil emphasized the economies of scale and claimed that transport costs of $28/tonne would allow Murmansk-exported oil to compete with supplies from the Persian Gulf.

However, Russia's oil transportation company Transneft remains cool to the idea. Instead, Transneft claims it would be more profitable to lay an oil pipeline from Angarsk to Nakhodka, giving impetus to development of oil supplies in Eastern Siberia and the Russian Far East.

Lukoil built its own oil export terminal in 2001 at Varandei Bay, which has a capacity of 1.5 million tpy. Varandei loads Timan-Pechora oil into ice-class tankers, which carry it to Murmansk, where larger amounts of crude are collected and reloaded into 300,000 dwt tankers for export to northwestern Europe. Plans now call for capacity expansion at Varandei, increasing Russia's oil exports to 20-25 million tpy from 5 million tpy.

Caspian deal

Russia and the other four nations with Caspian Sea coastlines — Azerbaijan, Iran, Kazakhstan, and Turkmenistan — failed to agree on surface water boundaries or seabed mineral rights at a summit in April, prompting individual countries to pursue bilateral agreements.

In May Russia and Kazakhstan agreed on a joint boundary dividing their interests in the northern Caspian Sea, which should encourage more foreign investment in the region. Petroleum reserves will be developed in previously disputed areas including Kurmangazi, Tsentralniye, and Khvalinskoye fields, and a joint state-controlled company will build ships, barges, and offshore drilling rigs to service foreign companies operating in the Caspian.

Sakhalin I development

The Sakhalin I project is developing three fields in the northeast shelf of Sakahlin Island in 10-60 m (30-200 ft) water depths in the Sea of Okhotsk. Phase 1 operation, which began in 2002, focuses on Chayvo oil and gas field, with first oil expected by year-end 2005, and Odoptu field, to begin producing in 2008. Eventually Phase 1 will also develop Arkutun Dagi field, reaching combined peak production of 250,000 b/d.

Chayvo field, 5-15 km (3-9 miles) offshore, is being developed from both onshore and offshore locations. The Sakhalin I consortium, led by an ExxonMobil subsidiary, began drilling the first well, in Chayvo's northern portion, from onshore Sakhalin Island in late 2002.

With planned horizontal displacements of 6-10 km (4-6 miles) and total vertical depth of 2,600 m (8,500 ft), the onshore wells represent the edge of industry experience in extended reach drilling. The 10 initial wells will be drilled from a single rig in a radial pattern with wellheads spaced about 10 m apart (30 ft).

SAKHALIN I PROJECT

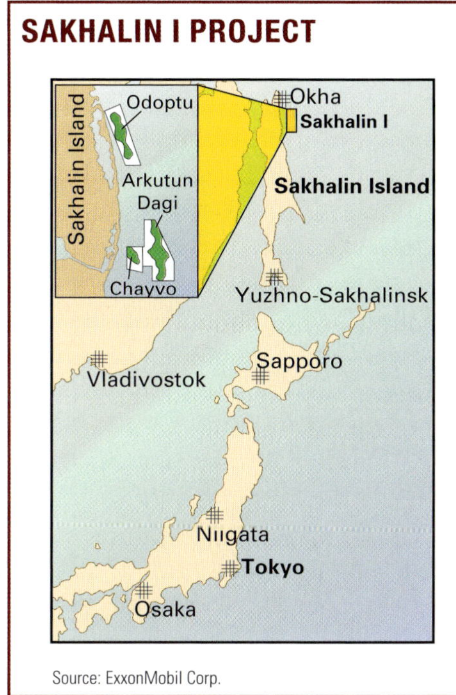

Source: ExxonMobil Corp.

The special-purpose, extended reach drilling rig will be built and operated by subsidiaries of Parker Drilling under two contracts. The Arctic-class rig will be designed to withstand earthquakes and operate in frigid winters where ice covers the Sea of Okhotsk for 6 months.

For offshore production, the 29 sq m (312 sq ft) concrete island drilling structure Orlan was moved during 2002 from Alaska's Prudhoe Bay to Sovietskaya Gavan, where it will be upgraded and begin drilling up to 20 wells in 2004. The modified concrete and steel structure will be installed 11 km (7 miles) off Sakhalin Island in about 14 m of water (45 ft). It will support existing living quarters, minimum production facilities, and a new drilling package designed for the extended-reach wells. Production will be moved by pipeline to a new onshore processing plant.

Detailed design of drilling facilities was subcontracted in late 2002 to KCA Deutag Drilling Inc. (Houston) by Hyundai Heavy Industries (South Korea), which holds the main contract for drilling package fabrication for Sakhalin I. Chayvo and Odoptu field support facilities were contracted in mid-2002. Two onshore production units and support facilities for three land-based well sites will be built by ABB Lummus Global, with most of the work to be subcontracted to Russian companies.

Although Phase 1 is an oil-only development, the Sakhalin I blocks could also produce gas. The consortium is proposing to deliver natural gas to Japan via a 190-km (120-mile) pipeline linking its fields with Sapporo on Japan's northernmost island of Hokkaido. A feasibility study of the pipeline was being conducted in 2002.

Far East exploration

Part of Russia's Sakhalin V area off Sakhalin Island will be explored by OJSC NK Rosneft, which received a 5-year license in July, on behalf of an alliance including BP. Seismic acquisition and exploratory drilling will be conducted on the Kaigansky-Vasuykansky blocks in the East Smidt (Shmidtovsky) area in the southern part of the Sakhalin V tract. The blocks cover 10,000 sq km (3,900 sq mi) in water depths up to 140 m (460 ft).

SAKHALIN V AREA

Rosneft postulates that the Sakhalin V structures identified thus far could hold resources of as much as 600 million tonnes of crude oil and 600 bcm (22 tcf) of natural gas. The partnership plans to purchase seismic data from a survey conducted over Sakhalin V in mid-2002 by Petroleum Geo-Services ASA and Dalmorneftegas. Drilling could begin as early as 2004, depending on results of the survey.

Sibneft operations

Gas from West Ozyornoye field is being delivered to a power station in Anadyr via a 104-km (65-mile) pipeline laid by Sibneft in mid-2002. West Ozyornoye is to produce 120 million cu m/yr (4.5 bcf/yr) from 6 bcm (220 bcf) of reserves. Sibneft is studying a range of projects to deliver pipeline gas and LNG and is assessing gas processing opportunities.

An operational center in the Noyabrsk area of Western Siberia will be established by Sibneft and Baker Hughes Inc. (Houston) at Sibneft's upstream base, where the companies will form a joint engineering group. The initiative is part of a plan to drill more than 100 sidetrack wells in 2003 to enhance production from existing Noyabrsk area oil wells, said Sibneft in September.

A full range of oil field services was being provided to Sibneft's upstream division by Halliburton International, beginning in spring 2002. Sibneft plans to improve reservoir management, increase production from existing wells, and work over wells that still have production potential.

In late 2002, majority interest in Russia's ninth-largest oil producer, OAO Slavneft, was won by Invest-Oil, acting for Sibneft and Tyumen Oil Co.

Other development and production

Western Siberia's Zapadno-Malobalysk field will be developed jointly by Yukos and MOL Rt. (Hungary) under an early 2002 agreement. The field, with estimated proven reserves of at least 20 million tonnes of crude oil, is near pipeline and other transportation infrastructure. It had been producing 10,000 b/d and was expected to peak at 55,000 b/d in 2005.

Production from Eguryak field in western Siberia increased to 6,500 b/d with completion of its 12th well, said Teton Petroleum in October. Teton also began pumping oil earlier in 2002 through its 40-km (25-mile) pipeline, enabling year-round production from Eguryak, which lies 30 km (20 miles) north of supergiant Samotlor oil field.

In the Komi region of the Timan-Pechora basin, 30,000 b/d was being produced from 27 wells, said Urals Energy in October.

Oil pipelines

Druzhba-Adria pipeline

At year-end 2002, Russian crude oil was to begin export from the deepwater Omisalj terminal on Croatia's Adriatic Sea coast. Russian Urals blend flowed 3,198 km (1,987 miles) via the Druzhba and Adria pipeline systems west to the Croatian port, the final link being a newly reversed 178-km (110.5-mile), 36-inch line from the inland terminal at Sisak, Croatia.

From the Adriatic port, the oil is trans-

MAIN RUSSIAN, CIS OIL PIPELINES

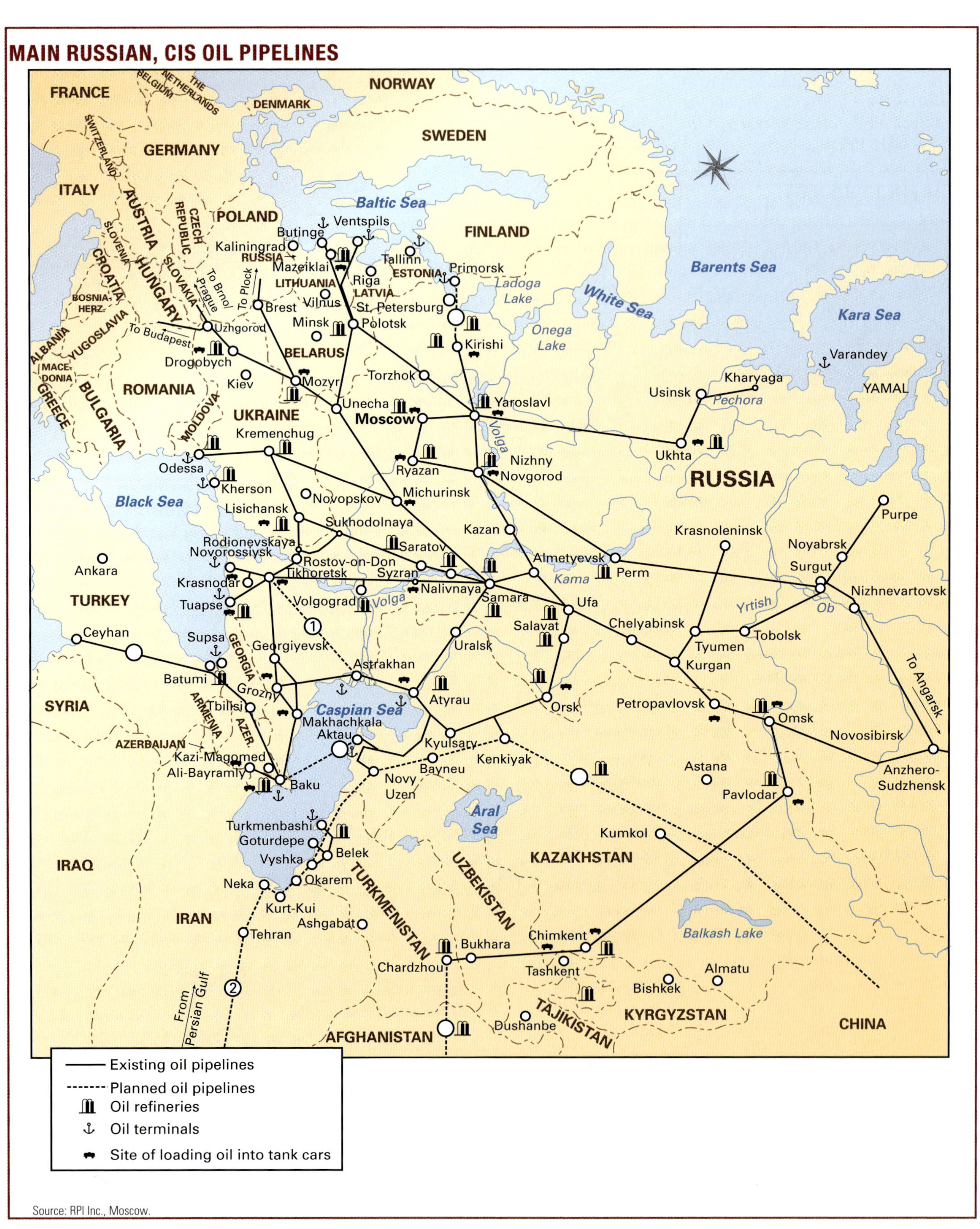

Source: RPI Inc., Moscow.

ported by tanker through the Mediterranean Sea to Asian and US markets, allowing Russia to bypass the Black Sea and its increasingly crowded Bosporus Straits. The entire pipeline will handle 100,000 b/d of crude in 2003, rising to 200,000 b/d after 5 years and to 300,000 b/d after 10 years.

Russia-China connections

Russia and China continued plans to build a 2,400-km (1,500-mile) crude oil pipeline linking Angarsk field in Western Siberia's Irkutsk region with refineries near China's top producing oil field complex at Daqing. State-owned China National Petroleum Corp. earmarked $700 million to invest in the pipeline, while Russian companies will invest another $1 billion, they reported in April.

The project, scheduled to start construction in 2003 and be completed in 2005, will transport an initial volume of 20 million tpy (400,000 b/d) by 2005. Design capacity would be expandable to 30 million tpy (600,000 b/d) by 2010 to accommodate production from new fields in the Angarsk region.

Also, a feasibility study for a 400,000-b/d pipeline linking eastern Siberia with northeastern China's Dalian port is due in mid-2003. Another pipeline project, proposed by Russia's pipeline operator Transneft, would deliver 1 million b/d of Russian crude from western and eastern Siberia to an export terminal at the Pacific coast port of Nakhodka, where oil could be shipped to China and other East Asian markets.

Baltic system

The capacity of the 457-km (284-mile) Baltic Pipeline System will be increased to 360,000 b/d from 240,000 b/d during the second stage of the project, which began construction in June. The first stage became operational at year-end 2001. The capacity increase includes building three more pump stations and eight storage tanks and upgrading the Yaroslavl-Kirishi pipeline. Completion is scheduled for year-end 2003.

Domestic plans

According to Russia's long-range development plans, 9,000 km (5,600 miles) of new oil pipelines could be built by 2010, said the US Commerce Department in March. Additionally, 10,000 km (6,000 miles), more than 20% of the existing network, could require some degree of refurbishing. State-owned AK Transneft operates, maintains, and develops Russia's pipeline system, which totals 46,800 km (more than 29,000 miles) of trunk pipelines.

During 2002 Transneft was planning to build and replace 1,044 km (649 miles) of trunklines, reinsulate 510 km (317 miles) of pipe, and commission seven pump stations. Transneft's design subsidiary, Giprotruboprovod Trunkline Design Institut (GTP), was also planning to develop a fuel and energy complex for the Timan-Pechora region and a pipeline from Tengiz field via Atyrau-Astrakhan-Grozny.

Gas pipelines

To offset declining production and meet export requirements, the Itera Group, Russia's second-largest gas supplier and the natural gas trader for state-owned OAO Gazprom, contracted in February to buy 9.5 bcm/yr (353 bcf/yr) of gas from Turkmenistan during 2002. Gazprom's gas transportation network was being opened to independent suppliers, which can use more than 15% of its capacity if available. However, 60% of Gazprom's existing gas pipelines need serious maintenance or outright replacement.

Blue Stream

The 1,218-km (757-mile) Blue Stream gas pipeline to Turkey was completed in October and became operational at year-end. The 378-km (235-mile) section was laid through the Black Sea in water depths approaching 2,100 m (7,000 ft) by the Saipem 7000, the world's only ship capable of pipelaying at that depth.

OIL EXPORT ROUTES FROM RUSSIA, CIS

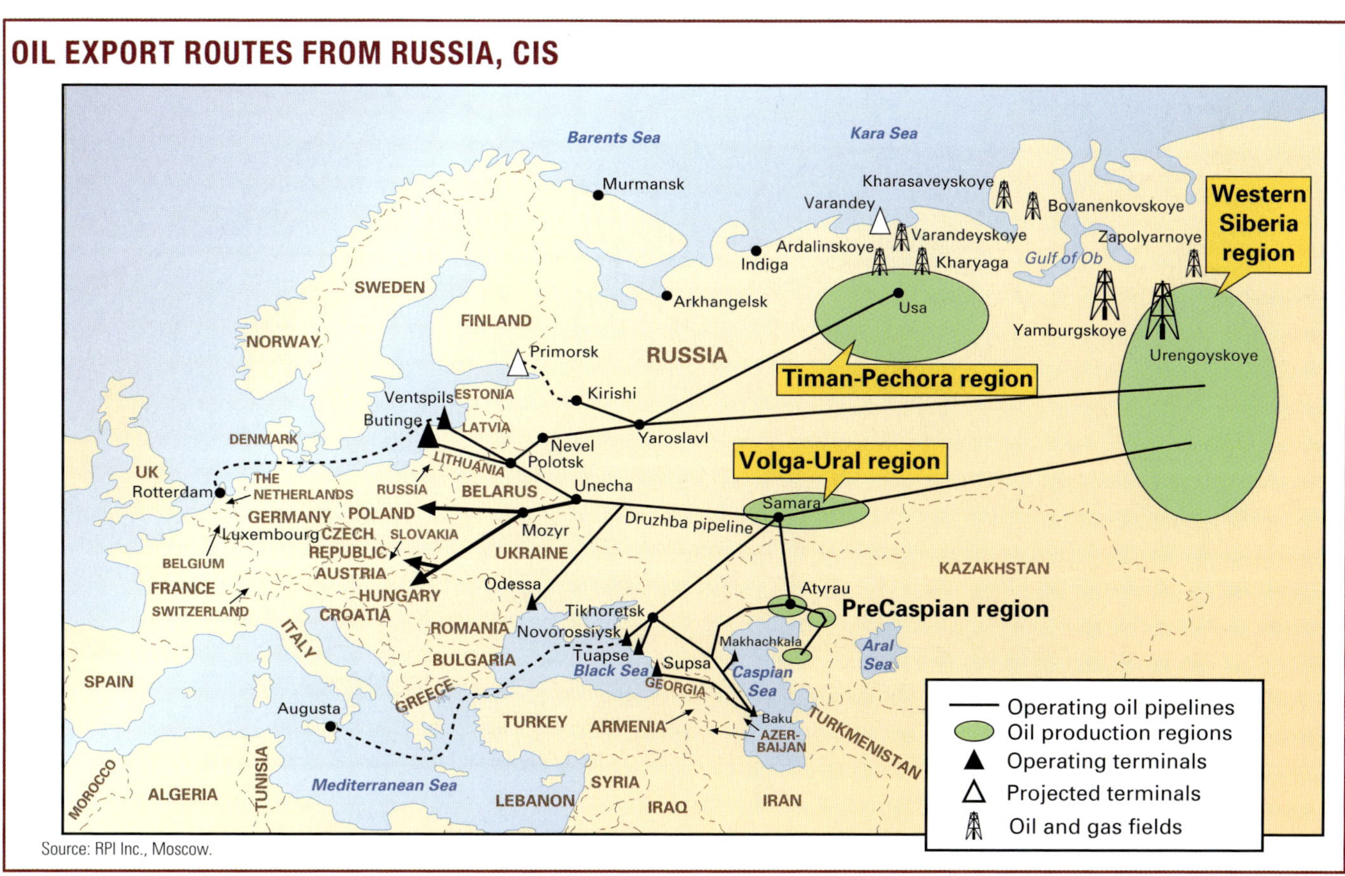

Source: RPI Inc., Moscow.

MAIN GAS PIPELINES IN RUSSIA AND CIS

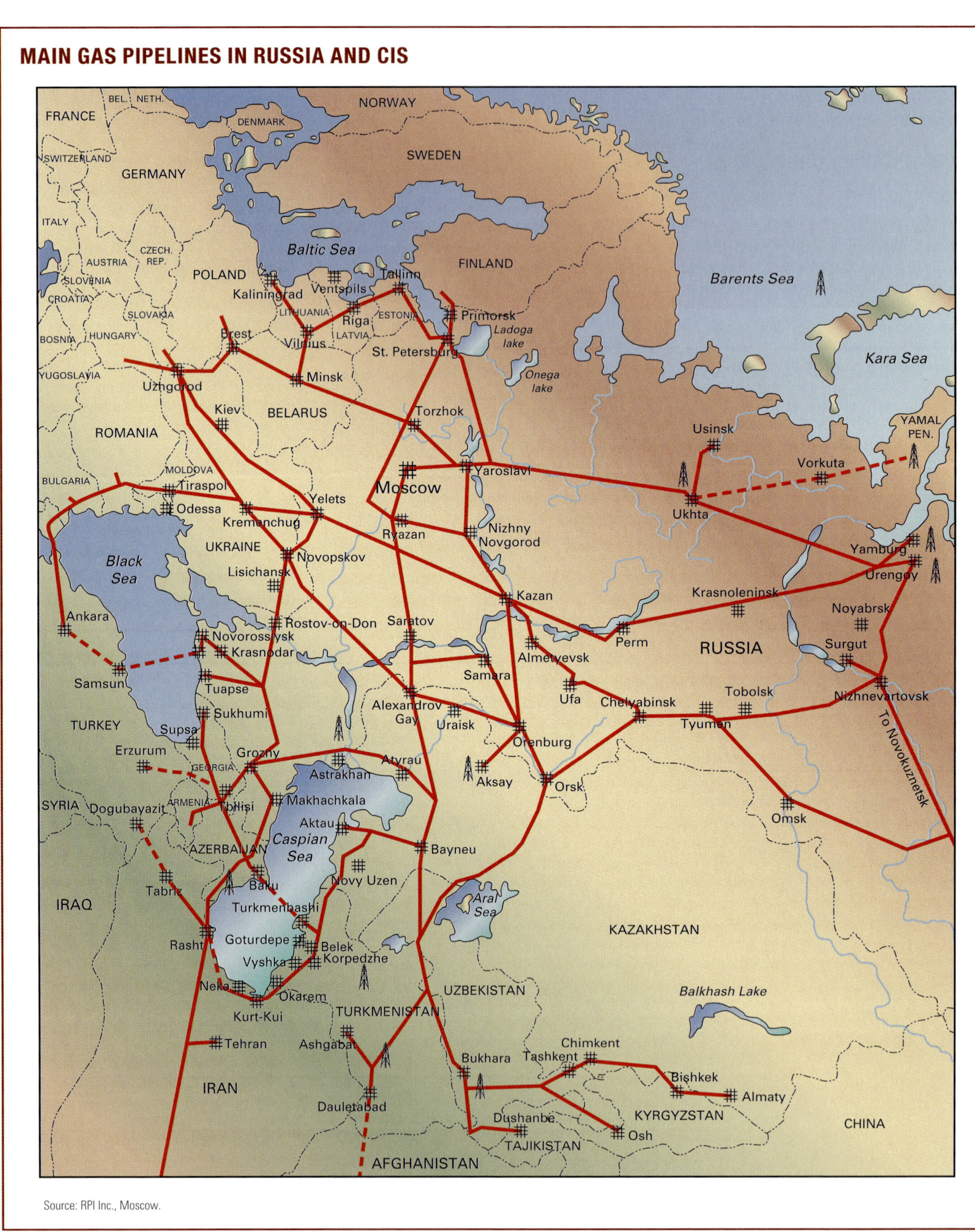

Source: RPI Inc., Moscow.

In 2003 Russia will send only about 2 bcm (70.6 bcf) of natural gas, owing to Turkey's current oversupply. Originally, Blue Stream deliveries were to increase to 15 bcm/yr (565 bcf/yr) by 2009, but Russia and Turkey renegotiated the agreement to lower expected volumes and prices.

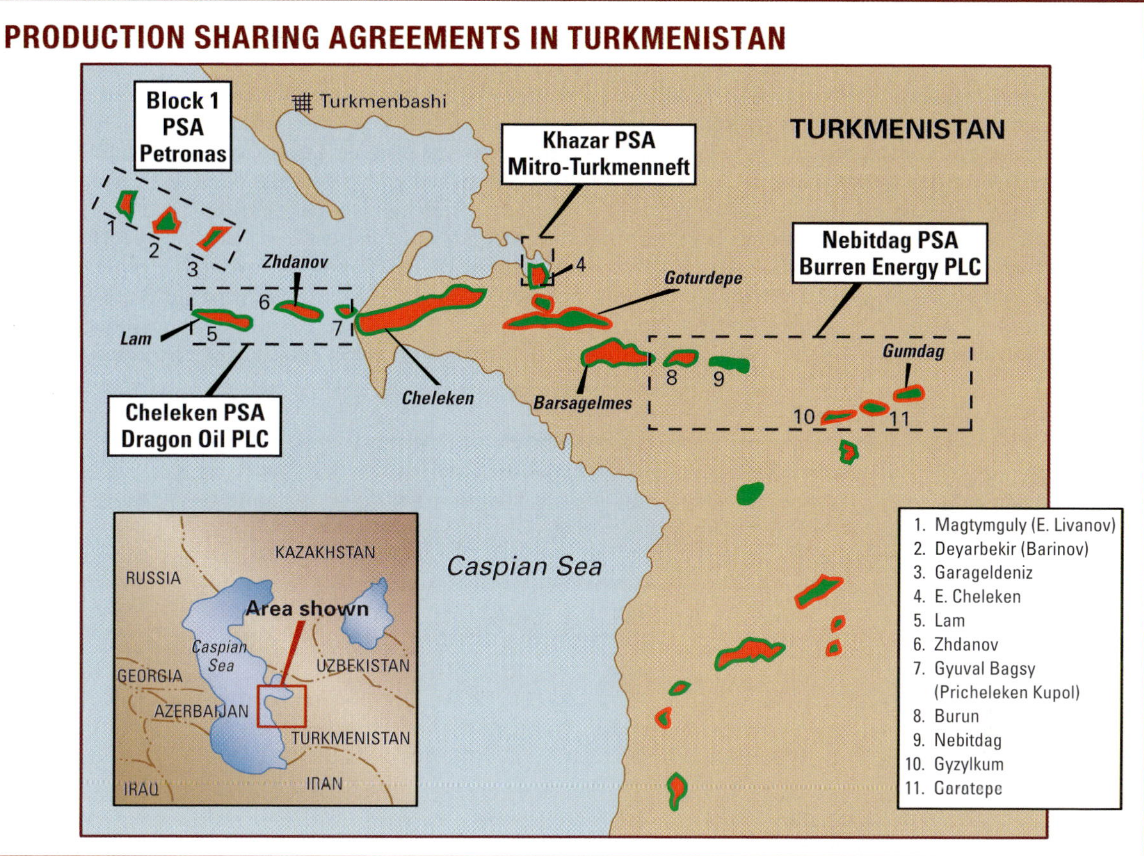

Ukraine, Yamal lines

Relations between Russia and Ukraine warmed considerably in mid-2002 when the two resolved long-term natural gas transit issues and agreed to form a consortium to modernize Ukraine's gas pipeline network. Russia will supply Ukraine with 25 bcm (918 bcf) of gas in 2003 as payment for transporting up to 107 bcm (4 tcf) of Russian gas to Europe. In addition, Gazprom will be allowed to operate Ukraine's underground gas storage facilities until 2013.

The second stretch of the Yamal-Europe gas pipeline through Poland was on hold during 2002, although Gazprom remains interested in diversifying its European export routes. The existing pipeline's capacity of 16 bcm/yr (600 bcf/yr) will handle up to 31.4 bcm/yr (1.17 tcf/yr) by 2003 when new compressor stations in Poland are completed. The second section would expand capacity to 56.4 bcm/yr (2.1 tcf/yr), but Russia and Poland have not agreed on a route.

Far East gas lines

In addition to pipelines proposed to supply Sakhalin Island gas to Japan and possibly to China, gas trunklines will be constructed, including a 1,000-km (600-mile) line to connect the island with the Russian mainland and a 700-km (450-mile) line across the island itself.

The China market is also the target for the East Siberian Kovykta gas project led by BP, which would build a 5,000-km (3,000-mile) gas trunkline to central China if Kovykta field's proven resources are sufficient to provide 25 bcm (930 bcf). Kovykta has estimated gas reserves of 49 tcf.

Another consortium, Sakha, proposes a 2,750-km (1,700-mile) gas pipeline from Chayandinovskoye field to northern China's Xinjiang region. Chayandinovskoye's estimated 1,200 bcm (43 tcf) of gas could supply the pipeline with 11.4-19.0 bcm/yr (423-706 bcf/yr).

Tanker traffic restricted

Turkey's concerns about the increasing tanker traffic through its Bosporus and Dardanelles Straits between the Black Sea and the Aegean Sea led to a policy banning supertanker traffic through the straits and severely restricting the passage of smaller tankers carrying cargo such as LPG and LNG.

Beginning in October, Turkey prevented supertankers from Russia and Kazakhstan from exporting oil through the straits, citing new regulations limiting tankers to 200 m in length (650 ft) to pass through the narrow, 30-km long (19-mile) Bosporus and restricting tanker length to 250 m (820 ft) for passage through the Dardanelles. The regulations also prohibit loaded oil tankers from passing through the straits at night and in other instances when visibility is obscured.

Tajikistan

CAPITAL: DUSHANBE

MONETARY UNIT: TAJIK RUBLE

REFINING CAPACITY: 0

OIL PRODUCTION: 0

OIL RESERVES: 12 MILLION BBL

GAS RESERVES: 200 BCF

Tajikistan relies heavily on Turkmen and Uzbek natural gas to meet domestic demand but has accumulated debts to suppliers for gas already consumed. During 2002 Tajikistan had already used up about 80% of its allotted annual gas volume by April.

A hydropower generation and distribution project was planned in a remote eastern province in Tajikistan, financed by the Aga Khan Fund for Economic Development and the International Finance Corp., with the Tajik government's support. The $26 million project involves creation of a new company, Pamir Energy Co., to generate and supply electricity in the Gorno-Badakhshan region that spans the Pamir mountains.

Pamir Energy will complete a partly constructed Soviet-era power plant, increasing its capacity from 14 MW to 28 MW, and operate another 8-MW hydroelectric plant in the city of Khorog, as well as several other smaller plants totaling 30 MW of installed capacity. Plans also include improving transmission and distribution facilities and adding a regulating structure to a nearby lake to ensure adequate flow in the winter months when demand is highest and water flow is at its lowest due to freezing.

As part of the arrangement, Pamir Energy will take over assets currently controlled by the state utility, Bark-i-Tajik, assuming responsibility for the 30,000 electricity customers and for improving and expanding the supply.

Turkmenistan

CAPITAL: ASHGABAT
MONETARY UNIT: MANAT
REFINING CAPACITY: 236,970 B/CD
OIL PRODUCTION: 180,000 B/D
OIL RESERVES: 546 MILLION BBL
GAS RESERVES: 71.0 TCF

Turkmenistan's economy remained healthy, with 13% GDP growth estimated for 2002. The country was seeking to boost its oil output to 200,000 b/d in 2002 and nearly 1 million b/d in 2010, plus about 110-120 bcm/yr (11.5 bcfd) of natural gas. President Niyazov survived an assassination attempt in November.

Turkmenistan and the other four nations with Caspian coastlines — Azerbaijan, Iran, Kazakhstan, and Russia — failed to agree on surface water boundaries or seabed mineral rights at a summit in April, prompting individual countries to pursue bilateral agreements.

Although its natural gas and oil reserves are abundant, Turkmenistan's export options are limited, and pipeline construction is at the top of the country's investment agenda, according to a series of reports by James P. Dorian (Honolulu). Although the projects have technically been approved, some have been temporarily shelved.

Turkmenistan's main export is natural gas, supplied to Russia and other neighbors including Ukraine, which was to import 37.9 bcm (1.41 tcf) in 2002 and 45.7 bcm (1.7 tcf) in 2003. The country also exports substantial oil volumes, which could increase to 320,000 b/d in 2005 and 660,000 b/d in 2010 if pipelines are rehabilitated and constructed.

A 1,460-km (900-mile) pipeline was proposed to carry up to 30 bcm/yr (2.9 bcfd) of natural gas through Afghanistan to Pakistan, probably to Sui field where existing infrastructure could be tapped to supply local markets. In mid-2002 financing for a feasibility study was secured from the Asian Development Bank after the three countries' governments agreed to move ahead with the project. Gas would come from one of Turkmenistan's largest fields, Dauletabad.

Turkmenistan also wants to build an 1,800-km (1,100-mile) line across the Caspian Sea to export up to 30 bcm/yr (2.9 bcfd) of gas to Erzurum, Turkey, via Azerbaijan and Georgia. Other gas pipelines being considered include a 1,500-km (900-mile) line connecting to the Trans-China gas pipeline, scheduled for 2005 completion, and a European connection through Iran and Turkey to Bulgaria.

For exporting oil, the most promising project would carry Turkmen oil to the Persian Gulf through the 2,500-km (1,600-mile) Kazakhstan-Turkmenistan-Iran pipeline. Also, the existing 42-km (26-mile) Vyshka-Belek trunkline could be rehabilitated, and a new, 135-km (84-mile) oil line from Okarem to Vyshka could be built.

To supply gas and oil export volumes, the government planned to install gas gathering facilities feeding the new 94-km (58-mile) Beshkyzyl-Yelkui-Uchadji pipeline and develop several oil fields including Eastern Cheleken, Northern Erdekli, and Southern Kamyshldja. Rehabilitation of Shatut, Ekerem, and Nebitlidje oil fields was also proposed.

Active PSAs in Turkmenistan include the offshore Cheleken PSA oil development,

GAS AND OIL PIPELINES IN TURKMENISTAN

where Dragon Oil plc (UAE, UK) was drilling three more production wells and expected to produce 11,400 b/d in 2002.

Also offshore, Block 1 was being developed by Petronas Carigali Sdn. Bhd. (Malaysia). Exploration well Ovez-IX tested at commercial production rates in early 2002 of 770,000 cu m/day (28.7 MMcfd) of gas and 2,190 b/d of condensate. In midyear the Makhtumkuli-2A exploratory well flowed more than 14,000 b/d and 539,000 cu m/day (about 20 MMcfd). Full-scale field development was being launched.

Onshore, rehabilitation of the Nebitdag area was under way, with anticipated 2002 production of 10,600 b/d, up from 8,200 b/d in 2001, according to Burren Energy plc (UK). The Khazar PSA area was also being rehabilitated and production wells drilled, with 2002 output expected to reach 8,000 b/d.

The government was planning to take bids in 2002 on building a 100,000-b/cd refinery, to be the country's third. The existing Turkmenbashi refinery was being upgraded and modernized, and upgrading of the Seidi facility was planned. Also at Seidi, the government wants to build a gas processing and petrochemicals complex that would produce 200,000 tpy of polyethylene.

Ukraine

CAPITAL: KIEV

MONETARY UNIT: HRYVNIA

REFINING CAPACITY: 1,024,759 B/CD

OIL PRODUCTION: 78,000 B/D

OIL RESERVES: 395 MILLION BBL

GAS RESERVES: 39.6 TCF

In 2002, for the first time, Ukraine received natural gas from Russia as payment for gas transportation services but did not buy any additional supplies. Instead, Ukraine imported gas from Turkmenistan and has a contract to purchase 237 bcm (8.83 tcf) during 2002-06. The two countries agreed in 2002 on a schedule for Ukraine to pay its past gas supply debts.

Relations between Ukraine and Russia warmed considerably in mid-2002 when the two resolved long-term natural gas transport issues and agreed to form a consortium to modernize Ukraine's gas pipeline network. Russia will supply Ukraine with 24.7 bcm (918 bcf) of gas in 2003 as payment for transporting up to 107 bcm (4 tcf) of Russian gas to Europe. In addition, Gazprom will be allowed to operate Ukraine's underground gas storage facilities until 2013.

Four gas fields in the Black Sea were being developed by Chornomornaftohaz. In mid-2002 Chornomornaftohaz invited foreign investors to participate in a joint venture to develop Odesa field, which holds proven reserves of 10.4 bcm (389 bcf).

Bugruvativske field in northeastern Ukraine's Dnieper-Donets basin has produced 41 million bbl of oil and still makes 5,000 b/d. Terms of the existing joint investment production activity (JIPA) regarding the oil field's development were revised with the state's Ukrnafta, said CanArgo Energy in September. Also, Gals-K Ltd. (Ukraine) will drill two wells under a farm-out agreement and will share incremental production with a CanArgo subsidiary.

Gas production began from Horodok field, which lies in western Ukraine's L'viv region, said Europa Oil & Gas Ltd. (London) and Zahidukrgeologia (Ukraine). Horodok field is estimated to hold 0.67-1.2 bcm (25-42 bcf) of gas reserves in Miocene sandstone at less than 1,000 m (3,300 ft). The field will be developed incrementally, with additional reserves potential in the northern extension to be tested in 2003.

Initially, three wells will produce gas into the nation's supply grid, with a fourth well being considered. An additional development well was being considered.

Uzbekistan

CAPITAL: TASHKENT

MONETARY UNIT: SOUM

REFINING CAPACITY: 222,271 B/CD

OIL PRODUCTION: 150,000 B/D

OIL RESERVES: 594 MILLION BBL

GAS RESERVES: 66.2 TCF

Several new gas fields will be developed in Uzbekistan, including the giant Kandym field, under a PSA signed in March 2002 between Uzbekneftegaz and Russia's Lukoil and Itera. The fields' reserves are estimated at 220 bcm (8.1 tcf), and combined natural gas production could rise from 4.25 bcm/yr (159 bcf/yr) to 7.5-9.4 bcm/yr (280-350 bcf/yr) at its peak.

A gas-condensate field was discovered in the dry bed of the Aral Sea in the Ustyurt basin, reported Uzbek sources through Caspian News Agency in August. Workers drilled a 2,700-m (8,900-ft) well and were preparing to begin commercial extraction at the deposit, named Uchsay.

IPE

CHINA

CHINA

CAPITAL: BEIJING

MONETARY UNIT: YUAN RENMINBI

REFINING CAPACITY: 4.528 MILLION B/D

OIL PRODUCTION: 3.4 MILLION B/D

OIL RESERVES: 18.25 BILLION BBL

GAS RESERVES: 53.3 TCF

Despite the global economic downturn, China's economy grew 11.2% in the year to fourth-quarter 2002, while real gross domestic product (GDP) increased during 2002 by an estimated 7%. The International Monetary Fund (IMF) projects that China's real GDP will increase 7.2% in 2003.

China is the world's second-largest energy consumer (after the US) and third-largest oil consumer, behind the US and Japan. The US Energy Information Administration expects China's oil consumption to surpass Japan's by 2012 and climb to 10.5 million b/d in 2020, making China a major factor in the world oil market.

As world energy demand increases through 2030 at a projected rate of 1.7%/yr, most of the growth will come from China and other developing countries, said the International Energy Agency (IEA), which called China "the new energy giant." By 2030 China's net oil imports will reach about 8% of projected world production.

China was considering opening a fuel oil futures exchange in Shanghai to make oil trading and pricing more transparent and to reflect market fundamentals and international standards. Fuel oil is one of the few oil products China allows to be imported.

Hu Jintao became the Chinese Communist Party's general secretary (head of the ruling party) in November 2002, signaling that he will likely become China's president in March 2003. Chinese officials are expected to proceed with many large energy projects, according to FACTS Inc. (Honolulu), but multinational oil companies should be on the lookout for changes. The supermajors are hoping to expand investment in China's domestically controlled industry, especially the retail market.

China relies on the Middle East for nearly half its oil imports, and tensions there have stimulated interest in other energy sources. China's oil imports from Russia increased dramatically throughout 2002. To improve its energy security as demand soars in the future, China was warming up its relations with Russia and welcomed a visit from President Putin in December.

Oil pipeline from Russia

China and Russia were expected to finalize plans for a 2,400-km (1,500-mile) oil pipeline from Angarsk field in western Siberia's Irkutsk region to refineries near China's top producing oil field complex at Daqing in northeastern China.

China National Petroleum Corp. (CNPC) and OAO Yukos (Russia) had already agreed on oil volumes and pricing. CNPC's PetroChina Co. Ltd. (Hong Kong) would develop the oil fields and operate the refineries. CNPC earmarked $700 million to invest in the oil pipeline, while Russian companies will invest another $1 billion. The route remained undecided, as China preferred a longer distance to bypass Mongolia for security reasons.

The project, scheduled to start construction in 2003 and be completed in 2005, will transport an initial volume of 20 million tonnes per year, or tpy (400,000 b/d) by 2005. Design capacity would be expandable to 30 million tpy (600,000 b/d) by 2010 to accommodate production from new fields in the Angarsk region. Yukos was shipping 1.5 million tpy (30,000 b/d) of crude by rail to China's state oil companies.

Also, a feasibility study for a 400,000 b/d pipeline linking eastern Siberia with northeastern China's Dalian port is due in 2003. Another pipeline project, proposed by Russia's pipeline operator Transneft, would deliver 1 million b/d of Russian crude from western and eastern Siberia to an export terminal at the Pacific coast port of Nakhodka, where oil could be shipped to China and other East Asian markets.

West-East gas pipeline

China's expansion of its natural gas

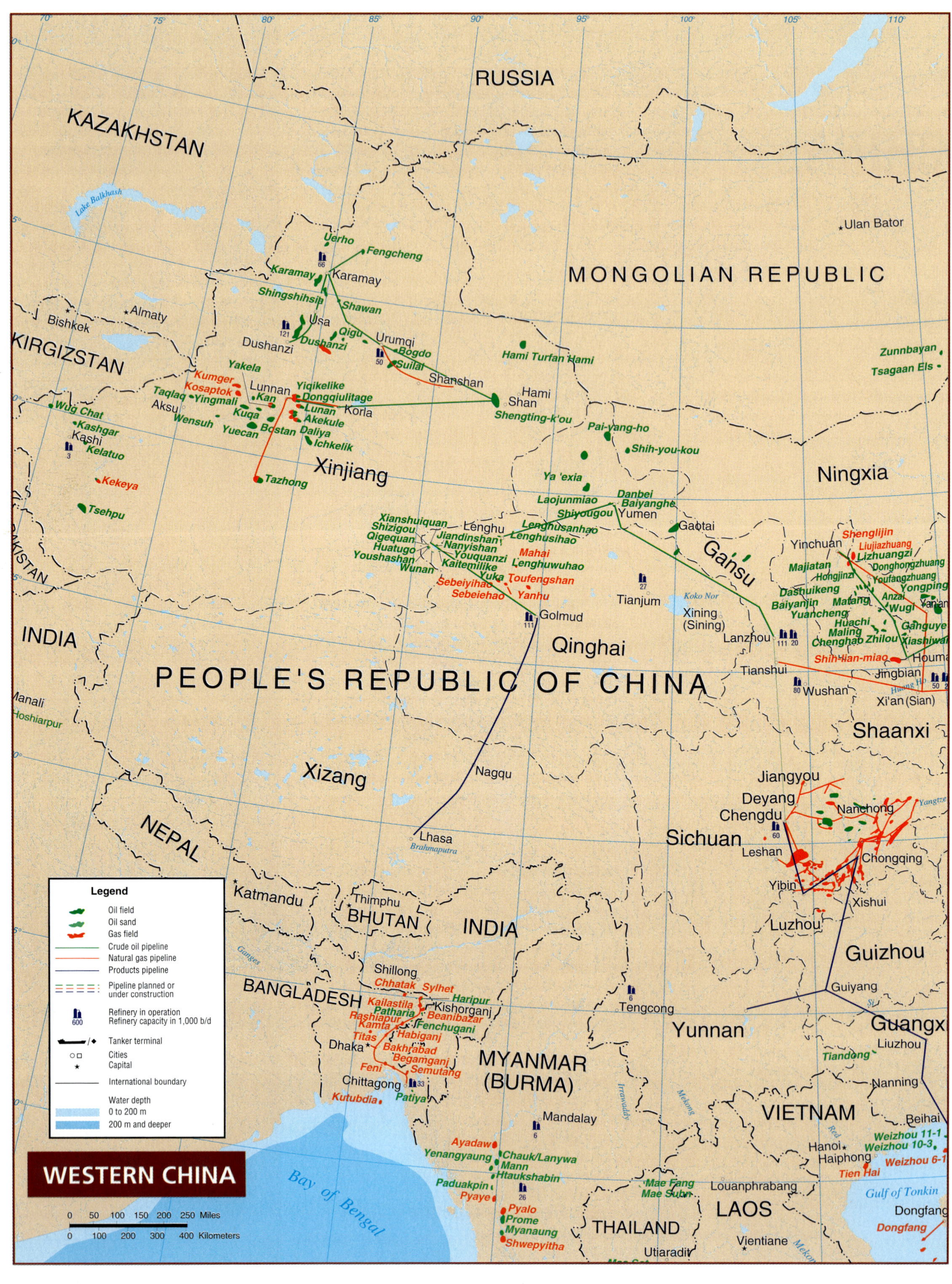

RUSSIA
KAZAKHSTAN
Lake Balkhash
MONGOLIAN REPUBLIC
Ulan Bator
Uerho
Fengcheng
Karamay
Karamay
Shingshihsia
Shawan
Almaty
Bishkek
KIRGIZSTAN
Usa
121
Qigu
Dushanzi
Dushanzi
Urumqi
Bogdo
50
Suilai
Hami Turfan Hami
Zunnbayan
Tsagaan Els
Yakela
Kumger
Kosaptok
Lunnan
Yiqikelike
Shanshan
Hami
Shan
Taqlaq
Yingmali
Kan
Dongqiulitage
Aksu
Lunan
Korla
Wug Chat
Kuqa
Akekule
Shengting-k'ou
Wensuh
Yuecan
Bostan
Daliya
Kashgar
Kashi
Ichkelik
Pai-yang-ho
Shih-you-kou
Kelatuo
3
Xinjiang
Tazhong
Kekeya
Ya 'exia
Ningxia
Tsehpu
Laojunmiao
Danbei
Baiyanghe
Shiyougou
Yumen
Xianshuiquan
Lenghu
Lenghusanhao
Gaotai
Shizigou
Jiandinshan
Lenghusihao
Qigequan
Nanyishan
Huatugo
Youquanzi
Mahai
Youshashan
Kaitemilike
Lenghuwuhao
Wunan
Yuka
Toufengshan
Sebeiyihao
Sebeiehao
Yanhu
Gansu
Yinchuan
Shenglijin
Liujiazhuang
Lizhuangzi
Majiatan
Donghongzhuang
Hongjinzi
Youfangzhuang
Dashuikeng
Yongping
Anzai
Baiyanjin
Matang
Wugi
Yanan
Yuancheng
Huachi
Ganguye
Maling
Chenghab
Zhilou
Xiashiwan
Shih-lian-miao
Houma
27
Tianjum
Koko Nor
Xining (Sining)
Lanzhou
111
20
Golmud
111
Qinghai
INDIA
PEOPLE'S REPUBLIC OF CHINA
Tianshui
Jingbian
80
Wushan
Huang He
50
Xi'an (Sian)
Manali
Hoshiarpur
Shaanxi
Xizang
Nagqu
Jiangyou
Deyang
Chengdu
Nanchong
Yangtze
60
NEPAL
Lhasa
Brahmaputra
Sichuan
Leshan
Chongqing
Yibin
Xishui
Katmandu
Thimphu
BHUTAN
INDIA
Luzhou
Guizhou
Ganges
Shillong
Chhatak
Sylhet
Guiyang
BANGLADESH
Haripur
Kailastila
Kishorganj
Patharia
Rashiapur
Beanibazar
Kamta
Fenchugani
6
Tengcong
Titas
Habiganj
Yunnan
Guangxi
Dhaka
Bakhrabad
Liuzhou
Begamganj
Tiandong
Feni
Semutang
MYANMAR (BURMA)
Chittagong
33
Kutubdia
Patiya
Irrawaddy
Mekong
Nanning
VIETNAM
Mandalay
6
Beihai
Red
Weizhou 11-1
Weizhou 10-3
Ayadaw
Hanoi
Haiphong
Weizhou 6-1
Yenangyaung
Chauk/Lanywa
Mann
Htaukshabin
Tien Hai
Paduakpin
Mae Fang
Mae Suhn
Louanphrabang
Gulf of Tonkin
Pyaye
26
Dongfang
Bay of Bengal
Pyalo
LAOS
Dongfang
Prome
Myanaung
THAILAND
Vientiane
Shwepyitha
Utiaradit
Legend
Oil field
Oil sand
Gas field
Crude oil pipeline
Natural gas pipeline
Products pipeline
Pipeline planned or under construction
600
Refinery in operation
Refinery capacity in 1,000 b/d
Tanker terminal
Cities
Capital
International boundary
Water depth
0 to 200 m
200 m and deeper
WESTERN CHINA
0 50 100 150 200 250 Miles
0 100 200 300 400 Kilometers
70°
75°
80°
85°
90°
95°
100°
105°
110°

EASTERN CHINA
RUSSIA
MONGOLIAN REPUBLIC
PEOPLE'S REPUBLIC OF CHINA
NORTH KOREA
SOUTH KOREA
JAPAN
OKINAWA
TAIWAN
PHILIPPINES
VIETNAM
LAOS
THAILAND
Heilongjiang
Jilin
Liaoning
Nei Mongol
Ningxia
Gansu
Shanxi
Hebei
Shandong
Henan
Shaanxi
Jiangsu
Anhui
Hubei
Sichuan
Hunan
Zhejiang
Jiangxi
Guizhou
Fujian
Yunnan
Guangxi
Guangdong
Beijing
Tianjin
Shanghai
Sea of Japan
Yellow Sea
Bohai Gulf
East China Sea
South China Sea
Gulf of Tonkin
HAINAN DAO
0 100 200 300 Miles
0 200 400 Km

infrastructure includes developing natural gas fields in the Tarim basin of western China's Xinjiang province and building the West-to-East pipeline to emerging markets in and around Shanghai on the eastern coast. The 4,167-km (2,589-mile) line would carry Tarim gas and pick up additional volumes from central China's Ordos basin along the way. Also, it could eventually be extended to tap Central Asian gas.

PetroChina is heading the project and selected Shell International Gas Ltd. (UK) as lead contractor of the pipeline consortium in early 2002. ExxonMobil Corp. (Irving, TX, US) and Petronas Carigali Sdn. Bhd. (Malaysia) were said to be seeking participation. Construction began in late 2001, and completion is scheduled for 2004.

Guangdong LNG project

China's first LNG project, which involves construction of an LNG import terminal and high-pressure gas pipelines, got a boost in November when Australia's North West Shelf venture participants signed agreements with the Guangdong LNG project companies for the purchase and supply of LNG from the North West Shelf in Western Australia.

China National Offshore Oil Corp. (CNOOC) had announced in August that the North West Shelf venture would be its sole supplier. The agreements signed by the six North West Shelf LNG sellers cover the supply of 3.3 million tpy of LNG in Phase I of the Guangdong LNG project for 25 years starting in late 2005. The contract is valued at $20-25 billion (Aus.).

Under Phase I of the Guangdong construction project in China, the grassroots LNG receiving terminal and regasification plant will be built along with 300 km (190 miles) of pipeline on the eastern side of the Pearl River delta in Guangdong province. A lateral pipeline will also be constructed to deliver natural gas to Hong Kong.

In addition to the Chinese facilities, Phase I calls for construction of additional LNG processing and a second trunkline from the North Rankin A platform to shore in western Australia. Each of three existing processing trains at the Karratha LNG liquefaction plant on the Burrup Peninsula produces 2.5 million tpy of LNG, and construction was under way on the fourth, which alone will have a capacity of 4.2 million tpy of LNG.

First LNG from the fourth train is scheduled for mid-2004. A fifth LNG liquefaction train at Karratha also was being designed. Completion of the fourth and fifth trains will more than double Karratha's current LNG processing capacity.

Phase II of the China construction, planned to start in 2008, is an extension of the pipeline around the western side of the Pearl River delta. Regasified LNG will be supplied to electric power generation plants and city gate distributors in Guangdong Province and in Hong Kong. In addition, two to three new LNG transport vessels will be required to service the China trade route. Under the Phase II proposal, the North West Shelf partners and the Chinese shipping companies would establish a joint venture to support LNG transport to Guangdong.

As a result of the sales and purchase

agreements, CNOOC's offshore oil and gas producing unit CNOOC Ltd. also will have the opportunity to acquire a participating interest in North West Shelf reserves and production that will supply gas to Guangdong.

Gas from the LNG development will be used primarily in China's southeastern coastal region, where Guangdong province has already launched a project to build six 320-MW gas-fired power plants and to convert existing oil-fired plants with a total capacity of 1,800 MW to gas.

Tangguh LNG plans

China was also planning a second new LNG terminal — this one at Fujian — that would be supplied with LNG from the BP plc (UK) Tangguh gas field in Indonesia. CNOOC, which has a 60% equity interest in the proposed terminal, plans to begin regasification operations in 2006-07 with an initial capacity of 2.5 million tpy of LNG. The Fujian terminal would be built on the coast of southern China, opposite Taiwan, with construction to start in 2004.

In October CNOOC Ltd. agreed to buy a stake in the reserves and upstream production of the Tangguh joint LNG project in Indonesia from BP, in what CNOOC called an acquisition "complementary to its natural gas strategy."

The Tangguh LNG project is operated by Indonesian state oil and gas enterprise Pertamina and BP Indonesia in Berau Bay, Irian Jaya. Tangguh consists of three offshore production-sharing contracts (PSCs) and the planned onshore LNG terminal. Pertamina will own the terminal, which will be operated by a company jointly owned by Pertamina and the PSC partners.

South China Sea E&P

A natural gas wildcat well, PY34-1-1, confirmed an earlier gas discovery in the PY34-1 area, reported CNOOC Ltd. in September. The well was tested to produce 350,000 cu m/d (13 MMcfd). PY34-1-1, at the Pearl River Mouth basin in the eastern South China Sea, is about 30 km (19 miles) southwest of the gas-bearing PY30-1 structure.

The discovery followed on the heels of a mid-2002 natural gas discovery in the western South China Sea. An appraisal well, YCH13-4-2 on the YCH13-4 discovery, produced about 540,000 cu m/d (20 MMcfd) on a 16-mm choke during drillstem tests.

In the Beibu Gulf basin of the South China Sea, Blocks 23/15 and 23/20 will be explored by a unit of Husky Energy Inc. (Calgary) under PSCs announced in September by CNOOC Ltd. Blocks 23/15 and 23/20, which cover 1,327 and 1,543 sq km, respectively (512 and 596 sq mi), are 80 km (50 miles) from Weizhou oil field.

Husky will drill and complete one wildcat well in each contract area in the first 3 years and will finance 100% of the exploration expenditures. CNOOC Ltd. will take a 51% participating interest in any commercial discoveries.

In December, Husky and CNOOC signed a production-sharing agreement (PSA) to explore for oil on deepwater Block 40/30 in the Pearl River Mouth basin of the South China Sea. The block is about 30 km (20 miles) southwest of the gas-bearing PY30-1 structure. The agreement is the first deepwater exploration PSA signed between CNOOC and foreign petroleum companies since it announced the tendering of deepwater areas in September. Other international companies are studying the geological data of the 12 deepwater blocks CNOOC offered.

Husky's Block 40/30, which lies about 100 km (60 miles) southeast of Hainan Island, covers an area of 6,704 sq km (2,588 sq mi) in water 600-1,500 m deep (2,000-4,900 ft). Under a three-phase exploration period, Husky must drill one wildcat at least 1,600 m deep (5,250 ft) in Phase I.

Wenchang Production

In July Husky produced first oil for CNOOC Ltd. from Wenchang field in the Pearl River Mouth basin about 140 km (90 miles) east of Hainan Island and 400 km (250 miles) southwest of Hong Kong. Production from Wenchang Blocks 13/1 and 13/2 surpassed 60,000 b/d of oil, exceeding the 50,000 b/d originally estimated for the blocks.

Cash flow from 13/1 and 13/2 oil sales will support exploration on the surrounding 230 sq km (57,000 acres, or 89 sq mi) on Block 39/05, which Husky also will operate under a 2001 agreement. Husky plans to initiate a drilling program for Block 39/05 involving two or three wells by early 2003. Wenchang facilities could be used to develop fields in the Block 39/05 lease area covering 5,700 sq km (2,200 sq mi).

Wenchang 13-1 and 13-2 fields in the western Pearl River Mouth basin are believed to hold 83 million bbl of proved reserves and 9 million bbl of probable reserves. The two fields are 7 km (4 miles) apart in 120 m (400 ft) of water about 140 km (90 miles) east of Hainan Island. Oil is produced into the Nanhai Endeavour turret-moored floating production, storage, and offloading (FPSO) vessel, which has storage capacity of 850,000 bbl. Husky estimated a 21-well development and a field life of 10-12 years, with production downtime at 35 days/yr.

S. CHINA SEA DEEPWATER BLOCKS

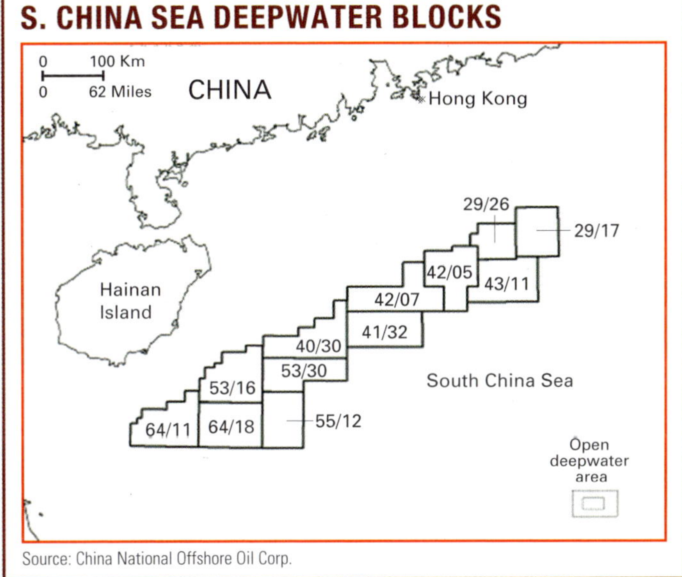

Source: China National Offshore Oil Corp.

Also in the South China Sea's Beibu Gulf, the 6/12-1 prospect will be explored by Bligh Oil & Minerals NL (Brisbane) and three other Australian companies. Plans included drilling a 1,750-m (5,750-ft) deep well 10 km (6 miles) east of the 12/1-1 producing field complex and less than 5 km (3 miles) from a pipeline. Primary target is the Weizhou formation at 1,400-1,750 m (4,600-5,750 ft). Other discoveries on the block are development candidates, the partners said in early 2002.

Bids were requested from foreign companies for oil and gas exploration and development of 12 virgin deepwater blocks in the South China Sea, said CNOOC in September. The area, south and east of Hainan Island, covers 76,000 sq km (29,000 sq mi) in waters 300-2,000 m (1,000-6,500 ft) deep.

Also in the South China Sea, a 15,400 sq km (5,950 sq mi) block southwest of Taiwan in the Tainan basin and Chaoshan Trough will be explored by CNOOC Ltd. and Taiwan's Chinese Petroleum Corp., which signed a long-negotiated agreement announced in June. The contract calls for reprocessing 500 km (300 miles) of seismic data, acquiring 4,000 km (2,500 miles), and drilling three wildcats. The companies will form a joint management committee and share the exploration costs equally.

Bohai Gulf E&P

An oil field was discovered on the Liaodong Block in the Bohai Gulf, said CNOOC Ltd. in October. JZ25-1s-1, drilled to 1,960 m (6,430 ft) in 24 m of water (80 ft), made a combined flow of 2,300 b/d of 27-29 degree gravity oil on drillstem tests. The find is 40 km (25 miles) north of SZ 36-1 oil field and an equal distance south of JZ 20-2 gas-condensate field.

Also in Bohai, two exploration wells discovered oil on Block 04/36, said Kerr-McGee Corp. (Oklahoma City) in June. The CFD 11-3-1 well, drilled in 30 m of water (90 ft), found high-quality oil in 14 gross m (45 ft) of sandstone reservoir. The well is 6 km (4 miles) east of CFD 11-1 field and 10 km (6 miles) northeast of CFD 11-2 field. The second exploration well, CFD 16-1-1, which lies 15 km (9 miles) southwest of CFD 11-1 field in 23 m (74 ft) of water, penetrated multiple zones of oil-bearing sandstones totaling nearly 20 gross m (70 ft).

The oil reservoirs found with these two wells are similar to oil-bearing sands that were drillstem-tested in numerous nearby wells on Block 04/36, so no additional testing was needed. Kerr-McGee planned to begin appraisal drilling during 2002. Its development strategy is to use CFD 11-1 field as a hub, with its spokes being CFD 11-2 and the discoveries.

In May, Kerr-McGee approved development of its discoveries on Block 04/36 in Bohai Gulf — CFD 11-1, CFD 11-2, and CFD 2-1 fields — using a centrally located FPSO vessel, along with fixed platforms that will serve as drillsites for dry wellheads. The FPSO, which is expected to be fabricated in China, will be located in 23 m of water (75 ft), 30 km (50 miles) from shore. It will be able to process more than 60,000 b/d of oil and will have a 1 million bbl storage capacity.

Kerr-McGee estimates area reserves at more than 150 million bbl. With government and partner approval of the development program, initial production could begin by year-end 2004, with peak production projected to exceed 50,000 b/d by mid-2005. The company has five Bohai Gulf discoveries to date and a number of additional prospects identified within its three operated licenses. Its exploration program includes discoveries CFD 12-1 and CFD 12-1S on the adjacent 05/36 block.

In the Bohai Gulf's Liaodong Bay, two independent appraisal wells were completed successfully, reported CNOOC Ltd. in June. LD 4-2-1, drilled near the LD 4-2 discovery made by wildcat SZ 36-1W-2 earlier, flowed 1,000 b/d of 24-26 degree gravity oil on two drillstem tests. It is 5 km (3 miles) southwest of SZ 36-1 oil field. LD 5-2-1, 2 km (1 mile) west of SZ 36-1 field, appraised the LD 5-2 discovery made by wildcat SZ 36-1-11. LD5-2-1 flowed more than 200 b/d of 14.7-16 degree gravity crude on drillstem tests.

More appraisal drilling was planned on Bohai Gulf Block 02/31, said CNOOC Ltd. and partners in June. The LD 27-2-3 appraisal well in Liaodong Bay 105 km (65 miles) southeast of Qinghuangdao encountered four oil zones that flowed at a combined 4,600 b/d. The well went to 2,700 m (8,900 ft) in 21 m of water (70 ft).

In early 2002 CNOOC and Shell signed a PSC for the Bohai Gulf's Bonan area, which Shell will operate on Block 11/26. The contract calls for at least two wells to be drilled in 2 years, with an option for drilling two additional wells during the subsequent 4 years. The block is in 25 m of water (80 ft) and 30 km (20 miles) northwest of Longkou.

Two more platforms came onstream in mid-2002 in Qinghuangdao 32-6 field, 130 km (80 miles) east of the city of Tianjin, said CNOOC Ltd. Platforms C and D added slightly more than 20,000 b/d of oil production to the output from first two platforms, A and B, which came on line in 2001 with an initial 25,000 b/d of oil. When platforms E and F start up, production is expected to peak at 65,000 b/d in 2002. Qinghuangdao 32-6 field's reserves are 103 million bbl.

Peng Lai onstream

Production began in early 2003 from Peng Lai 19-3 field in the Bohai Gulf, reported CNOOC and a subsidiary of ConocoPhillips (US). Phillips China Inc. has invested $2 billion at Peng Lai field, which analysts say may hold China's second-largest reserves behind PetroChina's onshore Daqing field. Peng Lai 19-3 eventually is expected to produce 120,000-150,000 b/d. The discovery well was followed by six successful appraisal wells.

Phase I development will utilize a single, 24-slot wellhead platform and an FPSO vessel with expected daily gross production rates of 35,000-40,000 b/d of oil.

Sichuan gas development

Gas resources in central China's Sichuan basin will be developed by Sunwing Energy Ltd. (Calgary) in partnership with China International Trust & Investment Corp. (CITIC) Energy, the companies said in October. The partnership will also expand its operations beyond China, acquiring interests in global oil and gas development projects and working to introduce gas-to-liquids and other technologies.

Sunwing, a subsidiary of Ivanhoe Energy Inc. (Vancouver), has a PSC with a PetroChina unit for joint venture development of gas reserves on Zitong Block in the western Sichuan basin. The 30-year PSC, announced in September, covers 3,600 sq km (900,000 acres, or 1,400 sq mi) with gross resource potential of 130 billion cu m, or bcm (5 tcf). The Sichuan basin is China's largest gas-producing region, with output of more than 21 million cu m/d (800 MMcfd) and an estimated 7 trillion cu m in place (260 tcf), Chinese officials have said. Ivanhoe also has the right to negotiate for a PSC on the 4,000 sq km (1 million acre, or 1,600 sq mi) Yudong Block on the eastern edge of the Sichuan basin.

In addition to developing existing producing structures, Ivanhoe will conduct exploration activities, which will include acquiring new 2D seismic data, reprocessing existing seismic data, and drilling at least four exploratory wells over 6 years.

Drilling could commence in late 2003. Interpretation of existing 2D seismic data has identified 16 structures with hydrocarbon-bearing potential. Four wells are producing, and larger potential could exist from other zones in the producing well bores.

Natural gas resources discovered in the Sichuan basin's Chuanzhong Block, which covers 24,000 sq km (1.8 million acres, or 9,400 sq mi), will be developed by a unit of Burlington Resources Inc. (Houston), which signed a gas purchase and sales deal with PetroChina in April — the first long-term agreement of its kind in China. The development-linked agreement covers a large natural gas field known to have several trillion cubic feet of potentially recoverable resources in tight sands reservoirs, with remaining potential awaiting evaluation.

Burlington expected to file a development plan for the first field, Bajiaochang, by year-end 2002, with work to start in 2003. Peak production would not occur for several years, pending expansion of the area's existing pipeline infrastructure and development of gas markets.

Burlington's other Chinese activities include a partnership with CNOOC in the Panyu oil development on Block 15/34, which contains Bootes and Ursa fields and an estimated 75 million bbl of reserves. Fabrication is under way on two offshore platforms as well as an FPSO at Panyu. Production is expected to reach a net peak of 17,500 b/d of oil beginning in late 2003.

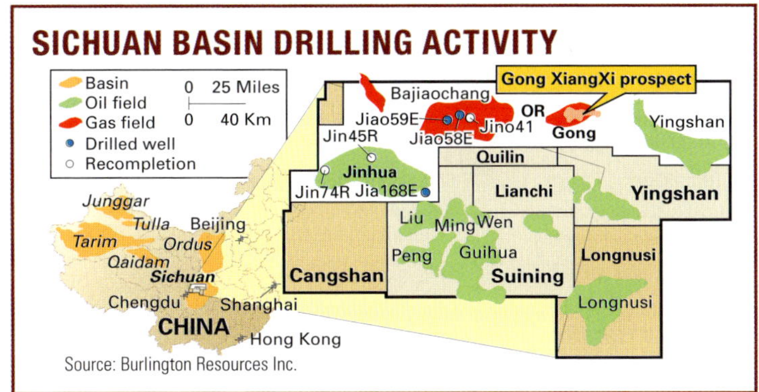

Source: Burlington Resources Inc.

Other E&D

East China Sea

Development of the Xihu Trough in the East China Sea was inaugurated in April by CNOOC Ltd. and China Petroleum & Chemical Corp. (Sinopec), with formation of a joint management committee and establishment of East China Sea Xihu Trough Oil & Gas Operating Co. The companies will develop a 59,565 sq km block (23,000 sq mi) 450 km (280 miles) southeast of Shanghai that contains the Chunxiao, Tianwaitian, and Duanqiao gas discoveries and the Canxue oil and gas discovery. They plan phased development of gas fields with first production by 2005.

Onshore Xinjiang

Oil and gas were discovered in central Junggar basin in northwestern China's Xinjiang region, said PetroChina in June. The Shidong 2 well, drilled to 2,683 m (8,802 ft), flowed 220 b/d of light oil and 2,780 cu m/d of gas (103.5 Mcf/d). PetroChina was assessing development and hopes to confirm additional reserves. PetroChina's Karamay oil field in the Junggar basin produced 23.9 million bbl during the first four months of 2002.

Qaidam basin

The first exploration well was drilled in spring 2002 on the natural gas-prone Sebei block in northwestern China's Qaidam basin, said Agip China BV. Sebei field, the largest in Qinghai Province's Qaidam basin, has proved and probable gas reserves of more than 90 bcm (3.3 tcf), according to CNPC estimates. A PetroChina unit, which was producing gas from shallow structures on the Sebei block, planned to raise gas production capacity there by 21% in 2002, to 1.15 bcm/yr (42.8 bcf/yr). Sebei gas is transported via a 953-km (592-mile) pipeline to Qinghai provincial capital Xining and to Gansu provincial capital Lanzhou.

Alternative-orientation hydraulic refracturing successfully stimulated low-permeability reservoirs in China's Maling oil field, located east of Lanzhou. Maling is a low-permeability sandstone reservoir, and the target formation for the refracturing was the Jurassic Yan'an group.

The treatments involved creating a new fracture azimuth by directionally perforating and plugging of existed fractures. Analysis indicated that some reasons for the low oil production rate prior to the retreatment of the wells included closure of the initial hydraulic fractures, reservoir sensitivity to water, wax and paraffin deposition, and scale buildup.

Tarim, Ordos geology

Five blocks prospective for oil and gas in northwestern China's Tarim and Ordos basins were the subject of a study by Sinopec and Royal Dutch/Shell Group (Netherlands), which was completed in 2002 as part of a partnership formed when Shell invested in Sinopec's initial public offerings in 2001. A Sinopec unit is producing oil and gas in the Tarim basin.

Two of the blocks in the Ordos basin, bordering the Inner Mongolia Autonomous Region and Shaanxi province, have potential for medium-sized gas fields. In the northern Tarim basin, the companies are studying geological data on three blocks, one of which is thought to contain gas and condensate.

Of the other two Tarim blocks, chances are high for confirming significant crude reserves, government sources said. After completion of the study, the two companies will decide whether to enter into PSCs to further explore and develop hydrocarbons on the five blocks.

Sulige field

Giant Sulige gas field in the Ordos basin in Inner Mongolia could be developed with investment by TotalFinaElf SA (France), if feasible, under a contract with PetroChina. Proved reserves exceed 230 bcm (8.5 tcf) of natural gas.

Songliao basin

A development well was completed in Daan oil field at 125 b/d from the Putaohua sand at 1,315 m (4,314 ft), said MI Energy Corp. (Houston) in June. The Daan 216-1 well, drilled to the Fuyang formation at 2,068 m (6,785 ft), was the field's eighth development well; an additional 30 were scheduled for 2002.

Putaohua is a major producing sand in the Songliao basin in northeastern China and is productive in Daan Bei field just north of Daan field. MI Energy operates Daan, Moliqing, and Miao 3 oil fields in the Songliao basin in partnership with PetroChina.

Petrochemicals plants

Construction began in 2002 on a world-scale, grassroots, integrated petrochemical complex that will produce 600,000 tpy of ethylene, plus derivatives. The facilities are being built at Nanjing, Jiangsu province. A 175-MW natural gas-fired, combined-cycle power plant at the complex will be built by Daelim Industrial Co. Ltd. (Seoul) under contract to Sinopec and BASF AG (Germany). The GE Power Systems turbines, due on-site in spring 2003, can use naphtha as a backup fuel. Commercial operation is set for mid-2004.

Another world-scale petrochemical complex is being built at Huizhou, Guangdong province, in southern China. The complex, expected to be onstream by the end of 2005, will manufacture about 2.3 million tpy of products including 560,000 tpy of styrene monomer, 250,000 tpy of propylene oxide, 135,000 tpy of polyols, 60,000 tpy of mono-propylene glycol, and 320,000 tpy of mono-ethylene glycol. Products sales could generate as much as $1.7 billion, primarily supplying customers in Guangdong and the high consumption areas of China's coastal economic zones.

The construction contract was awarded in November by a CNOOC-Shell joint venture to Technip-Coflexip Group (France) and partners. The complex is part of the $4.3 billion Nanhai petrochemical project, one of the largest Sino-foreign joint venture projects in China, which is based on an agreement between CNOOC and Shell Petrochemicals Co. Ltd. Construction will start in early 2003 at the 4.3 sq km site (1.7 sq mi) at the Daya Bay Economic and Technical Development Zone. The entire complex has been designed to international standards to protect the environment and use energy and material efficiently.

Production capacity at a carbon black facility in Shanghai is being increased by more than 50,000 tpy, said Shanghai Cabot Chemical Co. Ltd. in September, using advanced technology developed by Cabot Corp. (Boston). The plant expansion will be fully operational by early 2004.

Refining

China's Guangzhou refinery was slated for expansion under a proposed agreement between ExxonMobil and Guangzhou Petrochemical Corp., which would form a joint venture to increase capacity to 10 million tpy by 2005 and possibly to 18 million tpy in later years.

The companies also were considering expanding Guangzhou Petrochemical's ethylene cracker capacity to 300,000 tpy by 2005 and later to 800,000-1 million tpy. ExxonMobil earlier took a stake in Sinopec, parent of Guangzhou Petrochemical.

A 50,000 b/cd refinery in Ningxia province, along with a 36,000 b/d catalytic cracking unit, are now owned by CNPC through its acquisition of Ningxia Dayuan Refining & Chemical Corp. The refinery's crude comes from CNPC's Changqing oil field in Shaanxi province. The refinery, the only one in the province, operated at 50% capacity during 2001 due to prohibitive oil prices. With the acquisition, CNPC will bring the Ningxia operation under its unified crude allocation and oil products sales system.

PetroChina plans to buy the 1.5 million tpy Dongxing refinery, located at Zhanjiang port in southern China's Guangdong province, and expand its capacity to 5 million tpy by 2006. The acquisition would cut the cost of transporting products to PetroChina's 100 Guangdong gasoline stations. All of PetroChina's refineries are in the northeast and northwest, far from the large eastern and southern markets, so oil products have had to be shipped to Guangdong.

Product markets and imports

Sinopec planned to build 450 retail gasoline stations with foreign companies in 2002, as China lifts the ban on foreign investment in the retail fuels business. This will be the first wave of about 1,500 such sites to be built with Shell, ExxonMobil, and BP during 2002-2005. Sinopec also was building 1,000 independent retail outlets in 2002.

China has a total of 75,000 retail stations, about a third owned by Sinopec vs. only 8,000 a few years ago. China had banned foreign investment in the oil products retail business until it joined the World Trade Organization (WTO) in late 2001. Under WTO rules, foreign firms will be allowed to own up to 49% of retail marketing ventures for petroleum products within 3 years, and China agreed to open its oil wholesale markets 5 years after entry.

China also promised to lift its ban on imports of foreign gasoline and diesel fuel. Fuel oil has been one of the few oil products China allows to be imported. Under WTO rules, China has to increase its import quota for refined products each year by 15% and allocate 20% of the import quota to companies other than designated state oil companies, said FACTS in May.

However, Chinese officials waited 4 months before issuing a detailed 2002 import quota for refined products at the end of April. Of the total 420,400 b/d of imported petroleum products authorized under the year's quota, 80% will be brought into China by four firms designated by a government commission as the official state oil trading companies: China National Chemical Import & Export Co. (Sinochem), United International Petroleum & Chemicals Co. Ltd. (Unipec), China National United Oil Co. (Chinaoil), and Zhuhai Zhenrong Co.

However, competition for the remaining 20%, or about 84,000 b/d, also is open to any other Chinese state or local oil company other than the four designated state oil trading companies, said FACTS. That includes other large Chinese firms such as state-owned CNPC, its PetroChina subsidiary, and Sinopec. The industry had expected that the 84,000 b/d quota would be equally split between gasoline and diesel imports from foreign companies, which would have had a significant impact on China's product trade. As it turned out, however, most of the imports will go to fuel oil.

Because China already is the second largest importer of fuel oil in Asia, the 25% of fuel oil imports (79,000 b/d) that is allocated to non-state importers will have minimal impact on that market, especially if divided among many competitors. The 2002 gasoline import is restricted to 5,000 b/d, allocated entirely to the four designated state companies, which also get 70% of the allotted 20,000 b/d of light diesel imports and all of the 18,000 b/d of gas oils that are used as refinery feedstock in China.

As a result, imports of petroleum products into China could decrease during 2002. The four designated state companies need not import the full amount permitted by the quota, FACTS noted. Energy Security Analysis Inc. (Boston) expects Chinese refiners to increase their runs in order to prevent a jump in gas oil imports. An early growing season in China had already bolstered diesel demand in the agricultural sector, resulting in increased refinery runs in spring 2002. The firm also projected that China will increase its gasoline exports.

IPE

ASIA-PACIFIC

Australia

CAPITAL: CANBERRA
MONETARY UNIT: DOLLAR
REFINING CAPACITY: 848,250 B/CD
OIL PRODUCTION: 633,000 B/D
OIL RESERVES: 3.5 BILLION BBL
GAS RESERVES: 90.0 TCF

Australia weathered the world economic slowdown much better than expected, posting an estimated 3.7% growth in real gross domestic product (GDP) during 2002. The International Monetary Fund (IMF) projects that real GDP will grow by 3.8% in 2003.

As its economy expands, however, so will oil consumption. Despite substantial reserves, Australia has only 10 years of oil left, said the government, and the nation is using oil three times faster than it is finding it. By 2010, Australia's oil self-sufficiency is expected to slide to 40%. Also, a government report suggested that parts of Australia will run out of natural gas by 2020, but the country's gas industry sharply criticized the report.

Exploration

The first well was spudded as part of a three well exploration program in the oil-prone southern Cooper basin in South Australia, said 50-50 partners Beach Petroleum NL (Adelaide) and Magellan Petroleum Australia Ltd. (Brisbane) in August. Maslin-1 was being drilled on PEL 94 to 1,440 m (4,720 ft), while Aldinga-1 and Henley-1 were planned for drilling on PEL 95.

Maslin, 90 km (60 miles) south of Moomba oil and gas field, is one of the basin's largest untested structures, with a mean of 5.4 million bbl potentially recoverable. Primary objective is Jurassic-Cretaceous Namur sandstone. Secondary objectives are the Cretaceous Murta formation and the Jurassic Hutton sandstone.

Aldinga, 75 km (47 miles) southeast of Moomba, has potential mean recoverable volumes of 1 million bbl. Henley, 55 km (34 miles) east of Aldinga and also assessed at 1 million bbl, is a well-defined anticlinal trap on a prominent east-west structural trend 30 km (20 miles) south of Toolachee gas field.

Petroleum prospectivity could be more favorable than previously believed on the Naturaliste plateau seaward of the Perth basin off south Western Australia, concluded a study by Geoscience Australia (Canberra) in December 2002. The area covers 90,000 sq km (35,000 sq mi) and lies 350 km (220 miles) off the Cape Naturaliste-Whicher Range area in 2,000-5,000 m of water (6,600-16,400 ft). If oil has been generated, it likely migrated to the east where traps could occur in much shallower water, according to Aus-Geo News.

The western Mentelle basin, lying beneath the Naturaliste trough and west of the Perth basin and Vlaming subbasin, is the study area's largest depocenter with 4-5 km of sediment (13,000-16,000 ft). It includes about 1 km (3,000 ft) of Early Cretaceous sediments and at least 1.5 km (4,900 ft) of Jurassic section.

Permits awarded, sought

Australia's government awarded several exploration permits during 2002. Area VO1-4 in the center of the Central Deep of the Gippsland basin off Victoria will be explored by Australia Crude Oil Co. (Houston) under a permit announced in November. Area VO1-4 is sparsely explored, with eight wells drilled by 1990 based upon 2D seismic surveys, none of which was commercially successful. Australia Crude Oil plans a 3D seismic survey of the 740 sq km (290 sq mi) area to delineate prospects in the upper and lower Latrobe group and in deeper horizons.

In the Australian Timor Sea, the WA-316-P permit was awarded to West Oil NL (West Perth), the firm said in February. The 1,672 sq km (646 sq mi) WA-316-P is in the Bonaparte basin, 70 km (40 miles) west of Bayu-Undan gas and condensate field. West will reprocess 2D seismic in the first year and drill a well in the third year. The permit is in 60-100 m of water (200-300 ft) off

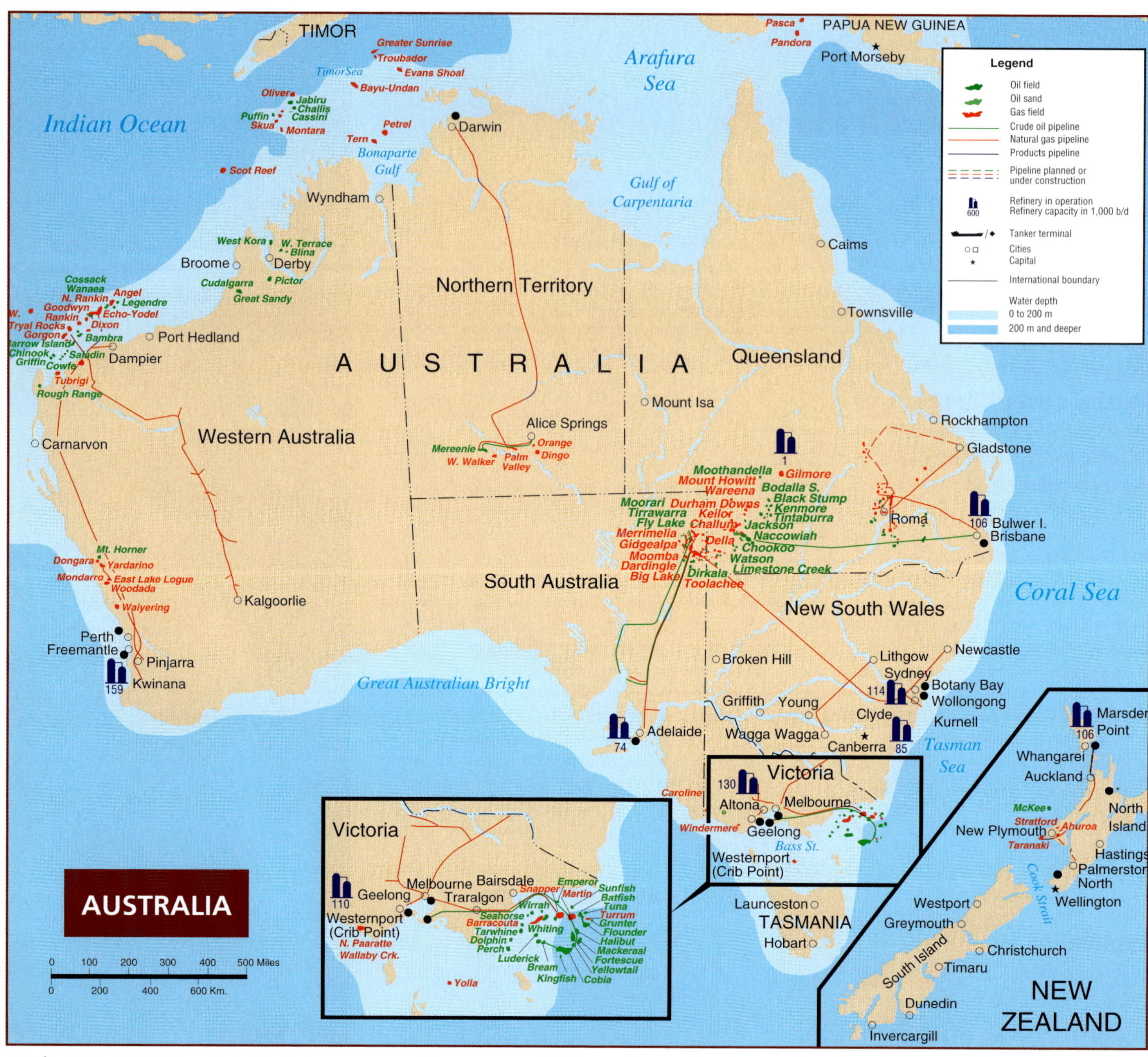

northernmost Western Australia between the Londonderry high and Sahul syncline.

Nine wells were drilled during 1972-98. WA-316-P has several Plover sandstone oil prospects and the Avocet Deep prospect with a P50 potential of 80 billion cu m, or bcm (3 tcf) in Permian limestones. Avocet Deep, to be offered for farmout, is about 85 km (50 miles) southwest of Corallina, Laminaria, and Buffalo oil fields.

The coalbed methane potential of brown coals in Victoria's Gippsland basin would be evaluated by Gastar Exploration Ltd. (Mount Pleasant, MI, US), if permits are granted on its licenses. Gastar planned to drill the first of several pilot programs of three to five wells each in 8,000 sq km (2 million acres, or 3,000 sq mi) covering the entire Gippsland basin onshore.

Gas in place on the licenses is estimated at 1-1.5 trillion cu m, or tcm (37-57 tcf), and the coals are nearly identical in age and characteristics to those in the Powder River basin. Gastar gathered data on the coals by participating in the shallow portion of two wells drilled by other operators. The company also has coal interests in the Gunnedah basin.

Bayu-Undan development

Natural gas to be produced from Bayu-Undan field would feed a proposed liquefaction plant and export terminal in Darwin, which would export 3 million tonnes/year (tpy) of LNG to Japan under a 17-year agreement announced in March 2002 between a unit of ConocoPhillips (Houston) and two Japanese customers, Tokyo Electric Power Co. Inc. and Tokyo Gas.

Bayu-Undan field lies in the Joint Petroleum Development Area in the Timor Sea between Australia and East Timor. The area is jointly administered by the two countries. About 90% of Bayu-Undan royalties will go to East Timor, which officially became an independent nation in 2002. ConocoPhillips also agreed separately to sell the Japanese firms a 10% interest in Bayu-Undan field.

Gas deliveries are expected to begin in late 2005, with fob shipments of the first LNG cargoes to begin in early 2006. The agreement commits nearly 100% of the field's 91 bcm (3.4 tcf) of natural gas, ConocoPhillips said. The Bayu-Undan gas project had been delayed over a tax dispute with the East Timor government as well as earlier partner differences over project approaches, but the pipeline to carry gas from the field to Darwin in northern Australia was approved by ConocoPhillips in late 2001. Australia still must approve the fiscal regime accepted in 2001 by East Timor.

One of the world's longest and heaviest flare barges was installed at Bayu-Undan field in late 2002, along with other development facilities, said McDermott International Inc. (New Orleans). The equipment will be used for the Bayu-Undan gas recycle project being developed in 80 m of water (24 ft). Facilities included two 10,000-tonne, 8-leg platform jackets — one for drilling, production, and processing and one for compression, utilities, and living quarters.

The Intermac 650 launch barge towed the two jackets 3,000 km (2,000 miles) from Batam Island where they were fabricated. The flare barge is 226 m long (741 ft) and weighs 2,000 tonnes. During the transportation stage, the jackets were moved on a single barge, in one tow. The gas project is operated by a unit of ConocoPhillips.

Also in the Timor Sea between Australia and East Timor, Greater Sunrise gas field contains larger reserves than Bayu-Undan, but development has been stalled by disputes. About 80% of Greater Sunrise lies within Australian waters as currently defined, but East Timor's new parliament declared a maritime border extending 370 km (200 nautical miles), which would encompass all of Greater Sunrise. Also, field partners disagreed on the development approach.

North West Shelf venture

The North West Shelf venture's $2.4 billion LNG expansion project was under way during 2002, including a fourth LNG train at the Karratha onshore liquefaction plant on Australia's Burrup Peninsula. First LNG from the fourth train is scheduled for mid-2004. The expansion includes construction of a 4.2 million tpy liquefaction facility and a 42-inch trunkline linking the plant and the venture's gas fields 130 km offshore (80 miles).

Several key LNG sales were finalized during 2002 to support the project. Under a 30-year supply agreement announced in March, the North West Shelf expansion will provide 1 million tpy of LNG to Osaka Gas Co. Ltd. beginning in 2004, said project operator Woodside Energy (UK) Ltd. Osaka Gas already was taking 790,000 tpy of LNG under an earlier, 20-year agreement. North West Shelf partners also had signed letters of intent to sell 1.37 million tpy of LNG to Tokyo Gas and Toho Gas Co. Ltd.

In May 2002, Kyushu Electric Power (Japan) agreed to purchase 500,000 tpy of LNG per year from Northwest Shelf beginning in 2006. In June a unit of Royal Dutch/Shell Group (Netherlands) finalized an agreement to purchase up to a total of 3.7 million tonnes during 2004-09 from the venture. This contract's actual volume will depend on the amount of LNG that North West Shelf partners commit to long-term customers in Japan, South Korea, China, and Taiwan. Shell will use the LNG to develop market opportunities outside those destinations.

North West Shelf also was selected as the sole Phase I supplier for China's first LNG import terminal, to be built in Guangdong province in southern China by China National Offshore Oil Corp. (CNOOC) and partners. The contract calls for initial deliveries from Western Australia of 3 million tpy of LNG for 25 years, beginning in 2005-06. CNOOC could become a partner in the North West Shelf venture.

A fifth liquefaction train for the Karratha plant was being designed in 2002, said North West Shelf operator Woodside in August. Completion of the fourth and fifth trains will more than double Karratha's current LNG processing capacity of 7.5 million tpy.

Two or three new LNG transport vessels will be required to service the China trade route. A fleet of eight LNG ships was already in service on the Australian project, with a ninth vessel under construction. Additional LNG transport to Guangdong could be supported by a joint venture between Chinese shipping companies and North West Shelf partners.

In late 2002 work on the second trunkline proceeded with a subsea installation contract award to Technip-Coflexip Oceania and Subsea 7 (Australia) Pty. Ltd. The trunkline, being installed by a unit of Saipem SpA (Italy), will connect onshore facilities to Goodwyn and Rankin gas condensate fields and will facilitate increased production from area fields when it comes online in April 2004, coinciding with completion of the fourth LNG train at Karratha.

The contract calls for tie-ins of the trunk-line into existing infield facilities near the North Rankin A platform. Following start-up of the second trunkline, extensive pipeline system modifications will be carried out on the existing trunkline at North Rankin A. Technip-Coflexip will utilize its CSO Venturer vessel to carry out subsea installations in three phases from mid-2003 through spring 2004.

North West Shelf development

Gibson and South Plato fields in the Carnarvon basin off northwestern Australia began producing oil, reported Apache Corp. (Houston) in June. The fields were discovered in 2001. Two wells were flowing at a combined rate of 22,265 b/d of oil. These Flag sandstone wells are notable for their high flow rates and their proximity to existing production infrastructure on Varanus Island, said Apache, meaning that they can be brought onstream at low cost. The company drilled 10 successful Flag sandstone wells from 2001 through mid-2002.

In October Apache and partners began production from two more Flag sandstone wells in the Carnarvon basin. A new discovery well, South Simpson-1, was flowing at 5,000 b/d, and an additional extension well, Tanami-6, at 2,500 b/d, should bring total production to the Varanus Island facilities close to 28,000 b/d. Apache was planning two more wells in late 2002.

A floating production, storage, and offloading (FPSO) vessel will be leased for development of Woollybutt oil field in the North West Shelf off Western Australia, said operator Agip Australia Ltd. and partners in July. The Four Vanguard vessel, formerly the 94,000-dwt Four Lakes tanker, will be converted into an FPSO as part of a project to develop the discovery, which offers preliminary reserves estimated at 100 million bbl of oil. The Vanguard unit also will connect the vessel to the offshore field and operate it for the partners.

Southeastern development

Development proposals were being evaluated for Thylacine and Geographe gas fields, discovered in 2001 in the Otway area off southwestern Victoria, said operator Woodside Energy in August. The Australian Commonwealth and Victorian governments would be jointly responsible for issuing environmental approvals for the gas fields' development.

The fields, which are 55-70 km (34-43 miles) south of Port Campbell, are expected to contain gas sufficient to provide more than 10% of current annual demand in southeastern Australia for at least 10 years. Reserves for the combined Geographe and Thylacine fields are estimated at 21 bcm (0.8 tcf) of gas and 9 million bbl of condensate. Initial screening studies gave the partners reason to believe the fields could be developed in time to provide gas to customers in Victoria and South Australia in 2006.

Gas pipelines

Australia's onshore gas pipeline network originally was designed to carry gas from centrally located fields to urban hubs. But as production from these fields declines and new gas is developed offshore, Australia will need massive investment in its pipeline infrastructure to bring energy into its grid.

The Bream natural gas pipeline started up in December carrying 2 bcm/yr (200 MMscfd) in Australia's Bass Strait, less than 1 year after it began construction in January 2002, said BHP Billiton Ltd. (Melbourne) and partner Esso Australia Resources Pty. Ltd.

The 51-km (32-mile), 14-inch line is the fourth gas pipeline extending from the existing Bream A offshore platform into the partners' Gippsland production network at the Longford processing facilities. The Bass Strait's Gippsland gas was supplying Victoria, New South Wales, and Tasmania. Beginning in 2004, the gas will flow into South Australia and the Australian Capital Territory.

Other new projects include the 681-km (423-mile) Sea Gas Pipeline, scheduled for 2003 completion, which will take gas from Otway basin off Victoria to the Quarintine power station in Adelaide. A gas pipeline to import gas from Papua New Guinea through the Gulf of Papua was under construction. In March 2002 a base load for the line was established when Australian Gas Light signed a conditional agreement with ExxonMobil Corp. (Irving, TX, US), which operates the Papua New Guinea fields, to buy up to 1.2 bcm/yr (43 bcf/yr) for 20 years.

Tasmanian gas project

Conversion of the 120-MW Bell Bay power station on the island of Tasmania from oil-fired to natural gas operation was a key factor in the Tasmanian gas project, which involves a new 753-km (455-mile), 350-mm subsea and underground pipeline from Longford, Victoria — Australia's longest subsea pipeline project. Of the two boiler-turbine units at the station, only Unit 1 was converted, while Unit 2 remains available for emergencies using oil fuel.

Pipeline construction and power plant conversion were conducted by a unit of Duke Energy Corp. (Charlotte, NC, US), and the pipeline was started up in September. The line links Tasmania to the Australian mainland gas network via connection to Duke's Eastern gas pipeline at Longford.

Duke also announced the start of construction of the VicHub project, which will enable gas to flow freely between Victoria, New South Wales, and Tasmania through construction of a link between the Eastern gas pipeline, the Tasmanian gas pipeline, and the GasNet pipeline system.

Downstream activities

A new hydrodesulfurization plant will be built at the Geelong, Victoria, refinery, said a Shell unit in July. The new plant, called HDS 2, will have a capacity of 6,000 tonnes/day (tpd). Work began in June and should be completed in late 2003. The plant will produce ultralow-sulfur diesel to meet Australia's new clean fuels specifications taking effect in January 2006. A turnkey contract for the work was awarded to a unit of Chicago Bridge & Iron Co. NV (Netherlands).

The Sweetwater gas-to-liquids (GTL) project in western Australia's Burrup Peninsula might get suspended, said Syntroleum Corp. (Tulsa) in October, although the company is continuing to explore alternatives.

Bangladesh

CAPITAL: DHAKA

MONETARY UNIT: TAKA

REFINING CAPACITY: 33,000 B/CD

OIL PRODUCTION: 5,000 B/D

OIL RESERVES: 56.9 MILLION BBL

GAS RESERVES: 10.6 TCF

Plans to export natural gas from Bangladesh to India were dealt a setback in November 2002 when a committee of geologists and economists declared that the country's gas reserves cannot support both domestic and foreign markets.

The Nagorik committee, a 7-person group formed by the Bangladesh Geological Society and Bangladesh Economic Association in June 2002, pegged Bangladesh's gas reserves at 166 bcm (6.2 tcf) proved and 156 bcm (5.8 tcf) probable. The committee recommended that the government refuse to authorize exports.

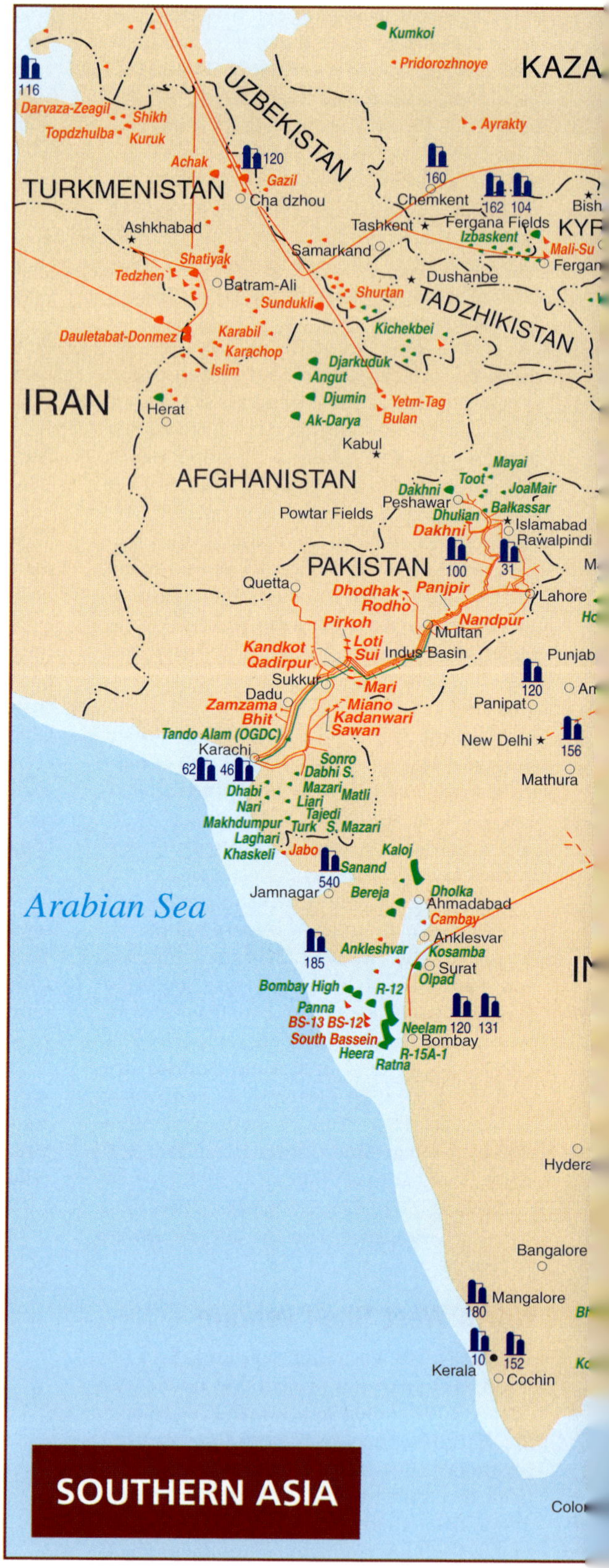

Export plans

Unocal Corp. (El Segundo, CA, US) had proposed to build an 1,360-km, 4.9 bcm/yr (847-mile, 500 MMcfd), 30-inch pipeline from Bibiyana gas-condensate field on Block 12 to the Delhi, India, area. Outside

engineers have assigned Bibiyana field proved and probable reserves of 64 bcm (2.4 tcf) plus 13.8 million bbl of condensate and found that it could hold a further 83 bcm (3.1 tcf) and 16.9 million bbl of possible reserves. Unocal had determined that the Bangladesh gas market is oversupplied for 5 years, might be saturated indefinitely, could not accept Bibiyana gas for more than 8 years, and could not take the field's maximum production volumes for more than 15 years.

The Nagorik committee said that Unocal's production-sharing contract (PSC) contains no provision for export of gas by pipeline. It argued that Unocal's share of reserves is not enough to support such exports and that Bangladesh Oil, Gas &

Mineral Corp. (Petrobangla) would have to supply some of its gas. Unocal's Block 12 PSC provides for LNG exports or an acceptable alternative.

Recent resource estimates

The US Geological Survey and Petrobangla had estimated that the nation holds undiscovered gas resources of 226 bcm (8.43 tcf) (95% probability), or 862 bcm (32.1 tcf) (mean probability). The Nagorik committee called these figures conjectural and unsuitable for gas utilization policy matters. The committee also argued that energy resources other than gas are not likely to penetrate Bangladesh markets in any significant way within 50 years and that gas consumed in Bangladesh would be of significantly greater economic value to the country than gas exported.

If Unocal were allowed the option to export, the committee said, other international oil companies operating in Bangladesh also would want the same option, hastening depletion. A pipeline that has been built across the Jamuna river was intended to supply gas to western Bangladesh, but this goal would be compromised if the line became part of any export project, the committee noted.

Unocal has spoken of a window of opportunity for Bangladesh gas that could close if Middle Eastern LNG or Iranian gas captures India's markets first. Gas was recently discovered in Indian waters a few miles offshore in the Bay of Bengal. These finds, in the Krishna-Godavari basin, are a similar distance from Delhi as the Bangladesh gas fields.

EXPLORATION ACTIVITY IN BANGLADESH

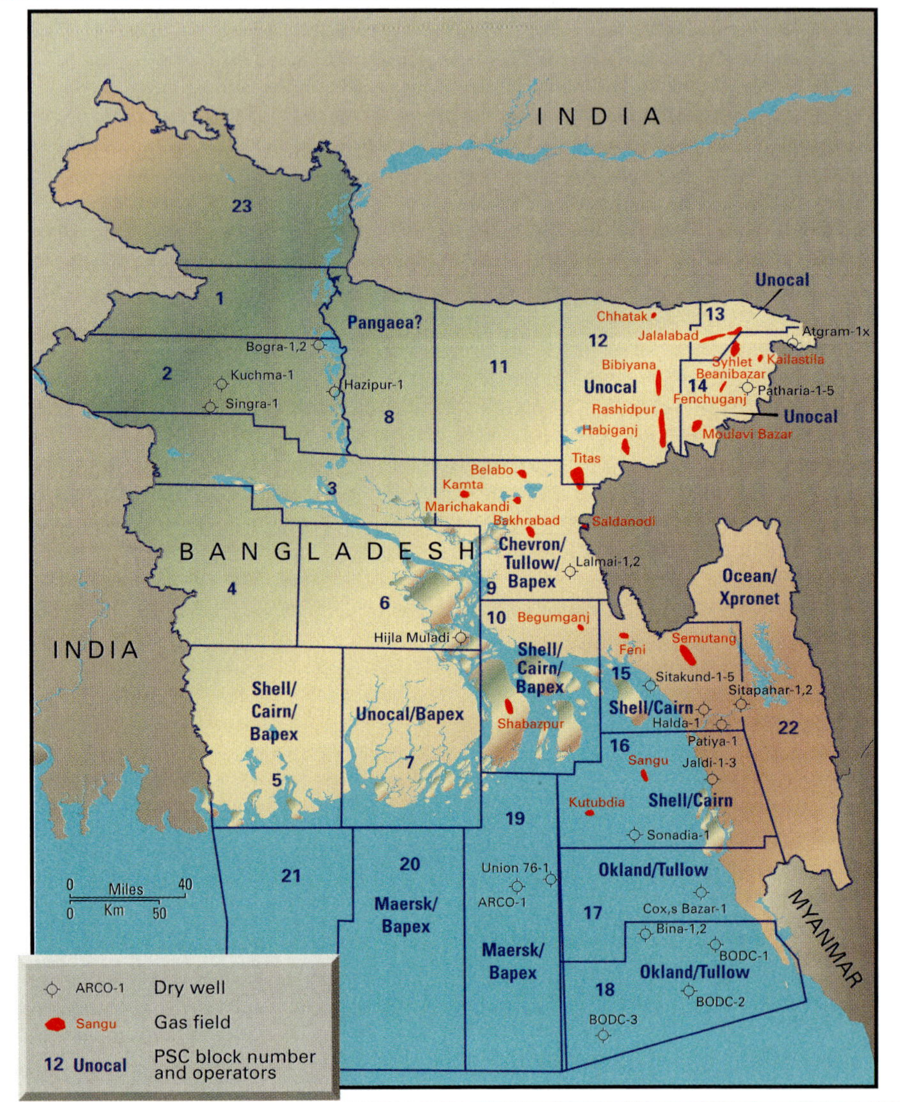

Other resource assessments

Bangladesh's gas resources have also been studied by a team from the Hydrocarbon Unit (HCU) of the Bangladesh Energy and Mineral Resources Division and the Norwegian Petroleum Directorate (NPD). The HCU/NPD study, released in early 2002, estimated that the resource ranges from 497 to 1,710 bcm (18.5 to 63.7 tcf), with a medium figure of 1,117 bcm (41.6 tcf). These estimates are significantly higher than the USGS/Petrobangla assessment.

In April 2002 Unocal presented its assessment of Bangladesh's gas potential. With an area of 207,000 sq km (79,900 sq mi) and only 64 exploration wells drilled as of early 2002, Bangladesh ranks as one of the world's most significant, underevaluated hydrocarbon provinces. Exploration has focused on eastern Bangladesh, while its western and offshore areas are relatively underexplored. Still, with 22 gas discoveries and 1 oil discovery, Bangladesh has a historical exploration success rate of about 35%, which is very strong by world standards.

Unocal evaluated Bangladesh's resource as divided into three categories: field discoveries, including prior production and reserves in existing fields; field growth, including revisions due to development/appraisal drilling and new technology applications; and new field discoveries, or the resource potential, which has been assessed multiple times.

Unocal's study identified proved plus probable reserves of 549 bcm (20.44 tcf) (ultimately recoverable gas), of which 433.4 bcm (16.14 tcf) remains unproduced; additional probable reserves of 344 bcm (12.8 tcf); and an undiscovered resource potential of more than 860 bcm (32 tcf) (mean countrywide estimate). Combined, these estimates translate to a hydrocarbon resource base of about 1.7-1.75 trillion cubic meters, or tcm (64-65 tcf).

Jalalabad production

In mid-2002 Unocal reported that production from Jalalabad field on Block 13 supplied 12% of Bangladesh's domestic gas demand in 2001. Jalalabad production averaged 2.2 million cu m/d of gas (83 MMcfd) and 1,000 b/d of liquids, including 1.5 million cu m/d (55 MMcfd) and 700 b/d net to operator Unocal Bangladesh Ltd. In addition to Jalalabad and Bibiyana, Unocal discovered and operates Moulavi Bazar gas field on Block 14.

Brunei

CAPITAL: BANDAR SERI BAGAWAN
MONETARY UNIT: DOLLAR
REFINING CAPACITY: 8,600 B/CD
OIL PRODUCTION: 185,000 B/D
OIL RESERVES: 1.35 BILLION BBL
GAS RESERVES: 13.8 TCF

Brunei announced preliminary awards in early 2002 from its first round of bidding for open exploration acreage (the Petroleum New Areas). The acreage consisted of two previously unexplored and unlicensed blocks in 1,000-2,800 m of water (3,300-9,200 ft). Block K went to Shell Deepwater Borneo and partners, and Block J went to BHP Billiton and Amerada Hess Corp. (New York). Contract negotiations were under way. Shell had drilled deepwater wells off Brunei (including Pungguk-1, Bilis-1, and Suit-2) but released little information due to their proximity to blocks on offer.

Development of Egret oil and gas field off Brunei proceeded during 2002, following a decision by Brunei Shell Petroleum Co. Sdn. Bhd. to move ahead with Phase I plans. Egret is expected to begin production in mid-2003, supplying natural gas to the Brunei LNG Sdn. Bhd. plant in Lumut as well as satisfying domestic demand. Without Egret's gas, Brunei LNG would encounter a shortfall in late 2003.

The Egret field is located 43 km (271 miles) offshore from Seria, approximately 7 km (4 miles) east-northeast of Shell's Fairley field on Block 2 in 60 m of water (200 ft). Egret is being developed in two phases, using two platforms that will support a total of 20 well slots, including spares for a possible Egret deep development and new wells. Both platforms will support a mixture of oil and gas wells.

Phase I work, awarded to Technip Far East Sdn. Bhd., involves construction and installation of drilling platform EGDP-1, which will produce from five gas wells. The gas will be transported by a new 25-km (15-mile), 20-inch multiphase subsea pipeline to Ampa-6 for separation. From Ampa an existing subsea pipeline will deliver the gas to the LNG plant. Phase I facilities including the jacket and topsides were being fabricated in 2002, with installation scheduled for spring 2003. A 15-km (9-mile) subsea cable also was being laid.

During Phase I, about 75% of Egret's recoverable gas reserves of 12.5 bcm (465 bcf) will be developed. Phase II of the project will focus on oil development, along with remaining gas from the field's eastern flank via a new well jacket (EGWJ-1), which will be installed 8 km (5 miles) northeast of the EGDP-1 platform. Phase II will potentially come onstream in 2006.

India

CAPITAL: NEW DELHI
MONETARY UNIT: RUPEE
REFINING CAPACITY: 2,134,625 B/CD
OIL PRODUCTION: 663,000 B/D
OIL RESERVES: 5.37 BILLION BBL
GAS RESERVES: 26.9 TCF

India was making major investments in its energy infrastructure to keep up with increasing demand, particularly for electric power and possible LNG imports to support power projects. Real GDP growth of about 5% during 2002 was expected to increase to 5.7% in 2003, projected the IMF. India's government revived a plan to build strategic petroleum reserves and initiated a review of its national hydrocarbon security in May 2002. A natural gas pipeline from Iran through Pakistan was being planned.

During 2002 reevaluation of India's resources led to large increases, with estimated proved reserves of oil rising by 11%, to 5.37 billion bbl from 4.84 billion, and natural gas by 18%, to 723.5 bcm from 614.0 bcm (to 26.943 tcf from 22.865 tcf), according to *Oil & Gas Journal*.

India moved a step closer toward privatizing its oil industry with a December 2002 announcement that the government will sell stakes in Hindustan Petroleum Corp. Ltd. and Bharat Petroleum Corp. However, the time frame and percentages to be sold were unspecified, and India's other state firms had not yet been barred from the bidding.

ONGC plans

India's state-owned Oil & Natural Gas Corporation (ONGC) said in November that it planned to invest $1.24 billion during the 2002-03 fiscal year to acquire oil equity abroad, mainly in Sudan, Iran, Libya, and the US. ONGC subsidiary ONGC Videsh Ltd. (OVL) will lead the acquisition effort, motivated by declining domestic oil output and the absence of any major oil discoveries at home in recent years. OVL floated a special-purpose vehicle, Nile-Ganga Pte., to buy the 25% equity stake in Sudan's Greater Nile Oil Project from Talisman Energy Inc. (Calgary), with payment scheduled for early 2003.

India launched its third licensing round in March 2002 under terms and conditions of the New Exploration Licensing Policy (NELP). The bidding, which closed in late August, covered 27 blocks including nine in deep water. During May and June the government held road shows in Singapore, London, Houston, and Calgary to promote bidding. A unit of BG Group plc (UK) indicated interest in the CB-OS-1 Block, off the Cambay basin, and the Panna, Mukta, and Tapti fields acquired from Enron India. BG was discussing partnerships with prominent Indian companies.

ONGC had called for tenders from global oil companies to help develop five deepwater blocks; however, due to lack of responses, ONGC decided to go it alone in its deepwater exploration program, the firm said in September. It also decided to call for tenders to develop 93 offshore marginal fields along India's western coast. Combined, these fields hold an estimated 200 million tonnes of oil and 120 bcm of gas reserves (4.5 tcf), said ONGC.

Within India, the company is restructuring its upstream activities and will implement enhanced oil recovery schemes in 15 fields, including Mumbai High. ONGC's downstream plans involve buying part of Hindustan Petroleum's stake in Mangalore Refinery & Petrochemicals Ltd., owned by the Aditya Birla Group, as well as upgrading ONGC's Hazira plant to produce aviation turbine fuel and converting naphtha to gasoline at the Uran plant.

Gas discoveries

A giant gas field was discovered in the Krishna-Godavari basin just off the central Andhra Pradesh coast, reported Reliance Industries Ltd. (Mumbai, India) in November. Four wells indicate that the A-1 accumulation has gas in place exceeding 190 bcm (7 tcf), which was being confirmed by outside consulting engineers. Individual well deliverability could exceed 2.7 million cu m/d (100 MMcfd).

Integrated 3D seismic and well interpretation indicates that the reserves could be significantly higher, said Reliance's partner Niko Resources Ltd. (Canada). Even at 190 bcm (7 tcf) in place, A-1 on KG-DWN-98/3 (Block D6) would be India's largest gas field. Cairn Energy plc (London) also has sizable gas and oil discoveries on KG-DWN-98/2, which adjoins KG-DWN-98/3 to the west. Block D6 covers 8,000 sq km (1.9 million acres, or 3,000 sq mi)

The wells and 3D seismic data indicate an anomaly of 177 sq km (68 miles) and a hydrocarbon column of 342 m (1,120 ft). The A-1 structure is 20 km (12 miles) off the Godavari River delta in an average 900 m of water (3,000 ft) in the Bay of Bengal. The Miocene reservoir is at 1,850-2,200 m subsea (6,070-7,200 ft). A-1 is positioned on the closest part of the block to shore. Two wells on the structure flowed at rates of 1.1 and 0.8 million cu m/d of gas (40 and 29 MMcfd) on cased-hole tests constrained by equipment capacity.

A-1 could go on line in 2004 and produce 38 million cu m/d of gas (1.4 bcfd) in 4 years, reported investment firm Canaccord Capital Corp. (Vancouver). Reliance would drill one well per month for the next several months, ultimately needing 20 wells for full development, along with subsea completions and a pipeline to shore.

Natural gas was also discovered in 2002 near Surat, Gujarat state, said Niko Resources in November. Two of three exploratory wells drilled on the CB-ONN-2000/2 onshore block, which covers 419 sq km (162 sq mi), encountered hydrocarbons, including a flow rate totaling 430,000 cu m/d of sweet dry gas (16 MMcfd) from Bheema-1. Niko planned to drill seven more wells in the block during the 2002-03 fiscal year.

Niko's 15-well onshore exploration and appraisal program, which started in April, targets Miocene reservoir sands analogous to those producing at Niko's 27-bcm (1-tcf) Hazira gas field, 20 km (12 miles) to the southwest. Niko was using a custom-built rig to reduce drilling time and enable multiwell pad drilling during development. The semi-automated, super single slant rig can drill in vertical and slant modes and is particularly suited for operations in environmentally sensitive and size-limited locations, said Precision Drilling Corp. (Calgary).

Under its PSC terms, Niko must also shoot 250 sq km (100 sq mi) of 3D seismic over 2 years. A second 2-year commitment, which depends on the first phase results, would require drilling five more wells and shooting an additional 50 sq km (20 sq mi) of 3D seismic. Niko also received government clearance to proceed with its platform at Hazira off India's west coast and planned to award the construction contract by year-end.

Lakshmi, Gauri development

Development wells in Lakshmi gas field on offshore Block CB/OS-2 showed strong flows of good quality crude oil, reported Cairn Energy in April. The gas development project was on schedule to produce first gas in late 2002. The two wells flowed on test at a combined rate of 10,446 b/d. The first well, on the field's southern edge, was directionally drilled to 1,460 m (4,790 ft) and encountered two oil-bearing zones below the main gas-bearing horizons. The wells' flow rates were constrained by surface test equipment limitations.

The flow rates demonstrate the high potential reservoir deliverability, said Cairn, though further appraisal is required to determine the reserve potential. Cairn was considering an accelerated appraisal program for the discovery, which follows three oil discoveries on the same block in 2001 and could lead to reassessment of the block's total oil potential.

Lakshmi gas will enter the industrialized Gujarat market under two gas sales contracts announced in September with local utilities Gujarat Powergen Energy Corp. and Gujarat Gas Co. Ltd. The Lakshmi facilities initially will have a maximum gas production and processing capacity of 4 million cu m/d (150 MMcfd).

Development of Gauri gas and oil field on the same block was being planned for 2003, after the field was declared commercial in July. Current reserve estimates for Gauri gas are 2-5 bcm (60-200 bcf). Sources revealed that some of the Gauri reservoirs are in communication with the adjacent producing Hazira field operated by Niko.

Development plan options, anticipating an initial 1.1 million cu m/d (40 MMcfd), were being evaluated by operator Cairn and block partners including ONGC, which boosted its stake in both Lakshmi and Gauri. Oil was discovered beneath the gas reservoirs at Gauri, for which further appraisal would be required to evaluate the potential. This is expected to coincide with the gas development drilling program in 2003.

LNG projects

An LNG receiving terminal at Dahej was under construction in 2002, scheduled to begin operating in 2003. Petronet, a consortium of four Indian state-owned companies, was building the terminal to receive LNG from Qatar. Petronet and Ras Laffan LNG reportedly amended the contract terms to set a price ceiling of $3.60-$3.80/MMBtu, and Petronet has an option to buy another 2.5 million tpy for a second planned terminal at Kochi that is seeking customers. The nearly completed Dabhol LNG project, which was on the market, has 20-year purchase agreements with Abu Dhabi Gas Liquefaction Co. for 0.5 million tpy and Oman LNG for 1.6 million tpy.

Power plants in northwestern and southern India will need 5 million tpy of LNG beginning in 2006-07 to support capacity expansions. National Thermal Power Corp. (India) received expressions of interest in supplying the LNG from 10 bidders, it said in April, including Shell, BG, and TotalFinaElf SA (France). The proposed expansions total 2,600 MW, consisting of 650-MW increases at each of four power plants: Anta in Rajasthan, Auraiya in Uttar Pradesh, and Kawas and Gandhar in Gujarat. These projects will require 3 million tpy, and another 2 million will be needed for a southern power plant.

Gas distribution, petrochemicals

A joint venture was being formed in early 2002 to distribute natural gas, LPG, and other gas liquids in Andhra Pradesh state. The company, tentatively called Bhagyanagar Gas Ltd., would be a partnership of Gas Authority of India Ltd. (GAIL), Hindustan Petroleum, and other interests. In addition to gas distribution, the company would pursue the development of infrastructure and would lay, operate, and maintain its own pipeline.

Rapid capacity growth in India's petrochemicals industry has surpassed market growth, resulting in a surplus of many key products, said Nexant Inc. (Singapore) in early 2002. The capacity buildup changed India into a net exporter of petrochemicals and refined petroleum products. However, continuing strong growth in demand will soon cause a supply deficit.

Indonesia

CAPITAL: JAKARTA
MONETARY UNIT: RUPIAH
REFINING CAPACITY: 992,745 B/CD
OIL PRODUCTION: 1.12 MILLION B/D
OIL RESERVES: 5.0 BILLION BBL
GAS RESERVES: 92.5 TCF

Indonesia established in mid-2002 a new national oil and gas agency, to be known by the Indonesian language acronym BALAK, which will assume all regulatory functions formerly performed by state-owned oil and gas entity Pertamina. BALAK will regulate Indonesia's oil and gas industry, award concessions, sign contracts, and manage oil and gas exploration and production companies operating in the country. It will be an independent agency accountable to the president.

Pertamina is being privatized as a commercial oil and gas entity. Officially relieved of oil and gas governing responsibilities, Pertamina is now free to concentrate on upstream exploration and development and downstream activities ahead of full privatization, scheduled for 2004.

Ranggas appraisal

Strong oil and natural gas flows from appraisal wells in the deepwater Ranggas oil field were reported by a Unocal subsidiary in spring 2002, which will lead toward commercializing Unocal's third deepwater oil field in Indonesia. Unocal

CHINA
RUSSIA
NORTH KOREA
SOUTH KOREA
JAPAN
Sea of Japan
Pacific Ocean
Vladivostok
Peter the Great Bay
42 Chongjin
Pohang
817 Ulsan
10 Pusan
Korean Channel
Tsushima Channel
Wakkanai
Soya
Katsu
Menas
Koetoi
Toyotomi
Kitatoyotomi
Teshio Fields
Iskikari Bay
HOKKAIDO
Ishikari
Atsuta
Barato
Tokachi Fields
Sapporo
Azuma
Fureoi
Kinausu
Tomakomai
133 180 Muroran
Oshamambe
Mori
Hakodate
Tsugaru Strait
Aomori
Nichi-Hachimori
Higashi-Hachimori
Sawame
Iwako
Sakake
Akita Fields
Surukawa
Funakawa
Kurokawa
Asakawa
Akita
Morioka
Yabase
Yuri
Kamekawa
Shonai
Innai
Kisagata
Okuoguni
Yokoka
Chokaizan
Yamagata Fields
Fukura
Sakata
Mogaami
Shonai
Nogouchi
Yamagata
Amarume
138 Sendai
Sendai Bay
Iwafune
Aga
Niitsu
Shibata
Niigata
Niigata-Plain
Okawatsu
Higashi
Mitsuke
Higashiyama
Echigo Fields
Nishiyama
Nagaoka
Kashiwazaki
Sekihara
Ojiya-Sanjo
Toyama Bay
Kubiki
Ubase
Gozu
Kubiki
Nadachi
Menji
Maki
Iwaki
Asakawa
Odagiri
Toyama
57
Kanazawa
HONSHU
Kashima
180 Choshi
Tokyo
Yokoshiba
Chiba
Shirasato
Kawasaki
Yokohama
Mobara
Goi
Negishi
Chiba Fields
Sodegaura
360
120 296 65
228 160 209 192
Fukui
Wakasa Bay
Kyoto
Nagoya
Shimizu
Shimizu
Yaizu
Sagara
Yokkaichi
Chita
152
Ise Bay
147 222
Himeyi
125
Hyogo
Osaka
Sakai
230 200
76
Mizushima
Hiroshima
Takamatsu
Yamaguchi
121 110
Marifu
114
Iwakuni
Sakaide
160
76
Wakayama
Kainan
Owase
Shimonoseki
Onoda
Ube
Tokuyama
Kikuma
114
Kudamatsu
Matsuyama
Yahata
Shimotsu
35
160
50
Kii Strait
Ehime
102
SHIKOKU
Tosa Bay
Fukuoka
Sasebo
Oita
130
Kashima
Bungo Strait
Nagasaki
Kumamoto
KYUSHU
Miyazaki
Osumi Strait
Okinawa
90 100
Off Map
JAPAN
0 50 100 Mi.
0 100 Km.

said in May that it has initiated engineering and development studies for the field. Ranggas lies in the southern part of the Rapak PSC off East Kalimantan.

The Ranggas-4 appraisal well, drilled in 1,587 m of water (5,208 ft), is 3.9 km (2.4 miles) north of the Ranggas-1 discovery well and 1.9 km (1.2 miles) south of the Ranggas-3 appraisal well. Ranggas-4 was tested at a rate of 8,158 b/d of oil and 172,000 cu m/d of gas (6.4 MMcfd) from a single interval at 3,101-3,116 m true vertical depth subsea (10,174-10,224 ft). Initial production rates from a well producing from this single zone were estimated at 10,000 b/d of oil and 215,000 cu m/d of gas (8 MMcfd). The Ranggas-4 well encountered 55 m of net oil pay and 17 m of net gas pay (181 ft and 57 ft, respectively).

Unocal's fifth successful well was drilled in 1,650 m of water (5,412 ft) to total vertical depth of 3,614 m (11,858 ft), encountering 62 m of net oil pay and 188 m of net gas pay (203 ft and 618 ft, respectively) — more than any other well in Unocal's history in Indonesia. The Ranggas-5 well is 2.4 km (1.5 miles) north of the Ranggas-1 discovery well and 1.1 km (0.7 mile) south of Ranggas-4. Ranggas-6 was drilled in August.

With these appraisal results, Unocal estimates the gross discovery volume for the main Ranggas structure at 200-350 million boe, with additional potential on other prospects on the same trend. Earlier, Unocal had estimated the unrisked exploration potential of the entire Ranggas complex at 350-650 million boe. Unocal has drilled six wells in the Ranggas complex, and all have found hydrocarbons.

In addition to appraisal drilling, two wells were drilled to test structures on the northern and western parts of the large central Ranggas prospect. The Ranggas Utara-1 well found an accumulation deemed subcommercial as an independent development, but demonstrated the potential for additional hydrocarbons to be found north of Ranggas field.

The Ranggas West-1 well, 4.7 km (2.9 miles) west of Ranggas-3, found an accumulation that could be tied back to future Ranggas development facilities via a single subsea well. Oil potential remains in the southern extension of this trend, said Unocal. Several additional prospects on trend or adjacent to the main Ranggas structure remain to be drilled, including the Api prospect where the Api-1 well was being drilled.

Other E&D

An exploration well was planned for late 2002 under a PSC work program submitted by GeoPetro Resources Co. (San Francisco) for the Yapen block, which covers 10,000 sq km (2.4 million acres, or 3,800 sq mi) off northern Irian Jaya. The 2003 work program also includes processing of 2D seismic data acquired in 2001 and reinterpreting seismic of old and new vintage. Prospects and leads identified on the block are in clastic and carbonate reservoirs with demonstrated structural closure, said GeoPetro.

Oil and gas were discovered by a wildcat well on the West Madura Block in the eastern Java Sea, said Kodeco Energy Co. Ltd. (Seoul) in July. The KE 40-1 well flowed 2,400 b/d of oil and 45,600 cu m/d of gas (1.7 MMcfd) from pay zones in the Oligocene Kujung III limestone.

An appraisal well to the KE 40 discovery also encountered gas and oil, said partner CNOOC Ltd. (China) in December. The KE 40-2 well, drilled to 2,202 m (7,225 ft), encountered 53 m of gas pay in the Kujung I carbonate buildup and 8.5 m of oil pay in Oligocene Kujung III limestone (174 ft and 28 ft, respectively). During drillstem tests, the pay zones flowed 1,430 bbl of oil and 184,000 cu m/d of gas (6.86 MMcfd).

The deepwater Gendalo-Gandang gas field complex contains estimated reserves of 54-67 bcm (2-2.5 tcf), said Unocal in spring 2002, after hefty natural gas and condensate flow rates from an appraisal well were reported by its subsidiary. The Gendalo-3 well was tested at a rate of 800,000 cu m/d of gas (30 MMcfd) and 2,200 b/d of condensate from a single interval at true vertical depth subsea of 3,523-3,547 m (11,559-11,638 ft). The well has an estimated potential production rate of 2.4 million cu m/d of gas (90 MMcfd) and 6,000 b/d of condensate.

In addition to the impressive gas rates, the appraisal well's high condensate rates demonstrate that this is a world-class gas-condensate resource, which could supply gas to Indonesia's Bontang LNG facilities by 2010 and produce significant liquids even sooner, said Unocal.

Natural gas was found in another appraisal well on the Senoro-Toili Block in central Sulawesi, said Pertamina in May. The well's discovery of 28.7 bcm (1.07 tcf) of reserves raises Pertamina's estimate of the block's gas reserves to 100 bcm (4 tcf). The Senoro No. 3 well flowed on test at a rate of 322,000 cu m/d of gas (12 MMscfd) and 237 b/d of condensate. Additional appraisal drilling was planned for 2002.

In the eastern Timor Sea in Indonesian waters close to the Indonesia-Australian Zone of Cooperation, Inpex was appraising its Abadi-1 gas discovery in the Masela PSC in 580 m of water (1,900 ft). According to industry sources, the large structure holds 130-270 bcm of gas (5-10 tcf) in Jurassic reservoirs.

Amerada Hess and partners drilled two wells in the Tanjung Aru PSC area. Halimun-1 and Papandayan-1 both encountered gas.

A 4,200-km (2,600-mile) nonexclusive 2D seismic survey was underway in September in the Makassar Straits off western Sulawesi and east of the Mahakam Delta. TGS-Nopec Geophysical Co. and WesternGeco were undertaking the survey in cooperation with Migas. This and an earlier Makassar survey were in preparation for licensing in late 2003.

ExxonMobil faced a large dilution of its interest in the 1,670 sq km (645 sq mi) Cepu block of east-central Java, where the company made significant oil discoveries including Banyu Urip-3 and the Sukowati prospect in 2001. In October Pertamina said it might offer ExxonMobil a 50-50 joint venture on Cepu rather than extend the technical assistance contract (TAC) in which the company's interest is 90%. The TAC expires in 2010.

Migas estimated that recoverable oil could reach 500 million bbl, about twice ExxonMobil's earlier estimate. Full-field development calls for gross output of 165,000 b/d starting in 2004 with a pipeline to an FPSO vessel off Surabaya. Prospects for commercializing the gas were unclear.

Gas and oil pipelines

Four pipeline projects were being planned in early 2002 by PT Perusahaan Gas Negara (PGN), Indonesia's state gas distributor. Construction began in June on the first line, which will ship gas to Singapore beginning in 2003 from Jabung field in Jambi Province and two other production areas in southern Sumatra. The other three projects, which would make up an Indonesian distribution network, include an 1,100-km (680-mile) East Kalimantan-East Java pipeline, to be ready by 2005-07, and a 680-km (420-mile) onshore line across Java.

PGN, which was distributing 3.4 bcm/yr (350 MMscfd), plans to increase its gas distribution volume to 10 bcm/yr by 2004 and 20 bcm/yr by 2005-07 (1 bcfd and 2 bcfd, respectively), when the four pipelines are completed. That volume could more than double by 2010 as the government removes fuel subsidies and encourages natural gas use. Industry sources say Indonesia's gas demand is expected to rise 5-8%/yr when the fuel subsidy is phased out in 2004.

Gas supply to Malaysia

Gas deliveries to Malaysia began in mid-2002 from the offshore South Natuna Block

B producing area, said Pertamina. Over a 20-year period, Pertamina will supply 40 bcm (1.5 tcf) from the block to Malaysia's state-owned oil and gas company Petronas Carigali Sdn. Bhd.

Operator ConocoPhillips began initial deliveries of 2.7 million cu m/d (100 MMcfd), which are expected to increase to 6.7 million cu m/d (250 MMcfd) by 2007. The natural gas is being produced from ConocoPhillips's Hang Tuah moveable offshore production unit and is flowing to the Petronas-operated Duyong field complex off peninsular Malaysia via a 96-km (60-mile) pipeline.

Pertamina was also negotiating in early 2002 to sell Petronas 10 bcm/yr (1 bcfd) for 20 years from Block Alpha-D fields operated by ExxonMobil in the Indonesian South China Sea. Sources in Kuala Lumpur said Petronas is short of gas for the fast-expanding Peninsular Malaysia market, as most of its reserves are near Borneo Island and are dedicated to LNG exports.

Alpha-D gas in place is estimated at more than 6 tcm (222 tcf), but recoverable gas is only 1.2 tcm (46 tcf) because of high levels of carbon dioxide, which also will make it very expensive to develop the block. Pertamina also reopened talks with the Petroleum Authority of Thailand for gas from the Alpha-D block.

LNG

The Tangguh LNG project, operated by Pertamina and BP Indonesia in Berau Bay, Irian Jaya, consists of three offshore PSCs and a planned onshore LNG terminal. Pertamina will own the terminal, to be operated by a company jointly owned by Pertamina and the PSC partners.

In October CNOOC Ltd. agreed to buy a stake in the reserves and upstream production of the Tangguh LNG project in Indonesia from BP. Tangguh project partners have a 25-year supply contract to provide up to 2.6 million tpy of LNG beginning in 2007 to an LNG terminal planned for Fujian on China's southern coast opposite Taiwan. The Fujian terminal is scheduled to begin construction in 2004.

Japan

CAPITAL: TOKYO
MONETARY UNIT: YEN
REFINING CAPACITY: 4,766,940 B/CD
OIL PRODUCTION: 12,000 B/D
OIL RESERVES: 58.5 MILLION BBL
GAS RESERVES: 1.4 TCF

Japan's economic growth remained slow during 2002, but the IMF projected that its real GDP would increase 1.1% in 2003. At a meeting with Russian officials in early 2003, Japan explored the possibility of an oil pipeline from eastern Siberia, which would reduce Japanese dependence on Middle East oil. However, China is competing for the same oil supplies, and its markets are closer.

Deregulation and liberalization of Japan's gas and power markets, including third-party access to pipelines and LNG terminals, has intensified gas-on-gas and interfuel competition and created uncertainty about market trends, said Gas Technology Institute, or GTI (Des Plaines, IL, US) in August. Japanese LNG demand is forecast to grow to 63-80 million tpy by 2015 from 54.5 million tpy in 2001. Other factors that affect demand are the pace of the development of nuclear power and the rigor of environmental regulations.

Japan is still the world's largest LNG importer, and its gas and power companies exerted their market strength by negotiating more flexible contract terms in 2002. For many years, Japanese companies were willing to pay premium prices for LNG in return for security of supply based on long-term ironclad contracts with stringent off-take obligations, said GTI. Shipping was left to the seller or the trading companies.

But deregulation is changing the LNG business in Japan, as elsewhere. Utilities now argue that they can no longer accept traditional 20-year contracts, which they wish to replace with a basket of short, medium, and long-term arrangements with more favorable prices as existing contracts expire. Also, said GTI, Tokyo Gas, Osaka Gas, and Tokyo Electric have acquired their own ships.

When Tokyo Gas and Tokyo Electric renewed baseload 20-year contracts with Malaysia's first two LNG export terminals for 7.4 million tpy, they were able to obtain new terms that reduced their prices by around 5%, GTI reported. The agreement also provides that 20% of the contract have terms set for only 1 year. The availability of LNG supplies enhances Japanese buyers' bargaining positions, with one official predicting prices could fall 10% by 2010.

Tokyo Gas, Toho Gas, and Osaka Gas also signed a binding contract in February to import natural gas from Malaysia's MLNG Tiga project, covering deliveries beginning in 2004. This contract also was much more flexible than traditional agreements. However, weak demand for electricity has undermined some LNG import projects, and Osaka Gas cancelled a planned LNG terminal in April.

Sulfur specs and GTL markets

As of April 2003 Japanese refiners will start manufacturing diesel fuel containing 50 ppm sulfur, though specifications don't officially change until January 2004, said FACTS Inc. (Honolulu) in February. All inventories, including distribution channels and gas stations, will be replaced with 50-ppm products by October 2003. A 10-ppm diesel sulfur content regulation will probably occur before 2010, and refiners seem set to meet this more stringent standard with additional upgrades and investment

A commercialization study was under way to investigate opportunities for utilization of GTL and NGL products in Japan, said Ivanhoe Energy Inc. (Vancouver) in March. The products would come from Ivanhoe's proposed Qatar project, which involves development of supergiant North field's gas reserves and transportation of the gas to plants capable of producing 155,000 boe/d of NGL products and 185,000 b/d of GTL fuels.

Ether production

Dimethyl ether (DME) could be produced by a new direct synthesis technology being developed under a government-subsidized project, said DME Development Co. Ltd., a consortium of eight Japanese firms and TotalFinaElf, in August. The project calls for a pilot plant to be built in Kushiro, Hokkaido, that will have the capacity to produce 100 tpd of DME from a feed of synthesis gas (comprising carbon monoxide and hydrogen) made from natural gas.

Partner NKK Corp. (Japan) had already constructed a bench-scale plant capable of producing 5 tpd of direct-synthesis DME. Based on the test results, the consortium said it would conduct a feasibility study for a commercial DME plant with a capacity of 2,500 tpd. DME Development will design, fabricate, construct, and perform commission tests on the pilot plant in 2002-03 and will run the plant on an experimental basis in 2004-06.

Malaysia

CAPITAL: KUALA LUMPUR
MONETARY UNIT: RINGGIT
REFINING CAPACITY: 516,000 B/CD
OIL PRODUCTION: 760,000 B/D
OIL RESERVES: 3.0 BILLION BBL
GAS RESERVES: 75.0 TCF

Following a long slowdown, Malaysia's economy began to recover during 2002, with growth estimated at 3.4% and forecast by the IMF at 5.3% for 2003.

Sarawak, Sabah exploration

Murphy Oil Co. (El Dorado, AR, US) completed a three-well exploration and appraisal drilling program on shallow-water Block SK 309 and neighboring Block SK 311 off Sarawak in March. West Patricia 5, the last appraisal well, confirmed the commercial viability of West Patricia field, with recoverable oil in excess of 30 million bbl. The field was on track for production by early 2003. More oil was found on Block SK 309 by the Congkak well in 41 m of water (136 ft), Murphy said in October. Congkak field is 3 km (2 miles) from West Patricia field platform.

The first deepwater oil discovery off Malaysia was reported by Murphy in August. The wildcat well was drilled on the Kikeh prospect in 1,300 m (4,400 ft) of water on southern Block K off Sabah close to deepwater Brunei J. It encountered several hundred feet of high quality oil reservoirs. Murphy called the discovery significant and said the structure covers a large area.

The Kikeh-2 appraisal well spudded in August using the Ocean Baroness semisubmersible was also successful, Murphy said in October. The well was drilled about 2 km (1 mile) from the discovery well and found the same five primary oil reservoirs, encountering over 120 m of net pay (400 ft). Murphy was moving the rig 3 km (2 miles) to drill Kikeh-3 to ascertain the structure's size. Based on information available, the Kikeh structure could extend into Brunei waters.

Bintang gas development

Plans for Bintang natural gas field, which lies 220 km (137 miles) off Terengganu in the South China Sea, were revealed in March by ExxonMobil, which will develop 27 bcm (1 tcf) of gas with partner Petronas. Peak production from Bintang is expected to reach 3.5 bcm/yr (355 MMcfd). The natural gas will be transported via pipeline to Kerteh and the Peninsular Malaysian gas pipeline grid to be used in the domestic market, mostly for power generation.

The Bintang development will use a minimum-facility design concept to shorten construction and installation times and improve operating efficien-

Pacific Ocean
Philippine Sea
PHILIPPINES
LUZON
MINDORO
LEYTE
PANAY
NEGROS
BOHOL
MINDANAO
PALAWAN
Sulu Sea
Celebes Sea
SABAH
Molucca Sea
Ceram Sea
Banda Sea
Flores Sea
Timor Sea
Arafura Sea
Bismarck Sea
Solomon Sea
Coral Sea
Gulf of Carpentaria
Gulf of Papua
Bonaparte Gulf
Strait of Makassar
SULAWESI (CELEBES)
BURU
CERAM
ARU
TANIMBAR
TIMOR
FLORES
SUMBA
SUMBAWA
Roti
IRIAN JAYA
PAPUA NEW GUINEA
AUSTRALIA
Yungan
Pingtung
Subic Bay
Quezon City
Manila
Limay
Batangas
Tabango
Calauit
Galoc
Octon
Saddle Rock
Malajon
West Linapacan
Malampaya
Camago
Linapacan
Matinloc
Cadlao
Destacado
Nido
Tacloban
Bacolod
Cebu
Davao
Kudat
Kinabalu
Sandakan
Begawan
Tawau
Bunyu Nibung
Tarakan
Manado
Batuputih
Soasiu
Panyilatan
Serang
Attaka
Tunu
Handil
Bekapi
Mutiara
Sepinggan
Wailawi
Palu
Kendari
Ujung Pandang
Ambon
Bula
Sorong
Klamono
1-X TBM
Kasim
Jaya
Salawati
Walio
Sele
Wiriagar
Mogo
Wasian
Manokwari
Steenkool
Fakfak
Hollandia
Wewak
Madang
Elvala
Ketu
Juha
Hides
Elevala
Agogo
S.E. Hedinia
Iagifu
2XGobe
4XGobe
S.E. Gobe
Barikewa
Kuru
Bwata
Omati
Puri
Lake Murray
Kikori
Baimuru
Uramu
Mt. Hagen
Goroka
Lae
Pasca
Pandora-B1X
Pandora-1X
Port Moresby
Raba
Dili
Matai
Ossulari
Aliambata
Kupang
Bayu
Jabiru
Challis
Cassini
Puffin
Darwin
SABAH
BRUNEI
SARAWAK
Samarang Kecil
Champion
Glayzer
Pelican
Iron Duke
Petrel
Magpie
Osprey
Egret
Fairley
Fairley Baram
Ampa
S.W. Ampa
Baram
Tali
Seria
W. Lutong
Kuala Belait
Rasau
Bandar Seri Begawan

cy. Two remotely operated platforms, Bintang A and B, were being installed in late 2002. These platforms, designed and built in Malaysia, have the capacity for 10 wells. Gas will be transported by a new 11-km (7-mile), 18-inch pipeline to Lawit A platform for processing. Drilling was to begin in late 2002.

ExxonMobil fields off Peninsular Malaysia

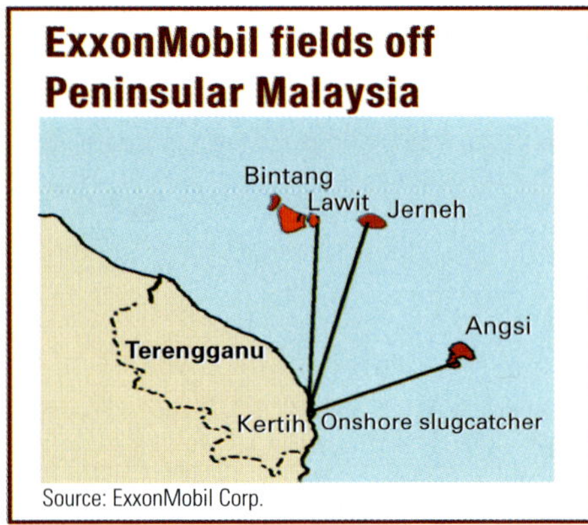

Source: ExxonMobil Corp.

Larut, Angsi production

Earlier in the year, ExxonMobil brought Larut field onstream, which is on Block PM5 about 200 km (125 miles) off Terengganu. Larut, the first development on PM5, is expected to reach peak production of 30,000 b/d of oil and 940,000 cu m/d of gas (35 MMcfd). Larut has an eight-legged steel jacket platform with slots for 36 wells. Production will flow into the existing Tapis oil and gas system through 200 km (125 miles) of new pipelines.

Larut and five other satellite fields lie near ExxonMobil's commercial Angsi field, which began production in late 2001. The satellites were considered economically marginal until ExxonMobil developed new cost-saving designs.

Total output from the six satellites and Angsi should approach 140,000 b/d of oil and 5.24 bcm/yr of gas (535 MMcfd). Over the next 25 years, production from the developments is expected to be more than 320 million bbl of oil and 40 bcm of gas (1.5 tcf). The satellite fields already had five platforms, the first of which began producing in late 2001.

ExxonMobil is Malaysia's largest oil producer, with operated production in September of about 280,000 b/d, representing nearly 50% of the nation's total. It also produces 13.7 bcm/yr of natural gas (1.4 bcfd), which supplies some 70% of Peninsular Malaysia's requirements.

Gas from Indonesia

Gas deliveries to Malaysia began in mid-2002 from the offshore South Natuna Block B producing area, said Pertamina, which will supply 40 bcm (1.5 tcf) to Petronas over a 20-year period. Operator ConocoPhillips began initial deliveries of about 1 bcm/yr (100 MMcfd), which are expected to increase by 2007 to 2.45 bcm/yr (250 MMcfd). The natural gas from ConocoPhillips's Hang Tuah moveable offshore production unit is flowing to the Petronas-operated Duyong field complex off Peninsular Malaysia via a 96-km (60-mile) pipeline.

Petronas was also negotiating with Pertamina in early 2002 to buy 10 bcm/yr (1 bcfd) for 20 years from Block Alpha-D fields operated by ExxonMobil in the Indonesian South China Sea. Sources in Kuala Lumpur said Petronas is short of gas for the fast-expanding Peninsular Malaysia market, as most of its reserves are near Borneo Island and are dedicated to LNG exports. Alpha-D gas in place is estimated at more than 6 tcm (222 tcf), but recoverable gas is only 1.2 tcm (46 tcf) because of high levels of carbon dioxide, which also will make it very expensive to develop the block.

Pipeline from Thailand

Thailand said in May, and again in July, that it would proceed with the newly rerouted natural gas pipeline to Malaysia and related gas separation plant project long stalled by local opposition. The announcement followed years of confusion, debates, and disputes over whether the major energy project should be built. In late 2002 Thai riot police clashed with protestors outside a meeting of the Malaysian and Thai cabinets, resulting in arrests and injuries.

The pipeline would deliver 3.8 bcm/yr of gas (390 MMcfd) to Malaysia from Cakerawala gas field on Block A-18 of the two countries' Joint Development Area (JDA). The main section of line would carry gas from the JDA to the separation plant, which will have capacity for more than 9 bcm/yr (1 bcfd). Pipelaying is planned for 2003 and commercial operation by 2005.

Mongolia

CAPITAL: ULAN BATOR
MONETARY UNIT: TUGHRIK
REFINING CAPACITY: 0
OIL PRODUCTION: N/A
OIL RESERVES: N/A
GAS RESERVES: N/A

Construction began in September 2002 on a 497-km (309-mile) gas pipeline connecting Changqing, China's largest gas field, with Hohhot, capital of north China's Inner Mongolia Autonomous Region, reported People's Daily. Completion is scheduled for late 2003. The pipeline is designed to carry 950 million cu m/yr (35 bcf/yr) but could increase capacity to 1.3 bcm/yr (48 bcf/yr) when fully pressurized.

Changqing field comprises five oil fields with proven reserves of 1 tcm (37 tcf) of natural gas in the Ordos basin, which covers Shaanxi, Gansu, and Shanxi provinces and Inner Mongolia and Ningxia Hui autonomous regions. The field is capable of providing gas to Beijing, Xi'an, and Yinchuan cities as well as meeting demand of 580 million cu m (21.6 bcf) from Inner Mongolia up to 2005.

Development of giant Sulige gas field, also in Inner Mongolia's Ordos basin, was being considered by TotalFinaElf under a contract with PetroChina for a feasibility study. Sulige's proved reserves exceed 230 bcm (8.5 tcf) of natural gas.

Myanmar

CAPITAL: YANGON
MONETARY UNIT: KYAT
REFINING CAPACITY: 57,000 B/CD
OIL PRODUCTION: 10,000 B/D
OIL RESERVES: 50 MILLION BBL
GAS RESERVES: 10 TCF

Myanmar's cabinet in 2002 endorsed amendments easing terms in the long-term contract for natural gas sales to Thailand from Yetagun field in Myanmar's Gulf of Martaban. The 78.4-bcm (2.92-tcf) gas-condensate field is 400 km (250 miles) south of Yangon.

The amendments involve certain adjustments in the gas pricing structure and deferment of the schedule to raise the daily contract quantity. The pact also would include a reduction in the minimum amount of gas that PTT plc must buy from Yetagun, reducing the take-or-pay obligation under which the gas is paid for regardless of whether the gas volume is taken. Myanmar agreed to reduce gas prices for additional volumes bought by Thailand during the stipulated contract's minimum period.

New Zealand

CAPITAL: WELLINGTON
MONETARY UNIT: DOLLAR
REFINING CAPACITY: 106,000 B/CD
OIL PRODUCTION: 34,000 B/D
OIL RESERVES: 189.7 MILLION BBL
GAS RESERVES: 3.09 TCF

During 2002 new data received by *Oil & Gas Journal* resulted in huge increases in New Zealand's estimated proved reserves, with oil more than doubling to 189.7 million bbl from 89.533 million bbl, and natural gas rising by 48%, to 82.87 bcm from 55.94 bcm (to 3.086 tcf from 2.083 tcf).

The New Zealand government launched a licensing round in late 2002 that covers the lightly explored Canterbury basin along the east coast of the South Island. Five blocks are available by competitive staged work program bidding, with a deadline of May 30, 2003. Crown Minerals held seminars related to the Taranaki basin deepwater bidding round in late 2002 in Singapore, London, and Houston.

An appraisal well was planned for the Taranaki basin on Petroleum Exploration Permit (PEP) 38734, said Westech Energy New Zealand (Denver) in December, to be drilled 2 km (1.2 miles) south of its Surrey 1 oil and gas discovery. Surrey 1, drilled to 1,935 m (6,348 ft), flowed 116 b/d of 30-degree gravity oil and 35,000 cu m/d of water-free gas (1.3 MMcfd) from 1,615-1,618 m (5,298-5,308 ft) in the Miocene Mount Messenger sand. The well cut 25 m (82 ft) of reservoir starting at 1,610 m (5,282 ft). Hookup involves a 2.5-km (1.5-mile) pipeline and surface equipment.

Surrey is 8 km (5 miles) south of Westech's Windsor 1 discovery on PEP 38732, which covers 39 sq km (15 sq mi), 20 km (12 miles) from New Plymouth. Westech continues to explore the East Coast basin, where it has drilled 14 wells the past 5 years.

Drilling of four offshore wells led to one gas-condensate discovery on each of two blocks. One well, Galleon-1, flowed 270,000 cu m/d of gas (10 MMcfd) and 2,300 b/d of condensate from Cretaceous Katiki coal measures at 2,752 m (9,029 ft). One group assigned the discovery a resource potential of 104-280 million bbl.

The Rimu discovery on PEP 38719 in the Taranaki basin will be explored by Swift Energy New Zealand Ltd. under a permit awarded in early 2002.

The southernmost part of New Zealand's South Island will be explored by a 50-50 partnership, which won the 3,124 sq km (1,206 sq mi) PEP 38223 in August. GeoSphere Exploration Ltd. (Wellington) and Thomasson International Ventures Inc. (Denver) were seeking partners to drill the Eastern Bush Tertiary prospect in the Waiau basin. Part of PEP 38223 also covers the Te Anau basin, where exploration is less advanced.

A sidetrack appraisal was to be drilled on the Taranaki basin's PEP 38736 from Kahili-1, which cut 44 m of net pay (144 ft) in a 58.5-m (192-ft) gross oil column in Oligocene Tariki sandstone in 2002. Kahili is 3 km (2 miles) north of the producing Tariki gas-condensate discovery. Tap Oil Ltd. (Perth), which took a share in the well, estimated potential recoverable volumes at 14 million bbl of oil and 0.5 bcm (18 bcf) of associated gas. Initially Tap will concentrate on the smaller, lower risk, oil-prone Miocene Mount Messenger formation at around 2,000 m (6,600 ft).

The results of a 6,200-km (3,900-mile) 2D seismic survey were reported in November by TGS-Nopec Geophysical Co. The survey was conducted over the deepwater Taranaki basin, the extension of the basin from the shelf edge to the 2,000-m (6,600-ft) isobath.

The study identified large traps (200 sq km, or 80 sq mi) and found that the deepwater Taranaki is a Cretaceous play, different from traditional (Kapuni) shallower water plays on the shelf. Thick sedimentary successions (up to 10 km, or 6 miles) also were established, and the results showed that Taranaki shelf wells have established source and reservoir facies for the deepwater areas.

Development of Kupe South field off the Taranaki coast was being sought by Genesis Power Ltd. (Auckland), which hopes to supply a 400-MW, combined-cycle gas turbine power station it plans to start up at Huntly by early 2005. Kupe South, lying in 35 m (115 ft) of water 30 km (20 miles) offshore, has reserves of 6.5 bcm (242 bcf), 15 million bbl of oil, and 300,000 tonnes of natural gas liquids in the Farewell formation of the Kapuni Group at 3,100 m (10,200 ft).

Coalbed methane is the main objective on 10 permits issued in various New Zealand locations. Resource Development Technology LLC (Morrison, CO, US) signed on PEP 38605 covering 2,860 sq km (1,200 sq mi) along the North Island west coast south of Auckland. The company also holds a permit on northern Taranaki coal fields. Kenham Holdings Ltd. (Christchurch) took PEP 38220 covering 57 sq km (22 sq mi) near Ohai in the western Southland coal field on the southernmost South Island. Kenham holds six other coalbed methane permits. The government granted Macdonald Investments Ltd. (Wellington) an extension on 70 sq km (27 sq mi) on northern PEP 38508 at the Greymouth coal field on the northwestern South Island.

Pakistan

CAPITAL: ISLAMABAD
MONETARY UNIT: RUPEE
REFINING CAPACITY: 233,850 B/CD
OIL PRODUCTION: 60,000 B/D
OIL RESERVES: 310.4 MILLION BBL
GAS RESERVES: 26.4 TCF

Pakistan reported higher proved reserves estimates during 2002, reflecting discoveries that boosted oil reserves by 4.1%, to 310.4 million bbl from 298.237 million bbl, and natural gas by 5.1% to 708.01 bcm from 673.45 bcm (to 26.365 tcf from 25.078 tcf), said *Oil & Gas Journal*.

Pakistan's current natural gas reserves can meet demand through 2018 without new discoveries, said a FACTS analysis in early 2002. Assuming 4%/yr growth in GDP, power generation will increase at 5%/yr. With moderate substitution of gas for fuel oil, natural gas consumption for power generation will rise at 8%/yr, which translates to overall gas consumption growth of 6%/yr. Pakistan's comfortable supply-demand picture and position as a transit route give the country leverage in negotiating gas deals.

Pakistan's privatization commission received high bids totaling $176.3 million in April 2002 for nine oil and gas fields, which were producing a total of more than 20,000 boe/d, or 9,700 b/d of oil and 1.7 million cu m/d of gas (62 MMcfd), and contain more than 50 million boe in reserves. The government's stake in the Badin-I concession received the highest bid of $131.5 million, which was offered jointly by BP Pakistan Exploration & Production Inc. and Occidental Oil & Gas Pakistan LLC.

Zamzama development

Pakistan's fourth largest gas field in terms of reserves, Zamzama, will undergo development by BHP Billiton, which was approaching the first phase of the project in mid-2002. Ultimate recovery is placed at 46 bcm (1.7 tcf) proved and probable. Plateau production is expected to be 2.9 bcm/yr (300 MMcfd) and 2,000 b/d of condensate for 10-12 years. Production start is set for late 2003, and field life is 15-25 years.

Zamzama field is on the Dadu concession in Sindh Province 200 km (125 miles) north of Karachi. Following an extended well test (EWT), production has averaged 2.7 million cu m/d (100 MMcfd) from the Zamzama-1/ST1 and Zamzama-2 wells. Pakistan has more than 500 bcm (20 tcf) of proved gas reserves and produces 19 bcm/yr (700 bcf/yr), all consumed domestically. The three largest producing fields are Sui at 7.4 bcm/yr, Mari 4.9 bcm/yr, and Qadirpur 2.9 bcm/yr (750 MMcfd, 500 MMcfd, and 300 MMcfd, respectively). The oil and gas ministry has forecast a 50% increase in internal gas demand by 2006, which the government hopes to satisfy via domestic developments such as Zamzama.

Zamzama field is BHP's first commercial production in Pakistan. Beyond the 46 bcm (1.7 tcf) in the core area, a further possible resource of 35 bcm (1.3 tcf) has been identified in the undrilled footwall to the east and the northern Phulji areas of Zamzama field.

Under BHP's EWT scheme, gas is treated at a minimum processing facility for dehydration, dewpoint control, and condensate stabilization before being compressed and delivered 8 km (5 miles) to the main Sui Southern Gas Co. Ltd. Sui-Karachi pipeline, where the gas is sold to Sui Southern under a temporary contract. BHP said the EWT was integral to managing risks associated with field development. It provided valuable long-term well and reservoir productivity data while securing a market position.

At least three development wells will be drilled in the core area. Development will involve a dual-train facility that largely incorporates the EWT plant. Stabilized condensate will be shipped via small-diameter pipeline connecting with the Pak-Arab Refinery Co., or Parco (Karachi) pipeline 40 km (25 miles) west of the Zamzama plant. Condensate will be refined at Parco's Muttar refinery in Punjab Province.

Nearby Bhit field reportedly was being developed by Eni SpA (Rome) during 2002 and expected to begin production in 2003.

GAS FIELDS AND FACILITIES IN PAKISTAN

Source: BHP Petroleum (Pakistan) Pty. Ltd.

Exploration

On the Sanghar block in Sindh province, production testing of the Chak 63 No. 1 well demonstrated commercial potential, said Pakistan's state owned Oil & Gas Development Co. Ltd. (OGDC) in April. An initial short duration test produced 2,000 b/d of oil and 107,000 cu m/d of natural gas (4 MMcfd). Chak 63 was drilled to 3,000 m (9,800 ft).

An oil exploration license for Block 2568-11 (Nawabshah) was issued in October to a joint venture of Tullow Oil plc (UK) subsidiaries and a unit of OMV AG (Austria). The block, which lies in Prospectivity Zone III, covers an area of 2,335 sq km (901 sq mi) in Sindh Province. The venture is obligated to carry out geological and geophysical studies, a 300-km (190-mile) seismic survey, and the drilling of one exploration well during Phase I of the initial 3-year term.

In September, another oil exploration license was granted to a joint venture for Manchar Block 2667.5 in the Dadu district, also in Sindh Province. The venture includes subsidiaries of Eni and other partners. The block is about 200 km (125 miles) north of Karachi near Bhit gas field and covers an area of 2,435 sq km (940 sq mi). It falls in Prospectivity Zone III and has the same sedimentary basin as that of adjacent Zamzama gas field.

FUTURE PIPELINES IN PAKISTAN

Segment	Distance, km	$ million	Year*
Port Bin Qasim to Mahmood Kot	817	480	2002
Morgah to Peshawar	160	80	2001-02
Morgah-Kharian-Machhike	300	141	2002-03
Faisalabad to Kharian	190	80—100	2003-04
Faisalabad to Sahiwal	90	40-50	2005-06
D.G. Khan to Peshawar	430	160-200	2006-07
Port Bin Qasim to Korangi	23	10-12	2001-02
Badin to Port Bin Qasim (crude)	140	33	2001-02

*Tentative completion dates.

Sawan, Miano gas

A gas sales and pricing agreement to supply 1.7 bcm/yr of gas (170 MMscfd) to Sui Northern Gas Pipeline Ltd. was finalized, said OMV's Pakistan unit in April. The gas is from Sawan field in Sindh Province, which holds reserves of 35 bcm (1.3 tcf).

Earlier in 2002, Phase 1 development of Sawan was initiated to bring onstream 1.7 bcm/yr (170 MMscfd) for delivery to Sui Southern Gas via the existing Kadanwari-Nawabshah transmission line by mid-2003. Sawan Phase 2 development will make available the same volumes of gas for Sui Northern by late 2003.

The OMV unit also started up Miano gas

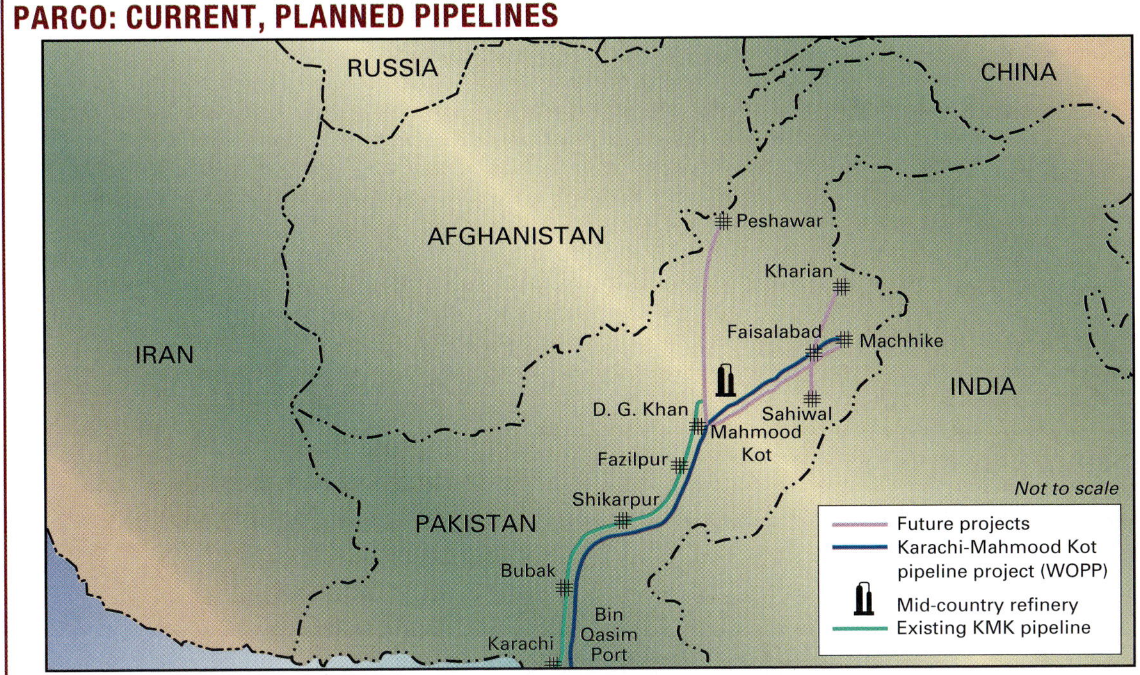

field in the Thar Desert in early 2002, producing close to 1 bcm/yr (100 MMcfd) for delivery to Sui Southern at the Kadanwari treatment plant.

Domestic pipeline expansion

The delivery capacity of Pakistan's crude oil and refined products pipelines was being expanded during 2002 by Parco. By year-end, as much as 980 km (610 miles) will be completed, including 140 km (90 miles) of crude oil pipeline and a major 817-km (508-mile) white-oil product line. Parco's plans call for finishing another 490 km of (300 miles) product lines by 2005 and an additional 520 km (320 miles) by 2008. Combined, new pipelines will total up to 2,150 km (more than 1,300 miles), mostly for refined products.

The 817-km (508-mile), 26-inch White Oil Pipeline Project follows conversion of the company's existing pipeline network for crude oil transportation. The line will transport 5 million tpy of refined products initially, building to full capacity of 12 million tpy, from Karachi to central and northern Pakistan, which account for almost 60% of total petroleum consumption. With up-country demand for petroleum products rising at 10%/yr, the company says total demand in the northern region will increase to 10 million tpy.

Import pipelines

Studies for a proposed 16 bcm/yr (1.6 bcfd) Qatar-to-Pakistan offshore gas pipeline were completed in early 2002 by Crescent Petroleum. The pipeline would travel mainly through Iranian waters, extending from Ras Laffan through Hamriyah, then Dibba on the Gulf of Oman, and terminating at Gadani, Pakistan. The section between Hamriyah and Dibba would be onshore.

Plans for a Turkmenistan-Afghanistan-Pakistan oil and gas pipeline were being finalized in mid-2002 by the three countries' oil ministers. In October Turkmenistan said it supports plans to build a 1,500-km (1,000-mile) gas pipeline linking Turkmenistan to Pakistan, with 764 km (475 miles) of the line in Afghanistan. The pipeline would have a throughput capacity of up to 30 bcm/yr (1.1 bcf/yr). In July the Asian Development Bank (Manila) granted $1.5 million for a prefeasibility study.

Downstream activities

Karachi might be the site of a new refining unit for Chinese companies to process crude from Sudan's Greater Nile oil project. The Chinese could also acquire a stake in the long-delayed Iran-Pak refinery project or build elsewhere. As a coastal city, Karachi has the advantage of port facilities, which can also be used to export naphtha to China, where plastics industry demand is high.

Engro Chemical Pakistan Ltd. (Karachi) will develop a world scale ammonia urea fertilizer complex in Oman under an agreement with Oman Oil Co. announced in November.

Papua New Guinea

CAPITAL: PORT MORESBY
MONETARY UNIT: KINA
REFINING CAPACITY: 0
OIL PRODUCTION: 46,000 B/D
OIL RESERVES: 240.0 MILLION BBL
GAS RESERVES: 12.2 TCF

A new oil system was discovered in the eastern Papuan basin of Papua New Guinea. As described by InterOil Corp. (Sydney) in March, the basin is proven to be prospective for oil, with the potential for world-class fields comparable to Kutubu (more than 300 billion bbl of oil).

The eastern Papuan basin has a good quality reservoir sand and two sources of oil, said InterOil, which won the PPL 230 license covering 16,000 sq km (4 million acres, or 6,000 sq mi) in early 2002 and has become the largest net onshore acreage holder in the country. The new license includes all the identified Pale sandstone prospects.

InterOil's initial work included field coring of stratigraphic wells and reprocessing of original gravity and seismic data. The company identified 12 Pale leads and prospects, plus 20 others. Four wells were planned for 2002 and eight in 2003 to test large structures. A successful drilling program could revitalize exploration in Papua New Guinea and reverse its decline in oil production, said InterOil.

The pipeline project to move natural gas from Papua New Guinea to Queensland, Australia, was reorganized under a new agreement excluding Santos Ltd. (Australia), which quit the project in early 2002. The 3,200-km (2,000-mile) pipeline had been delayed because of a dispute among the partners on how to split revenue from the gas and because of a lack of political support.

Philippines

CAPITAL: MANILA
MONETARY UNIT: PESO
REFINING CAPACITY: 419,500 B/CD
OIL PRODUCTION: 14,000 B/D
OIL RESERVES: 152.0 MILLION BBL
GAS RESERVES: 3.77 TCF

The Philippines government expected to increase domestic production of natural gas by 3.9 bcm/yr (146.4 bcf/yr), reaching 40 bcm (1.5 tcf) by 2011. The impetus behind this dramatic change in the country's gas sector is the Malampaya deepwater offshore gas field, which began producing gas in 2001 for three onshore power plants at Batangas, southern Luzon Island. Malampaya is the largest gas development project in Philippine history, and one of the largest ever foreign investments in the country.

Philippine National Oil Co. had an indicated gas discovery in early 2002 at the Victoria-3 well in Tarlac Province, Luzon, 100 km (60 miles) north of Manila. The well, drilled to 2,700 m (9,000 ft), reportedly flared gas continuously for several days from 1,577-1,678 m (5,175-5,505 ft). The drillsite is a close offset to Victoria-2/2A, at 5,593 m (18,349 ft) the Philippines' deepest hole.

In April the company encountered water and plugged a wildcat in Victoria, Tarlac. The well initially yielded as much as 27,000 cu m/d (1 MMcfd), but flow could not be sustained. Philippine National Oil said it is still optimistic on gas prospects in the Central Luzon basin.

Shell Philippines Exploration BV took a farmout on the Geophysical Survey and Exploration Contract 99 offshore, east of northern Palawan Island, covering 5,300 sq km (1.3 million acres, or 2,000 sq mi). The northern Sulu Sea block lies south of the gas pipeline connecting Malampaya field to Luzon Island. The tract is on the Mindoro-Cuyo platform, the Philippines' largest basin at 110,000 sq km (27 million acres, or 42,000 sq mi).

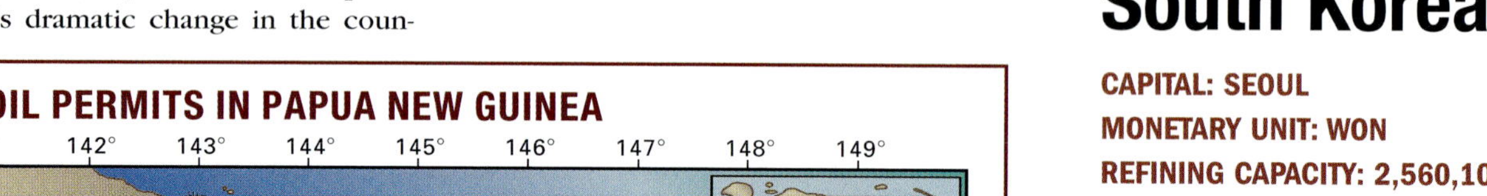
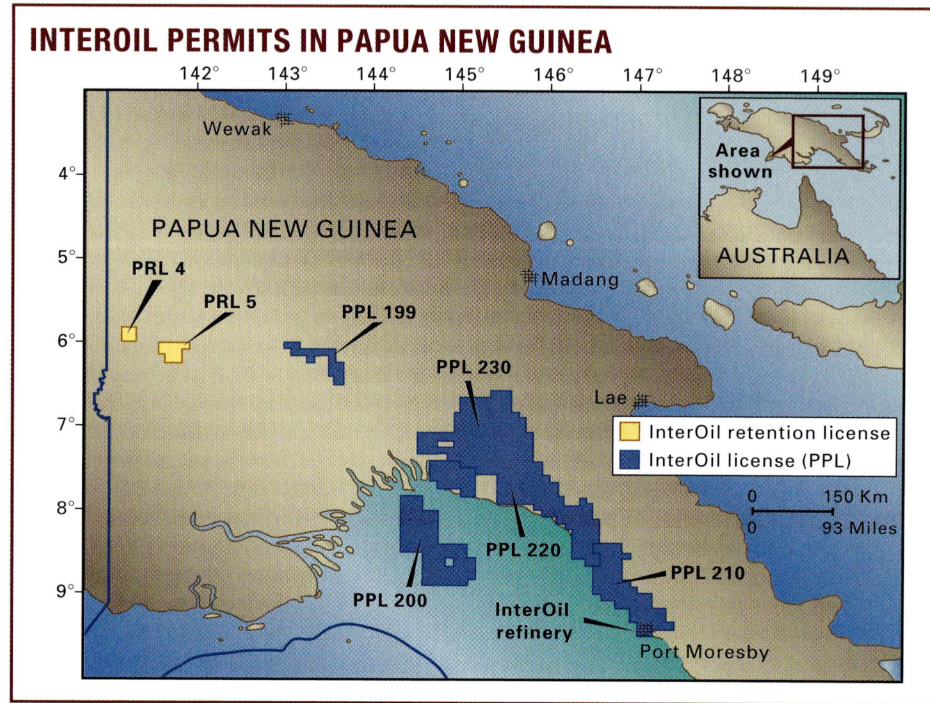

Singapore

CAPITAL: SINGAPORE
MONETARY UNIT: DOLLAR
REFINING CAPACITY: 1,258,500 B/CD
OIL PRODUCTION: 0
OIL RESERVES: 0
GAS RESERVES: 0

Singapore's economy began to rise from the doldrums, after negative growth in 2001. Although GDP was expected to grow only 2.2% in 2002, an increase of 4.2-4.8% was projected for 2003. Despite a short-term recovery in refining margins during 2002, the longer term outlook for the refining sector was uncertain, and some operators were considering selling their stakes. Singapore's refining capacity is nearly double its rate of petroleum product consumption.

Construction began in June on a subsea pipeline to ship natural gas to Singapore beginning in 2003 from Jabung field in Jambi Province, Sumatra, Indonesia. Singapore has an ideal location to become Southeast Asia's regional gas hub.

Singapore uses gas mainly for power generation and as a petrochemical feedstock, but has launched a program to convert public buses in some areas to CNG. The first SembGas CNG filling station opened in April, and the program could later expand to taxis.

South Korea

CAPITAL: SEOUL
MONETARY UNIT: WON
REFINING CAPACITY: 2,560,100 B/CD
OIL PRODUCTION: N/A
OIL RESERVES: N/A
GAS RESERVES: N/A

Like other Southeast Asian nations, South Korea is recovering from the global economic slowdown, with GDP growth estimated at 5.8% in 2002 and projected at 5.9% in 2003. However, profit margins in its refining sector remained very weak. South Korea officially joined the International Energy Agency in March 2002.

The country's Korea National Oil Co. could provide 10% of South Korea's energy needs by 2010, the government hopes. The firm anticipates becoming a minor natural gas producer with its stake in the Rong Doi concession off Viet Nam, with estimated reserves of 27 bcm (1 tcf).

Korea National is also planning development of Donghae-1 natural gas field in the Ulleung basin 60 km (40 miles) off Ulsan. Once onstream, the field would mark the country's first commercial oil and gas production. Plans call for drilling three subsea wells, to be tied back to a production platform and pipeline to an onshore gas processing plant. Multiple drilling contracts were awarded to a unit of Halliburton Co. (Dallas) in April. Donghae-1 contains an estimated 5.4 bcm (200 bcf) of reserves.

Meanwhile, South Korea continues to rely exclusively on imported LNG for its gas requirements. Korea Gas Corp. (Kogas) was expanding its LNG terminals at Pyeong-Taek and Inchon, and construction of a third terminal at Tong Yung was near completion. In June construction began on a 1.7 million tpy LNG terminal at Kwangyang south of Seoul, to be completed by mid-2005. The facility was being built by South Korea's Pohang Iron and Steel Corp. to cut costs and secure fuel for its power plants.

After slowing in recent years, LNG

demand in South Korea has again accelerated and could reach 26 million tpy by 2010 and 33 million tpy by 2015 from the current level of 17 million tpy, said GTI in August. The government was planning to split Kogas's import and wholesale divisions into three parts, two of which would be privatized by year-end 2002, subject to parliamentary approval. ExxonMobil and Petronas have submitted bids to buy a 15% share in Kogas, GTI said.

Taiwan

CAPITAL: TAIPEI
MONETARY UNIT: NEW TAIWAN DOLLAR
REFINING CAPACITY: 920,000 B/CD
OIL PRODUCTION: 800 B/D
OIL RESERVES: 4.0 MILLION BBL
GAS RESERVES: 2.7 TCF

A long-negotiated agreement was signed in June to explore a 15,400 sq km (5,900 sq mi) block southwest of Taiwan in the Tainan basin and Chaoshan Trough in the South China Sea. The contract between Taiwan's state-run Chinese Petroleum Corp. (CPC) and CNOOC calls for reprocessing 500 km (300 miles) of seismic data, acquiring 4,000 km (2,500 miles), and drilling three wildcats. The companies will form a joint management committee and share the exploration costs equally.

Plans to build Taiwan's eighth naphtha cracking plant were revived in October by CPC. The facility would have projected refining capacity of 450,000 b/cd of oil and an output of 2.6 million tpy of ethylene, 1.63 million tpy of propylene, and 420,000 tpy of butadiene. The plan has received the backing of Taiwan's Ministry of Economic Affairs, which supports Nanchou in Pingtung County as the site for building the cracker.

In another deal, CPC agreed in early 2002 to refine 600,000 bbl of crude oil for China National Petroleum Corp. (China). Middle East crude oil began arriving in 2002 at Taiwan's 300,000 b/cd Ta-Lin facility. If the short-term arrangement proves satisfactory, the companies may enter a longer-term agreement.

ExxonMobil applied for a license to market its refined products in Taiwan and was negotiating in March to use CPC's facilities to process imported oil, once ExxonMobil gains a certain share of the market. According to CPC, ExxonMobil plans to buy products from CPC for resale in Taiwan until ExxonMobil secures a foothold in that market. Once established, ExxonMobil plans to begin local production using CPC's facilities.

ExxonMobil moved quickly during 2002 to establish a presence in the Taiwanese market, which was recently opened to foreign competitors. It established a joint venture with Taiwan's Pan Overseas Corp. to import ExxonMobil products into Taiwan. The venture has set up seven storage tanks at Taichung Harbor in central Taiwan.

Thailand

CAPITAL: BANGKOK
MONETARY UNIT: BAHT
REFINING CAPACITY: 703,100 B/CD
OIL PRODUCTION: 130,000 B/D
OIL RESERVES: 583.35 MILLION BBL
GAS RESERVES: 13.34 TCF

Capital and operating expenditures of $1.93 billion (85.27 billion baht) were planned for 2002-06 by Thailand's PTT Exploration & Production plc (PTTEP), a subsidiary of partially privatized PTT plc. During 2002 PTTEP will focus on Bongkot and Pailin fields and the S1 onshore tract projects. These will account for 79% of the year's capital expenditures of $233 million (10.3 billion baht). During 2002 Bongkot produced an estimated 5.56 bcm of gas (567 MMcfd). In addition, future projects in the PTTEP budget will include Arthit field and areas of the Thai-Malay JDA.

Terms were eased in the long-term contract for natural gas sales to PTT from Yetagun field in Myanmar's Gulf of Martaban, as the Myanmar cabinet endorsed amendments in 2002. The 78.4-bcm (2.92-tcf) gas-condensate field is 400 km (250 miles) south of Yangon.

The amendments involve certain adjustments in the gas pricing structure and deferment of the schedule to raise the daily contract quantity. The pact also would include a reduction in the minimum amount of gas that PTT plc must buy from Yetagun, reducing the take-or-pay obligation under which the gas is paid for regardless of whether the gas volume is taken. Myanmar agreed to reduce gas prices for additional volumes bought by Thailand during the stipulated contract's minimum period.

E&P awards

Thai authorities granted five new production licenses and awarded seven exploration blocks in October 2002, after a 2-year delay. The production approvals went to existing concessionaire groups led by Shell, Pacific Tiger Energy Inc. (Singapore), Harrods Natural Resources Inc., and PTTEP. The licenses, covering a total production area of 168 sq km (65 sq mi) both offshore and onshore, will add about 1 bcm/yr of natural gas (100 MMcfd), 15,000 b/d of crude oil, and 1,000 b/d of condensate to Thailand's rising domestic petroleum output.

The following production licenses were granted:

- PTTEP, 1.93 sq km (0.75 sq mi) in Pikun field on Block B13/38 and 18.28 sq km (7.1 sq mi) in Tong Rang field, on Block 15, both in the Gulf of Thailand
- Harrods, 48.64 sq km (18.78 sq mi) in Jasmine field on Block B5/27, also in the Gulf
- Pacific Tiger, 8.69 sq km (3.39 sq mi) in Wichian Buri License II field, on onshore Block SW1A
- Shell, 90.43 sq km (34.92 sq mi) in South Pratu Tao prospect, part of onshore Block S1.

The Thai government is expected to receive $65 million in royalties and $80 million in income tax based on production from these sources.

The exploration blocks were awarded to six firms, pending formal endorsement from the Thai Cabinet:

- A subsidiary of ChevronTexaco Corp. (San Francisco), Block G4/43 in the Gulf of Thailand
- Shell's Thai unit, onshore Block L22/43
- Pacific Tiger, onshore blocks L33/43 and L44/43
- China National Petroleum, onshore Block L21/43
- SVS Energy Resources, onshore Block L71/43
- Nucoastal Corp., offshore Block G5/43.

ChevronTexaco and Shell provided bonuses and other incentives during the negotiations, and the concessionaires must spend more than $32 million exploring the tracts over the first 3-year obligation.

Unocal deals

Unocal, Thailand's largest natural gas producer, delivers about 10 bcm/yr (1 bcfd) from its 13 fields in the Gulf of Thailand, which accounts for about a third of the kingdom's overall indigenous gas supplies. Over the decade from 2002, Unocal plans to spend more than $4 billion in Thailand.

In April Unocal agreed to cut the price

of natural gas it sells to PTT by about $1.30/cu m (5 cents/Mcf) or 2% of the prevailing wellhead prices. The discount covers gas supplies from sales agreements that account for 82% of the daily contract minimum for Unocal-operated Thai fields. Separately, Unocal was also seeking a license to produce crude oil from Yala field, part of a large project that also includes Plamuk, Surat, and Kaphong fields.

Unocal planned to spend $440 million in Thailand in 2002, mostly to maintain gas output by arresting production declines and starting up North Pailin field, as well as ramping up crude production. Unocal's Thai unit was to drill up to 230 wells in 2002, double 2001's count. Most are development wells, with only 11-12 exploration wells. Condensate output was expected to increase slightly.

In October Unocal Thailand Ltd. boosted its hydrocarbon portfolio in the region with plans to buy out a BP unit's share of Blocks 5 and 6 in the gas-prone Thai-Cambodian Overlapping Claims Area in the Gulf of Thailand. The BP Thai unit was operator and 50% owner of the concession, and the two governments have agreed to divide the area.

The blocks, which cover 10,155 sq km (3,921 sq mi), offer high expectation of substantial gas and condensate reserves and will complement Unocal's existing gas-producing areas, Blocks 10 and 11 and parts of 12 and 13. A 3D seismic campaign and exploratory drilling are planned after the boundary dispute is settled.

Other E&D

The first three wells were drilled in a planned six-well exploration program on Block B8/32 in the Gulf of Thailand, said Pogo Producing Co. (Houston) in July.

In central Thailand, Pacific Tiger spudded the WB-N3 well in January at Wichian Buri oil field to test the productive F-sand downdip from earlier wells. Reserves are 10 million bbl, and output will exceed 1,500 b/d by year-end 2002, said partner Carnarvon Petroleum NL (South Yarra, Victoria, Australia).

The well was part of a multi-well exploitation program in Wichian Buri field, which received fresh funding from Gemini Oil & Gas Ltd. in April. The partners were drilling up to four development wells on the PL I Wichian Buri production license in mid-2002 and planned to drill the PL II license later in the year. Two wells were on production, and one was shut in pending the PL II license award, which was approved in October.

Shell's planned revival of production from Nang Nuan field in 2002 was a failure, raising questions about the productivity of the field off the Chumphon coast.

Bongkot, Arthit development

The 13th wellhead platform will be installed at Bongkot gas field in the Gulf of Thailand 600 km (400 miles) from Bangkok in mid-2003, under terms of a contract with Nippon Steel Corp. (Japan) awarded in July by a Thai-European consortium led by PTTEP.

The platform is part of the Phase 3C development of Bongkot, aimed at sustaining its gas production at current levels of 5.4-6.2 bcm/yr (550-630 MMcfd) plus about 13,000 b/d of condensate and crude oil. Twelve wellhead platforms, a production platform, and a platform for living quarters are in place at Bongkot, one of Thailand's largest gas fields.

About 54 bcm (2 tcf) of proven gas reserves remain in the structure, where 32 bcm (1.2 tcf) has already been recovered since production began in 1993. The potential for sustaining Bongkot production is supported by continued efforts to improve recovery and by success in finding new reserves in satellite prospects, said PTTEP.

One of the Gulf of Thailand's major gas discoveries, the 4,000 sq km (1,500 sq mi) Arthit structure, will be developed by PTTEP, which applied for a production license in July. Twenty of the 21 exploration and appraisal wells drilled on the prospect have yielded gas, and reserves are estimated at 40 bcm (1.5 tcf). Arthit's structure and reservoir behavior are similar to Bongkot's, the gulf's biggest gas field.

In addition to Arthit's commercial potential, Thailand's gas demand is increasing, due in part to public concern over pollution from coal-fired power plants. Arthit development plans call for initial production of 2.45 bcm/yr (250 MMcfd) in 2006.

Arthit gas will be piped to the country's central plains through a third trunkline planned in the Gulf of Thailand by PTT. The line would extend northward, splitting into two spurs to reach Petchaburi on the western side and Rayong on the eastern seaboard.

Pailin production

Natural gas and condensate production started in mid-2002 from the northern extension of Pailin field in the Gulf of Thailand, said Unocal's Thai unit. Phase II development facilities were producing 1.62 bcm/yr of gas (165 MMcfd), boosting gross contracted gas sales from Pailin to 3.2 bcm/yr (330 MMcfd) under the agreement with PTT. North Pailin produced at rates as high as 2.25 bcm/yr (230 MMcfd) during commissioning in June, demonstrating the capacity of Pailin field to provide more than 3.7 bcm/yr (380 MMcfd) if warranted by market demand.

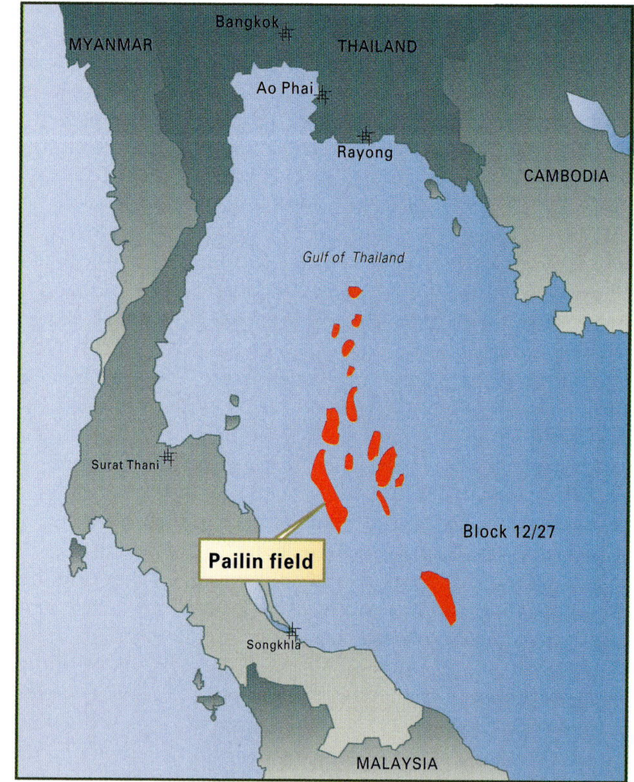

In addition, North Pailin produces 6,800 b/d of condensate, raising total condensate production from Pailin, on Block B12/27, to more than 15,300 b/d. With additional production from Pailin Phase II, gross natural gas production from Unocal-operated fields totals 10.5 bcm/yr (1.07 bcfd), all of which is sold to PTT for domestic consumption.

Unocal has installed 11 wellhead platforms and two processing platforms to serve Pailin field, 160 km (100 miles) east of the southern province of Nakon Si Thammarat. Unocal estimates proved gas reserves in the field at 54 bcm (2 tcf).

Pipeline to Malaysia

Thailand said in May, and again in July, that it would proceed with the newly rerouted Thai-Malay natural gas pipeline and related gas separation plant project long stalled by local opposition. The announcement followed years of confusion, debates, and disputes over whether the major energy project should be built. In late 2002 Thai riot police clashed with protestors outside a meeting of the Thai and Malaysian cabinets, resulting in arrests and injuries.

The pipeline would deliver 3.8 bcm/yr of gas (390 MMcfd) to Malaysia from Cakerawala gas field on Block A-18 of the two countries' JDA. The main section of line would carry gas from the JDA to the separation plant, which will have capacity for more than 9 bcm/yr (1 bcfd). Pipelaying is planned for 2003 and commercial operation by 2005.

To avoid the area where most of the project's opponents live and yet skirt another environmental impact assessment, the onshore section through the southern Thai province of Songkhla on the Malay Peninsula will be rerouted only slightly. It would come ashore at Ban Nairat in Tambon Taling Chan, 4.8 km (3 miles) from the original route at Ban Kobsak in Tambon Sakom. The pipeline rerouting would mean relocation of the two 11.4 million cu m/d (425 MMcfd) gas separation units to Ban Nairat.

Oil line to northern Asia

An oil pipeline across southern Thailand was proposed in August to expedite shipments of Middle East crude oil to northern Asia. When completed in 2005, the pipeline would provide a shortcut and divert oil traffic that normally goes through the congested Malacca Straits and at a lower transportation cost. Taksina Pipeline Co., a consortium of Thai, Omani, and Canadian firms, is undertaking the $450 million project.

The scheme essentially calls for laying a 46-inch pipeline 160 km (100 miles) from Satun on the Andaman Sea coast to Songkhla on the Gulf of Thailand coast. It will include two storage terminals, each with crude storage capacity of 10 million bbl, on both ends of the pipeline that will be laid underground along Thai highways.

Also, two single-buoy moorings would be installed about 15-20 km (9-12 miles) off the coasts, to facilitate loading and offloading of crude oil from 200,000-300,000 dwt very large crude carriers. Currently, only tankers with maximum capacity of 150,000 dwt are allowed to transit the Malacca Straits.

The pipeline would have crude oil throughput capacity of about 1 million b/d, compared with 12 million b/d of crude oil shipped through the Malacca Straits every day. By using the pipeline, the shipment of crude oil from Middle East to countries such as Japan, South Korea, China, and Taiwan can be cut by a few days or so.

Petrochemicals from domestic gas

A 14 million cu m/d (530 MMcfd) natural gas processing plant, PTT's fifth, will be built by Foster Wheeler International Corp. (Clinton, NJ, US) following a feasibility study, under a contract announced in March. The facility would be sited alongside PTT's three existing gas processing trains at Mab Ta Phud, Rayong, on Thailand's eastern coast about 220 km (140 miles) southeast of Bangkok. The plant will produce 500,000 tpy of ethane, 170,000 tpy of propane, and 480,000 tpy of LPG.

Liquids produced from domestic natural gas will be supplied by PTT to Thailand's largest olefins producer, Rayong Olefins Co., under a long-term accord announced in August. The Mab Ta Phud plant in Rayong will receive 50,000-100,000 tpy of LPG and 35,000-70,000 tpy of NGL for 15 years starting in December 2002. Previously, the olefins complex had used naphtha. The plant produces up to 800,000 tpy of ethylene and 400,000 tpy of propylene.

Another olefins plant in Rayong was being expanded by Thai Olefins Co., which awarded construction contracts in early 2002. The work will boost ethylene production capacity by 300,000 tpy to 685,000 tpy; propylene capacity will remain unchanged at 190,000 tpy. Construction, beginning by year-end, should be completed by late 2004, coinciding with an upturn anticipated for the petrochemicals market in 2004-05 and an expected rise in petrochemical product prices. The new unit will crack ethane, separated from indigenous natural gas, into ethylene.

Thailand's National Petrochemical plc is building a 250,000 tpy polyethylene plant in Rayong, which will add to the company's existing production of 437,000 tpy of ethylene and 170,000 tpy of propylene. National Petchem's production is based on ethane and propane extracted from natural gas piped from the Gulf of Thailand. The plant is expected to begin operation in 2004.

Production capacity at the aromatics complex in Rayong will be increased by 13.27%, to a total of about 1 million tpy, said Aromatics (Thailand) plc in September. The firm recently completed first-phase debottlenecking, boosting the plant's output by 57.28% to 972,000 tpy from 618,000 tpy and reducing unit production costs by 20%.

The next phase, due onstream by spring 2004, will further improve the plant's cost-effectiveness and increase production capacities of benzene to 467,000 tpy from 333,000, paraxylene to 495,000 tpy from 394,000, and orthoxylene to 78,000 tpy from 44,000.

Aromatics (Thailand) also plans to build its own condensate splitter, which will convert excess condensate residue from aromatics production and other sources into light oil products, the company said in December 2002. Installation could start in 2003 for completion within a few years.

Viet Nam

CAPITAL: HANOI
MONETARY UNIT: DONG
REFINING CAPACITY: 0
OIL PRODUCTION: 304,000 B/D
OIL RESERVES: 600.0 MILLION BBL
GAS RESERVES: 6.8 TCF

Two new wellhead platforms in Rang Dong field produced first oil in November, boosting production by 40% to an average gross level of 65,000 b/d, which is the facilities' design capacity. Block 15-2 is in 50 m of water (165 ft) in the Cuu Long basin, 180 km (120 miles) southeast of Ho Chi Minh City. The block covers 1,600 sq km (400,000 acres, or 600 sq mi) and is near Bach Ho and Ruby fields. Earlier in 2002, the 15P development well had boosted field output to 55,000 b/d, a production milestone.

On neighboring Block 15-1, ConocoPhillips and partners completed their appraisal well for the Sutu Den discovery made last year and were testing it in April. ConocoPhillips holds more acreage than any other foreign energy company in Viet Nam, with interests in five offshore blocks totaling more than 23,000 sq km (5.75 million acres, or 9,000 sq mi).

Other development

Three natural gas fields will be developed off southwestern Viet Nam by Unocal subsidiaries, which filed a declaration of commercial discovery in May following 10 successful exploration wells drilled on Blocks B and 52/97. The declaration, which paves the way toward a long-term gas sales agreement, covers resources discovered in Kim Long, Ac Quy, and Ca Voi natural gas fields.

Unocal plans to develop the three fields in a single program using shared facilities.

The fields could have a combined gross discovery volume of 67 bcm (2.5 tcf), sufficient to support major new power development projects. Four wells drilled in 2001 all found net pay, including the 52/97-CV-3X well, which extended Ca Voi field and flowed on test at a rate of 990,000 cu m/d (37 MMcfd), and appraisal well 52/97-AQ-4X, which extended Ac Quy field and flowed at 1.3 million cu m/d (48 MMcfd).

Exploration

The first exploration well drilled on the 1,370 sq km (530 sq mi) Block 9-2 discovered oil and gas in the Cuu Long basin off Vung Tau City south of Viet Nam, said SOCO International plc (London) in September. The 09-2-CNV-1X well flowed 2,500 b/d of crude oil and of natural gas 177,000 cu m/d (6.6 MMcfd) without stimulation during preliminary tests in a basement granite formation, and oil shows were found in upper sandstone.

SOCO and partners planned to drill a total of four wells on Blocks 9-2 and 16-1, including a second well on Block 9-2 by late 2002. SOCO's Viet Nam unit farmed out part of its interest in these blocks to PTTEP in February.

Offshore Block 46/02 will be explored by a consortium including units of Talisman Energy, Petronas, and state-owned PetroVietnam under a license awarded in May, pending government approval and contract finalization. The block lies close to Talisman's Malaysia-Vietnam PM-3 development and could contain a play extension, offering potential for small field discoveries that could be tied back quickly and economically.

Block 46/02 covers 12,000 sq km (3 million acres, or 5,000 sq mi) and is 160 km (100 miles) off the southern coast of Viet Nam in 40-50 m (120-150 ft) of water. Previous discoveries flowed at rates up to 5,500 b/d of oil and 698,000 cu m of gas (26 MMcfd) with low carbon dioxide content. The consortium's commitment on Block 46/02 includes conducting a 3D seismic survey and drilling four wells during a 3-year exploration phase.

Blocks 10 and 11.1 off Viet Nam were being explored by a joint venture of the state oil companies of Viet Nam, Malaysia, and Indonesia. The blocks are in the Nam Con Son basin on the Viet Nam's southern continental shelf, 210 km (130 miles) southeast of Vung Tau. Seismic data reprocessing and other geological and geophysical studies were ongoing in 2002, and the venture plans to drill two exploration wells in 2003.

Gas power project

BP's Nam Con Son integrated gas-to-power project was underway in 2002 off the southern province of Ba Ria-Vung Tau. Commissioning and testing services were contracted in June for the 400-km (250-mile) pipeline that will carry natural gas to shore from Lan Tay and Lan Do fields on Block 6/1. The project also entailed construction of a gas-fired power station, an offshore steel production platform, and an onshore gas processing terminal.

The new BP platform is 360 km (220 miles) off Vung Tau, with the primary pipeline surfacing with a safety valve at Long Hai and continuing 8 km (5 miles) over land to the new terminal at Dinh Co.

IPE

KEY STATS

LNG Trade and Technology

The LNG shipping business has seen unprecedented growth while other shipping industry sectors have suffered (Drewry Shipping Consultants Ltd.)

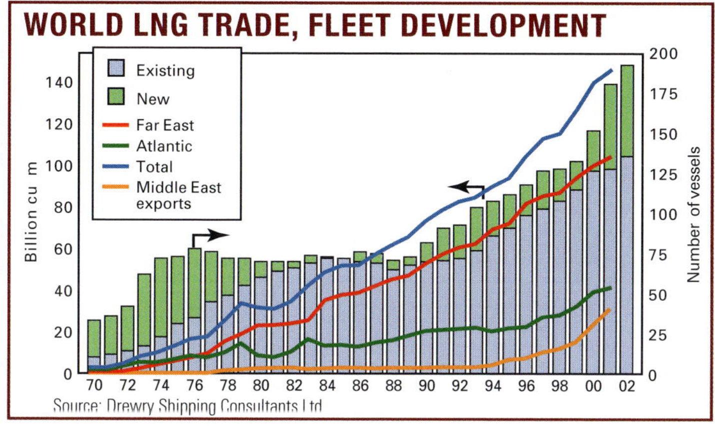

Source: Drewry Shipping Consultants Ltd

US LNG imports are needed to compensate for declining gas imports from Canada and growing gas exports to Mexico (Raymond James & Associates Inc.)

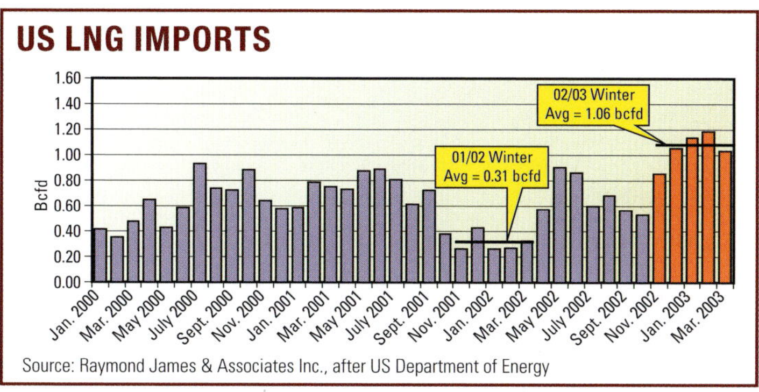

Source: Raymond James & Associates Inc., after US Department of Energy

Exports of LNG, one of the world's most rapidly growing fuels, have increased 6.4%/yr since the mid-1980s (Gas Technology Institute, GTI)

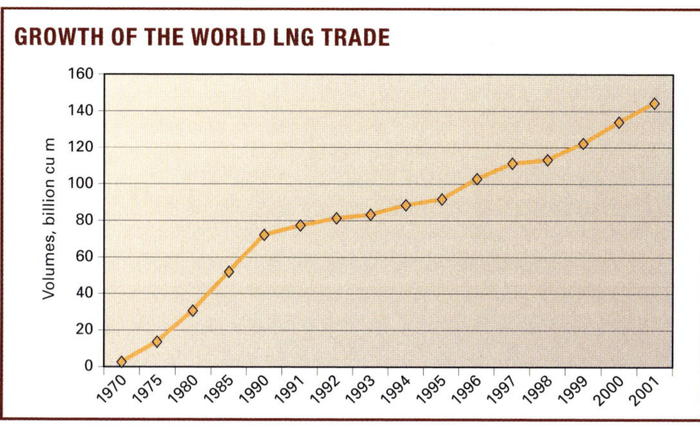

Although Asia remains the dominant player, LNG exports are rising from the Middle East, Africa, and Trinidad and Tobago (GTI)

LNG TRADES BY COUNTRY, 2001

Importing	US (incl. Puerto Rico)	Belgium	France	Greece	Italy	Spain/ Portugal	Turkey	Japan	Korea	Taiwan	Total
	billion cu m										
Exporting											
US, Alaska								1.79			1.79
Trinidad	3.20					0.45					3.65
Western Hemisphere (total)	**3.20**					**0.45**		**1.79**			**5.44**
Algeria	1.84	2.32	9.80	0.50	2.25	5.20	3.63				25.54
Libya						0.77					0.77
Nigeria	1.08	0.08	0.50		3.00	1.97	1.20				7.83
Africa (total)	**2.92**	**2.40**	**10.30**	**0.50**	**5.25**	**7.94**	**4.83**				**34.14**
Abu Dhabi						0.02		6.89	0.17		7.08
Oman	0.39					0.91		0.83	5.30		7.43
Qatar	0.64		0.15			0.78		8.30	6.67		16.54
Middle East (total)	**1.03**		**0.15**			**1.71**		**16.02**	**12.14**		**31.05**
Australia	0.07							10.05	0.08		10.13
Brunei								8.20	0.80		9.0
Indonesia								22.74	5.77	3.70	32.21
Malaysia								15.27	3.04	2.60	20.91
Australia-Asia (total)	**0.07**							**56.26**	**9.69**	**6.30**	**72.25**
Total	**7.22**	**2.40**	**10.45**	**0.50**	**5.25**	**10.10**	**4.83**	**74.07**	**21.83**	**6.30**	**142.95**

Source: Cedigaz.[1]

Note: 1 billion cu m of regasified LNG = 0.73 million tonnes of LNG.

Expansions could add 60 million tonnes/year (mty) to the world's LNG production capacity (GTI)

ANNOUNCED, PLANNED EXPANSIONS

Exporter	Country	Train No.	Volumes, mty	Contractor	Start-up date	Status
Atlantic LNG	Trinidad	2& 3	6.6	Bechtel	2002, 2003	Approved by govt. Construction has begun.
		4&5	6.4			Under discussion.
Nigeria LNG	Nigeria	3	3.3	TKJS	2002	Construction has begun on first train.
		4& 5	8		2005,2006	EPC contract awarded.
Ras Laffan LNG	Qatar	3	4.7	Chiyoda, Mitsui, Snamprogetti	2004	Contract awarded.
		4	4.7			Planned.
Qatargas	Qatar	4	4.8		2005-2006	Feasibility study.
Oman LNG	Oman	3	3.3 initial	—	2005	SPA for 1.65 mty with Union Fenosa.
Malaysia LNG Tiga	Malaysia	1 & 2	Up to 7.6	Kellogg Brown & Root, JGC, Sime Eng.	Late 2002, third quarter 2003	
PT Badak NGL	Indonesia	Train I	3	—	2004	Proposed; looking for customers.
North West Shelf project	Australia	3	4.2	Kellogg Joint Venture (Halliburton Aust., Hatch-Kaiser, Clough, JGC Corp.)	2004	Letters of intent.
Total		**16 trains**	**60.4**			

Source: GTI, World LNG Source Book 2001.[2]

Construction of new terminals in the US, Mexico, and Caribbean will depend on resolution of economic and regulatory issues (GTI)

Terminals announced for the Western Hemisphere

Site	Capacity, bcf/year	Company	Potential source of LNG
US and Canada			
Bahamas-Florida	200	El Paso	
Radio Island, NC	200	El Paso	
Vallejo, Calif.	N/a	Bechtel, Shell	
Tampa, Fla.	200	BP	
Gulf of Mexico, offshore	365	Texaco	
Brownsville, Tex.	365	Cheniere	
Freeport, Tex.	365	Cheniere	
Sabine Pass, Tex.	365	Cheniere	
Hackberry, La.	275	Dynegy	
New Brunswick, Canada	275	Irving Oil	Hibernia, Nfld.
Mexico			
Altamira, Mex. (Gulf Coast)	475	El Paso, Shell	
Ensenada, Baja California, Mex.	10 mty	Shell	Shell: Australia, Timor Sea
"	365	CMS/Sempra	Bolivia
"		ChevronTexaco	Australia
"		Marathon	Australia, Indonesia
"		El Paso/Phillips	Australia, Indonesia

Plenty of LNG could become available to supply markets in both Asia and the Atlantic Basin (GTI)

Potential supplies for the Atlantic Basin

Project & Train	Volume, mty
Firm expansions	
Atlantic LNG, 2&3	6.6
Nigerian LNG, 3	3.3
Ras Laffan LNG, 3	4.7
Possible expansions	
Atlantic LNG, 4&5	6.4
Nigerian LNG, 4&5	7.6
Oman LNG, 3	3.5
Ras Laffan LNG, 4	4.7
Proposed new projects	
Angola	3
Egypt, Damietta	3-4.5
Egypt, Union Fenosa	4-8
Egypt, Idku	3-4.5
Egypt, Shell	3.2
Iran	8-10
Nigeria, new plant	3-6
Venezuela, N. Paria	4
Venezuela, Jose	2.1
Norway, Snovhit	4

Potential supplies for Asia, US-Mexican West Coast

Project, train	Capacity, mty
Firm expansions	
Malaysia Tiga, 2003	6.8
Northwest Shelf, Train 4	4.2
Ras Laffan, Train 3	4.7
Bayu Undan, 2006	3.0
Qatargas, debottlenecking	3.2
Possible expansions	
Oman LNG, 3	3.5
Bontang, I, J	6
Ras Laffan, 4	4.7
Brunei LNG, 6	4.0
Qatargas 4 & 5	7+
North West Shelf, Train 5	4.2
Proposed new projects	
Gorgon	7+
Bayu-Undan/Darwin	4.8
Tangguh	3.5-7.0
Papua New Guinea	4
Sakhalin Island	9.6
Ras Laffan 5 & 6	7
Iran, South Pars	10
Sakhalin II	4.8-9.6

The LNG "spot" market includes agreements to deliver multiple cargoes over a 6-month period or longer, as opposed to single-cargo oil spot-market deals (GTI)

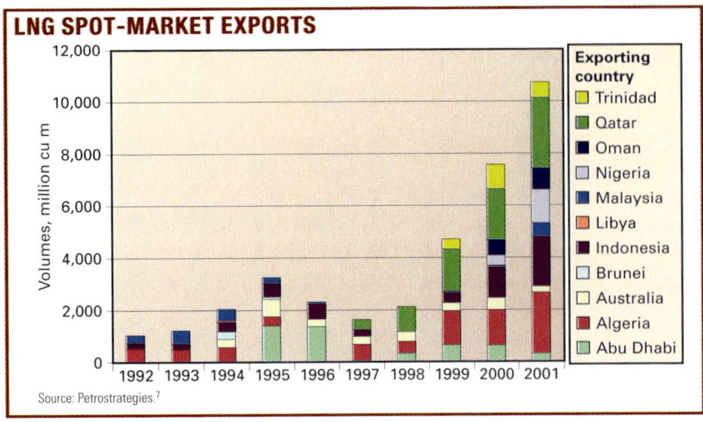

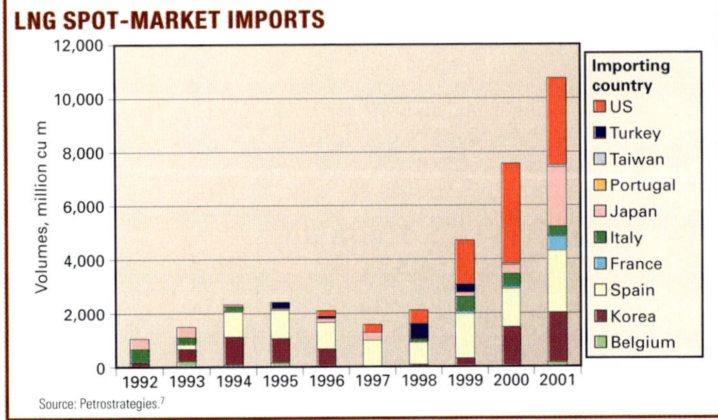

The Asian LNG market has all the elements of a buyer's market, with strong competition among potential suppliers to secure any new opportunity (Andy Flower)

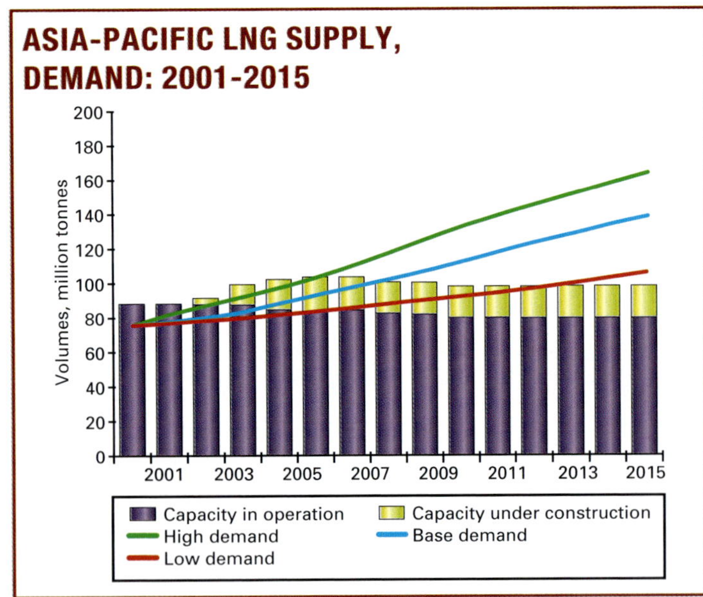

Liquefaction costs have been reduced mainly by building ever larger LNG trains, requiring sellers to bring together several buyers to accept the output from a single train (Flower)

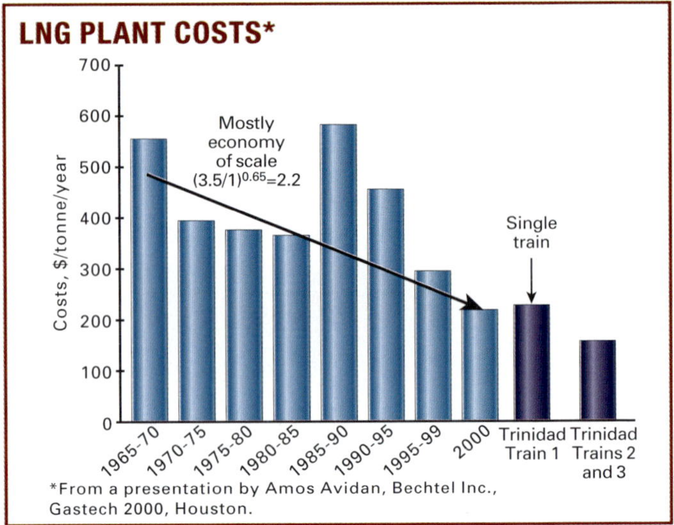

Advances in compressor and turbine technologies have contributed to a steady decrease in specific power requirement for liquefaction (Bechtel Corp.)

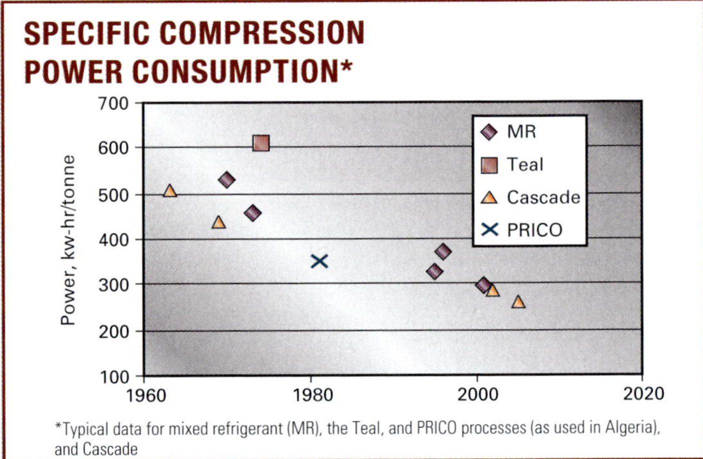

The new Liquefin process uses a mixed refrigerant instead of propane for the prerefrigeration cycle, operating at lower temperatures and eliminating phase separation (IFP-Axens)

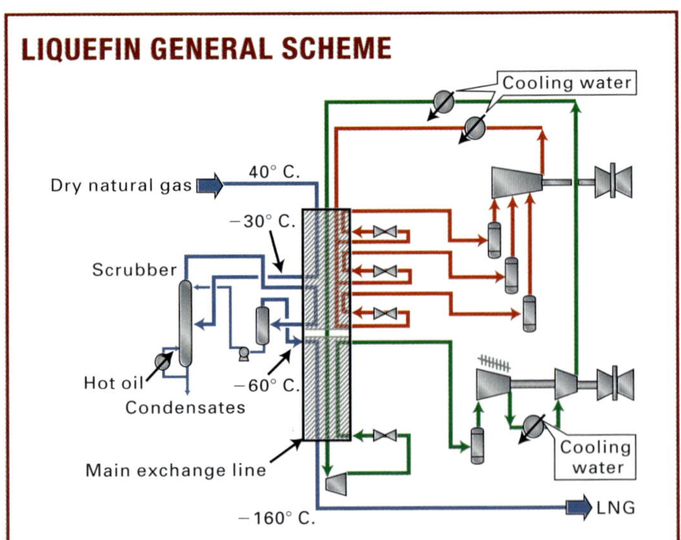

Production of LNG from offshore gas sources could become economically viable with a new FPSO design (Costain Oil, Gas & Process Ltd.)

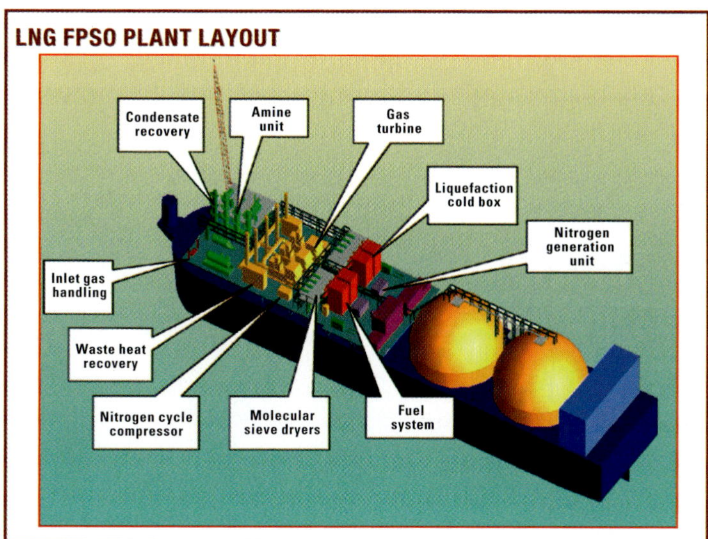

One of the most important aspects of an LNG FPSO design is the storage tank (Costain)

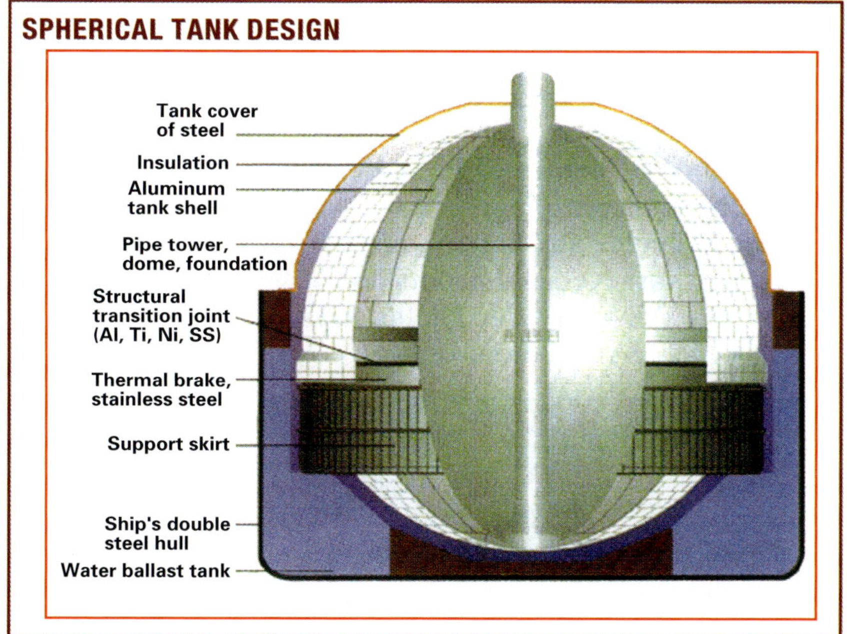

The LNG FPSO concept has also been applied to the design of a floating storage and regasification unit, or FSRU (Costain)

The proposed LNG FPSO design concept (left) was applied to a FSRU vessel (right) (Fig. 1).

Canada's Gas Potential

A detailed survey by the Canadian Gas Potential Committee (CGPC) includes some gas that could be difficult to market

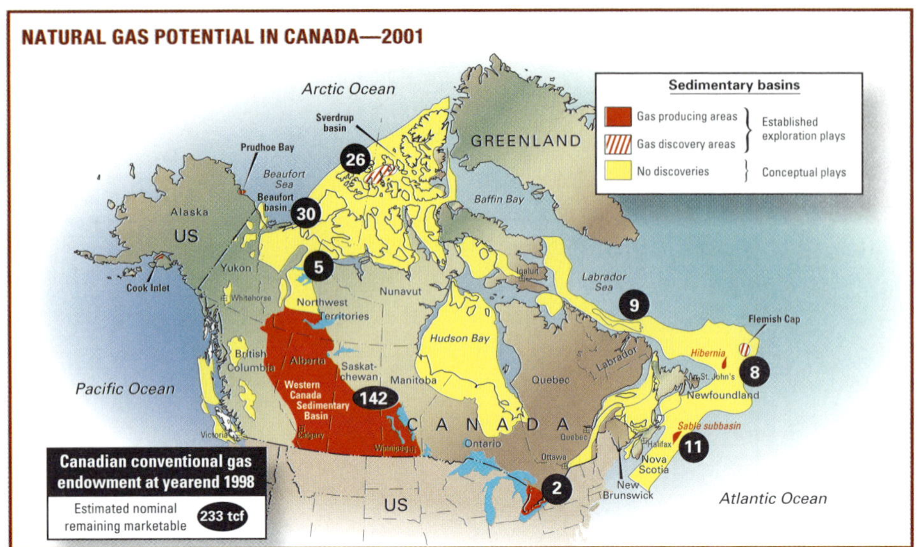

The Western Canada Sedimentary Basin (WSCB) will continue to be the mainstay of Canada's gas supplies

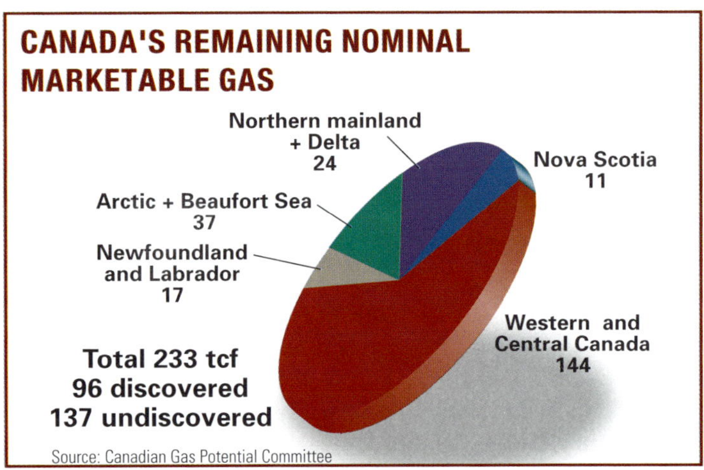

Up to 200,000 exploration wells, twice the number drilled so far, will be needed to tap the WSCB's undiscovered potential

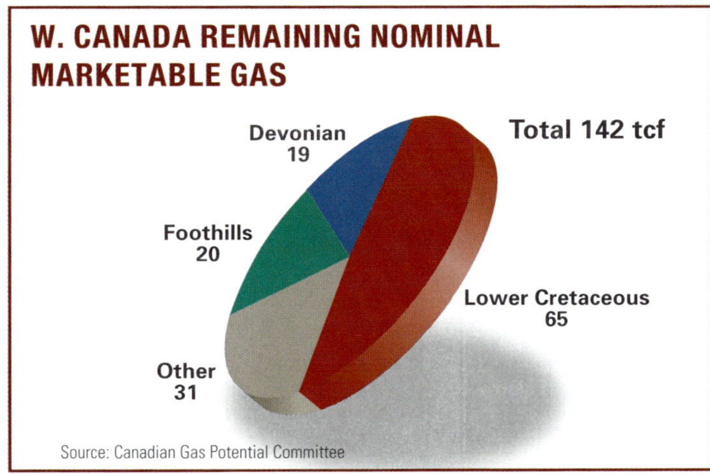

Canada's undiscovered gas represents 61% of its remaining marketable gas

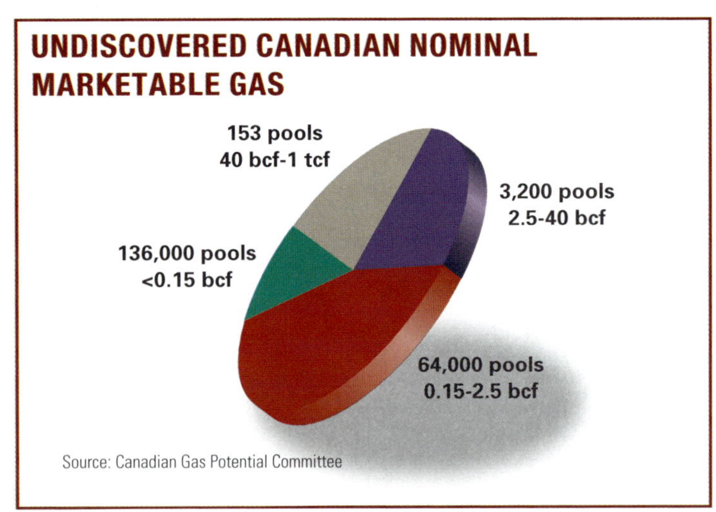

North America's gas demand is expected to increase 30% through 2010 due to economic growth and the proliferation of gas-fired power projects

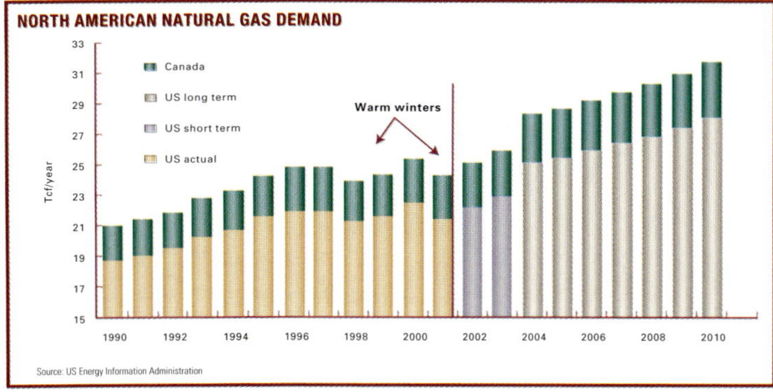

Finding and development costs will rise as gas is found in smaller accumulations, requiring more drilling

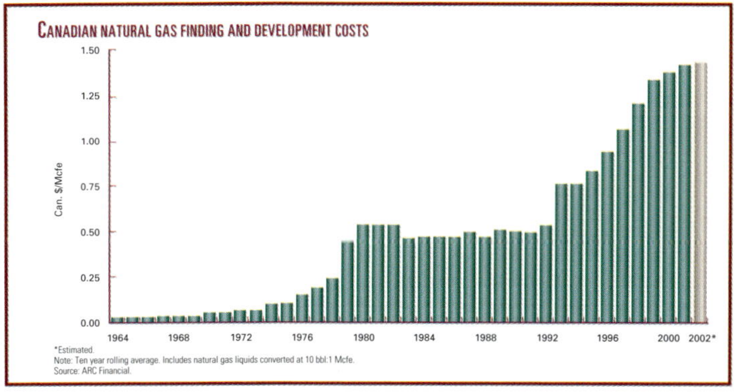

Moving Arctic gas from Alaska and from Canada's Mackenzie Delta will become a priority as WCSB production declines and economic activity rebounds

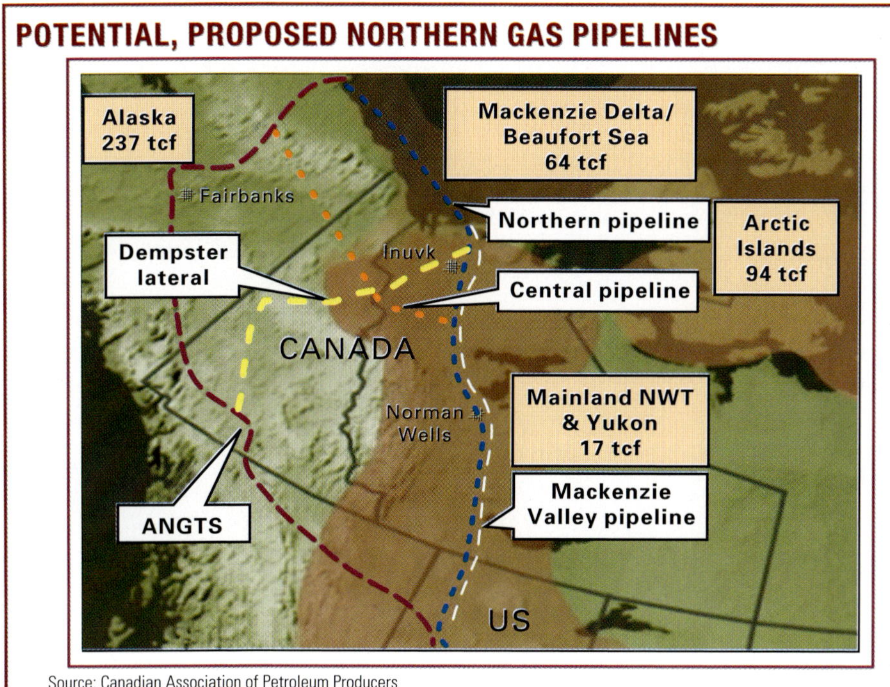

The Canadian Association of Petroleum Producers (CAPP) anticipates a gas drilling rebound in 2003, given strong price signals

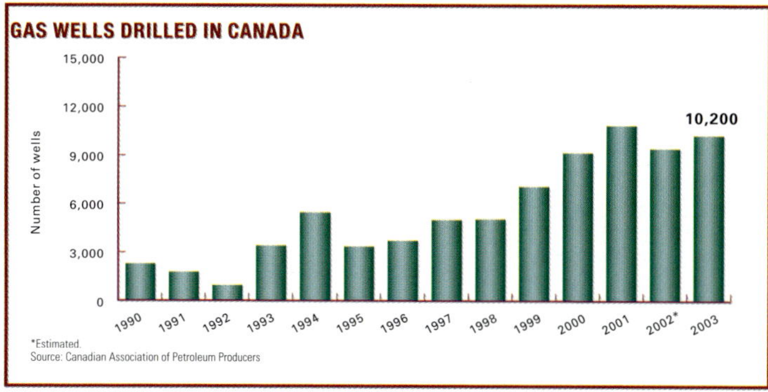

Refining Trends

Tightening inventories at US refineries should help boost refining margins in 2003 (Prudential Securities Inc.)

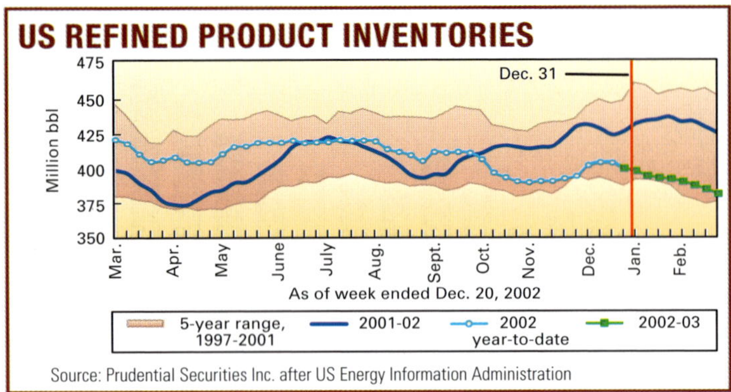

Source: Prudential Securities Inc. after US Energy Information Administration

Crudes processed in US refineries have continued to decline in quality, though less rapidly than before (Edward J. Swain)

Sulfur content of the lower quality crudes has increased moderately, about 0.032 wt%/yr

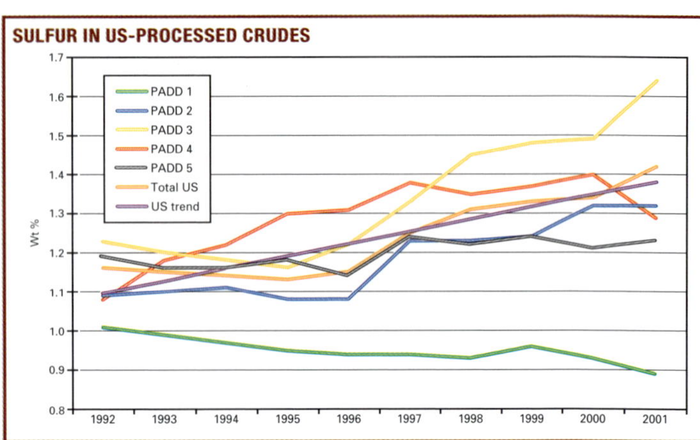

Crudes Processed in US Refineries

	1992	1993	1994	1995	1996	1997	1998	1999	2000	2001
PADD 1										
Gravity, °API	31.25	31.36	31.79	31.68	31.63	33.03	33.42	33.38	33.00	31.83
Sulfur, wt %	1.01	0.99	0.97	0.95	0.94	0.94	0.93	0.96	0.93	0.89
PADD 2										
Gravity, °API	33.94	33.74	34.25	34.01	33.92	33.31	32.94	33.37	33.03	32.78
Sulfur, wt %	1.09	1.10	1.11	1.08	1.08	1.23	1.23	1.24	1.32	1.32
PADD 3										
Gravity, °API	32.37	32.32	32.21	31.93	31.67	31.32	31.49	31.33	31.09	30.32
Sulfur, wt %	1.23	1.20	1.18	1.16	1.22	1.33	1.45	1.48	1.49	1.64
PADD 4										
Gravity, °API	35.16	34.51	34.33	34.97	33.64	33.10	33.18	34.10	33.16	33.19
Sulfur, wt %	1.08	1.18	1.22	1.30	1.31	1.38	1.35	1.37	1.40	1.29
PADD 5										
Gravity, °API	24.95	25.39	25.26	25.51	25.69	26.08	25.22	26.67	26.45	26.82
Sulfur, wt %	1.19	1.16	1.16	1.18	1.14	1.24	1.22	1.24	1.21	1.23
Total US										
Gravity, °API	31.32	31.30	31.39	31.30	31.14	31.07	30.98	31.31	30.99	30.49
Sulfur, wt %	1.16	1.15	1.14	1.13	1.15	1.25	1.31	1.33	1.34	1.42

Source: EIA Petroleum Supply Annuals

Gravities have decreased by 0.07 degrees API per year over the past 10 years

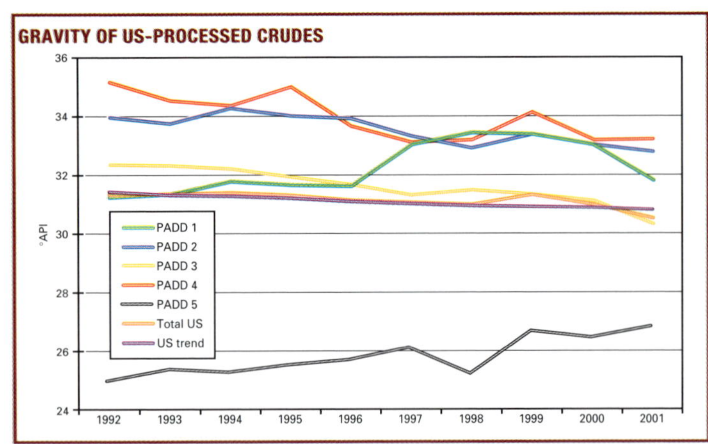

To meet upcoming ultra-low sulfur specifications, US refiners will need more hydrogen, and the most economical source is to build a hydrogen plant, vs. buying sweet crude (Stone & Webster, Inc.)

Raw Materials Required

	Today	With hydrogen	No hydrogen
Arabian Light, b/cd	95,000	90,275	41,016
Brent crude, b/cd	—	4,725	53,984
Isobutane, b/cd	2,249	2,275	3,244
MTBE, b/cd	2,929	2,655	2,382
Hydrogen, MMcfd	—	5,540	—

Refinery products

Sales, b/d	Today	With hydrogen	No hydrogen
LPG	4,138	3,929	5,939
Unleaded premium	10,051	3,310	10,857
Unleaded regular	20,102	26,250	21,714
RFG regular	20,102	13,163	21,714
RFG premium	—	6,543	—
Diesel	22,966	17,840	20,865
No. 2 fuel oil	5,742	4,460	5,216
1% fuel oil	—	9,759	1,894
3% fuel oil	11,865	9,627	7,898
Coke	3,887	3,584	2,165
Total	**98,853**	**98,465**	**98,262**

Unit capacity utilized

	Today	With hydrogen	No hydrogen
		b/cd	
Crude distillation	95,000	95,000	95,000
Vacuum distillation	28,552	27,255	26,499
FCC	24,000	24,000	24,000
FCC naphtha splitter	14,780	15,038	15,435
Delayed coker	12,735	12,105	11,233
Continuous catalyst regeneration reformer	11,400	11,400	11,400
Semiregenerative reformer	14,515	14,851	16,391
C_5-C_6 isomerization	2,214	4,714	5,207
HF alkylation	5,943	6,023	8,350
Distillate hydrotreater	24,002	20,432	21,580
Naphtha hydrotreater	25,965	26,251	28,651
FCC feed hydrodesulfurizer	16,667	24,256	24,254
Gasoline desulfurizer	—	11,199	5,423

Economic comparison

	Prices, $/bbl	Today	With hydrogen	No hydrogen
			$/cd	
Raw materials				
Arabian light crude	21.99	2,089,050	1,985,147	901,942
Brent crude	25.26	—	119,354	1,363,636
Isobutane	24.36	54,786	55,419	79,024
MTBE	47.04	137,780	124,891	112,049
Hydrogen	2.50 $/MMcf	—	13,850	—
Total		**2,281,616**	**2,298,661**	**2,456,651**
Product sales				
LPG	16.80	69,518	66,007	99,775
Unleaded premium	35.30	354,800	116,843	383,252
Unleaded regular	31.98	642,862	839,475	694,414
RFG regular	35.13	706,183	462,416	762,813
RFG premium	36.89	—	241,371	—
Diesel	27.30	626,972	487,032	569,615
No. 2 fuel oil	26.64	152,967	118,814	138,954
1% fuel oil	19.65	—	191,764	37,217
3% fuel oil	19.15	227,215	184,357	151,247
Total		**2,780,517**	**2,708,080**	**2,837,286**
Gross margin		498,902	409,420	380,636
Utilities		47,842	50,932	53,867
EBITDA*		451,060	358,488	326,769

*EBITDA = Earnings before interest, taxes, depreciation, and amortization.

Cost comparison

	Year 0	Year 1-14	Year 15
		Million $	
Build option			
Salvage	—	—	0.60
Raw materials	—	(29.059)	(29.059)
Fuel required	—	(18.889)	(18.889)
Power	—	(0.318)	(0.318)
Depreciation	—	(4.800)	(4.800)
Taxable income	—	(53.066)	(52.466)
Tax, 40%	—	21.226	20.986
Net income	—	(31.839)	(31.479)
Cash cost	—	(27.039)	(26.679)
Capital cost	(80.000)	—	—
Cash flow	(80.000)	(27.039)	(26.679)
Net present value	(224.085)	—	—
Purchase option			
Cost of hydrogen	(87.5)	(87.5)	(87.5)
Taxable income	(87.5)	(87.5)	(87.5)
Tax, 40%	35.0	35.0	35.0
Net income	(52.5)	(52.5)	(52.5)
Cash flow	(52.5)	(52.5)	(52.5)
Net present value	(302.349)	—	—

Discount rate = 10%

Catalytic refining processes account for about 82% of total crude capacity worldwide (Université Catholique de Louvain)

Capacity of catalytic processes

Process	1997	1998	1999	2000	2001
	1,000 b/d				
FCC	13,260	13,367	13,550	13,702	13,886
Hydrotreating	34,780	34,615	36,600	36,582	37,018
Hydrocracking	3,550	4,008	3,941	4,254	4,302
Naphtha reforming	11,071	11,144	11,052	11,038	11,004
Total	**62,661**	**63,134**	**65,143**	**65,576**	**66,210**

Source: OGJ refining surveys

Construction, revamps, and expansions in 2001-05 will add about 11 million b/d of catalytic processing capacity (UCL)

New capacity additions, 2001-05

Region	Number of process units: capacity, b/d			
	Hydrotreating	Hydrocracking	FCC	Naphtha reforming
North America	11: 602,300	6: 283,000	6: 143,000	—
Western Europe	14: 404,700	4: 70,900	4: 132,000	1: 23,000
Russia, Eastern Europe	5: 68,347	4: 145,000	6: 137,600	3: 22,350
Africa	6: 121,200	—	—	2: 27,900
Middle East	12: 351,165	1: 1,300	2: 105,000	3: 42,400
Asia Pacific	17: 4,370,000	2: 1,520,000	4: 193,000	6: 1,564,000
South America	10: 329,750	5: 233,249	5: 87,798	2: 17,850
Total	**75: 6,247,462**	**22: 2,253,449**	**27: 798,398**	**17: 1,697,500**

Source: Reference 11

The catalyst market shows signs of recovery, with stronger demand expected for hydrotreating catalyst in 2003-04 (UCL)

Projected processing capacities, catalyst market

	Capacity, million b/d		Catalyst market, $ million		2005 market distribution, %
	2001	2005 estimate	2001	2005	
Hydrotreating	36.6	45.1	789	968	35
FCC	13.7	15.1	696	787	28
Hydrocracking	4.3	7.3	116	222	8
Naphtha reforming	11.0	12.0	139	133	5
Total	**65.6**	**79.5**	**1,740**	**2,110**	**76**
World crude capacity	81.3	95.4			
Worldwide refining catalyst market, $ million			2,320	2,776	100

Asia, North America, and western Europe continue to dominate the refining market, while most of 2002's capacity increases occurred in North America and the Middle East (*Oil & Gas Journal*)

REGIONAL LOOK AT WORLDWIDE REFINING OPERATIONS

Region	No. of refineries	Crude distillation, b/cd	Vacuum distillation, b/cd	Catalytic cracking, b/cd	Catalytic reforming, b/cd	Catalytic hydrocracking, b/cd	Catalytic hydrortreating, b/cd	Coke, tonnes/day
Africa	45	3,213,262	491,844	195,000	387,273	28,708	835,449	781
Asia	202	20,204,810	3,841,586	2,720,395	2,019,390	782,173	8,244,163	12,810
Eastern Europe	95	10,617,099	3,649,826	894,334	1,476,831	212,090	4,061,323	12,498
Middle East	46	6,318,615	1,873,695	371,500	599,049	603,810	1,853,210	3,100
North America	160	20,290,751	8,857,554	6,546,810	4,145,437	1,754,360	13,006,745	115,993
South America	69	6,650,364	2,798,066	1,301,020	418,662	165,000	1,827,352	15,877
Western Europe	105	14,582,745	5,164,683	2,166,460	2,139,209	891,148	8,506,925	10,731
Total	**722**	**81,877,646**	**26,677,254**	**14,195,519**	**11,185,851**	**4,437,289**	**38,335,167**	**171,790**

Processing capability is defined as conversion capacity (cracking) and fuels production (reforming and alkylation) divided by crude distillation capacity (OGJ)

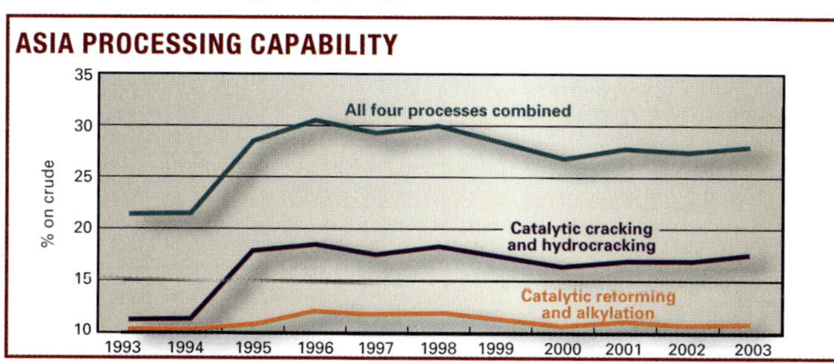

ASIA PROCESSING CAPABILITY

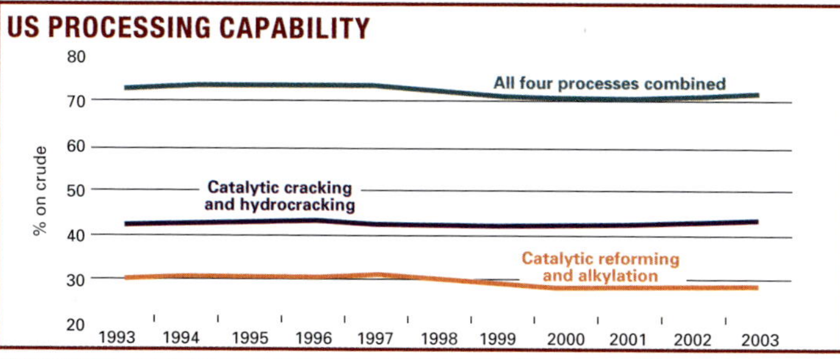

US PROCESSING CAPABILITY

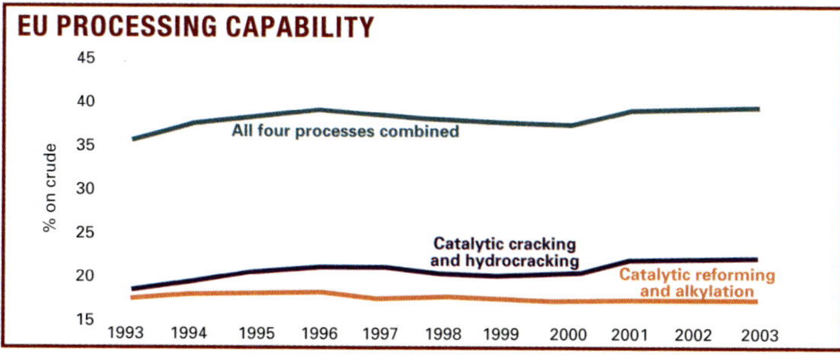

EU PROCESSING CAPABILITY

The Flying J refinery saved $900,000 in the first year of implementing recommendations from a best practices program (Energetics Inc., Flying J Inc., UOP LLC)

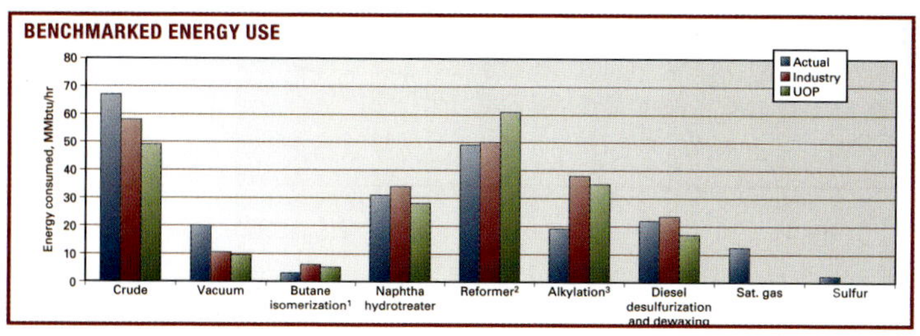

BENCHMARKED ENERGY USE

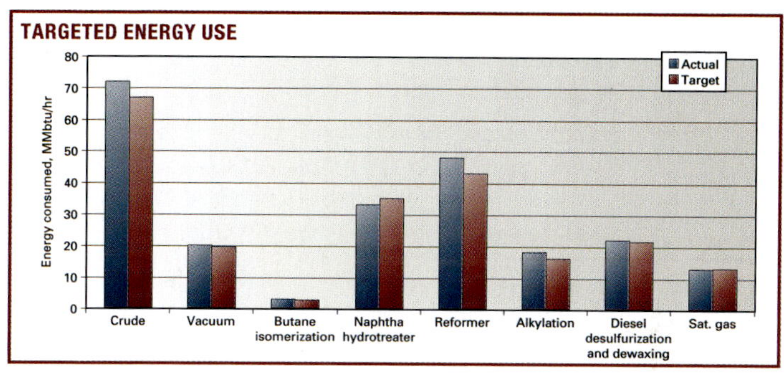

TARGETED ENERGY USE

ENERGY SAVINGS AREAS

System	Study area	Potential savings, $
Process heating	Use oxygen control for flue gases	100,000/year
	Recover waste heat	1,100,000/year
Pumps	Minimize throttle losses on two 200-hp charge pumps	39,000/year
	Shut down 250-hp pump during low-flow operation and minimize throttle losses during high flow	28,000/year
	Reduce or eliminate 75-hp wax crude recirculation flow	10,000/year
Steam	Improve boiler efficiency (two boilers)	120,000/year
	Replace cracking unit off-gas compressor steam turbine with electric motor	500,000/year
Insulation	Main steam line, 8 in. at 365° F.	139/year-lin ft
	97 lin ft of ducting, 30 in. at 320° F.	37,000/year
	Top of 80,000-bbl storage tank at 225° F.	148,000/year
	Reactor head, 44 in.	8,600/year
	Crude unit exchanger, 24 in.	4,000/year
	Condensate receiver at 210° F.	1,800/year
	Flanged valve, 10 in. at 600° F.	1,900/year
Compressed air	Modify control valves	8,600/year
	Repair leaks	5,900/year
	Excessive operating pressure	2,400/year
	Implement central drying	3,700/year

STUDY RECOMMENDATIONS

Item	Potential savings, $
Repair leaking steam traps	147,000/year
Install or repair insulation on piping and heat exchanger in reformer unit	47,000/year
Use premium-efficient motors	27,000/year
Install fixed capacitance to increase power factor	13,000/year
Reschedule operation of butane isomerization drier regenerations to avoid high-demand times	12,000/year
Use synthetic lubricants on all electric motors	10,000/year
Repair steam leaks	3,400/year
Retrofit exit sign lamps	1,200/year

Russia's Position

Of Russia's nine oil majors, five are essentially integrated, private companies (International Center for Petroleum Business Studies, ICPBS)

Russian Oil Majors Status in 2001[1]

	Oil reserves, billion bbl[2]	Oil output, 1,000 b/d[3]	Refinery runs, 1,000 b/d[4]	Oil exports, 1,000 b/d[5]	Retail stations[6]	Employees, 1,000	Capital expen-diture, billion $[7]
Lukoil[8]	14.24	1,263.5	443.1	440.2	850	128.0	1.12
Yukos[9]	12.18	1,167.0	577.6	455.8	1,280	86.3	0.59
SNG	9.10	884.2	317.2	325.4	470	86.4	1.37
TNK[10]	11.47	998.9	392.2	349.1	370	60.9	0.80
Sibneft	4.64	413.5	266.3	144.3	860	38.3	0.14
Total private majors	**51.64**	**4,727.1**	**1,996.4**	**1,714.8**	**3,830**	**400**	**4.02**
Rosneft	5.16	300.1	154.7	111.1	1,100	54.6	0.42
Slavneft	1.67	299.8	229.8	107.1	200	33.6	0.25
Total federal majors	**6.83**	**599.9**	**384.5**	**218.2**	**1,300**	**88**	**0.67**
Tatneft	6.65	494.3	6.2	182.2	100	50.2	0.61
BTK	1.74	238.3	407.6	79.3	90	57.9	0.33
Total regional majors	**8.39**	**732.6**	**413.8**	**261.5**	**190**	**108**	**0.94**
Total majors	**66.86**	**6,059.6**	**2,794.7**	**2,194.5**	**5,320**	**596**	**5.62**
% of Russia's total	***72.7***	***86.7***	***78.0***	***86.8***	***23.2***	***90.3***	***84.4***

[1]Excluding operations outside Russia. [2]Proved reserves of crude oil and condensate as of Jan. 1, 2001 – Western-audited or estimated based on typical 1:0.7 ratio between proved and Russian ABC_1 (explored) reserves. [3]Including field condensate diluted with crude oil. [4]Including minirefineries; excluding processing deals with other companies. [5]To non-Commonwealth of Independent States destinations via Transneft-controlled system. [6]As of Jan. 1, 2001. [7]E&P investments only. Converted from Russian rubles at 2001's average rate of 29.2 Rbl/$10. [8]Including Komitek. [9]Including VNK and VSNK. [10]Including Onaco and Sidanco. [11]Bashneft and Bashneftekhim only.

Sources: Roskomstat, Russian Federation Ministry of Energy, Troika Dialog, company reports, and author's estimates

Four of these five majors control about half of Russia's oil reserves and account for more than half of its E&P spending (ICPBS)

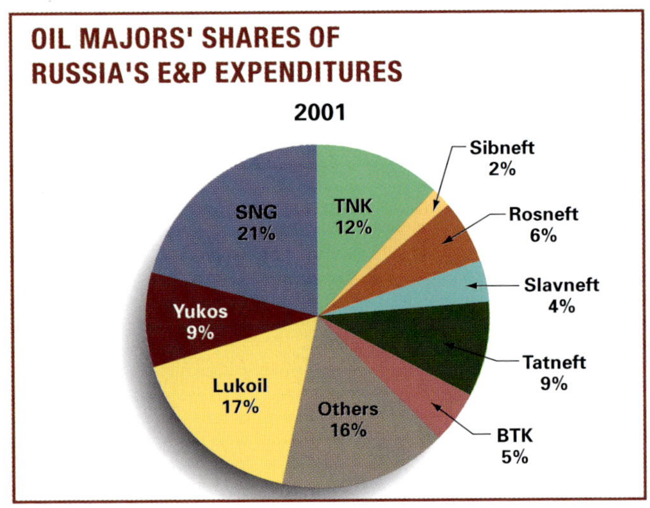

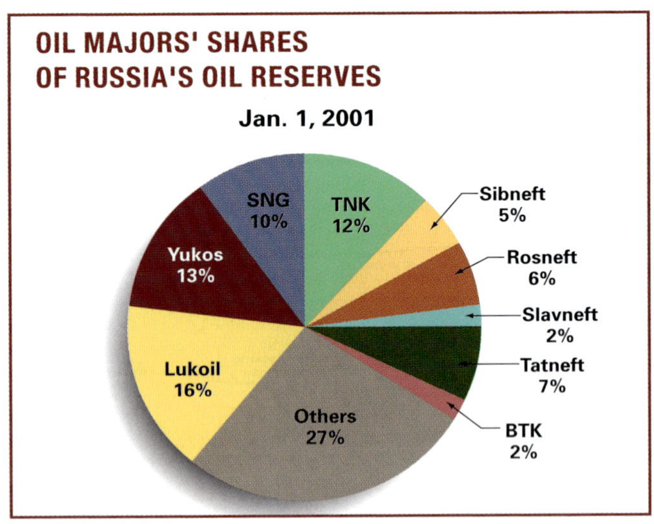

Russian majors' reserves and production are comparable with their international peers, though their downstream and sales performance lags (ICPBS)

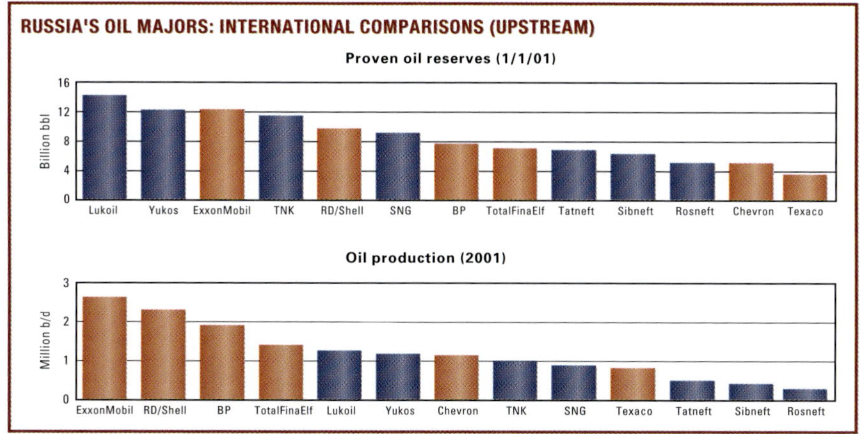

Russia and other FSU countries will produce more oil for a few years, but output could peak as early as 2007-08 (National Iranian Oil Co.)

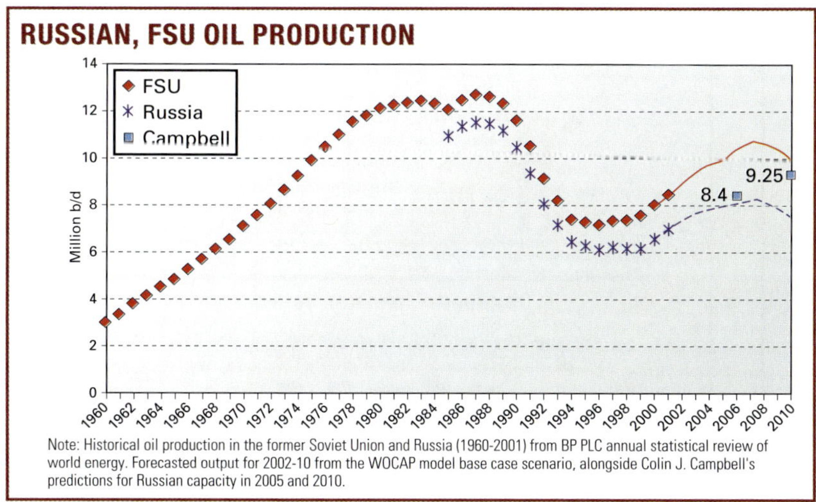

Note: Historical oil production in the former Soviet Union and Russia (1960-2001) from BP PLC annual statistical review of world energy. Forecasted output for 2002-10 from the WOCAP model base case scenario, alongside Colin J. Campbell's predictions for Russian capacity in 2005 and 2010.

Russia's 2002 tax reform reduced the profits tax rate, but overall has failed to make the system both more transparent and effective (Energy Charter Secretariat)

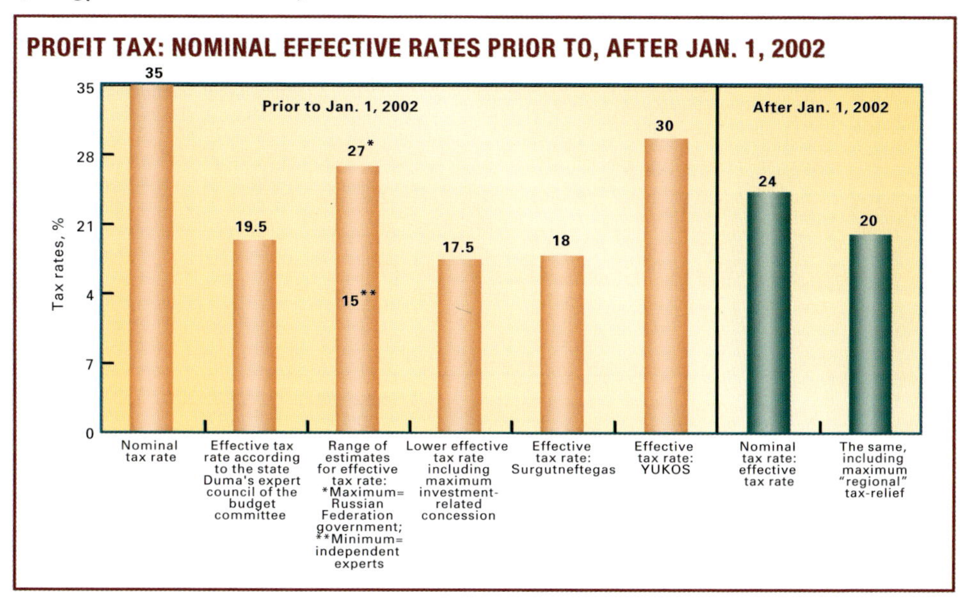

Middle East Oil Trade

The Middle East is the world oil market's center of gravity, and the Asia-Pacific region is the natural home for Middle East exports (FACTS Inc.)

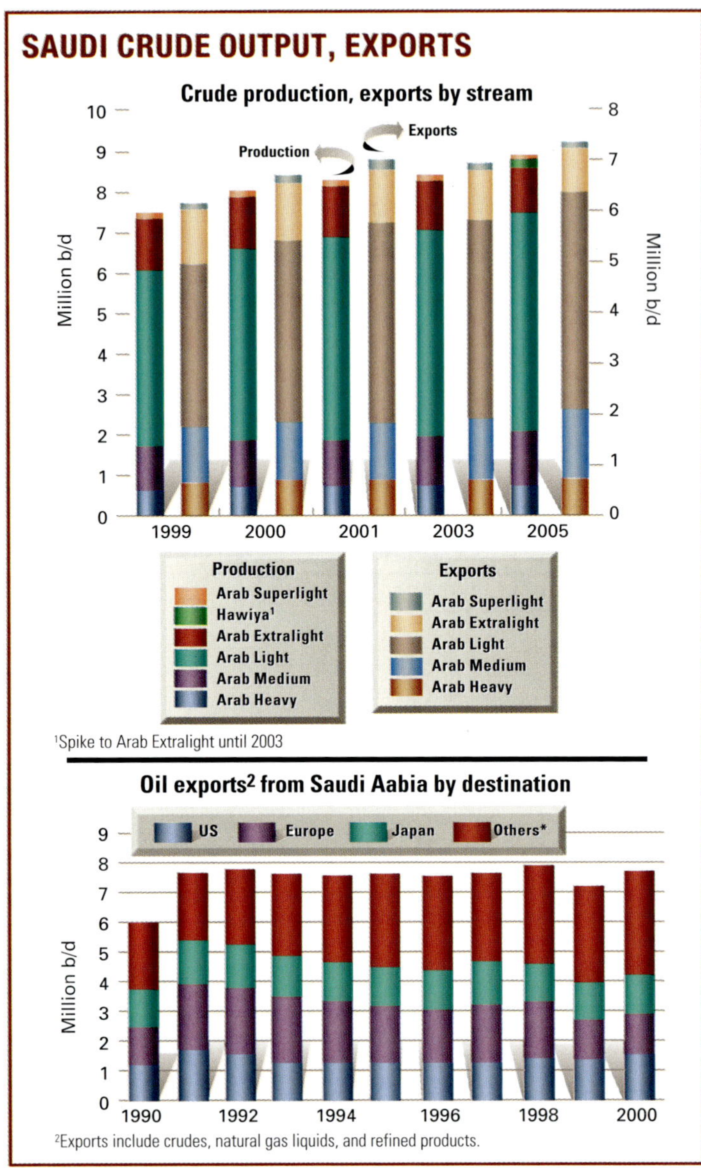

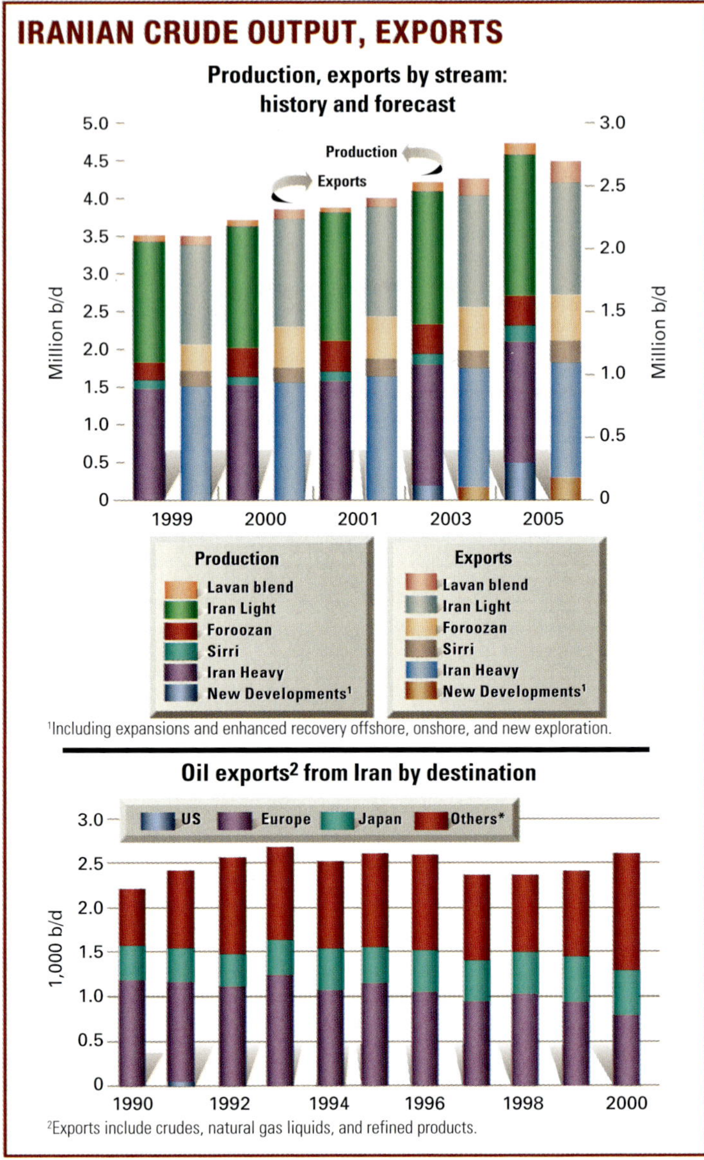

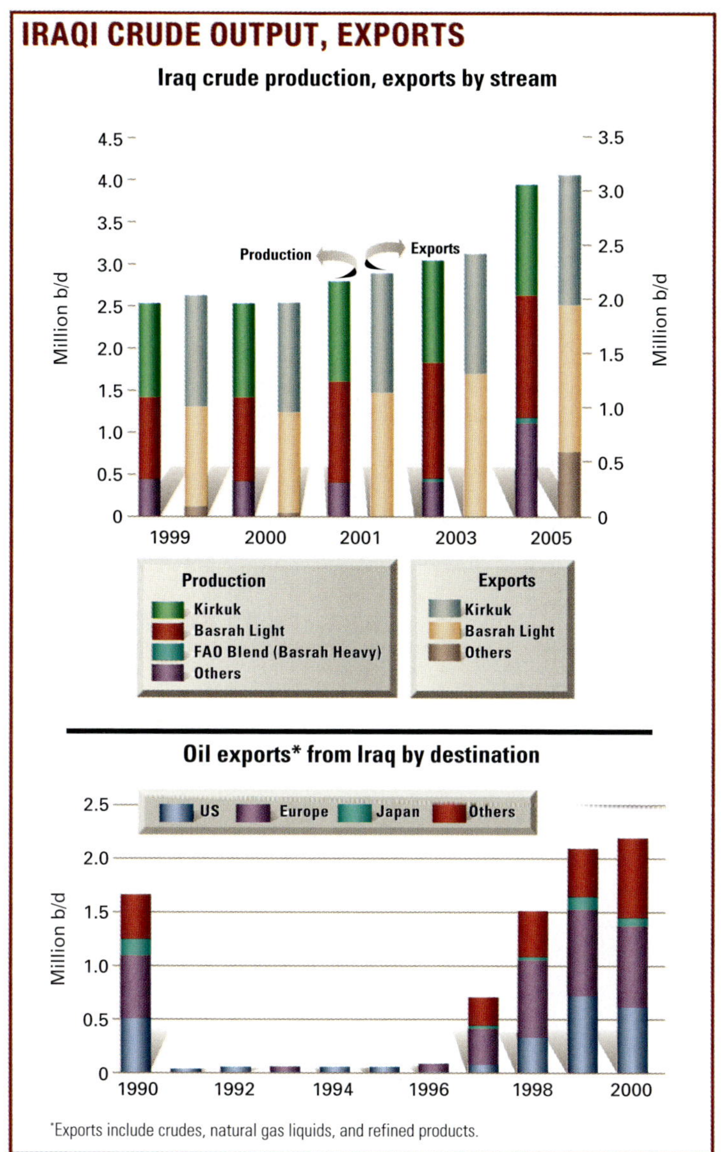

*Exports include crudes, natural gas liquids, and refined products.

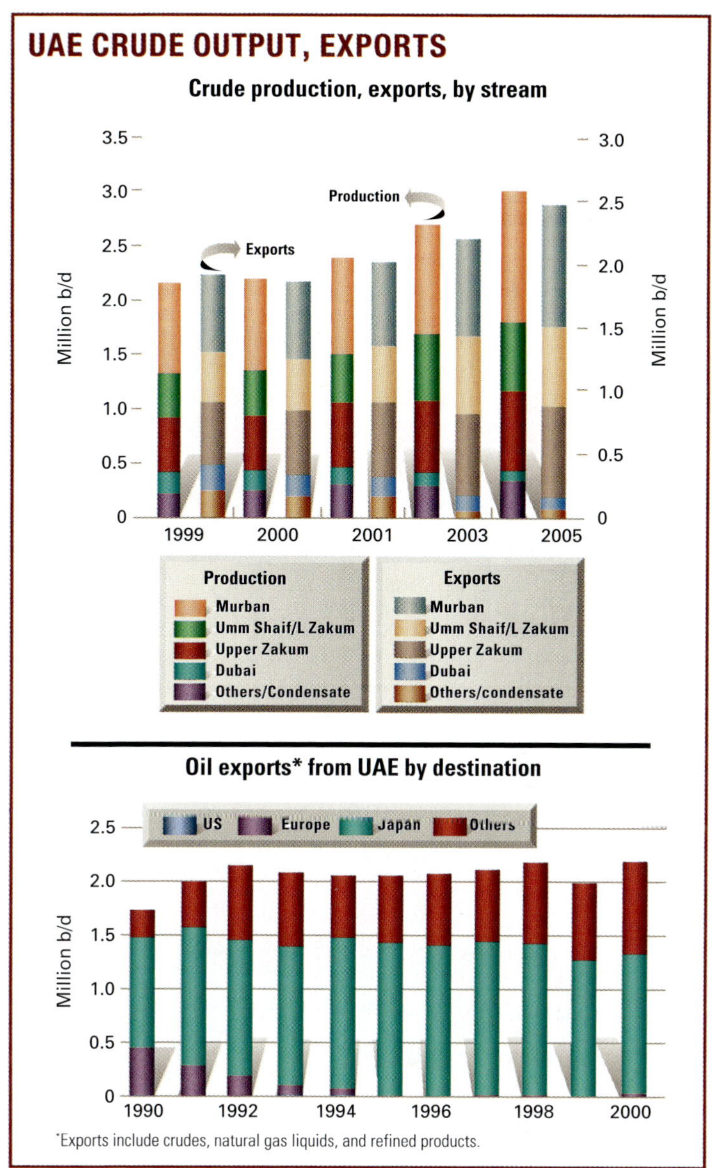

*Exports include crudes, natural gas liquids, and refined products.

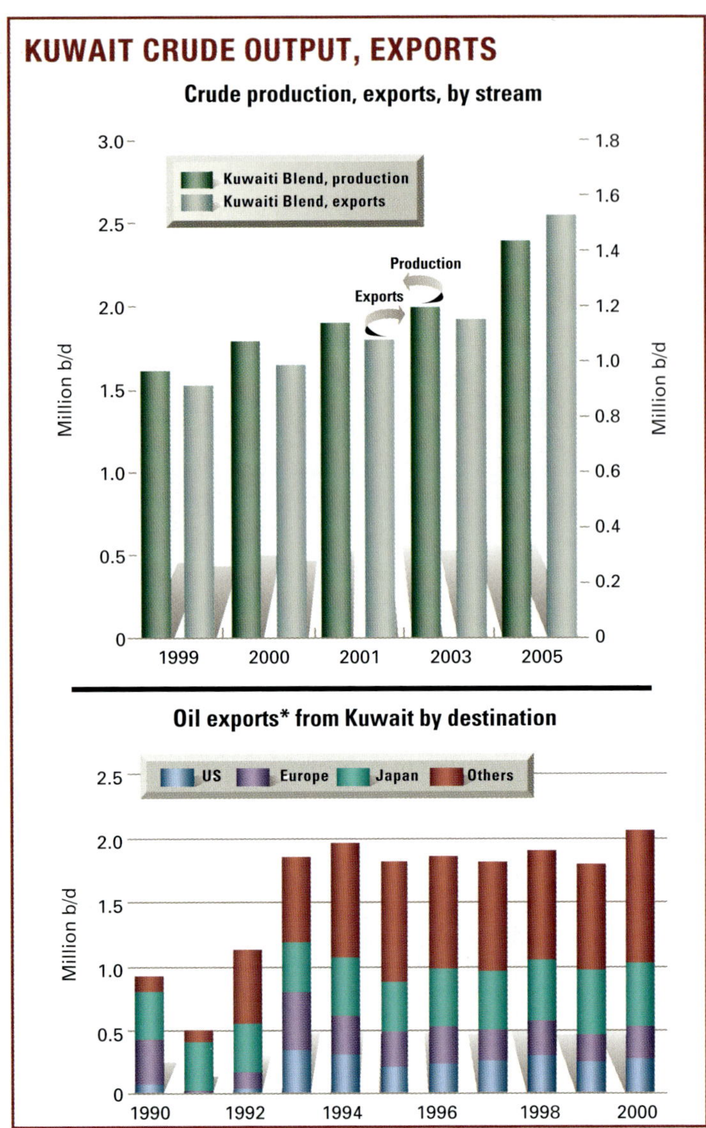

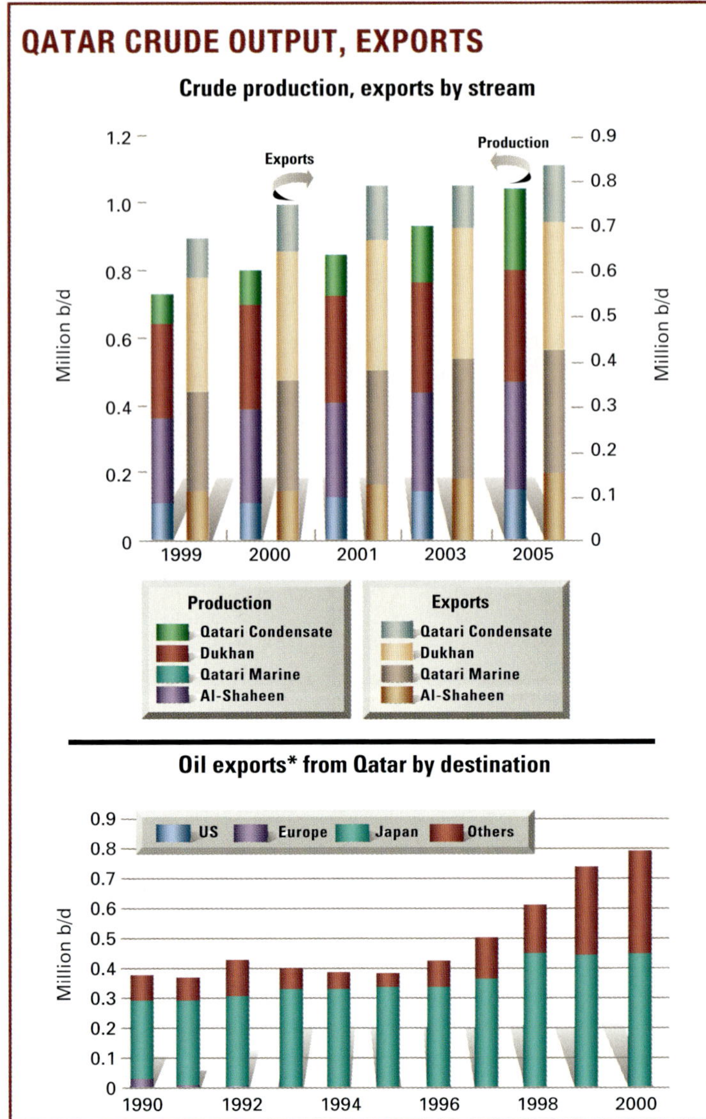

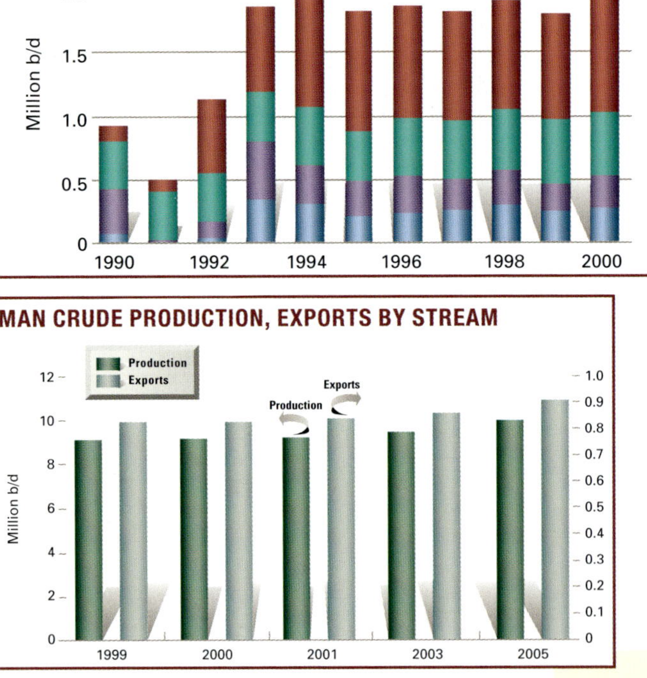

The Middle East might add 6 million b/d of oil production capacity by 2005, but an increase of 3.6 million b/d is more likely (FACTS)

FACTS PRODUCTION CAPACITY FORECAST

	Actual	Base case		High case		Low case	
	2000	2001	2005	2001	2005	2001	2005
	1,000 b/d						
Saudi Arabia	10,000	10,000	*10,500	10,000	*11,500	10,000	10,000
Iran	3,700	3,950	4,750	4,050	5,000	3,700	4,000
Iraq	2,812	3,100	4,000	3,300	4,500	3,000	3,300
UAE	2,430	2,550	3,050	2,750	3,300	2,430	2,700
Kuwait	1,925	2,125	2,650	2,175	2,750	1,925	2,125
Neutral Zone	650	650	800	650	1,000	650	650
Qatar	756	850	1,100	950	1,200	800	1,020
Total Middle East	**22,273**	**23,225**	**26,850**	**23,875**	**29,250**	**22,505**	**23,795**

*Includes condensate.

Asia-Pacific Imports

The Asia-Pacific region's oil consumption, mainly for refining, will grow rapidly through 2010 (East-West Center)

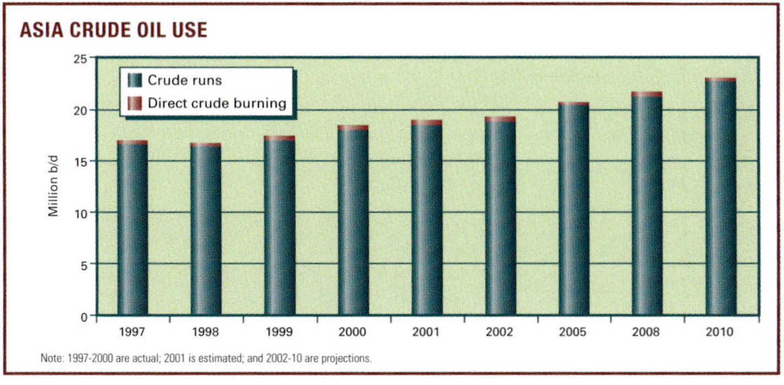

The region's export availability (domestic production minus consumption) will begin to fall sharply by 2005 (EWC)

Asia-Pacific Crude Production, Export Availability

	1999	2000	2001	2002	2005	2008	2010
	1,000 b/d						
Crude production	7,532	7,690	7,692	7,700	7,493	7,212	6,902
Export availability	2,237	2,222	2,110	2,056	1,574	1,265	946

Asia Pacific's reliance on imported oil will increase from 61% in 2000 to an unprecedented level of 72% by 2010 (EWC)

Net Oil Imports, Asia-Pacific Region

	1999	2000	2001	2002	2005	2008	2010
	1,000 b/d						
Oil production	7,814	7,953	7,968	8,001	7,849	7,573	7,293
Net oil import requirements	11,860	12,564	12,728	13,005	14,716	16,694	18,668

Note: Oil production = crude output plus nonrefinery NGLs and condensate.

The Middle East accounted for 74% of Asia-Pacific imports in 2000 and is expected to supply nearly 84% by 2010 (EWC)

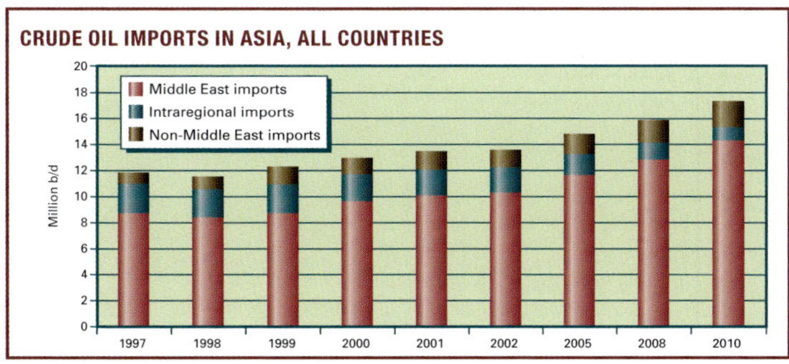

Central Asian oil and gas look tempting to Asia-Pacific markets, but Central Asia's energy potential is limited, and European markets are easier to access (EWC)

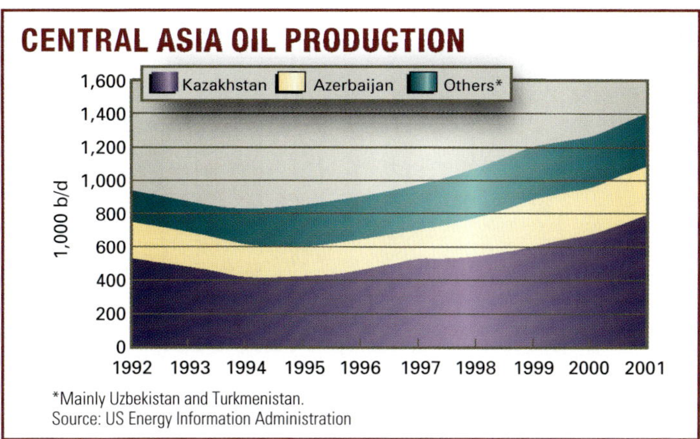

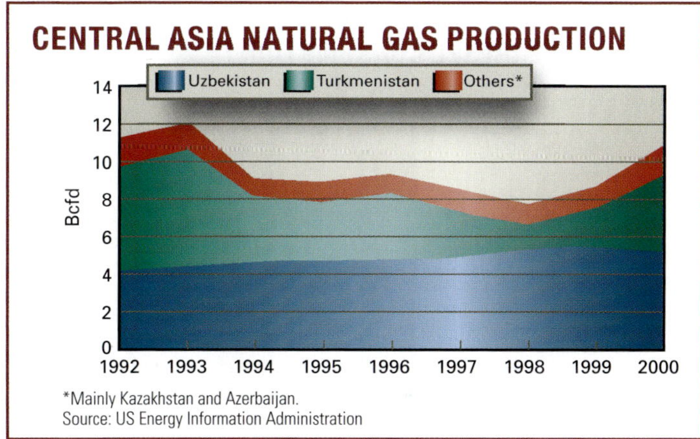

Asia-Pacific's refining industry could start to recover in 2005 from its current overcapacity, contributing to continued reliance on imports (FACTS)

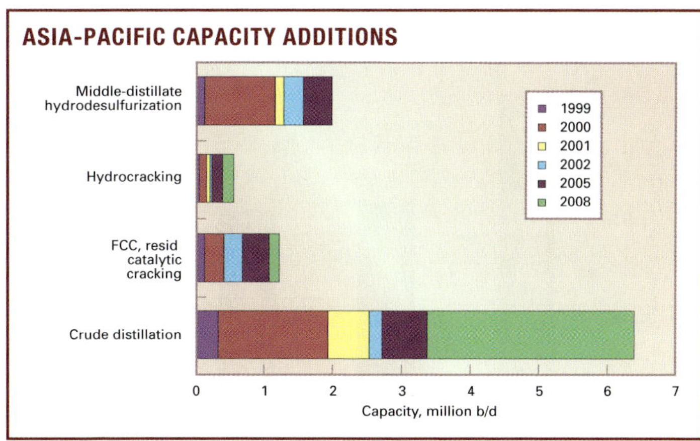

The region imports significant amounts of LPG and naphtha (FACTS)

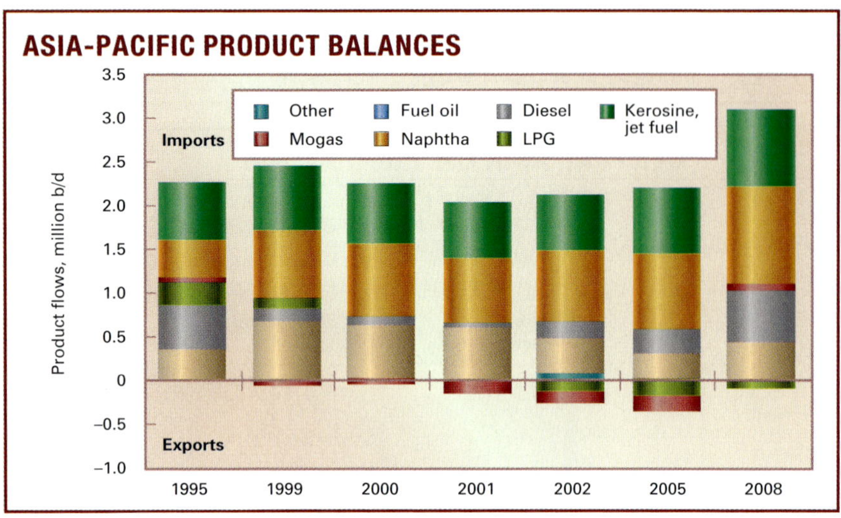

Refiners have been adding desulfurization capacity to meet future demand for cleaner fuels (FACTS)

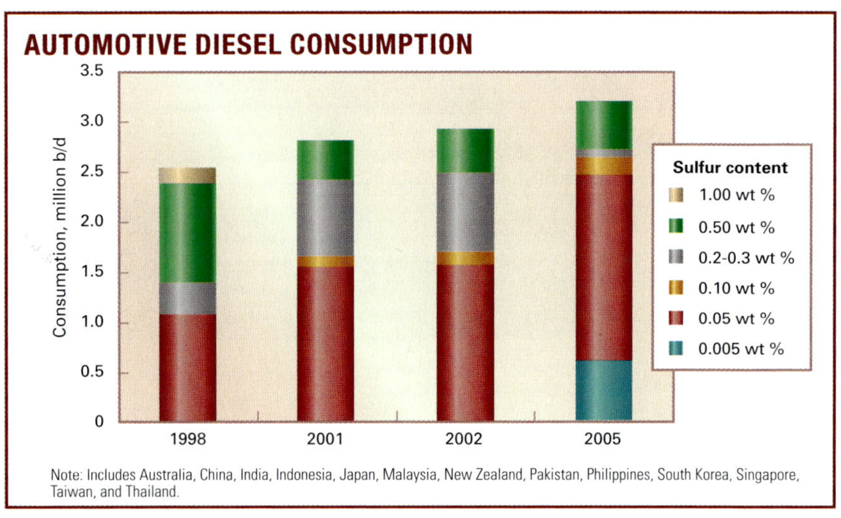

Note: Includes Australia, China, India, Indonesia, Japan, Malaysia, New Zealand, Pakistan, Philippines, South Korea, Singapore, Taiwan, and Thailand.

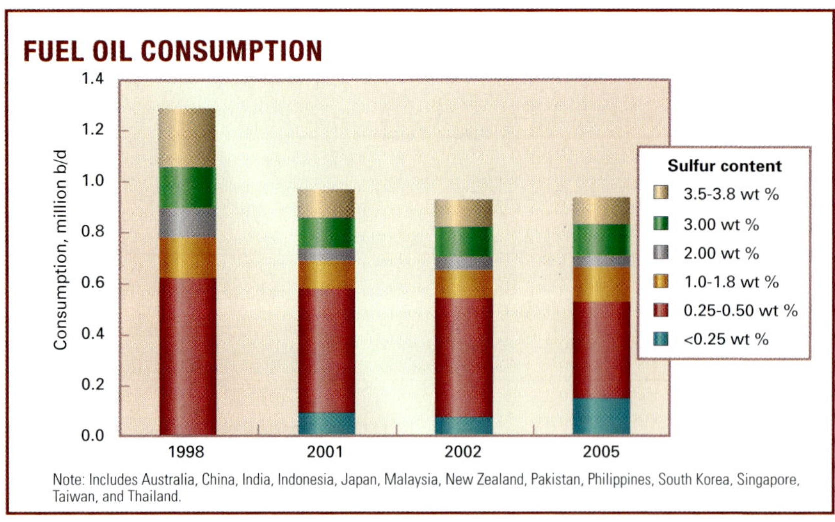

Note: Includes Australia, China, India, Indonesia, Japan, Malaysia, New Zealand, Pakistan, Philippines, South Korea, Singapore, Taiwan, and Thailand.

Together, the Asia-Pacific and Middle East regions will have net surpluses of gasoline and kerosine jet fuel by 2005 (FACTS)

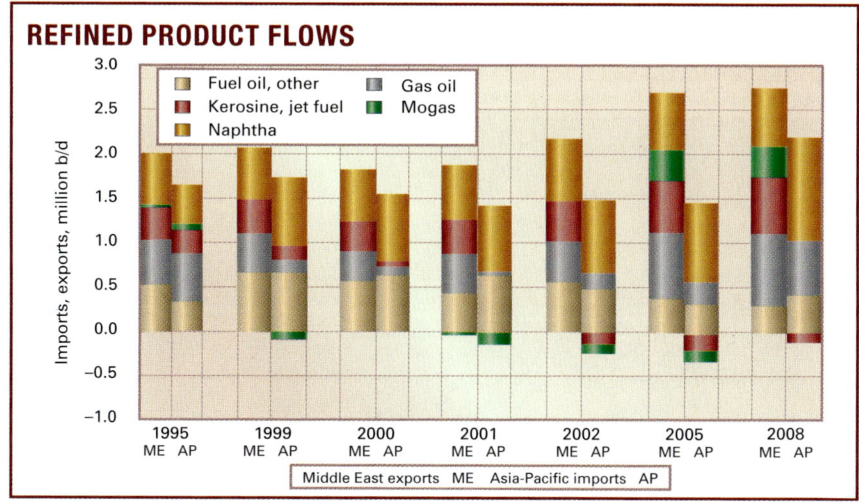

If demand recovers and refinery utilization rates are rationalized, refining margins could begin a mild recovery in 2005 (FACTS)

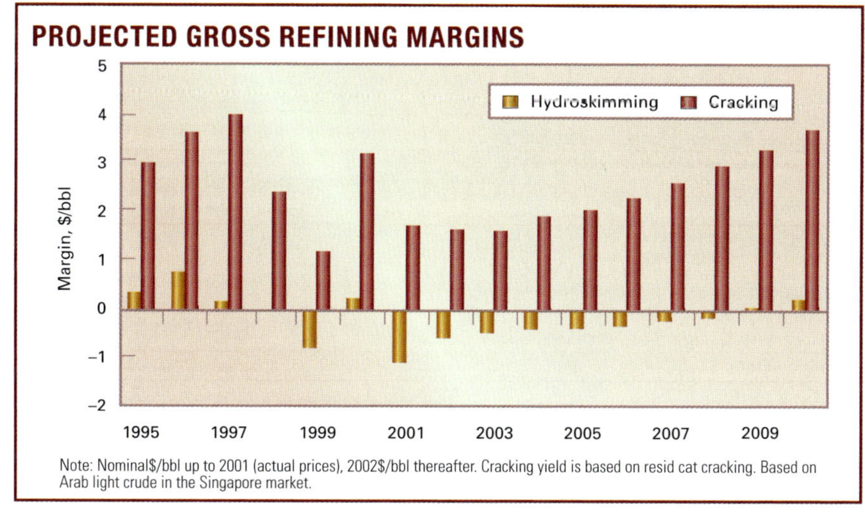

World Oil Flow Map

Global oil trade movements for 2001-02
Oil and gas reserves, oil production, and refining capacity as of year-end 2002

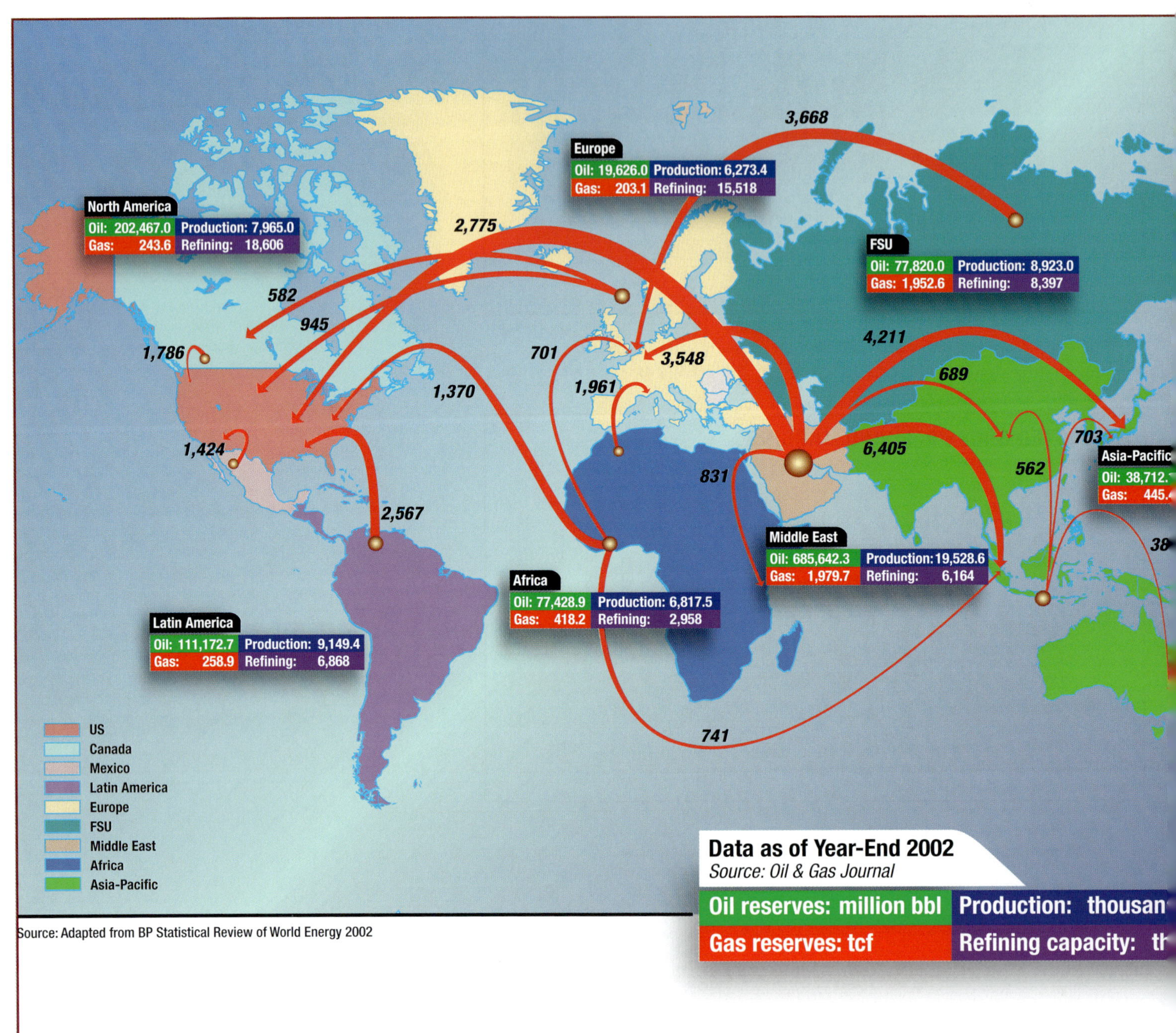

Source: Adapted from BP Statistical Review of World Energy 2002

North America
Oil: 202,467.0
Gas: 243.6
Production: 7,965.0
Refining capacity: 18,606

Latin America
Oil: 111,172.7
Gas: 258.9
Production: 9,149.4
Refining capacity: 6,868

Europe
Oil: 19,626.0
Gas: 203.1
Production: 6,273.4
Refining capacity: 15,518

Middle East
Oil: 685,642.3
Gas: 1,979.7
Production: 19,528.6
Refining capacity: 6,164

Africa
Oil: 77,428.9
Gas: 418.2
Production: 6,817.5
Refining capacity: 2,958

Former Soviet Union
Oil: 77,820.0
Gas: 1,952.6
Production: 8,923.0
Refining capacity: 8,397

Asia-Pacific
Oil: 38,712.1
Gas: 445.4
Production: 7,376.8
Refining capacity: 20,088

GLOBAL OIL TRADE MOVEMENTS, THOUSANDS B/D, 2001-02

	To												
	North America		Latin America		Europe	Africa	Asia-Pacific						
From	US	Canada	Mexico	South & Central America			Australasia	China	Japan	Other Asia-Pacific	Rest of world	Unidentified	Total
US	—	130	253	165	232	4	6	6	13	82	19	—	910
Canada	1,786	—	—	4	10	—	—	—	—	4	—	—	1,804
Mexico	1,424	26	—	185	197	4	—	—	22	20	4	—	1,882
South and Central America	2,567	122	31	—	281	12	—	6	8	114	—	—	3,143
Europe	945	582	6	45	—	148	—	22	2	108	88	—	1,947
Former Soviet Union	90	—	—	143	3,668	10	—	109	14	179	47	418	4,679
Middle East	2,775	145	23	237	3,548	831	183	689	4,211	6,405	52	—	19,098
North Africa	286	72	17	86	1,961	79	—	6	10	142	64	—	2,724
West Africa	1,370	20	—	227	701	30	—	76	16	741	—	—	3,182
East and Southern Africa	—	—	—	—	—	—	—	100	28	18	—	—	147
Australasia	45	—	—	—	—	—	—	20	80	285	—	—	430
China	23	—	—	6	4	—	6	—	85	174	—	—	298
Japan	8	—	—	—	2	4	4	23	—	52	—	—	94
Other Asia-Pacific	193	4	4	—	48	6	391	562	703	222	17	—	2,151
Unidentified	107	47	—	—	878	—	25	175	10	22	—	—	1,265
Total	**11,618**	**1,149**	**334**	**1,098**	**11,531**	**1,130**	**616**	**1,796**	**5,202**	**8,569**	**291**	**418**	**43,754**

Source: BP Statistical Review of World Energy 2002

Stratigraphic Charts

NORTH AMERICA

Canada: Cumberland subbasin

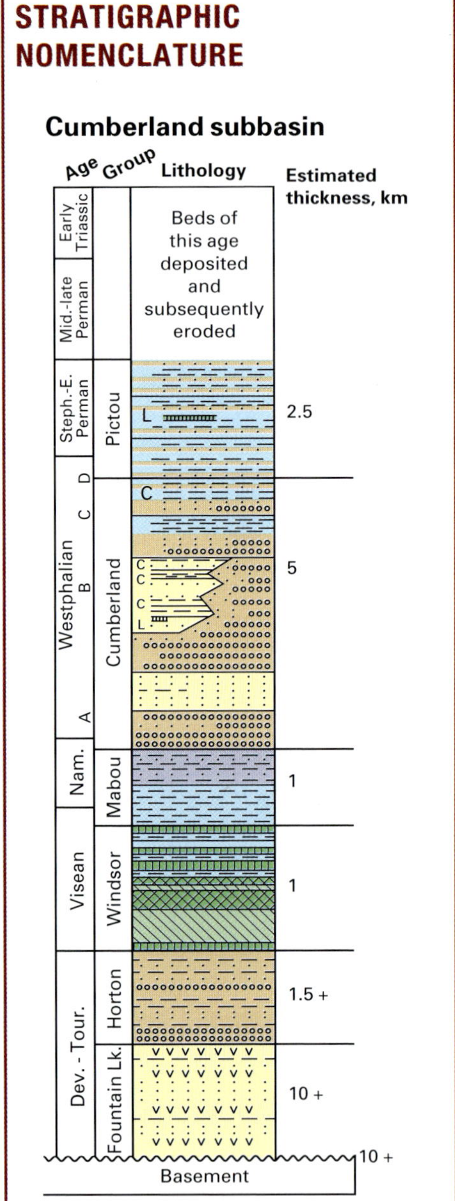

US: Middle Devonian rocks, New York

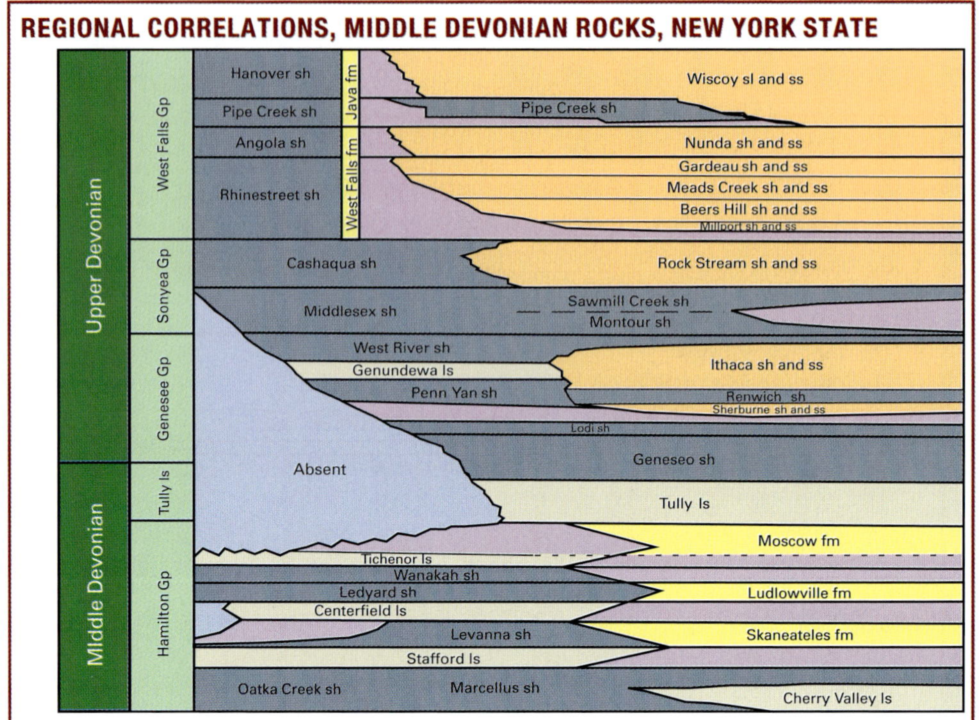

US: Desmoinesian series, Kansas

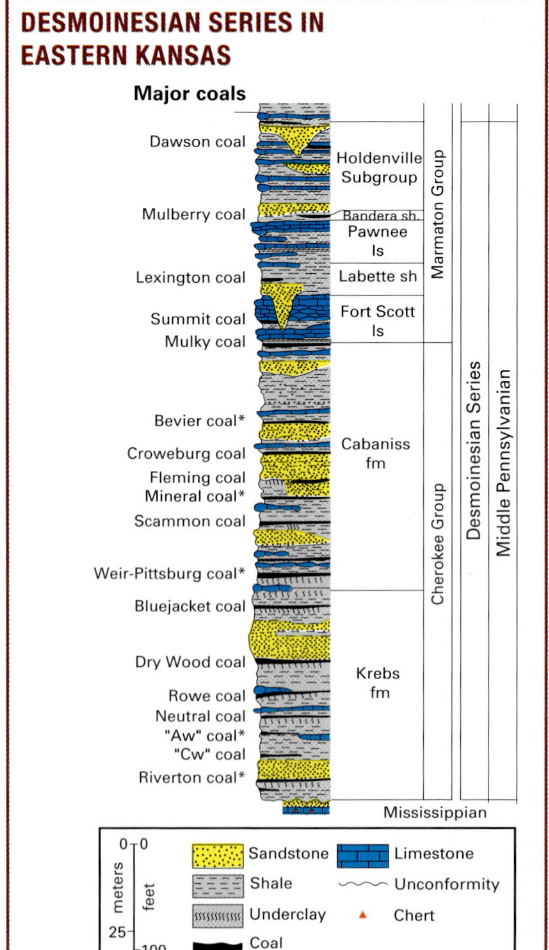

US: West Florida shelf-slope, Florida

ONSHORE AND OFFSHORE STRATIGRAPHY, WEST FLORIDA SHELF-SLOPE

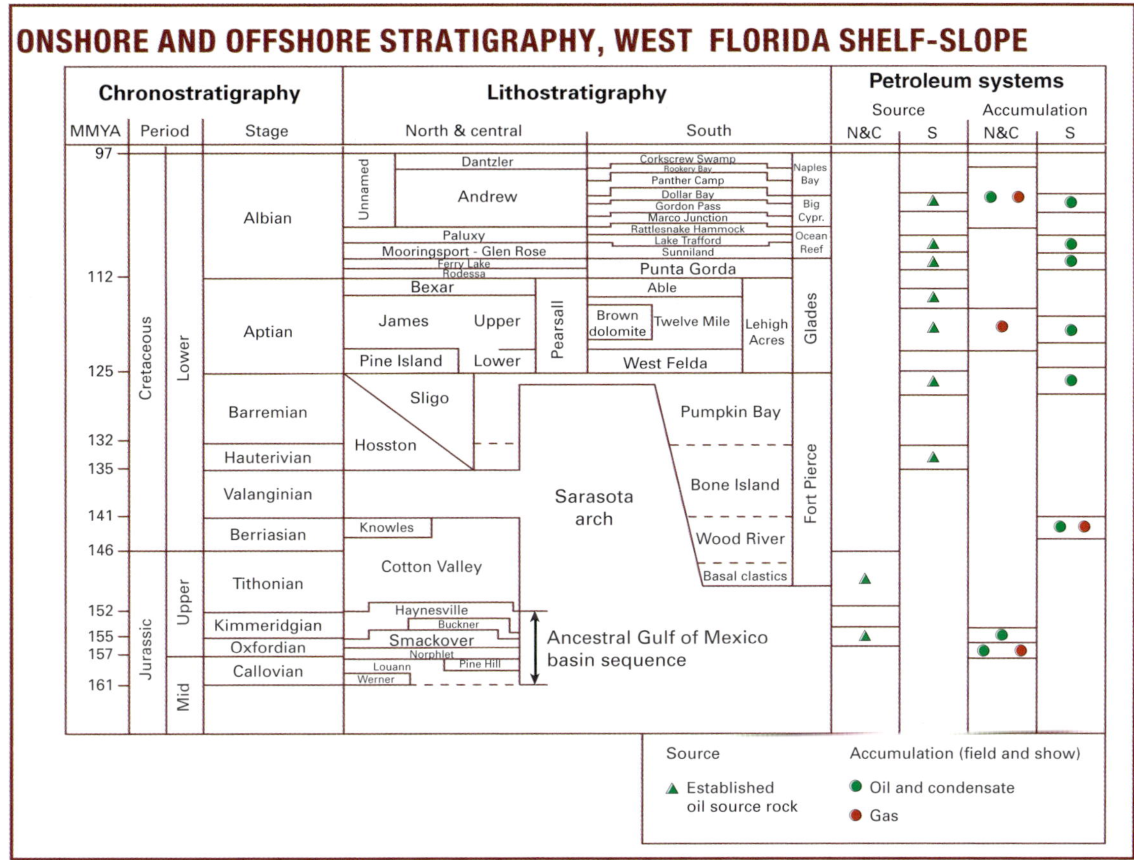

LATIN AMERICA

Brazil: Potiguar basin

DIAGRAMMATIC ILLUSTRATION OF STRUCTURAL AND STRATIGRAPHIC FRAMEWORK

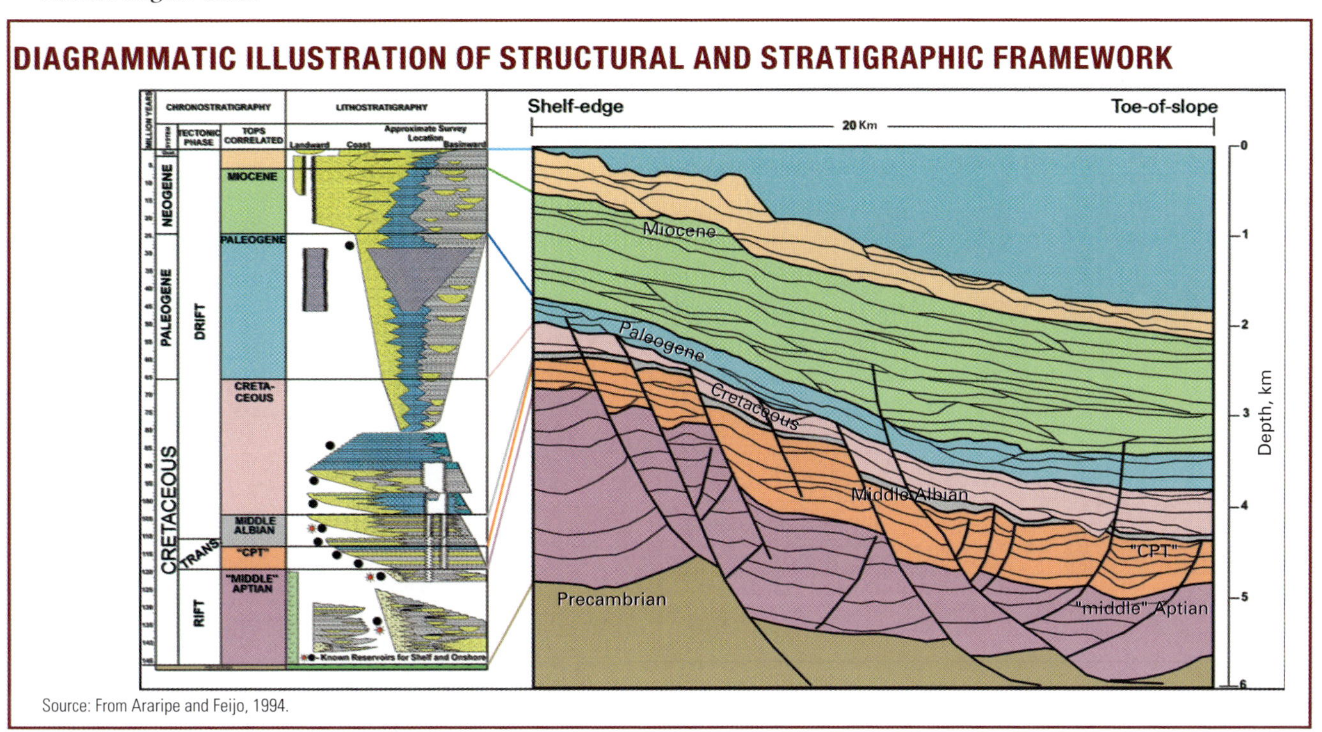

Source: From Araripe and Feijo, 1994.

Brazil: Potiguar basin (Cretaceous)

CRETACEOUS STRATIGRAPHY OF POTIGUAR BASIN[1]

[1]Cenozoic from Berggren et al., 1995. Cretaceous from Bralower et al., 1995. Triassic/Jurassic from Gradstein et al., 1995.

Source: From Araripe and Feijo, 1994.

Brazil: Potiguar basin (Tertiary)

TERTIARY STRATIGRAPHY OF POTIGUAR BASIN[1]

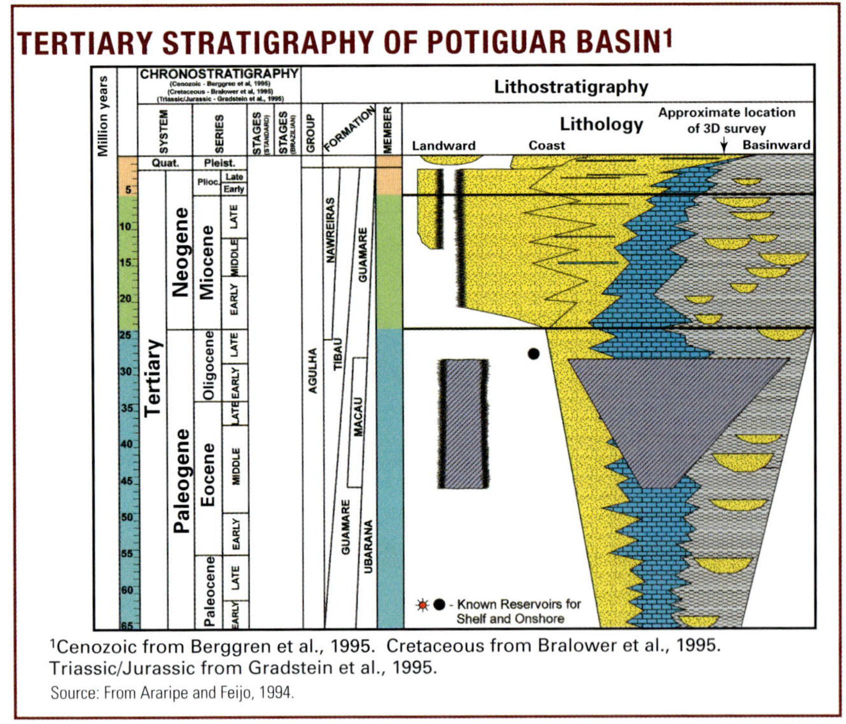

[1]Cenozoic from Berggren et al., 1995. Cretaceous from Bralower et al., 1995. Triassic/Jurassic from Gradstein et al., 1995.

Source: From Araripe and Feijo, 1994.

Mexico/US: Burgos basin

BURGOS BASIN STRATIGRAPHIC COLUMN

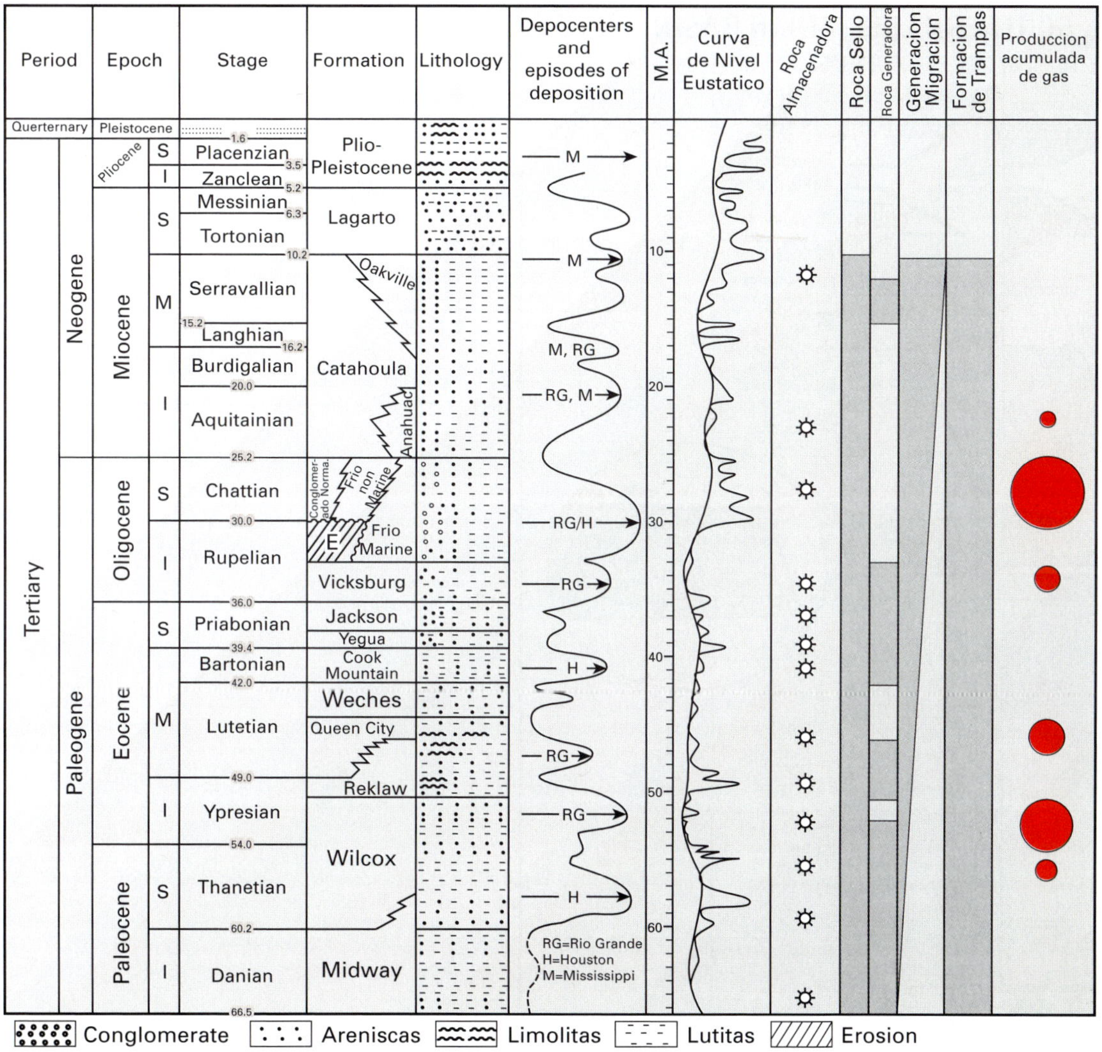

Period	Epoch		Stage	Base (M.A.)	Formation	Depocenters and episodes of deposition
Querternary	Pleistocene			1.6	Plio-Pleistocene	
Tertiary / Neogene	Pliocene	S	Placenzian	3.5	Plio-Pleistocene	M
Tertiary / Neogene	Pliocene	I	Zanclean	5.2	Plio-Pleistocene	
Tertiary / Neogene	Miocene	S	Messinian	6.3	Lagarto	
Tertiary / Neogene	Miocene	S	Tortonian	10.2	Lagarto	
Tertiary / Neogene	Miocene	M	Serravallian	15.2	Oakville	M
Tertiary / Neogene	Miocene	M	Langhian	16.2	Catahoula	M, RG
Tertiary / Neogene	Miocene	I	Burdigalian	20.0	Catahoula	RG, M
Tertiary / Neogene	Miocene	I	Aquitainian	25.2	Anahuac	
Tertiary / Paleogene	Oligocene	S	Chattian	30.0	Conglomerado Norma; Frio non Marine	RG/H
Tertiary / Paleogene	Oligocene	I	Rupelian	36.0	E; Frio Marine; Vicksburg	RG
Tertiary / Paleogene	Eocene	S	Priabonian	39.4	Jackson; Yegua	
Tertiary / Paleogene	Eocene	M	Bartonian	42.0	Cook Mountain	H
Tertiary / Paleogene	Eocene	M	Lutetian	49.0	Weches; Queen City	RG
Tertiary / Paleogene	Eocene	I	Ypresian	54.0	Reklaw; Wilcox	RG
Tertiary / Paleogene	Paleocene	S	Thanetian	60.2	Wilcox	H
Tertiary / Paleogene	Paleocene	I	Danian	66.5	Midway	

Other column headings: Lithology; M.A. (10, 20, 30, 40, 50, 60); Curva de Nivel Eustatico; Roca Almacenadora; Roca Sello; Roca Generadora; Generacion Migracion; Formacion de Trampas; Produccion acumulada de gas

RG=Rio Grande
H=Houston
M=Mississippi

Conglomerate | Areniscas | Limolitas | Lutitas | Erosion

Source: Modified from Roman, 1996.

Portugal: Southern Lusitanian basin

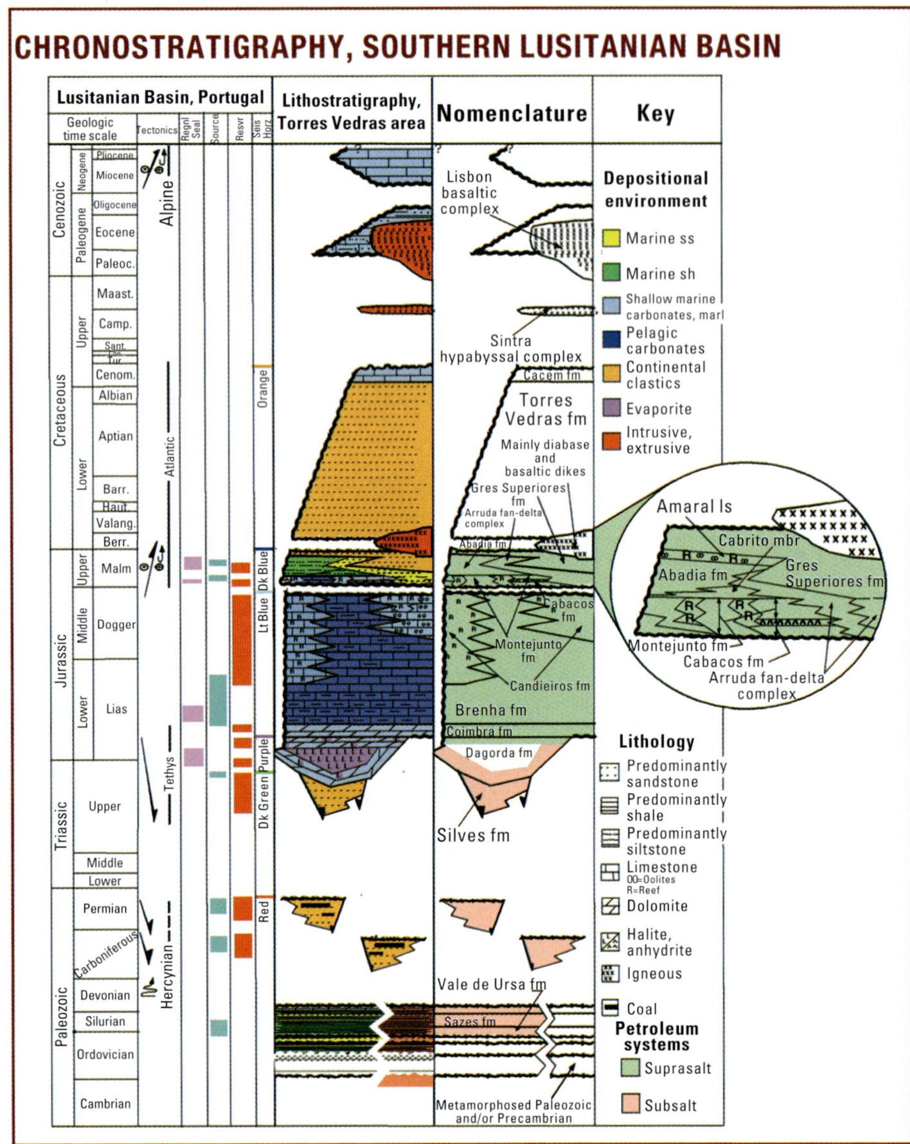

FSU

Armenia: Garni-Shorakhpur area

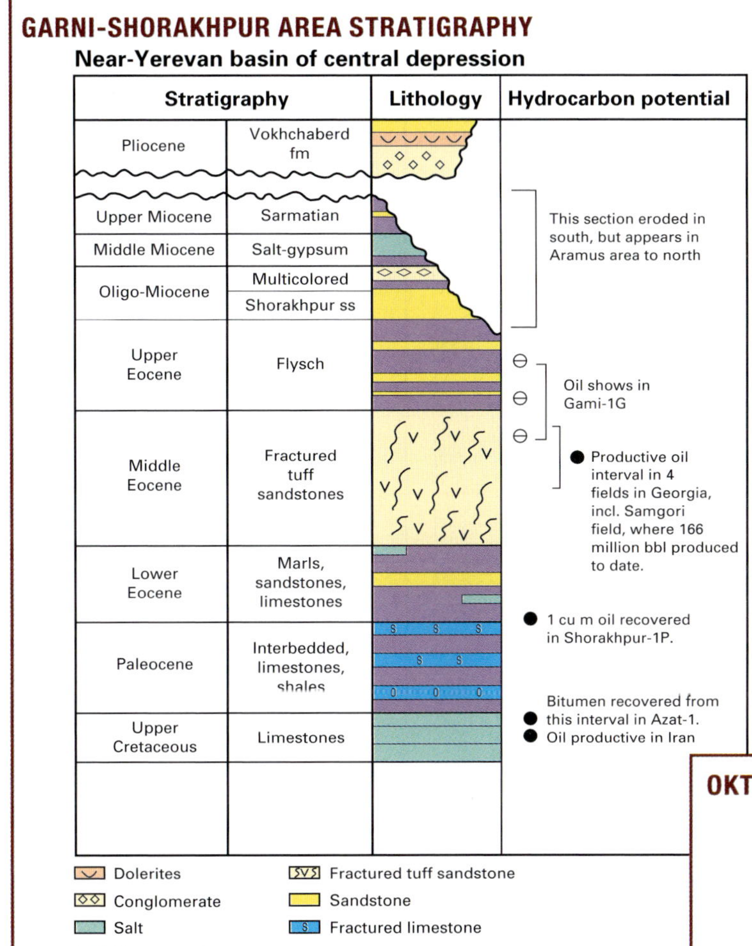

Armenia: Oktemberyan basin

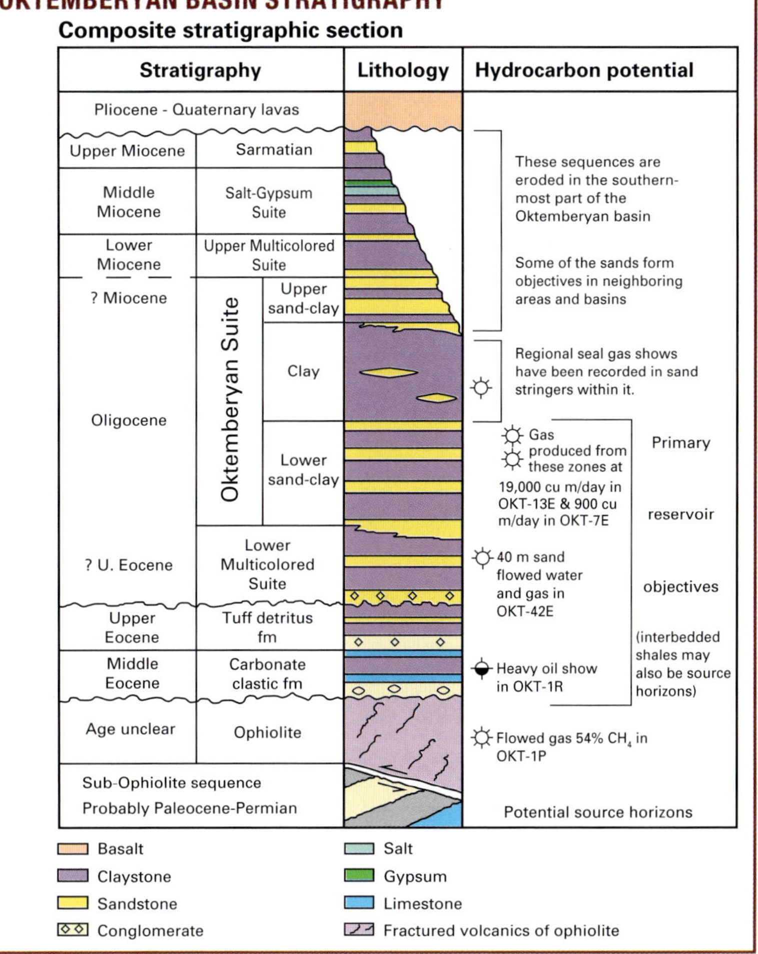

Bangladesh: Western and eastern areas

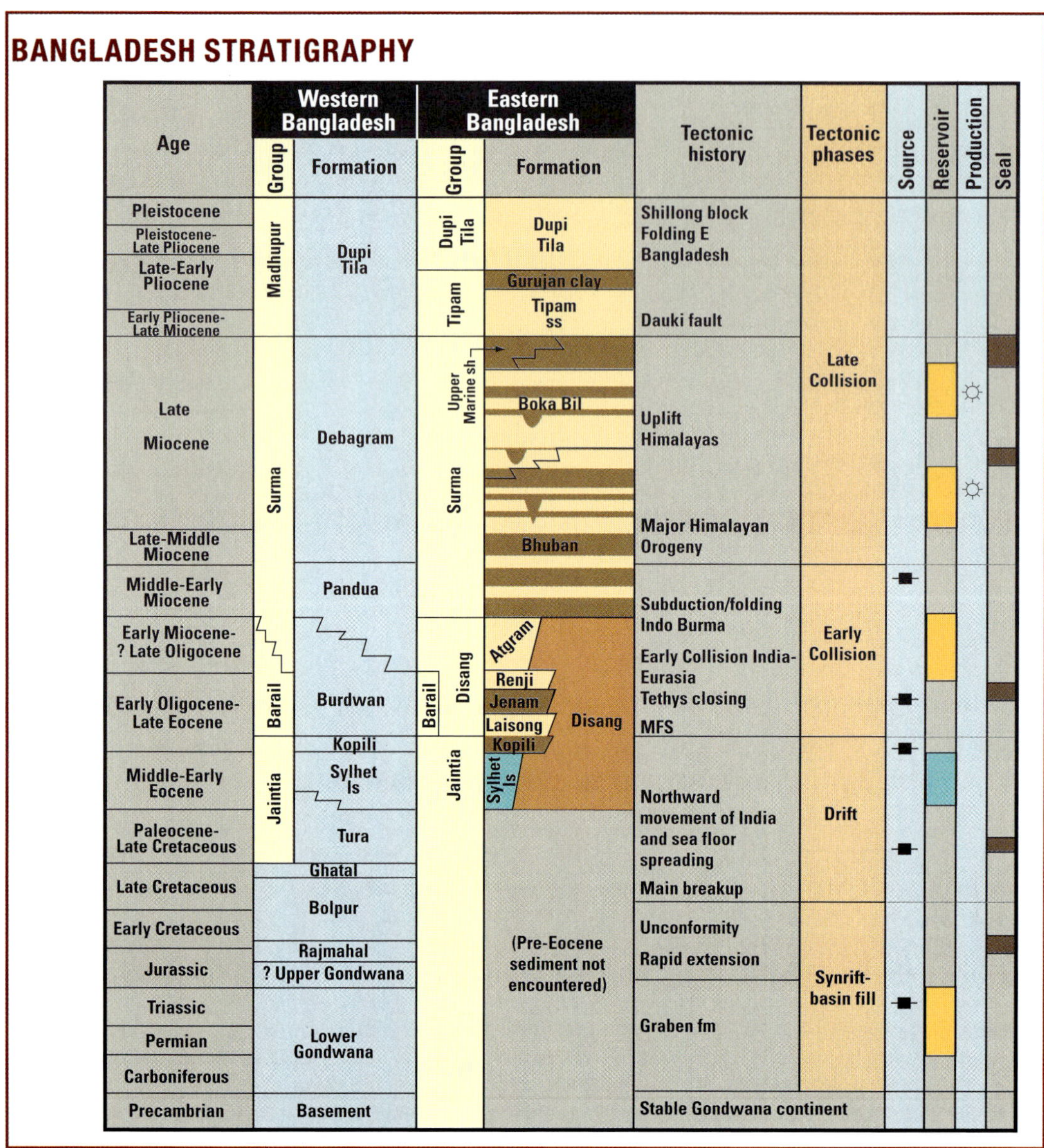

India: Bikaner-Nagaur basin

GENERAL STRATIGRAPHY, BIKANER-NAGAUR BASIN

Era	Period	Southern part / Description	Northern part
Quarter-nary	**Recent:**	Aeolian sand, alluvium, kankar and soil (not always differentiable from Pleistocene) but normally incoherent and not much of $CaCO_3$, Concentration	
	Pleistocene	Aeolian sand, alluvium, kankar, grit, gravel, and clay	
		-----------Unconformity-----------	
Tertiary	**Lower Paleocene to Lower Eocene**	Marh fm: Poorly sorted ferruginous sandstone with clay with plant remains and siltstone. Palana fm: Carbonaceous shales, lignite, grey and yellow sandstones with occasional limestone (rich in marine fossils at Palana) and clay.	
		-----------Unconformity-----------	
Mesozoic	**Cretaceous**	In the southern part (Jodhpur-Nagaur) of the basin: Dark grey claystone with interbedded sandstone with rich formaniferal assemblage.	In the northern part (Nagaur-Ganganagar) of the basin: Not reported.
	Jurassic	In the southern part (Jodhpur-Nagaur) of the basin and as recorded in Baghewala area: Fine to medium grained micaceous sandstone with thin bands of coal in upper part. Red, brittle claystone and pinkish brown ferruginous sandstone in the lower part.	In the northern part (Nagaur-Ganganagar) of the basin: Not reported.
	Permo-Triassic	Bap-Badhura fm In southern part (Jodhpur-Nagaur) of the basin: Upper unit has thin pebble bed containing pebbles of granite, basalt, dolerite, chert, and quartzites embedded in a red clayey matrix. Lower unit has red-yellow and grey calcareous clay stone with interbeds of light grey siltstone and occasional sandstone conglomerate containing pebbles and grains of dolomite, chert, quartz, and sandstone cemented in pink to brown sandy matrix.	Bap-Badhura fm In the northern part (Nagaur -Ganganagar) of the basin: Encountered in Pugal Well. Boundary and areal extent is not clear in this basin.
Paleozoic	**Upper Cambrian**	Upper Carbonate fm Developed locally in the southern part (Jodhpur-Nagaur) of the basin: Upper part comprises of dolostone/dolomite (occasionally anhydritic) with minor interbeds of grey claystone. Lower part comprises of inter-bedded dolostone, reddish brown, grey claystone and siltstones.	Upper Carbonate fm Not developed in the northern part (Nagaur-Ganganagar) of the basin.
	Lower Cambrian (Marwar Supergroup)	Nagaur Group Tunklian fm: Sandstone mostly gritty and pebbly Nagaur fm: Siltstone, clay, claystone, sandstone with gypsum veins.	
		-----------Gradational Contract-----------	
		Bilara Group Developed in the southern part (Jodhpur – Nagaur) of the basin: Pondle fm Dolomite, dolomitic limestone, cherty dolomite and stromatolitic limestone Gotan fm Limestone and dolomitic limestone Dhanapa fm Dolomite, dolomitic limestone, chert and cherty dolomite with stromatolites	Hanseran Evaporite Group Developed in the northern part (Nagaur-Ganganagar) of the basin: HEG has been classified into 8 formations on the basis of lithological characters and order of superposition and they from bottom to top are: Lakhasar, Kalu, Chhatergarh, Malkisar, Kupli, Harsinghpura, Satiyan, and Lakhusar. Each of the formations broadly made up of (in the order of deposition) reddish colored clay,stromatolitic dolomite, dolomite/limestone, anhydritic dolomite and halite, and also occasionally magnesite and rare potash salts.
		-----------Transitional contact-----------	
		Jodhpur Group Girbhaker fm: Light brown to buff and grey colored coarse-medium grained sandstone, gritty to pebbly with interbeds of light grey dolomite. Sonia fm: Sandstone, siltstone, interbedded with chert and cherty dolomite, light grey dolomitic limestone.	
		-----------Unconformity-----------	
	Precambrian	Malani Igneous Suite: Malani rhyolite, Jalore Siwana granite Delhi Supergroup: Ajabgarh slate and phyllite	

Source: Based on the published literature and data of wells/boreholes in India.

Pakistan: Lower Indus basin

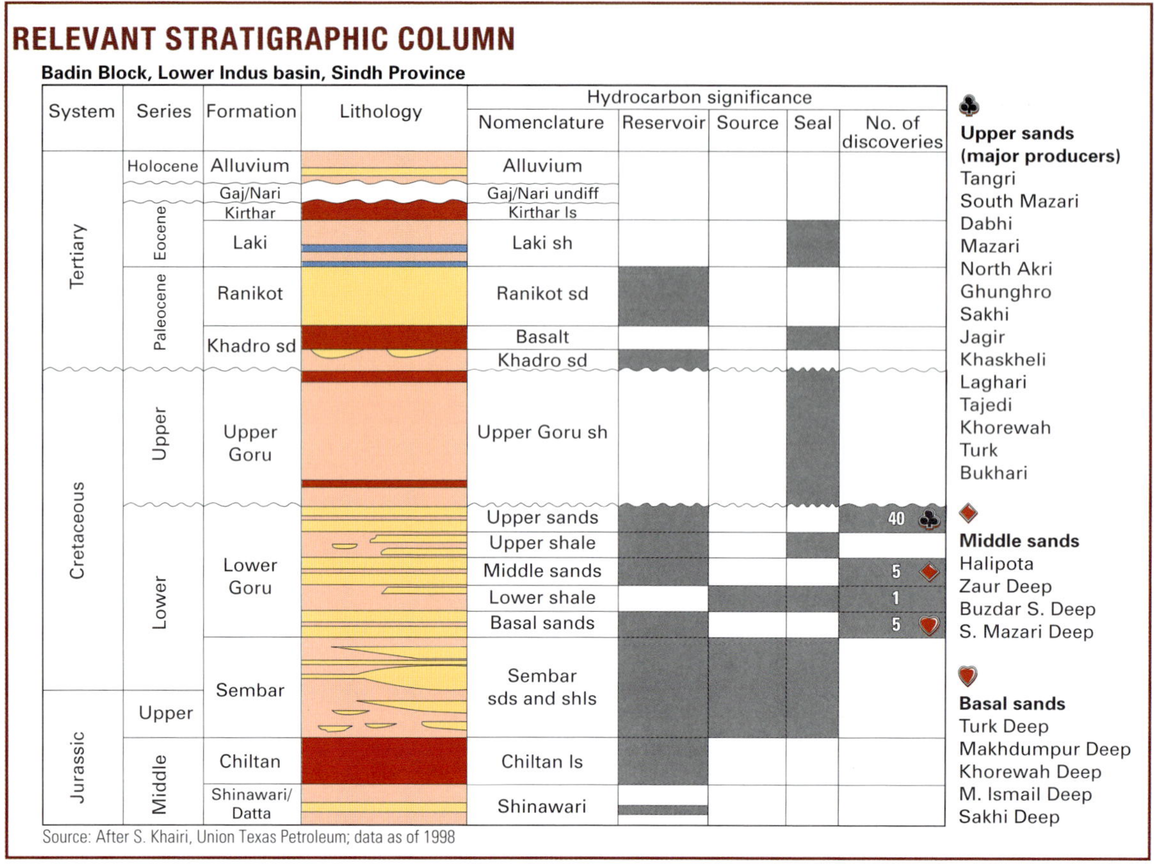

Source: After S. Khairi, Union Texas Petroleum; data as of 1998

Papua New Guinea: Eastern Papuan basin

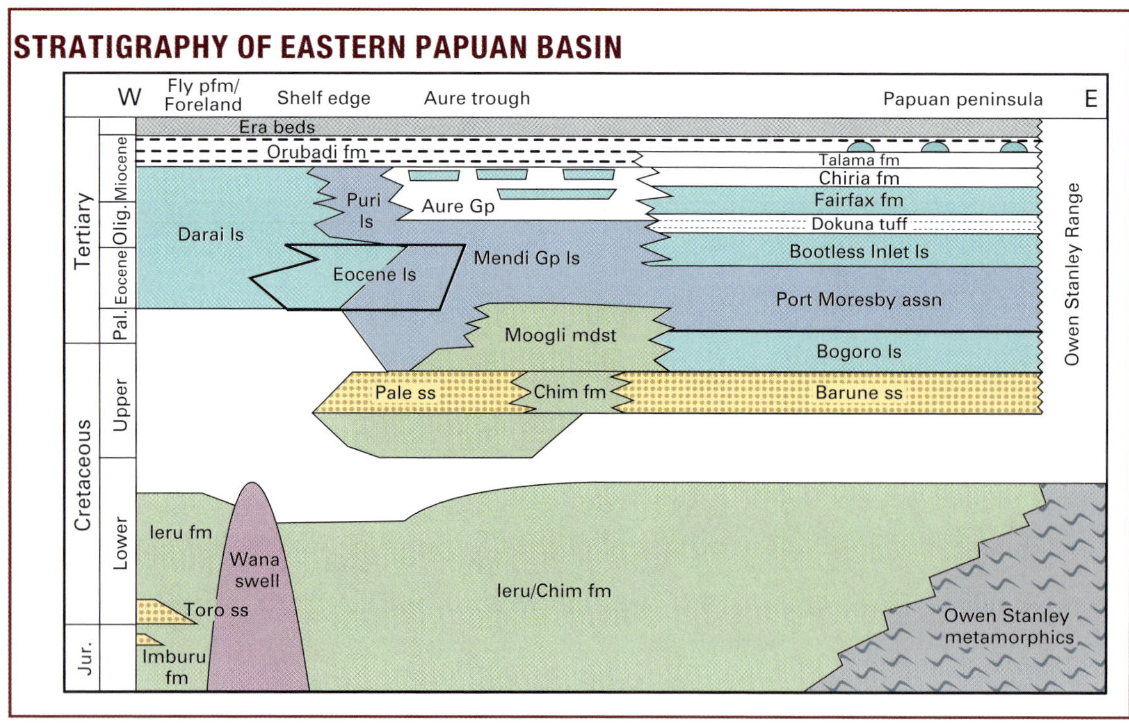

STATISTICAL TABLES

OIL PRODUCTION/CONSUMPTION

World Oil Consumption

COUNTRY	1999	2000	2001	1999	2000	2001	Change 2001 over 2000	2001 share of total
	million tonnes/year			1000 b/d			%	
North America								
Canada	87.2	88.1	88.0	1926	1937	1941	—0.1	2.5
US	888.9	897.6	895.6	19519	19701	19633	—2.0	25.5
Total	976.1	985.7	983.6	21445	21638	21573	—0.2	28.0
Latin America								
Argentina	21.0	20.3	19.0	445	431	404	—6.2	0.5
Brazil	85.7	85.4	85.1	1879	1867	1865	—0.4	2.4
Chile	11.7	11.8	12.0	253	256	262	2.0	0.3
Colombia	10.6	10.5	9.9	238	232	220	—5.6	0.3
Ecuador	6.0	5.8	5.9	131	129	132	2.3	0.2
Mexico	80.8	84.1	82.7	1765	1835	1813	—1.7	2.4
Peru	7.4	7.3	6.8	157	153	145	—5.8	0.2
Venezuela	21.3	22.5	22.2	474	496	491	—1.3	0.6
Other	57.3	57.4	57.5	1168	1168	1175	0.2	1.6
Total	301.8	305.1	301.1	6510	6567	6506	—1.3	8.6
Europe								
Austria	12.1	11.8	12.4	250	244	257	5.3	0.4
Belgium/Luxembg.	32.4	33.9	32.3	670	702	672	—4.6	0.9
Bulgaria	4.5	4.5	4.6	93	95	96	1.5	0.1
Czech Republic	8.2	7.9	8.3	174	169	178	5.1	0.2
Denmark	10.6	10.4	10.1	222	215	211	—2.4	0.3
Finland	10.7	10.7	10.5	224	224	222	—1.5	0.3
France	96.4	94.9	95.8	2044	2007	2032	0.9	2.7
Germany	132.4	129.8	131.6	2824	2763	2804	1.4	3.7
Greece	18.7	19.9	19.4	383	406	396	—2.8	0.6
Hungary	7.1	6.8	6.8	151	145	144	—1.2	0.2
Ireland	8.3	8.2	8.7	172	170	181	6.3	0.2
Italy	94.4	93.5	92.8	1980	1956	1946	—0.8	2.6
Netherlands	40.6	41.7	43.9	880	899	948	5.3	1.3
Norway	10.1	9.4	9.7	216	201	213	2.9	0.3
Poland	19.9	20.0	19.0	431	427	407	—4.8	0.5
Portugal	15.4	14.9	15.2	321	312	318	1.7	0.4
Romania	9.5	10.0	10.1	195	203	207	1.4	0.3
Slovakia	3.4	3.4	3.4	73	73	73	—0.9	1.0
Spain	68.4	70.0	72.7	1423	1452	1508	3.9	2.1
Sweden	16.1	15.2	15.6	337	318	326	2.5	0.4
Switzerland	12.6	12.2	13.1	271	263	280	6.9	0.4
Turkey	29.5	31.6	30.4	638	695	662	—3.9	0.9
UK	79.7	77.9	76.1	1727	1684	1649	—2.3	2.2
Other Europe	16.7	16.3	16.8	345	336	346	2.8	0.5
Total	758.6	755.8	760.2	16063	15975	16093	0.6	21.7
Middle East								
Iran	57.3	56.1	54.2	1192	1158	1131	—3.4	1.5
Kuwait	10.3	10.4	10.5	202	202	206	1.6	0.3
Qatar	1.1	1.2	1.4	24	25	30	19.7	
Saudi Arabia	60.9	62.4	62.7	1306	1333	1347	0.5	1.8
UAE	13.0	14.2	14.3	257	280	282	0.6	0.4
Other	63.0	63.5	63.3	1301	1309	1309	—0.3	1.8
Total	205.6	207.8	206.4	4283	4307	4306	—0.6	5.9

Source: BP Statistical Review of World Energy 2002

World Oil Consumption-cont'd

COUNTRY	1999	2000	2001	1999	2000	2001	Change 2001 over 2000	2001 share of total
	million tonnes/yr			1000 b/d			%	
Africa								
Algeria	8.1	8.5	8.8	187	192	200	3.5	0.3
Egypt	27.8	27.2	26.2	573	564	551	—3.5	0.7
South Africa	21.8	22.5	23.0	462	475	488	2.1	0.7
Other	57.4	57.9	59.0	1216	1224	1251	1.9	1.7
Total	115.1	116.1	117.0	2439	2455	2490	0.8	3.3
Former Soviet Union								
Azerbaijan	7.4	6.2	4.6	149	124	92	—25.8	0.1
Belarus	6.9	6.1	5.9	139	122	118	-3.3	0.2
Kazakhstan	6.6	7.0	7.7	133	140	155	10.0	0.2
Lithuania	3.1	2.4	2.8	63	49	57	15.9	0.1
Russia	126.2	123.5	122.3	2534	2474	2456	—1.0	3.5
Turkmenistan	2.5	2.3	2.4	50	46	48	4.3	0.1
Ukraine	12.7	12.0	12.7	255	240	255	5.8	0.4
Uzbekistan	7.1	6.4	6.5	143	128	131	1.6	0.2
Other	4.5	4.4	4.7	91	88	94	7.2	0.1
Total	177.0	170.3	169.6	3556	3412	3407	—0.4	4.8
Asia-Pacific								
Australia	38.0	37.7	38.1	843	837	845	1.0	1.1
Bangladesh	3.4	3.4	3.4	70	70	71	1.0	0.1
China	207.2	230.1	231.9	4416	4985	5041	0.8	6.6
India	95.2	97.5	97.1	2016	2067	2072	—0.4	2.8
Indonesia	46.8	50.4	52.3	980	1053	1095	3.7	1.5
Japan	257.3	255.4	247.2	5618	5576	5427	—3.2	7.0
Malaysia	20.3	20.4	18.6	439	441	407	—8.8	0.5
New Zealand	6.3	6.3	6.2	134	134	134	—0.4	0.2
Pakistan	18.2	18.8	18.9	363	373	377	0.7	0.5
Philippines	18.0	16.6	16.5	375	348	347	—0.5	0.5
Singapore	31.6	33.5	36.9	619	654	726	10.2	1.1
South Korea	100.7	103.2	103.1	2178	2229	2235	—0.1	2.9
Taiwan	39.9	39.8	37.7	820	816	776	—5.2	1.1
Thailand	35.4	34.8	33.8	734	725	714	—2.8	1.0
Other	19.0	20.6	21.5	400	432	453	4.4	0.6
Total	946.6	978.2	972.7	20200	20941	20916	—0.5	27.7
Total World	3480.8	3519.0	3510.6	74495	75295	75291	—0.2	100.0

World Petroleum Product Consumption

REGION/ PRODUCT	1996	1997	1998	1999	2000	2001	Change 2001 over 2000	2001 share of total
	1000 b/d						%	
US								
Gasoline	8167	8324	8579	8716	8813	8883	0.8	45.2
Middle distillates	5342	5502	5545	5700	5852	5843	—0.2	29.8
Fuel oil	831	777	869	814	893	829	—7.2	4.2
Others	3969	4017	3924	4290	4143	4078	—1.6	20.8
Total	18309	18621	18917	19519	19701	19633	—0.3	100.0
North America incl. US, Mexico								
Gasoline	9389	9560	9849	9998	10106	10200	0.9	43.6
Middle distillates	6194	6398	6450	6628	6811	6767	—0.6	28.9
Fuel oil	1348	1351	1506	1415	1518	1451	—4.4	6.2
Others	4805	4882	4788	5169	5037	4969	—1.4	21.3
Total	21736	22191	22593	23210	23473	23386	—0.4	100.0
South & Central America								
Gasoline	1197	1266	1283	1396	1374	1346	—2.0	28.7
Middle distillates	1538	1625	1700	1680	1690	1705	0.9	36.3
Fuel oil	788	845	868	776	749	725	—3.2	15.4
Others	804	831	857	893	919	917	—0.3	19.6
Total	4327	4568	4709	4745	4732	4693	—0.8	100.0
Europe								
Gasoline	4227	4279	4318	4334	4217	4121	—2.3	25.6
Middle distillates	6356	6421	6631	6694	6722	6939	3.2	43.1
Fuel oil	2260	2206	2191	2088	1982	1981	—0.1	12.3
Others	2790	2933	2952	2946	3053	3052	0	19.0
Total	15633	15839	16092	16063	15975	16093	0.7	100.0
Middle East								
Gasoline	774	814	830	841	840	843	0.4	19.6
Middle distillates	1468	1465	1459	1482	1483	1517	2.2	35.2
Fuel oil	1206	1205	1173	1251	1281	1241	—3.1	28.8
Others	662	678	699	709	703	704	0.2	16.4
Total	4110	4161	4161	4283	4307	4306	0	100.0
Africa								
Gasoline	552	559	568	574	580	591	2.1	23.8
Middle distillates	893	919	949	985	1007	1032	2.4	41.5
Fuel oil	445	466	491	494	474	456	—3.6	18.3
Others	349	363	376	386	395	410	3.8	16.4
Total	2240	2307	2385	2439	2455	2490	1.4	100.0
China								
Gasoline	986	1111	1098	1164	1313	1245	—5.2	24.7
Middle distillates	1080	1202	1277	1455	1633	1698	4.0	33.7
Fuel oil	719	750	725	694	725	728	0.4	14.4
Others	888	871	946	1103	1314	1371	4.3	27.2
Total	3672	3935	4047	4416	4985	5041	1.1	100.0

Source: BP Statistical Review of World Energy 2002

NOTE: Gasolines include aviation, motor, and light distillate feedstock.
Middle distillates include jet and heating kerosenes and gas and diesel oils.
Others include refinery gas, LPGs, solvents, petroleum coke, lubricants, bitumen, wax, and refinery fuel and loss.

World Petroleum Product Consumption-cont'd

WORLD PETROLEUM PRODUCT CONSUMPTION, cont d.

REGION/ PRODUCT	1996	1997	1998	1999	2000	2001	Change 2001 over 2000	2001 share of total
	1000 b/d						%	
Japan								
Gasoline	1576	1646	1611	1702	1735	1711	—1.4	31.5
Middle distillates	2027	1991	1949	1978	1958	1958	0	36.1
Fuel oil	1067	977	879	861	804	690	—14.2	12.7
Others	1142	1147	1085	1077	1079	1067	—1.1	19.7
Total	5812	5761	5525	5618	5576	5427	—2.7	100.0
Asia-Pacific incl. China, Japan								
Gasoline	4549	4973	5018	5311	5549	5525	—0.4	26.4
Middle distillates	7160	7430	7186	7596	7798	7878	1.0	37.7
Fuel oil	3795	3829	3563	3548	3505	3366	—4.0	16.1
Others	3364	3447	3483	3745	4089	4147	1.4	19.8
Total	18868	19680	19250	20200	20941	20916	—0.1	100.0
World excl. Former Soviet Union								
Gasoline	20688	21451	21866	22454	22665	22626	—0.2	31.5
Middle distillates	23609	24258	24375	25064	25512	25839	1.3	35.9
Fuel oil	9842	9902	9794	9572	9510	9221	—3.0	12.8
Others	12775	13133	13155	13848	14196	14199	0	19.8
Total	66913	68745	69190	70939	71883	71884	0	100.0

World Oil Production

First line = 1,000 bbl per day (b/d); second line = million tonnes per year (mtpy)

COUNTRY	1950	1960	1970	1980	1990	1996	1997	1998	1999	2000	Actual 2001	Est. 2002
North America												
Canada	79.6	525.6	1263.6	1412.0	1508.0	1834.0	1910.3	2017.0	1900.8	2035.0	2052.4	2195.0
	4.0	26.2	62.9	70.3	75.1	91.3	95.1	100.4	94.7	101.4	102.2	109.3
US	5407.1	7054.6	9630.0	8569.0	7220.0	6464.5	6452.0	6252.0	5881.5	5822.0	5801.2	5770.0
	269.3	351.3	479.6	426.7	359.6	321.9	321.3	311.3	292.9	289.5	288.9	287.3
Total	5486.7	7580.2	10893.6	9981.0	8728.0	8298.5	8362.3	8269.0	7782.3	7857.0	7853.6	7965.0
	273.3	377.5	542.5	497.0	434.7	413.2	416.4	4111.7	387.6	390.9	391.1	396.7
Latin America												
Argentina	64.0	171.9	382.9	487.0	473.0	783.2	834.6	846.9	801.8	750.7	763.0	750.0
	3.2	8.6	19.1	24.3	23.6	39.0	41.6	42.2	39.9	37.4	38.0	37.4
Bolivia	1.7	9.8	16.3	30.0	19.2	29.3	31.0	28.0	32.5	27.6	31.3	31.0
	0.1	0.5	0.8	1.5	1.0	1.5	1.5	1.4	1.6	1.4	1.6	1.5
Brazil	0.9	80.9	160.5	182.0	633.0	780.6	841.5	956.1	1085.9	1128.0	1302.5	1488.0
	--	4.0	8.0	9.1	31.5	38.9	41.9	47.6	54.1	56.2	64.9	74.1
Chile	1.7	19.8	34.6	29.0	20.3	9.3	8.8	9.0	9.0	7.0	7.0	7.0
	0.1	1.0	1.7	1.4	1.0	0.5	0.4	0.4	0.4	0.3	0.3	0.3
Colombia	93.3	152.7	214.0	125.0	445.0	623.2	652.2	754.3	815.6	687.3	604.3	583.0
	4.6	7.6	10.7	6.2	22.2	31.0	32.5	37.6	40.6	34.2	30.1	29.0
Cuba	--	--	--	--	15.0	33.0	35.6	27.0	27.0	44.0	40.3	40.0
	--	--	--	--	0.7	1.6	1.8	1.3	1.3	2.2	2.0	2.0
Ecuador	7.2	7.7	4.1	222.0	287.0	383.7	388.2	377.9	375.7	401.1	407.1	398.0
	0.4	0.4	0.2	11.1	14.3	19.1	19.3	18.8	18.7	20.0	20.3	19.8
Guatemala	--	--	--	5.0	4.0	13.0	19.5	25.5	23.2	20.7	21.1	23.5
	--	-	--	0.2	0.2	0.6	1.0	1.3	1.2	1.0	1.1	1.2
Mexico	198.5	270.6	430.2	1936.0	2633.0	2961.1	3022.1	3071.1	2906.4	3012.0	3127.0	3180.0
	9.9	13.5	21.4	96.4	131.1	147.5	150.5	152.9	144.7	150.0	155.7	158.4
Peru	41.1	52.6	72.2	191.0	132.0	119.8	105.7	113.5	102.5	98.0	93.1	93.0
	2.0	2.6	3.6	9.5	6.6	6.0	5.3	5.7	5.1	4.9	4.6	4.6
Trinidad	56.5	115.7	139.8	211.0	151.0	126.9	123.7	122.6	125.0	119.0	113.3	127.0
	2.8	5.8	7.0	10.5	7.5	6.3	6.2	6.1	6.2	5.9	5.6	6.3
Venezuela	1498.0	2846.1	3708.0	2167.0	2118.0	2938.0	3181.8	3121.7	2786.7	3028.0	2685.0	2415.0
	74.6	141.7	184.7	107.9	105.5	146.3	158.5	155.5	138.8	150.8	133.7	120.3
Total	1962.9	3727.8	5162.6	5585.0	6935.7	8809.7	9255.1	9465.6	9105.2	9323.4	9209.1	9149.4
	97.7	185.7	257.2	278.1	345.5	438.7	461.0	471.4	453.3	464.3	458.6	455.6

Source: Oil & Gas Journal
NOTE: Some totals include countries with low production that are not listed.

World Oil Production-cont'd

First line = 1,000 bbl per day (b/d); second line = million tonnes per year (mtpy)

COUNTRY	1950	1960	1970	1980	1990	1996	1997	1998	1999	2000	Actual 2001	Est. 2002
Europe												
Albania	--	--	--	--	--	--	--	--	--	5.9	6.0	6.2
	--	--	--	--	--	--	--	--	--	0.3	0.3	0.3
Austria	27.9	45.4	53.3	29.0	24.0	20.2	20.4	20.7	20.2	19.1	18.9	18.4
	1.4	2.3	2.7	1.4	1.2	1.0	1.0	1.0	1.0	1.0	0.9	0.9
Bulgaria	--	4.0	6.7	6.0	5.0	0.7	0.8	1.0	1.0	1.0	1.0	1.0
	--	0.2	0.3	0.3	0.2	--	--	--	--	--	--	--
Croatia	--	--	--	--	--	30.0	30.8	26.7	24.6	23.0	21.7	21.0
	--	--	--	--	--	1.5	1.5	1.3	1.2	1.1	1.1	1.0
Czech Rep.*	0.8	2.7	4.1	1.9	3.0	9.0	9.0	9.0	8.9	3.2	3.3	5.0
	--	0.1	0.2	0.1	0.1	0.4	0.4	0.4	0.4	0.2	0.2	0.2
Denmark	--	--	--	6.0	118.0	207.2	230.5	238.1	299.3	363.0	348.4	365.0
	--	--	--	0.3	5.9	10.3	11.5	11.9	14.9	18.1	17.4	18.2
France	2.5	38.6	45.8	26.0	62.0	42.3	35.8	34.3	30.9	28.4	27.8	26.2
	0.1	1.9	2.3	1.3	3.1	2.1	1.8	1.7	1.5	1.4	1.4	1.3
Germany	23.2	108.3	148.4	100.0	78.0	56.8	55.6	57.9	54.3	62.5	68.9	71.7
	1.2	5.4	7.4	5.0	3.9	2.8	2.8	2.9	2.7	3.1	3.4	3.6
Greece	--	--	--	--	15.0	10.3	8.5	5.6	0.3	5.1	4.0	3.2
	--	--	--	--	0.7	0.5	0.4	0.3	--	0.3	0.2	0.2
Hungary	10.1	24.3	38.7	40.6	40.0	30.0	27.6	25.8	24.2	25.8	23.6	21.7
	0.5	1.2	1.9	2.0	2.0	1.5	1.4	1.3	1.2	1.3	1.2	1.1
Italy	0.2	37.5	26.3	39.0	97.0	108.3	110.4	108.0	99.9	90.8	65.2	87.0
	--	1.9	1.3	1.9	4.8	5.4	5.5	5.4	5.0	4.5	3.2	4.3
Netherlands	13.4	36.0	35.8	25.0	71.0	57.0	56.7	55.5	54.7	29.3	27.0	42.0
	0.7	1.8	1.8	1.2	3.5	2.8	2.8	2.8	2.7	1.5	1.3	2.1
Norway	--	--	--	528.0	1574.0	3092.0	3148.8	3021.0	3017.6	3212.0	3244.6	3150.0
	--	--	--	26.3	78.4	154.0	156.8	150.4	150.3	160.0	161.6	156.9
Poland	3.3	3.9	8.5	6.6	3.0	6.6	5.0	4.8	4.2	14.3	17.4	16.5
	0.2	0.2	0.4	0.3	0.1	0.3	0.2	0.2	0.2	0.7	0.9	0.8
Romania	87.7	231.0	269.6	230.9	160.0	135.0	138.7	130.1	122.9	122.9	120.0	118.0
	4.4	11.5	13.4	11.5	8.0	6.7	6.9	6.5	6.1	6.1	6.0	5.9
Spain	--	--	3.1	32.0	14.3	11.4	7.5	13.0	6.1	4.6	6.9	6.5
	--	--	0.2	1.6	0.7	0.6	0.4	0.6	0.3	0.2	0.3	0.3
Turkey	0.3	7.0	67.9	44.0	70.0	66.9	66.5	68.0	56.3	56.0	48.54	47.0
	--	0.3	3.4	2.2	3.5	3.3	3.3	3.4	2.8	2.8	2.4	2.3
UK	0.9	1.7	1.7	1619.0	1860.0	2553.0	2543.0	2632.7	2724.9	2513.7	2330.8	2250.0
	--	0.1	0.1	80.6	92.6	127.1	126.6	131.1	135.7	125.2	116.1	112.1
Total	170.3	540.4	709.9	2734.0	4194.3	6436.7	6495.6	6452.2	6550.3	6580.6	6383.9	6256.4
	8.5	26.9	35.4	136.0	208.7	320.3	323.3	321.2	326.0	327.8	317.9	311.6

* Slovakia production totaled with Czech Republic before 2000.

World Oil Production-cont'd

First line = 1,000 bbl per day (b/d); second line = million tonnes per year (mtpy)

COUNTRY	1950	1960	1970	1980	1990	1996	1997	1998	1999	2000	Actual 2001	Est. 2002
Middle East												
Bahrain	30.2	45.1	76.6	48.0	42.0	103.8	103.9	102.6	102.2	102.3	173.6	174.0
	1.5	2.2	3.8	2.4	2.1	5.2	5.2	5.1	5.1	5.1	8.6	8.7
Iran	664.3	1067.6	3328.8	1467.0	3120.0	3668.0	3632.7	3607.5	3504.2	3681.7	3695.8	3450.0
	33.1	53.2	165.8	73.1	155.4	182.7	180.9	179.7	174.5	183.4	184.1	171.8
Iraq	136.2	1004.2	1548.6	2638.0	2083.0	576.0	1147.0	2110.0	2525.0	2566.7	2355.0	2030.0
	6.8	50.0	77.1	131.4	103.7	28.7	57.1	105.1	125.7	127.8	117.3	101.1
Kuwait	344.4	1628.2	2734.5	1382.0	1080.0	1807.0	1836.3	1800.0	1653.3	1765.0	1715.0	1600.0
	17.2	81.1	136.2	68.8	53.8	90.0	91.4	89.6	82.3	87.9	85.4	79.7
Neutral Zone	--	136.2	500.6	540.0	315.0	483.0	533.0	550.0	590.0	630.0	565.0	535.0
	--	6.8	24.9	26.9	15.7	24.1	26.5	27.4	29.4	31.4	28.1	26.6
Oman	--	--	332.4	283.0	658.0	885.0	904.5	900.0	832.2	933.3	964.2	895.0
	--	--	16.6	14.1	32.8	44.1	45.0	44.8	41.4	46.5	48.0	44.6
Qatar	33.6	175.1	362.4	472.0	387.0	487.0	622.9	660.8	647.7	688.3	671.7	640.0
	1.7	8.7	18.0	23.5	19.3	24.3	31.0	32.9	32.3	34.3	33.5	31.9
Saudi Arabia	546.7	1247.1	3548.9	9630.0	6215.0	7838.0	8082.9	8021.7	7521.3	7995.0	7695.0	7380.0
	27.2	62.1	176.7	479.6	309.5	390.3	402.5	399.5	374.6	398.2	383.2	367.5
Syria	--	--	83.1	165.0	385.0	577.0	570.0	553.3	542.0	522.5	518.3	490.0
	--	--	4.1	8.2	19.2	28.7	28.4	27.6	27.0	26.0	25.8	24.4
UAE	--	--	693.8	1709.0	2101.0	2292.5	2254.0	2296.7	2044.5	2230.5	2153.5	1984.5
	--	--	34.6	85.1	104.6	114.2	112.2	114.4	101.8	111.1	107.2	98.8
Yemen	--	--	--	--	179.0	365.0	370.0	385.0	407.5	354.0	350.0	350.0
	--	--	--	--	8.9	18.2	18.4	19.2	20.3	17.6	17.4	17.4
Total	1755	5306	13280	18335	16566	19083	20058	20988	20370	21469	20857	19529
	87.5	264.2	661.3	913.1	825.0	950.4	998.7	1045.2	1014.4	1069.2	1038.7	972.5

World Oil Production-cont'd

First line = 1,000 bbl per day (b/d); second line = million tonnes per year (mtpy)

COUNTRY	1950	1960	1970	1980	1990	1996	1997	1998	1999	2000	Actual 2001	Est. 2002
Africa												
Algeria	--	182.7	1029.1	1016.0	797.0	818.0	849.2	824.2	745.0	808.0	835.8	850.0
	--	9.1	51.2	50.6	39.7	40.7	42.3	41.0	37.1	40.2	41.6	42.3
Angola	--	1.3	13.7	150.0	480.0	691.0	713.6	734.5	761.8	740.0	696.0	900.0
	--	0.1	0.7	7.5	23.9	34.4	35.5	36.6	37.9	36.8	34.7	44.8
Cameroon	--	--	--	58.0	164.0	110.0	124.3	125.0	100.0	85.0	80.0	69.0
	--	--	--	2.9	8.2	5.5	6.2	6.2	5.0	4.2	4.0	3.4
Congo (Brz.)	--	--	--	56.0	161.0	220.0	244.0	238.4	264.3	265.0	265.0	250.0
	--	--	--	2.8	8.0	11.0	12.2	11.9	13.2	13.2	13.2	12.5
Congo (Zr.)	--	--	--	20.0	28.0	29.3	28.3	26.2	23.7	25.0	24.0	23.0
	--	--	--	1.0	1.4	1.5	1.4	1.3	1.2	1.2	1.2	1.1
Egypt	44.9	61.6	327.3	596.0	873.0	855.0	860.0	865.8	815.0	810.0	760.0	750.0
	2.2	3.1	16.3	29.7	43.5	42.6	42.8	43.1	40.6	40.3	37.8	37.4
Eq. Guinea	--	--	--	--	--	22.0	51.7	83.0	94.0	167.5	183.6	135.0
	--	--	--	--	--	1.1	2.6	4.1	4.7	8.3	9.1	6.7
Gabon	--	16.1	108.8	180.0	289.0	366.5	365.2	355.0	340.0	325.0	300.8	294.0
	--	0.8	5.4	9.0	13.4	18.3	18.2	17.7	16.9	16.2	15.0	14.6
Ivory Coast	--	--	--	--	2.1	28.0	15.0	10.0	20.0	12.0	5.2	5.0
	--	--	--	--	0.1	1.4	0.7	0.5	1.0	0.6	0.3	0.2
Libya	--	--	3318.0	1785.0	1369.0	1393.0	1420.4	1391.7	1346.7	1410.0	1365.0	1300.0
	--	--	165.2	88.9	68.2	69.4	70.7	69.3	67.1	70.2	68.0	64.7
Nigeria	--	17.3	1083.3	2057.0	1808.0	2246.2	2281.9	2132.1	1964.2	2030.0	2083.3	1930.0
	--	0.9	53.9	102.4	90.0	111.9	113.6	106.2	97.8	101.0	103.7	96.1
South Africa	--	--	--	--	--	10.0	16.3	17.9	10.0	25.9	22.9	21.5
	--	--	--	--	--	0.5	0.8	0.9	0.5	1.3	1.1	1.1
Sudan	--	--	--	--	--	--	--	--	--	185.0	200.0	210.0
	--	--	--	--	--	--	--	--	--	9.2	10.0	10.5
Tunisia	--	3.0	87.7	110.0	93.0	87.6	78.4	80.5	81.6	75.9	69.6	71.0
	--	0.1	4.4	5.5	4.6	4.4	3.9	4.0	4.1	3.8	3.5	3.5
Total	45.7	285.0	5969.4	6032.0	6048.4	6887.2	7066.2	6904.5	6586.5	6971.6	6901.3	6817.5
	2.2	14.2	297.2	300.5	301.2	343.0	351.9	343.8	328.1	347.2	343.7	339.5

World Oil Production-cont'd

First line = 1,000 bbl per day (b/d); second line = million tonnes per year(mtpy)

COUNTRY	1950	1960	1970	1980	1990	1996	1997	1998	1999	2000	Actual 2001	Est. 2002
Former Soviet Union												
Azerbaijan	--	--	--	--	--	--	--	--	--	275.0	298.5	300.0
	--	--	--	--	--	--	--	--	--	13.7	14.9	14.9
Belarus	--	--	--	--	--	--	--	--	--	35.0	35.0	35.0
	--	--	--	--	--	--	--	--	--	1.7	1.7	1.7
Georgia	--	--	--	--	--	--	--	--	--	2.0	2.0	2.0
	--	--	--	--	--	--	--	--	--	0.1	0.1	0.1
FSU*	760.6	2969.1	7089.2	12031.0	11500.0	6752.0	7003.2	7008.2	7125.7	--	--	--
	37.9	147.9	353.0	599.1	572.7	336.2	348.8	349.0	354.9	--	--	--
Kazakhstan	--	--	--	--	--	--	--	--	--	675.0	707.5	800.0
	--	--	--	--	--	--	--	--	--	33.6	35.2	39.8
Kyrgyzstan	--	--	--	--	--	--	--	--	--	1.0	2.0	2.0
	--	--	--	--	--	--	--	--	--	--	0.1	0.1
Russia	--	--	--	--	--	--	--	--	--	6325.0	6780.8	7385.0
	--	--	--	--	--	--	--	--	--	315.0	337.7	367.8
Turkmenistan	--	--	--	--	--	--	--	--	--	140.0	160.0	180.0
	--	--	--	--	--	--	--	--	--	7.0	8.0	9.0
Ukraine	--	--	--	--	--	--	--	--	--	73.0	80.0	78.0
	--	--	--	--	--	--	--	--	--	3.6	4.0	3.9
Uzbekistan	--	--		--	--	--	--	--	--	155.0	145.0	150.0
	--	--	--	--	--	--	--	--	--	7.7	7.2	7.5
Total	760.6	2969.1	7089.2	12031.0	11500.0	6752.0	7003.2	7008.2	7125.7	7896.2	8210.8	8932.0
	37.9	147.9	353.0	599.1	572.7	336.2	348.8	349.0	354.9	393.2	408.9	444.8

* Total of all Former Soviet Union countries before 2000.

World Oil Production-cont'd

First line = 1,000 bbl per day (b/d); second line = million tonnes per year(mtpy)

COUNTRY	1950	1960	1970	1980	1990	1996	1997	1998	1999	2000	Actual 2001	Est. 2002
Asia-Pacific												
Australia	--	--	175.6	379.0	582.0	535.3	566.6	534.9	526.8	700.0	632.6	633.0
	--	--	8.7	18.9	29.0	26.7	28.2	26.6	26.2	34.9	31.5	31.5
Bangladesh	--	--	--	--	0.9	1.6	1.7	2.0	2.8	3.3	3.3	5.0
	--	--	--	--	--	0.1	0.1	0.1	0.1	0.2	0.2	0.2
Brunei	84.8	92.9	148.0	230.0	143.0	155.5	146.9	141.3	165.7	177.9	180.5	185.0
	4.2	4.6	7.4	11.5	7.1	7.7	7.3	7.0	8.3	8.9	9.0	9.2
China	4.0	100.0	500.0	2119.0	2755.0	3128.5	3188.9	3199.9	3224.5	3237.3	3296.7	3400.0
	0.2	5.0	24.9	105.5	137.2	155.8	158.8	159.4	160.6	161.2	164.2	169.3
India	5.1	9.0	138.6	182.0	679.0	656.0	680.0	658.2	651.8	646.0	643.8	663.0
	0.3	0.4	6.9	9.1	33.8	32.7	33.9	32.8	32.5	32.2	32.1	33.0
Indonesia	132.6	411.2	853.6	1576.0	1274.0	1392.0	1364.2	1315.4	1280.0	1267.3	1214.2	1120.0
	6.6	20.5	42.5	78.5	63.4	69.3	67.9	65.5	63.7	63.1	60.5	55.8
Japan	5.6	10.2	16.3	10.0	10.5	13.9	14.5	13.7	11.6	12.7	13.1	12.0
	0.3	0.5	0.8	0.5	0.5	0.7	0.7	0.7	0.6	0.6	0.7	0.6
Malaysia	--	--	--	288.0	605.0	645.8	752.2	720.0	723.3	690.0	744.2	760.0
	--	--	--	14.3	30.1	32.2	37.5	35.9	36.0	34.4	37.1	37.8
Myanmar	1.5	11.1	16.4	30.0	13.0	11.0	11.0	15.0	15.0	8.0	8.0	10.0
	0.1	0.6	0.8	1.5	0.6	0.5	0.5	0.7	0.7	0.4	0.4	0.5
New Zealand	--	--	--	7.0	39.0	43.0	58.5	50.0	45.0	36.0	34.0	34.0
	--	--	--	0.3	1.9	2.1	2.9	2.5	2.2	1.8	1.7	1.7
Pakistan	3.5	7.2	9.9	10.0	60.0	58.0	50.9	53.6	53.4	53.3	59.7	60.0
	0.2	0.4	0.5	0.5	3.0	2.9	2.5	2.7	2.7	2.7	3.0	3.0
Pap.N.Guinea	4.8	4.2	--	--	--	115.1	75.9	81.1	88.6	69.7	57.0	46.0
	0.2	0.2	--	--	--	5.7	3.8	4.0	4.4	3.5	2.8	2.3
Philippines	--	--	--	15.0	5.0	1.4	0.8	0.8	0.9	1.1	7.1	14.0
	--	--	--	0.7	0.2	0.1	--	--	--	--	0.4	0.7
Taiwan	--	0.4	1.8	5.0	2.5	0.9	0.9	0.9	0.8	0.6	0.8	0.8
	--	--	0.1	0.2	0.1	--	--	--	--	--	--	--
Thailand	--	--	0.4	--	41.0	61.2	77.0	75.0	81.9	111.0	113.1	130.0
	--	--	--	--	2.0	3.0	3.8	3.7	4.1	5.5	5.6	6.5
Viet Nam	--	--	--	--	40.0	174.0	180.2	190.0	295.2	304.0	304.8	304.0
	--	--	--	--	2.0	8.7	9.0	9.5	14.7	15.1	15.2	15.1
Total	241.9	646.2	1860.6	4851.0	6249.9	6993.2	7170.2	7051.8	7167.3	7318.2	7312.7	7376.8
	12.1	32.2	92.6	241.5	310.9	348.2	356.9	351.1	356.8	364.4	364.2	367.4
Total World	10428	21088	45059	59684	60317	63290	65441	66164	64711	67234	66747	66043
	519.4	1050.2	2243.9	2972.0	3003.4	3151.6	3258.4	3294.7	3222.2	3348.2	3324.0	3288.9

World Reserves and Production

COUNTRY	ESTIMATED PROVED RESERVES				OIL PRODUCTION			
	Jan. 1, 2003		Jan 1., 2002					
	Oil, 1,000 bbl	Gas, bcf	Oil, 1,000 bbbl	Gas, bcf	Producing oil wells, Dec. 31, 2001	Estimated 2002, 1,000 b/d	Change from 2001, %	Actual 2001, 1,000 b/d
Canada	180,021,000	60,118	4,858,000	59,733	54,061	2,195.0	6.9	2,052.4
US	22,446,000	183,460	22,045,000	177,427	521,070	5,770.0	—0.5	5,801.2
N. America	202,467,000	243,578	26,903,000	237,160	575,131	7,965.0	1.4	7,853.6
Argentina	2,878,680	26,960	2,973,700	27,460	15,094	750.0	—1.7	763.0
Bolivia	440,500	24,000	440,500	24,000	328	31.0	—0.9	31.3
Brazil	8,321,700	8,092	8,464,744	7,805	11,983	1,488.0	14.2	1,302.5
Chile	150,000	3,460	150,000	3,460	315	7.0	--	7.0
Colombia	1,842,290	4,507	1,750,000	4,322	7,641	583.0	—3.5	604.3
Cuba	750,000	2,500	750,000	2,500	245	40.0	—0.6	40.3
Ecuador	4,629,600	345	2,115,000	3,670	1,044	398.0	—2.2	407.1
Guatemala	526,000	109	526,000	109	20	23.5	11.4	21.1
Mexico	12,622,000	8,776	26,941,000	29,505	2,991	3,180.0	1.7	3,127.0
Peru	323,393	8,655	323,393	8,655	4,915	93.0	—0.1	93.1
Trinidad-Tob.	716,000	23,450	716,000	23,450	3,911	127.0	12.1	113.3
Venezuela	77,800,000	148,000	77,685,000	147,585	15,395	2,415.0	—10.1	2,685.0
Latin America	111,172,663	258,859	122,911,337	282,525	64,522	9,149.4	—0.1	9,209.1
Albania	165,000	100	165,000	100	2,275	6.2	3.3	6.0
Austria	85,680	844	85,680	915	950	18.4	—2.5	18.9
Bulgaria	15,000	210	15,000	210	100	1.0	--	1.0
Croatia	92,196	1,237	92,196	1,237	723	21.0	—3.0	21.7
Czech Repub.	15,000	140	15,000	140	--	5.0	49.7	3.3
Denmark	1,347,000	2,975	1,113,300	2,719	213	365.0	4.8	348.4
France	148,473	506	140,040	403	463	26.2	—5.7	27.8
Germany	342,311	11,294	364,300	12,088	991	71.7	4.1	68.9
Greece	9,000	18	9,000	18	10	3.2	—20.0	4.0
Hungary	102,484	1,210	110,919	1,282	944	21.7	—8.0	23.6
Ireland	--	700	--	700	--	--	--	--
Italy	621,700	8,000	621,763	8,072	208	87.0	33.4	65.2
Netherlands	106,000	62,000	106,927	62,542	203	42.0	55.6	27.0
Norway	10,265,000	77,300	9,447,290	44,037	833	3,150.0	—2.9	3,244.6
Poland	96,375	5,829	114,883	5,119	1,404	16.5	—5.1	17.4
Romania	955,620	3,556	955,620	3,556	6,000	118.0	—1.7	120.0
Serbia	77,500	1,700	77,500	1,700	646	16.0	—5.9	17.0
Slovakia	9,000	530	9,000	530	--	1.0	—9.1	1.1
Spain	157,626	94	21,009	18	21	6.5	—6.2	6.9
Turkey	300,000	300	295,760	310	846	47.0	—2.9	48.4
UK	4,715,000	24,600	4,930,000	25,956	1,387	2,250.0	—3.5	2,330.8
Europe	19,625,965	203,143	18,690,187	170,652	2,254	6273.4	—2.0	6402.0
Bahrain	124,560	3,250	124,560	3,249	496	174.0	0.2	173.6
Iran	89,700,000	812,300	89,700,000	812,300	1,120	3,450.0	—6.7	3,695.8
Iraq	112,500,000	109,800	112,500,000	109,800	1,685	2,030.0	—13.8	2,355.0
Israel	3,810	1,375	3,840	1,470	7	0.1	25.0	0.1
Jordan	890	230	890	230	4	--	--	--
Kuwait	94,000,000	52,200	94,000,000	52,200	790	1,600.0	—6.7	1,715.0
Neutral Zone	5,000,000	1,000	5,000,000	1,000	578	535.0	—5.3	565.0
Oman	5,506,000	29,280	5,506,000	29,280	2,298	895.0	—7.2	964.2
Qatar	15,207,000	508,540	15,207,000	508,540	417	640.0	—4.7	671.7
Saudi Arabia	259,300,000	224,200	259,250,000	219,000	1,560	7,380.0	—4.1	7,695.0
Syria	2,500,000	8,500	2,500,000	8,500	132	490.0	—5.5	518.3
UAE	97,800,000	212,100	97,800,000	212,100	1,456	1984.5	—7.8	2,153.5
Yemen	4,000,000	16,900	4,000,000	16,900	302	350.0	--	350.0
Middle East	685,642,260	1,979,675	685,592,290	1,974,569	10,845	19,528.6	—6.4	20,857.1

Source: Oil & Gas Journal

NOTES: All reserves figures are reported as prove d reserves recoverable with current technology and prices except FSU, some eastern European countries, and Canada s gas reserves; these figures include proved plus some probable. Some totals include countries with low reserves and production that are not listed.

World Reserves and Production-cont'd

COUNTRY	ESTIMATED PROVED RESERVES				OIL PRODUCTION			
	Jan. 1, 2003		Jan 1., 2002					
	Oil, 1,000 bbl	Gas, bcf	Oil, 1,000 bbl	Gas, bcf	Producing oil wells, Dec. 31, 2001	Estimated 2002, 1,000 b/d	Change from 2001, %	Actual 2001, 1,000 b/d
Algeria	9,200,000	159,700	9,200,000	159,700	1,312	850.0	1.7	835.8
Angola	5,412,000	1,620	5,412,000	1,620	561	900.0	29.3	696.0
Cameroon	400,000	3,900	400,000	3,900	255	69.0	—13.8	80.0
Congo (Braz.)	1,505,910	3,200	1,505,913	3,200	445	250.0	—5.7	265.0
Congo (Zaire)	187,000	35	187,000	35	150	23.0	—4.2	24.0
Egypt	3,700,000	58,500	2,947,560	35,180	1,258	750.0	—1.3	760.0
Eq. Guinea	12,000	1,300	12,000	1,300	38	135.0	—26.5	183.6
Gabon	2,499,000	1,200	2,499,000	1,200	375	294.0	—2.3	300.8
Ivory Coast	100,000	1,050	100,000	1,050	9	5.0	—3.2	5.2
Libya	29,500,000	46,400	29,500,000	46,400	1,470	1,300.0	—4.8	1,365.0
Morocco	1,600	43	1,800	47	8	3.0	--	3.0
Nigeria	24,000,000	124,000	24,000,000	124,000	2,586	1,930.0	—7.4	2,083.3
S. Africa	15,680	1	15,680	1	22	21.5	—6.1	22.9
Sudan	563,000	3,000	563,000	3,000	9	210.0	5.0	200.0
Tunisia	307,560	2,750	307,560	2,750	211	71.0	2.0	69.6
Africa	77,428,898	418,162	76,676,661	394,846	8,270	6,817.5	—1.2	6,901.3
Azerbaijan	7,000,000	30,000	1,178,000	4,400	2,102	300.0	0.5	298.5
Belarus	198,000	100	198,000	100	--	35.0	--	35.0
Georgia	35,000	300	35,000	300	281	2.0	--	2.0
Kazakhstan	9,000,000	65,000	5,417,000	65,000	11,676	800.0	13.1	707.5
Kyrgyzstan	40,000	200	40,000	200	--	2.0	--	2.0
Russia	60,000,000	1,680,000	48,573,000	1,680,000	41,192	7,385.0	8.9	6,780.8
Tajikistan	12,000	200	12,000	200	--	--	--	--
Turkmenistan	546,000	71,000	546,000	101,000	2,460	180.0	12.5	160.0
Ukraine	395,000	39,600	395,000	39,600	1,353	78.0	—2.5	80.0
Uzbekistan	594,000	66,200	594,000	66,200	2,190	150.0	3.4	145.0
FSU	77,820,000	1,952,600	56,988,000	1,957,000	61,254	8,932 .0	8.8	8,210.8
Australia	3,500,000	90,000	3,500,000	90,000	1,417	633.0	0.1	632.6
Bangladesh	56,900	10,615	56,902	10,615	41	5.0	51.5	3.3
Brunei	1,350,000	13,800	1,350,000	13,800	779	185.0	2.5	180.5
China	18,250,000	53,325	24,000,000	48,300	72,255	3,400.0	3.1	3,296.7
India	5,367,173	26,943	4,840,150	22,865	3,300	663.0	3.0	643.8
Indonesia	5,000,000	92,500	5,000,000	92,500	8,373	1,120.0	—7.8	1,214.2
Japan	58,500	1,400	58,577	1,414	157	12.0	—8.4	13.1
Malaysia	3,000,000	75,000	3,000,000	75,000	788	760.0	2.1	744.2
Myanmar	50,000	10,000	50,000	10,000	450	10.0	25.0	8.0
New Zealand	189,700	3,086	89,533	2,083	73	34.0	--	34.0
Pakistan	310,443	26,365	298,237	25,078	250	60.0	0.4	59.7
Pap. N.Guin.	240,000	12,230	238,345	12,230	39	46.0	—19.3	57.0
Philippines	152,000	3,772	178,060	3,693	8	14.0	98.4	7.1
Taiwan	4,000	2,700	4,000	2,700	73	0.8	4.4	0.8
Thailand	583,350	13,341	515,690	12,705	749	130.0	14.9	113.1
Viet Nam	600,000	6,800	600,000	6,800	28	304.0	—0.3	304.8
Asia-Pacific	38,712,066	445,407	43,779,494	433,312	88,780	7,376.8	0.9	7,312.7
Total World	1,212,880,852	5,501,424	1,031,553,477	5,451,065	827,469	66,042.7	—1.1	66,746.7
Total OPEC	819,007,000	2,490,740	818,842,000	2,485,125	36,742	25,235	—6.7	27,034

Historical Oil Reserves

COUNTRY	January 1												
	1970	1980	1990	1994	1995	1996	1997	1998	1999	2000	2001	2002	2003
	billion bbl												
North America													
Canada	10.7	6.4	6.1	5.1	5.0	4.9	4.9	4.8	4.9	4.9	4.7	4.9	180.0
US	37.0	26.4	25.9	23.7	22.9	22.5	22.4	22.0	22.5	21.0	21.8	22.0	22.4
Latin America													
Argentina	4.5	2.5	2.3	1.6	2.2	2.2	2.4	2.6	2.6	2.8	3.0	3.0	2.9
Brazil	--	1.3	2.8	3.6	3.8	4.2	4.8	4.8	7.1	7.4	8.1	8.5	8.3
Colombia	1.7	1.0	2.0	1.9	3.4	3.5	2.8	2.8	2.6	2.6	2.0	1.8	1.8
Ecuador	--	1.1	1.5	2.0	2.0	2.1	2.1	2.1	2.1	2.1	2.1	2.1	4.6
Mexico	3.2	44.0	56.4	50.9	50.8	49.8	48.8	40.0	47.8	28.4	28.3	26.9	12.6
Venezuela	1.4	17.9	58.5	63.3	64.5	64.5	64.9	71.7	72.6	72.6	76.9	77.7	77.8
Europe													
Norway	1.0	5.5	11.5	9.3	9.4	8.4	11.2	10.4	10.9	10.8	4.4	9.4	10.3
UK	1.0	14.8	4.2	4.6	4.5	4.3	4.5	5.0	5.2	5.2	5.0	4.9	4.7
Middle East													
Iran	70.0	57.5	92.8	92.9	89.3	88.2	93.0	93.0	89.7	89.7	89.7	89.7	89.7
Iraq	32.0	30.0	100.0	100.0	100.0	100.0	112.0	112.5	112.5	112.5	112.5	112.5	112.5
Kuwait	67.1	64.9	94.5	94.0	94.0	94.0	94.0	94.0	94.0	94.0	94.0	94.0	94.0
Neutral Zone	25.7	6.0	5.2	5.0	5.0	5.0	5.0	5.0	5.0	5.0	5.0	5.0	5.0
Oman	--	2.3	4.2	4.7	4.8	5.1	5.1	5.2	5.3	5.3	5.5	5.5	5.5
Qatar	3.5	3.6	4.5	3.7	3.7	3.7	3.7	3.7	3.7	3.7	13.2	15.2	15.2
Saudi Arabia	128.5	165.0	254.9	258.7	258.7	258.7	259.0	259.0	259.0	261.0	259.2	259.3	259.3
Syria	1.0	1.3	1.7	1.7	2.5	2.5	2.5	2.5	2.5	2.5	2.5	2.5	2.5
UAE	11.8	30.4	96.2	96.2	96.5	96.5	96.2	96.2	96.2	96.2	96.2	97.8	97.8
Yemen	--	--	3.0	4.0	4.0	4.0	4.0	4.0	4.0	4.0	4.0	4.0	4.0
Africa													
Algeria	7.0	8.2	9.2	9.2	9.2	9.2	9.2	9.2	9.2	9.2	9.2	9.2	9.2
Angola	--	1.2	2.0	1.5	5.4	5.4	5.4	5.4	5.4	5.4	5.4	5.4	5.4
Egypt	4.5	2.9	4.5	6.3	3.3	3.9	3.7	3.8	3.5	2.9	2.9	2.9	3.7
Gabon	--	--	0.7	0.7	1.3	1.3	1.3	2.5	2.5	2.5	2.5	2.5	2.5
Libya	29.2	23.0	22.8	22.8	22.8	29.5	29.5	29.5	29.5	29.5	29.5	29.5	29.5
Nigeria	9.3	16.7	16.0	17.9	17.9	20.8	15.5	16.8	22.5	22.5	22.5	24.0	24.0
FSU													
Azerbaijan	--	--	--	--	--	--	--	--	--	--	1.2	1.2	7.0
Kazakhstan	--	--	--	--	--	--	--	--	--	--	5.4	5.4	9.0
Russia	--	--	--	--	--	--	--	--	--	--	48.6	48.6	60.0
FSU*	77.0	63.0	24.0	57.0	57.0	57.0	59.1	59.1	59.1	59.0	--	--	--
Asia-Pacific													
Australia	2.0	2.4	1.7	1.6	1.6	1.6	1.8	1.8	2.9	2.9	2.9	3.5	3.5
Brunei	1.0	1.7	1.4	1.4	1.4	1.4	1.4	1.4	1.4	1.4	1.4	1.4	1.4
China	--	--	24.0	24.0	24.0	24.0	24.0	24.0	24.0	24.0	24.0	24.0	18.3
India	--	2.5	7.5	5.9	5.8	5.8	4.3	4.3	3.9	4.8	4.7	4.8	5.4
Indonesia	10.0	9.5	8.2	5.8	5.8	5.8	4.9	4.9	4.9	4.9	4.9	5.0	5.0
Malaysia	--	3.0	3.0	4.3	4.3	4.3	4.0	3.9	3.9	3.9	3.9	3.0	3.0
Total World	611.4	648.5	1001.6	999.1	999.8	1007.5	1018.8	1019.5	1034.3	1016.0	1031.6	1031.6	1212.9

* USSR before 1992; Former Soviet Union plus Eastern Europe 1993-2000.

NOTE: Some totals include countries with low reserves that are not listed.
Canada's large increase was due to including its oil sands.

World Oil Balance

	Avg. 1998	Avg. 1999	Avg. 2000	2001 1Q01	2Q01	3Q01	4Q01	Avg. 2001	2002 1Q02	2Q02	3Q02	4Q02	Avg. 2002
	million b/d												
DEMAND													
OECD													
North America	23.1	23.9	24.1	24.2	23.7	23.8	23.6	23.8	23.6	23.9	24.2	N/A	N/A
Europe	15.3	15.2	15.2	15.2	14.8	15.5	15.6	15.3	15.2	14.7	15.2	N/A	N/A
Asia-Pacific	8.4	8.7	8.7	9.4	8.0	8.1	8.8	8.6	9.1	7.7	8.1	N/A	N/A
Total OECD	46.9	47.7	47.9	48.9	46.5	47.4	48.0	47.7	47.9	46.2	47.4	N/A	N/A
Non-OECD													
FSU	3.8	3.8	3.7	3.8	3.6	3.6	3.6	3.6	3.8	3.6	3.6	N/A	N/A
Europe	0.7	0.7	0.6	0.6	0.6	0.6	0.6	0.6	0.6	0.6	0.6	N/A	N/A
China	4.1	4.4	4.8	4.9	4.9	4.8	4.9	4.9	5.1	5.0	5.0	N/A	N/A
Other Asia	6.8	7.2	7.3	7.4	7.4	7.1	7.4	7.3	7.4	7.4	7.2	N/A	N/A
Other Non-OECD	11.5	11.6	11.7	11.7	11.9	12.0	11.8	11.9	11.7	12.0	12.0	N/A	N/A
Total Non-OECD	26.9	27.6	28.1	28.4	28.4	28.1	28.3	28.3	28.6	28.7	28.4	N/A	N/A
Total Demand	73.8	75.3	76.0	77.3	74.9	75.6	76.3	76.0	76.5	74.8	75.8	N/A	N/A
SUPPLY													
OECD													
North America	15.5	15.0	15.3	15.2	15.3	15.3	15.6	15.4	15.7	15.7	15.5	15.6	15.6
North Sea	6.2	6.3	6.2	6.4	6.1	6.1	6.5	6.3	6.3	6.3	5.8	6.3	6.2
Other OECD	1.7	1.6	1.7	1.7	1.6	1.6	1.6	1.6	1.6	1.6	1.6	1.6	1.6
Total OECD	23.3	22.9	23.2	23.2	23.0	23.1	23.7	23.3	23.7	23.6	23.0	23.6	23.5
Non-OECD													
OPEC	30.5	29.3	30.9	31.1	29.9	30.2	29.2	30.1	27.9	27.4	28.3	29.1	28.1
FSU	7.3	7.7	8.2	8.6	8.7	8.9	9.1	8.8	9.0	9.2	9.6	9.8	9.4
China	3.2	3.2	3.3	3.3	3.3	3.3	3.3	3.3	3.3	3.4	3.4	3.4	3.4
Other Non-OECD	10.9	11.2	11.3	11.3	11.1	11.3	11.3	11.3	11.5	11.6	11.4	11.5	11.5
Total Non-OECD	51.8	51.3	53.6	54.3	53.0	53.7	52.9	53.5	51.7	51.4	52.7	53.7	52.4
Total Supply	75.1	74.2	76.7	77.5	76.0	76.8	76.6	76.7	75.4	75.0	75.6	77.3	75.8
Stock Change	1.3	—1.1	0.7	0.2	1.1	1.2	0.3	0.7	—1.1	0.2	—0.2	N/A	N/A

Source: U.S. Energy Information Administration
NOTES: North America includes Mexico in addition to the U.S. and Canada
OECD Europe excludes Slovakia
OECD Asia-Pacific includes Australia, Japan, New Zealand, and South Korea
Some columns do not add up due to rounding.

OIL REFINING

Summary of Operating Refineries Worldwide

COUNTRY	No. refineries	Charge capacity, bbl per calendar day (b/cd)			COUNTRY	No. refineries	Charge capacity, bbl per calendar day (b/cd)		
		Crude	Catalytic cracking	Catalytic reforming			Crude	Catalytic cracking	Catalytic reforming
Albania	2	26,300	0	3,500	Kuwait	3	889,200	41,400	13,500
Algeria	4	450,000	0	88,900	Kyrgyzstan	1	10,000	0	0
Angola	1	39,000	0	1,900	Libya	3	343,400	0	20,250
Argentina	10	639,075	168,610	58,800	Malaysia	6	516,000	0	77,400
Australia	9	848,250	234,060	198,660	Mexico	6	1,684,000	375,000	284,300
Austria	1	208,600	26,250	32,725	Morocco	2	154,901	5,040	24,359
Azerbaijan	2	441,808	57,750	24,466	Myanmar	2	57,000	0	0
Bahrain	1	248,900	41,400	14,175	Netherlands	6	1,206,842	113,600	150,223
Bangladesh	1	33,000	0	1,800	New Zealand	1	106,000	0	27,672
Belgium	5	791,013	113,500	98,300	Nigeria	4	438,750	82,700	70,070
Bolivia	3	63,000	0	14,560	Norway	2	310,000	54,000	37,600
Brazil	13	1,865,140	483,953	24,386	Oman	1	85,000	0	16,000
Brunei	1	8,600	0	5,700	Pakistan	4	233,850	0	26,215
Bulgaria	1	115,240	23,300	4,060	Peru	6	190,950	24,700	0
Cameroon	1	42,000	0	7,000	Philippines	4	419,500	24,500	61,900
Canada	21	1,983,450	494,455	348,900	Poland	5	350,000	46,000	39,000
Chile	3	204,849	52,257	16,497	Portugal	2	304,172	31,500	50,182
China	95	4,528,100	892,000	157,000	Qatar	1	200,000	60,000	29,400
Colombia	5	285,850	90,000	0	Romania	10	501,182	103,478	63,063
Congo (Bz.)	1	21,000	0	2,000	Russia	42	5,435,480	330,817	775,201
Congo (Zr.)	1	15,000	0	3,500	Saudi Arabia	8	1,745,000	103,600	193,360
Croatia	3	260,337	51,000	49,514	Singapore	3	1,258,500	65,000	144,500
Cuba	4	301,400	14,700	20,000	Slovakia	1	115,000	18,000	21,000
Czech Rep.	4	198,000	53,740	28,140	S.Africa	4	489,547	107,260	78,892
Denmark	2	176,400	0	21,500	S.Korea	6	2,560,100	168,000	230,970
Ecuador	3	176,000	18,000	12,800	Spain	9	1,321,500	184,150	194,660
Egypt	9	726,250	0	40,540	Sudan	3	121,700	0	1,900
Finland	2	251,800	51,000	44,430	Sweden	5	423,500	29,700	69,900
France	13	1,903,493	372,510	270,850	Switzerland	2	132,000	0	28,000
Gabon	1	17,300	0	1,400	Syria	2	239,865	0	31,242
Georgia	1	106,436	0	10,276	Taiwan	4	920,000	123,000	115,000
Germany	17	2,267,100	343,865	397,227	Thailand	4	703,100	77,530	91,911
Greece	4	406,500	72,300	54,700	Trinidad-Tob.	1	160,000	28,000	19,000
Guatemala	1	16,000	0	3,000	Tunisia	1	34,000	0	3,300
Hungary	2	161,000	24,000	29,600	Turkey	6	719,275	28,935	64,762
India	17	2,134,625	167,305	41,673	Turkmenistan	2	236,970	15,151	52,540
Indonesia	8	992,745	101,450	92,970	Ukraine	6	1,024,759	70,100	144,711
Iran	9	1,474,000	30,000	167,555	UAE	5	514,250	34,350	25,875
Iraq	8	417,500	0	43,500	UK	11	1,788,500	438,050	329,200
Ireland	1	71,250	0	10,800	US	133	16,623,301	5,677,355	3,512,237
Israel	2	220,000	49,500	26,500	Uzbekistan	3	222,271	0	23,487
Italy	17	2,300,800	307,100	284,150	Venezuela	5	1,282,100	231,800	49,500
Ivory Coast	1	65,200	0	12,827	Yemen	2	130,000	0	14,500
Japan	34	4,766,940	867,550	733,950	Total	722	81,877,646	1,4195,514	11,185,852
Kazakhstan	3	427,093	38,356	59,452					

Source: Oil & Gas Journal
NOTE: Some totals include countries with low capacity that are not listed.

World Refining Capacity

First line = 1,000 bbl per calendar day (b/cd); second line = million tonnes per year (tpy)

COUNTRY	1950	1960	1970	1980	1990	1997	1998	1999	2000	2001	2002	2003
North America												
Canada	329	920	1400	2222	1852	1852	1851	1873	1912	1906	1944	1983
	16.4	45.8	69.7	110.7	92.2	92.2	92.2	93.3	95.2	94.9	96.8	98.8
US	6696	10250	12600	17720	16244	15432	15898	16423	16541	16539	16564	16623
	333.5	510.5	627.5	882.5	809.0	768.5	791.7	817.9	823.7	823.6	824.9	827.8
Total	7025	11170	14000	19942	18096	17284	17749	18296	18453	18445	18508	18606
	349.9	556.3	697.2	993.2	901.2	860.7	883.9	911.2	918.9	918.5	921.7	926.6
Latin America												
Argentina	152	238	457	676	689	665	666	653	662	639	639	639
	7.6	11.9	22.8	33.7	34.3	33.1	33.2	32.5	33.0	31.8	31.8	31.8
Bolivia	7	12	12	74	58	48	48	48	48	63	63	63
	0.3	0.6	0.6	3.7	2.9	2.4	2.4	2.4	2.4	3.1	3.1	3.1
Brazil	12	156	502	1205	1397	1256	1662	1772	1783	1918	1756	1865
	0.6	7.8	25.0	60.0	69.6	62.5	82.8	88.2	88.8	95.5	88.9	92.9
Chile	4	24	91	139	147	192	205	205	205	205	205	205
	0.2	1.2	4.5	6.9	7.3	9.6	10.2	10.2	10.2	10.2	10.2	10.2
Colombia	24	51	137	193	227	249	249	249	286	286	286	286
	1.2	2.5	6.8	9.6	11.3	12.4	12.4	12.4	14.2	14.2	14.2	14.2
Cuba	7	87	93	160	160	301	301	301	301	301	301	301
	0.3	4.3	4.6	8.0	8.0	15.0	15.0	15.0	15.0	15.0	15.0	15.0
Ecuador	5	7	33	86	145	148	168	176	176	176	176	176
	0.2	0.3	1.6	4.3	7.2	7.4	8.4	8.8	8.8	8.8	8.8	8.8
Guatemala	--	--	21	16	16	20	20	20	20	22	16	16
	--	--	1.0	0.8	0.8	1.0	1.0	1.0	1.0	1.1	0.8	0.8
Mexico	160	357	495	1394	1514	1520	1520	1525	1525	1525	1525	1684
	8.0	17.8	24.7	69.4	75.4	75.7	75.7	75.9	75.9	75.9	75.9	83.9
Peru	35	49	92	170	172	182	182	182	182	182	182	191
	1.7	2.4	4.6	8.5	8.6	9.1	9.1	9.1	9.1	9.1	9.1	9.5
Trinidad-Tob.	104	182	430	456	300	245	245	160	160	160	160	160
	5.2	9.1	21.4	22.7	14.9	12.2	12.2	8.0	8.0	8.0	8.0	8.0
Venezuela	254	886	1313	1446	1201	1177	1177	1187	1239	1282	1282	1282
	12.6	44.1	65.4	72.0	59.8	58.6	58.6	59.1	61.7	63.8	63.8	63.8
Total	1397	2841	5114	8620	7315	7452	7916	7795	7973	8202	7942	6868
	69.4	141.5	254.5	429.2	364.0	371.1	394.2	388.1	396.9	408.0	395.2	342.0

Source: Oil & Gas Journal
NOTE: Some totals include countries with low capacity that are not listed.

World Refining Capacity-cont'd

First line = 1,000 bbl per calendar day (b/cd); second line = million tonnes per year (tpy)

COUNTRY	1950	1960	1970	1980	1990	1997	1998	1999	2000	2001	2002	2003
Europe												
Albania	--	--	--	--	--	40	26	26	26	26	26	26
	--	--	--	--	--	2.0	1.3	1.3	1.3	1.3	1.3	1.3
Austria	26	47	94	280	204	210	210	209	209	209	209	209
	1.3	2.3	4.7	13.9	10.2	10.5	10.5	10.4	10.4	10.4	10.4	10.4
Bulgaria	--	--	--	--	--	300	325	133	115	115	115	115
	--	--	--	--	--	14.9	16.2	6.6	5.7	5.7	5.7	5.7
Croatia	--	--	--	--	--	294	294	236	236	253	293	260
	--	--	--	--	--	14.6	14.6	11.8	11.8	12.6	14.6	12.9
Czech Repub.	16	46	220	455	455	187	187	198	186	198	198	198
	0.8	2.3	11.0	22.7	22.7	9.3	9.3	9.9	9.3	9.9	9.9	9.9
Denmark	1	1	186	214	187	189	132	135	135	176	176	176
	--	--	9.3	10.7	9.3	9.4	6.6	6.7	6.7	8.8	8.8	8.8
Finland	--	25	174	336	241	200	200	200	200	200	239	252
	--	1.2	8.7	16.7	12.0	10.0	10.0	10.0	10.0	10.0	11.9	12.4
France	306	769	2317	3385	1820	1786	1865	1947	1902	1895	1896	1903
	15.2	38.3	115.4	168.6	90.6	88.9	92.9	97.0	94.7	94.4	94.4	94.8
Germany	81	573	2359	2986	1507	2108	2184	2246	2275	2259	2259	2267
	4.0	28.5	117.5	148.7	75.0	105.0	108.8	111.9	113.3	112.5	112.5	112.9
Greece	--	30	102	431	385	395	391	387	383	407	407	407
	--	1.5	5.1	21.5	19.2	19.7	19.5	19.3	19.1	20.2	20.2	20.2
Hungary	20	58	130	290	220	232	232	232	232	232	161	161
	1.0	2.9	6.5	14.4	11.0	11.6	11.6	11.6	11.6	11.6	8.0	8.0
Ireland	--	40	55	56	56	65	62	66	71	71	71	71
	--	2.0	2.7	2.8	2.8	3.2	3.1	3.3	3.5	3.5	3.5	3.5
Italy	107	73	2963	4131	2804	2262	2453	2446	2341	2359	2283	2300
	5.3	38.5	147.6	205.7	139.6	112.6	122.2	121.8	116.6	117.5	113.7	114.5
Netherlands	77	359	1362	1828	1381	1187	1188	1188	1188	1204	1206	1207
	3.8	17.9	67.8	91.0	68.8	59.1	59.2	59.2	59.2	60.0	60.1	60.1
Norway	1	2	123	264	295	307	307	312	358	305	310	310
	--	0.1	6.1	13.1	14.7	15.3	15.3	15.5	17.8	15.2	15.4	15.4
Poland	4	20	156	385	385	352	365	382	382	382	382	350
	0.2	1.0	7.8	19.2	19.2	17.5	18.2	19.0	19.0	19.0	19.0	17.4
Portugal	8	23	37	378	313	304	304	304	304	304	304	304
	0.4	1.1	1.8	18.8	15.6	15.1	15.1	15.1	15.1	15.1	15.1	15.1
Romania	182	260	356	608	617	559	542	522	499	504	504	501
	9.1	12.9	17.7	30.3	30.7	27.8	27.0	26.0	24.9	25.1	25.1	24.9
Slovakia	--	--	--	--	--	115	115	115	115	115	115	115
	--	--	--	--	--	5.7	5.7	5.7	5.7	5.7	5.7	5.7
Spain	17	146	685	1455	1293	1296	1294	1315	1316	1294	1294	1322
	0.8	7.3	34.1	72.5	64.4	64.5	64.4	65.5	65.5	64.4	64.4	65.8
Sweden	26	48	230	458	428	427	427	427	427	423	424	424
	1.3	2.4	11.5	22.8	21.3	21.3	21.3	21.3	21.3	21.1	21.1	21.1
Switzerland	--	--	106	137	132	132	132	132	132	132	132	132
	--	-	5.3	6.8	6.6	6.6	6.6	6.6	6.6	6.6	6.6	6.6
Turkey	1	7	149	356	725	683	708	688	691	694	719	719
	0	0.3	7.4	17.7	36.1	34.0	35.3	34.3	34.4	34.6	35.8	35.8
UK	197	963	2301	2527	1831	1941	1826	1854	1785	1771	1784	1789
	9.8	48.0	114.6	125.8	91.2	69.7	90.9	92.3	88.9	88.2	88.8	89.1
Total	1070	3490	14105	20960	15279	15571	15769	15700	15508	15528	15507	15518
	53.0	208.5	702.6	245.6	761	748.3	785.6	782.1	772.4	773.4	772.1	772.8

World Refining Capacity-cont'd

First line = 1,000 bbl per calendar day (b/cd); second line = million tonnes per year (tpy)

COUNTRY	1950	1960	1970	1980	1990	1997	1998	1999	2000	2001	2002	2003
Middle East												
Bahrain	150	187	205	250	243	249	249	249	249	249	249	249
	7.5	9.3	10.2	12.4	12.1	12.4	12.4	12.4	12.4	12.4	12.4	12.4
Iran	502	495	645	921	530	1242	1358	1448	1474	1484	1484	1474
	25.0	24.7	32.1	45.9	26.4	61.9	67.6	72.1	73.4	73.9	73.9	73.4
Iraq	9	56	104	168	319	348	348	348	348	418	418	418
	0.4	2.8	5.2	8.4	15.9	17.3	17.3	17.3	17.3	20.8	20.8	20.8
Israel	83	87	107	195	180	220	220	220	220	220	220	220
	4.1	4.3	5.3	9.7	9.0	11.0	11.0	11.0	11.0	11.0	11.0	11.0
Kuwait	25	270	569	645	819	824	886	886	865	764	773	889
	1.2	13.5	28.4	32.1	40.8	41.0	44.1	44.1	43.1	38.0	38.5	44.3
Oman	--	--	--	--	--	--	--	--	--	--	85	85
	--	--	--	--	--	--	--	--	--	--	4.2	4.2
Qatar	--	1	1	11	62	58	58	58	58	58	58	200
	--	--	--	0.5	3.1	2.9	2.9	2.9	2.9	2.9	2.9	10.0
Saudi Arabia	140	189	377	487	1484	1656	1651	1685	1710	1745	1745	1745
	7.0	9.4	18.8	24.3	73.9	82.5	82.2	83.9	85.2	86.9	86.9	86.9
Syria	--	--	59	223	244	246	242	242	242	242	242	240
	--	--	2.9	11.1	12.2	12.3	12.1	12.1	12.1	12.1	12.1	12.0
UAE	--	--	--	14	180	213	213	213	429	444	515	514
	--	--	--	0.7	9.0	10.6	10.6	10.6	21.4	22.1	25.6	25.6
Yemen	--	120	178	175	162	120	120	120	120	130	130	130
	--	6.0	8.9	8.7	8.1	6.0	6.0	6.0	6.0	6.5	6.5	6.5
Total	916	1429	2290	3163	4447	5399	5636	5757	5928	5965	5532	6164
	45.5	71.2	114.0	151.4	221.6	269.0	280.7	286.8	295.4	297.2	275.6	307.0

World Refining Capacity-cont'd

First line = 1,000 bbl per calendar day (b/cd); second line = million tonnes per year (tpy)

COUNTRY	1950	1960	1970	1980	1990	1997	1998	1999	2000	2001	2002	2003
Africa												
Algeria	--	--	48	122	465	465	483	503	503	503	450	450
	--	--	2.4	6.1	23.2	23.2	24.1	25.0	25.0	25.0	22.4	22.4
Angola	--	2	14	36	32	32	32	39	39	39	39	39
	--	0.1	0.7	1.8	1.6	1.6	1.6	1.9	1.9	1.9	1.9	1.9
Cameroon	--	--	--	--	43	42	45	35	35	42	42	42
	--	--	--	--	2.1	2.1	2.2	1.7	1.7	2.1	2.1	2.1
Congo (Braz.)	--	--	14	--	21	21	21	21	21	21	21	21
	--	--	0.7	--	1.0	1.0	1.0	1.0	1.0	1.0	1.0	1.0
Congo (Zaire)	--	--	21	16	17	17	17	17	17	15	15	15
	--	--	1.0	0.8	0.8	0.8	0.8	0.8	0.8	0.7	0.7	0.7
Egypt	39	87	175	234	489	546	546	578	578	726	726	726
	1.9	4.3	8.7	11.7	24.4	27.2	27.2	28.8	28.8	36.2	36.2	36.2
Gabon	--	--	13	20	24	17	17	17	17	17	17	17
	--	--	0.6	1.0	1.2	0.8	0.8	0.8	0.8	0.8	0.8	0.8
Ivory Coast	--	--	19	50	69	64	73	69	69	65	65	65
	--	--	0.9	2.5	3.4	3.2	3.6	3.4	3.4	3.2	3.2	3.2
Libya	--	--	10	138	329	348	348	348	348	343	343	343
	--	-	0.5	6.9	16.4	17.3	17.3	17.3	17.3	17.1	17.1	17.1
Morocco	--	2	35	72	155	156	152	157	157	157	155	155
	--	0.1	1.7	3.6	7.7	7.8	7.6	7.8	7.8	7.8	7.7	7.7
Nigeria	--	--	46	160	433	433	439	439	439	439	439	439
	--	--	2.3	8.0	21.6	21.6	21.9	21.9	21.9	21.9	21.9	21.9
S. Africa	--	25	184	478	434	414	465	468	467	474	469	490
	--	1.2	9.2	23.8	21.6	20.6	23.2	23.3	23.3	23.6	23.3	24.4
Sudan	--	--	21	26	25	22	22	22	82	122	122	122
	--	--	1.0	1.3	1.2	1.1	1.1	1.1	4.1	6.1	6.1	6.1
Tunisia	--	--	23	34	34	34	34	34	34	34	34	34
	--	--	1.1	1.7	1.7	1.7	1.7	1.7	1.7	1.7	1.7	1.7
Total	39	116	787	1665	2824	2849	2932	3007	3046	3265	3169	2958
	1.9	5.7	38.8	82.9	140.4	141.6	145.7	149.2	151.2	162.3	157.1	147.3

World Refining Capacity-cont'd

First line = 1,000 bbl per calendar day (b/cd); second line = million tonnes per year (tpy)

COUNTRY	1950	1960	1970	1980	1990	1997	1998	1999	2000	2001	2002	2003
Former Soviet Union												
Azerbaijan	--	--	--	--	--	--	-	--	--	--	442	442
	--	--	--	--	--	--	-	--	--	--	22.0	22.0
Belarus	--	--	--	--	--	--	-	--	--	--	493	493
	--	--	--	--	--	--	-	--	--	--	24.6	24.6
Georgia	--	--	--	--	--	--	-	--	--	--	106	106
	--	--	--	--	--	--	-	--	--	--	5.3	5.3
FSU*	900	2800	5640	10950	12300	10100	10171	9749	9762	8400	--	--
	44.8	139.4	280.9	545.3	612.5	503.0	506.5	485.5	486.1	418.3	--	--
Kazakhstan	--	--	--	--	--	--	-	--	--	--	427	427
	--	--	--	--	--	--	-	--	--	--	21.3	21.3
Kyrgyzstan	--	--	--	--	--	--	-	--	--	--	10	10
	--	--	--	--	--	--	-	--	--	--	0.5	0.5
Russia	--	--	--	--	--	--	-	--	--	--	5435	5435
	--	--	--	--	--	--	-	--	--	--	270.7	270.7
Turkmenistan	--	--	--	--	--	--	-	--	--	--	237	237
	--	--	--	--	--	--	-	--	--	--	11.8	11.8
Ukraine	--	--	--	--	--	--	-	--	--	--	1206	1025
	--	--	--	--	--	--	-	--	--	--	60.1	51.0
Uzbekistan	--	--	--	--	--	--	-	--	--	--	222	222
	--	--	--	--	--	--	-	--	--	--	11.1	11.1
Total	900	2800	5640	10950	12300	10100	10171	9749	9762	8400	8578	8397
	44.8	139.4	280.9	545.3	612.5	503.0	506.5	485.5	486.1	418.3	427.4	418.2

* Total of all Former Soviet Union countries before 2002.

World Refining Capacity-cont'd

First line = 1,000 bbl per calendar day (b/cd); second line = million tonnes per year (tpy)

COUNTRY	1950	1960	1970	1980	1990	1997	1998	1999	2000	2001	2002	2003
Asia-Pacific												
Australia	14	237	594	725	675	771	762	807	812	847	848	848
	0.7	11.8	29.6	36.1	33.6	38.4	37.9	40.2	40.4	42.2	42.2	42.2
Bangladesh	--	--	--	31	31	31	31	33	33	33	33	33
	--	--	--	1.5	1.5	1.5	1.5	1.6	1.6	1.6	1.6	1.6
Brunei**	--	--	--	--	--	--	--	--	--	--	9	9
	--	--	--	--	--	--	--	--	--	--	0.4	0.4
China	8	98	300	1600	2200	2867	2967	4347	4347	4347	4528	4528
	0.4	4.9	14.9	79.7	109.6	142.8	147.8	216.5	216.5	216.5	225.5	225.5
India	6	112	446	557	1080	1086	1086	1141	1858	2113	2135	2135
	0.3	5.6	22.2	27.7	53.8	54.1	54.1	56.8	92.5	105.2	106.3	106.3
Indonesia	203	274	268	528	714	805	930	930	993	993	993	993
	10.1	13.6	13.3	26.3	35.6	40.1	46.3	46.3	49.5	49.5	49.5	49.5
Japan	34	640	3140	5509	4198	4989	4966	5059	4998	4962	4786	4767
	1.7	31.9	156.4	274.3	209.1	248.5	247.3	251.9	248.9	247.1	238.3	237.4
Malaysia**	--	--	--	--	--	--	--	--	--	--	515	516
	--	--	--	--	--	--	--	--	--	--	25.6	25.7
Myanmar	--	9	26	26	32	32	32	32	32	32	32	57
	--	0.4	1.3	1.3	1.6	1.6	1.6	1.6	1.6	1.6	1.6	2.8
New Zealand	--	--	66	74	95	91	91	98	98	106	106	106
	--	--	3.3	3.7	4.7	4.5	4.5	4.9	4.9	5.3	5.3	5.3
Pakistan	5	6	107	98	121	137	137	139	143	239	239	234
	0.2	0.3	5.3	4.9	6.0	6.8	6.8	6.9	7.1	11.9	11.9	11.7
Philippines	--	22	182	253	284	323	323	389	401	420	420	420
	--	1.1	9.1	12.6	14.1	16.1	16.1	19.4	20.0	20.9	20.9	20.9
Singapore**	10	46	320	1093	1049	1496	1536	1654	1788	1792	1259	1259
	0.5	2.3	15.9	54.4	52.2	74.5	76.5	82.4	89.0	89.3	62.7	62.7
S. Korea	--	--	180	601	867	2210	2540	2540	2540	2560	2560	2560
	--	--	9.0	29.9	43.2	110.1	126.5	126.5	126.5	127.5	127.5	127.5
Taiwan	--	28	119	425	570	770	770	770	770	920	920	920
	--	1.4	5.9	21.2	28.4	38.3	38.3	38.3	38.3	45.8	45.8	45.8
Thailand	--	1	62	186	215	558	704	713	713	682	682	703
	--	--	3.1	9.3	10.7	27.8	35.1	35.5	35.5	34.0	34.0	35.0
Total	284	1472	5866	11986	12181	16285	16994	18771	19645	20166	20185	20088
	14.1	73.3	292.1	596.9	606.6	811.0	846.2	934.7	978.2	1004.3	1005.1	1000.4
Total World	11658	24239	48982	79514	74535	76066	78317	80310	81556	81252	81166	81878
	580.0	1206.9	2438.8	3959.5	3711.5	3787.8	3900.0	3999.0	4060.8	4045.7	4041.4	4077.5

** Brunei and Malaysia were totaled with Singapore before 2002.

World Refinery Throughput

REGION	1996	1997	1998	1999	2000	2001	Change 2001 over 2000	2001 share of total
	1000 bbl per calendar day (b/cd)						%	
US	14195	14662	14889	14804	15067	15130	0.4	21.6
Canada	1644	1694	1709	1714	1765	1823	3.3	2.6
Mexico	1491	1438	1451	1389	1364	1400	2.6	2.0
South & Central America	4994	5238	5443	5405	5401	5535	2.5	7.9
Europe	14697	14859	15336	14838	14989	14849	—0.9	21.2
Former Soviet Union	4624	4735	4482	4468	4558	4900	7.5	7.0
Middle East	5408	5448	5595	5672	5635	5616	—0.3	8.0
Africa	2381	2539	2408	2448	2393	2442	2.0	3.5
Australasia	844	872	865	881	881	885	0.5	1.3
China	2850	3084	3060	3686	4218	4210	—0.2	6.0
Japan	4168	4319	4212	4149	4145	4107	—0.9	5.9
Other Asia Pacific	7544	8183	8010	8309	8914	9061	1.6	13.0
Total World	64840	67072	67460	67763	69330	69958	0.9	100.0

Source: BP Statistical Review of World Energy 2002

World Refining Margins

MARKET	1990	1991	1992	1993	1994	1995	1996	1997	1998	1999	2000	2001	Dec. 2002
	$/bbl*												
US Gulf Coast													
W.TX sour	4.03	3.69	2.47	2.33	2.20	2.19	3.37	3.83	3.31	2.15	5.54	2.91	3.81
W.TX sr-RFG	--	--	--	--	--	--	--	--	--	--	--	3.70	4.63
Arab. light	5.80	5.91	4.25	4.15	4.40	3.56	3.63	4.57	4.04	3.14	7.27	5.02	5.71
Bonny light	4.11	4.61	3.11	2.73	2.64	2.37	2.41	2.90	2.83	1.74	6.40	0.97	1.54
US PADD II (Chicago)													
W.TX Interm.	6.06	4.50	3.35	3.22	3.70	3.00	3.74	3.78	3.17	2.47	4.60	2.52	4.30
US East Coast (New York Harbor)													
Arab. med.	2.98	3.48	1.71	2.53	3.94	3.40	2.89	3.89	3.62	2.50	6.38	4.15	4.49
E.Comp.-RFG	--	--	--	--	--	--	--	--	--	--	--	2.88	4.58
US West Coast (Los Angeles)													
AK N.Slope	7.56	6.63	6.11	6.98	5.00	4.33	5.13	5.58	5.60	6.55	12.48	4.33	3.57
NW Europe (Rotterdam)													
Brent	4.68	5.16	3.16	3.33	2.59	2.48	2.78	3.26	3.26	2.19	4.71	0.88	2.76
Mediterranean (Italy)													
Urals	2.36	3.58	2.52	2.80	1.33	1.67	1.82	2.24	2.73	1.67	3.28	0.05	—0.62
Far East (Singapore)													
Dubai	4.85	5.92	3.43	3.69	2.52	2.52	2.57	2.50	1.61	0.78	3.89	0.63	1.96

Source: The Pace Consultants, Inc.

NATURAL GAS

World Natural Gas Production

COUNTRY	1996	1997	1998	1999	2000	2001	Change 2001 over 2000	2001 share of total
	billion cubic meters						%	
North America								
Canada	153.6	156.2	160.5	162.2	167.8	172.0	2.5	7.0
US	542.2	543.1	549.2	541.6	544.9	555.4	1.9	22.5
Total	695.8	699.3	709.7	703.8	712.7	727.4	2.1	29.5
Latin America								
Argentina	26.6	27.4	29.6	34.6	37.4	38.4	2.6	1.6
Bolivia	3.2	3.3	3.1	2.5	3.4	4.1	20.5	0.2
Brazil	5.5	6.0	6.3	6.7	6.8	7.7	13.3	0.3
Colombia	4.7	5.9	6.3	5.2	5.9	6.1	3.0	0.2
Mexico	31.2	33.8	36.6	38.5	37.1	34.7	—6.3	1.4
Trinidad & Tobago	7.1	7.4	8.6	10.9	13.0	12.9	—0.5	0.5
Venezuela	29.7	30.8	32.3	27.4	27.9	28.9	3.3	1.2
Other	2.3	2.4	2.5	2.1	2.1	2.0	—2.6	0.1
Total	110.3	117.0	125.3	127.9	133.6	134.8	0.1	5.5
Europe								
Denmark	6.4	7.9	7.6	7.8	8.1	8.4	3.5	0.3
Germany	17.4	17.1	16.7	17.8	16.9	17.0	0.9	0.7
Hungary	4.0	3.7	3.3	2.9	2.7	2.7	1.0	0.1
Italy	20.0	19.3	19.0	17.5	16.2	15.5	—4.6	0.6
Netherlands	75.8	67.1	63.6	59.3	57.3	61.4	7.1	2.5
Norway	41.0	46.7	47.8	51.0	54.0	57.5	6.4	2.3
Romania	17.2	15.0	14.0	14.0	13.8	12.6	—8.5	0.5
UK	84.2	85.9	90.2	99.1	108.3	105.8	—2.3	4.3
Other	13.3	12.8	12.3	11.6	11.8	11.6	—1.4	0.5
Total	279.3	275.5	274.5	281.0	289.1	292.5	1.2	11.9
Middle East								
Bahrain	7.4	8.0	8.4	8.7	8.8	8.9	1.4	0.4
Iran	39.0	47.0	50.0	57.8	60.2	60.6	0.6	2.5
Kuwait	9.3	9.3	9.5	8.6	9.6	9.5	—1.0	0.4
Oman	4.4	5.0	5.2	5.5	8.4	13.4	59.7	0.5
Qatar	13.7	17.4	19.6	22.1	29.1	32.5	11.7	1.3
Saudi Arabia	44.4	45.3	46.8	46.2	49.8	53.7	7.8	2.2
UAE	33.8	36.3	37.1	38.5	39.8	41.3	3.8	1.7
Other	6.0	7.2	7.5	7.9	7.9	8.1	2.0	0.3
Total	158.0	175.5	184.1	195.3	213.6	228.0	6.7	9.3

Source: BP Statistical Review of World Energy 2002
NOTE: Some columns do not add up due to rounding.

World Natural Gas Production-cont'd

COUNTRY	1996	1997	1998	1999	2000	2001	Change 2001 over 2000	2001 share of total
	billion cubic meters						%	
Africa								
Algeria	62.3	71.8	76.6	86.0	84.4	78.2	—7.3	3.2
Egypt	11.5	11.6	12.2	14.7	18.3	21.0	14.4	0.9
Libya	5.8	6.0	5.8	5.5	5.4	5.4	0	0.2
Nigeria	5.4	5.1	5.1	6.0	10.8	13.4	23.6	0.5
Other	4.2	4.9	5.0	5.4	5.6	6.0	5.8	0.2
Total	89.2	99.4	104.7	117.6	124.5	124.0	-0.5	5.0
Former Soviet Union								
Azerbaijan	5.9	5.6	5.2	5.6	5.3	5.2	—2.0	0.2
Kazakhstan	6.1	7.6	7.4	9.3	10.8	10.8	0.3	0.4
Russia	561.1	532.6	551.3	551.0	545.0	542.4	—0.5	22.0
Turkmenistan	32.8	16.1	12.4	21.3	43.8	47.9	9.1	1.9
Ukraine	17.2	17.4	16.8	16.9	16.7	17.1	2.3	0.7
Uzbekistan	45.7	47.8	51.1	51.9	52.6	53.5	1.8	2.2
Other	0.3	0.3	0.4	0.4	0.4	0.4	—8.7	--
Total	669.1	627.4	644.6	656.4	674.6	677.3	0.4	27.5
Asia-Pacific								
Australia	30.6	30.0	30.4	30.6	31.1	32.7	5.3	1.3
Bangladesh	7.6	7.6	7.8	8.3	10.0	10.8	8.2	0.4
Brunei	11.7	11.7	10.8	11.2	11.3	11.4	1.0	0.5
China	19.9	22.2	22.3	24.3	27.2	30.3	11.5	1.2
India	20.4	20.7	24.6	24.9	26.1	26.4	1.1	1.1
Indonesia	67.1	67.6	64.7	71.4	67.3	62.9	—6.4	2.6
Malaysia	33.6	38.6	38.5	40.8	45.3	47.4	4.7	1.9
Pakistan	15.4	15.6	16.0	17.3	18.9	19.9	5.2	0.8
Thailand	11.8	14.1	15.5	16.9	17.9	18.1	1.0	0.7
Other	10.2	10.8	10.6	11.6	18.6	20.1	8.3	0.8
Total	228.3	238.9	241.2	257.3	273.7	280.0	2.4	11.4
Total World	2230.0	2233.0	2284.1	2339.3	2421.8	2464.0	1.7	100.0

World Natural Gas Consumption

COUNTRY	1996	1997	1998	1999	2000	2001	Change 2001 over 2000	2001 share of total
	billion cubic meters						%	
North America								
US	631.7	630.9	614.3	621.7	647.1	616.2	—4.8	25.6
Canada	74.3	74.8	70.3	72.7	77.5	72.6	—6.4	3.0
Total	706.0	705.7	684.6	694.4	724.6	688.8	—4.9	28.6
Latin America								
Argentina	28.6	28.5	30.5	32.4	33.2	33.2	‹	1.4
Brazil	5.5	6.0	6.3	7.1	9.1	10.9	19.3	0.5
Chile	1.7	2.8	3.3	4.6	5.2	5.6	6.5	0.2
Colombia	4.7	5.9	6.2	5.2	5.9	6.1	2.6	0.3
Ecuador	0.1	0.1	0.1	0.1	0.1	0.1	0	--
Mexico	31.0	31.7	34.4	33.8	34.9	33.7	—3.3	1.4
Peru	0.4	0.2	0.4	0.4	0.3	0.4	7.1	--
Venezuela	29.7	30.8	32.3	27.4	27.9	28.9	3.3	1.2
Other	8.2	8.9	9.7	10.3	11.2	11.8	5.2	0.5
Total	109.9	114.9	123.2	121.3	127.8	130.7	2.3	5.4
Europe								
Austria	7.3	7.7	7.6	7.7	7.3	7.4	0.6	0.3
Belgium/Luxembg.	13.1	12.5	13.8	14.7	14.9	14.7	—1.4	0.6
Bulgaria	5.2	4.1	3.4	3.0	2.9	2.6	—9.3	0.1
Czech Republic	8.4	8.5	8.5	8.6	8.3	8.9	7.0	0.4
Denmark	4.1	4.4	4.8	5.0	4.9	5.1	4.4	0.2
Finland	3.3	3.2	3.7	3.7	3.7	4.1	8.5	0.2
France	36.1	34.6	37.0	37.7	39.7	40.7	2.5	1.7
Germany	83.6	79.2	79.7	80.2	79.5	82.9	4.3	3.4
Greece	--	0.2	0.8	1.4	1.9	1.9	1.1	0.1
Hungary	11.4	10.8	10.9	11.0	10.7	11.9	11.1	0.5
Ireland	3.0	3.1	3.1	3.3	3.8	4.0	4.3	0.2
Italy	51.5	53.2	57.2	62.2	64.9	64.5	—0.6	2.7
Netherlands	41.7	39.1	38.7	37.9	39.2	39.3	0.2	1.6
Norway	3.2	3.7	3.8	3.6	4.0	4.5	10.6	0.2
Poland	10.6	10.5	10.6	10.3	11.1	11.4	2.7	0.5
Portugal	--	0.1	0.8	2.3	2.4	2.5	7.6	0.1
Romania	24.2	20.0	18.7	17.2	17.1	17.5	2.3	0.7
Slovakia	6.2	6.3	6.4	6.4	6.5	7.4	14.3	0.3
Spain	9.3	12.3	13.1	15.0	16.9	18.2	7.7	0.8
Sweden	0.9	0.8	0.9	0.8	0.7	0.8	4.8	--
Switzerland	2.6	2.5	2.6	2.7	2.7	2.8	4.2	0.1
Turkey	9.0	9.4	9.9	12.0	14.1	15.5	9.7	0.6
UK	82.1	83.8	87.2	92.5	96.0	95.4	—0.6	4.0
Other	6.0	6.1	5.9	5.3	5.6	6.1	7.8	0.3
Total	422.8	416.1	429.1	444.5	458.8	470.1	2.4	19.5
Middle East								
Iran	38.9	47.1	51.8	59.8	63.0	65.0	3.2	2.7
Kuwait	9.3	9.3	9.5	8.6	9.6	9.5	—1.0	0.4
Qatar	13.7	14.6	14.8	14.0	15.1	16.0	6.0	0.7
Saudi Arabia	44.4	45.3	46.8	46.2	49.8	53.7	7.8	2.2
UAE	27.2	29.0	30.4	31.4	32.9	34.3	4.2	1.4
Other	17.3	19.6	20.5	21.5	22.3	23.0	2.9	1.0
Total	150.8	164.9	173.8	181.5	192.7	201.5	4.5	8.4

Source: BP Statistical Review of World Energy 2002

NOTE: Differences between consumption and production are due to variations in storage and liquefaction plant stocks as well as other disparities.

World Natural Gas Consumption-cont'd

COUNTRY	1996	1997	1998	1999	2000	2001	Change 2001 over 2000	2001 share of total
	billion cubic meters						%	
Africa								
Algeria	21.4	20.2	20.9	21.2	21.0	21.6	2.7	0.9
Egypt	11.3	11.6	12.0	14.3	18.3	21.0	14.4	0.9
Other	14.3	14.4	14.9	14.6	16.2	17.6	8.4	0.7
Total	47.0	46.2	47.8	50.1	55.5	60.2	8.2	2.5
Former Soviet Union								
Azerbaijan	5.9	5.6	5.2	5.6	5.4	8.4	55.2	0.3
Belarus	13.0	14.8	15.0	15.3	16.2	16.1	—0.6	0.7
Kazakhstan	9.0	7.1	7.3	7.9	9.7	10.1	3.8	0.4
Lithuania	2.5	2.6	2.3	2.4	2.7	2.8	3.8	0.1
Russia	379.9	350.4	364.7	363.6	377.2	372.7	—1.2	15.5
Turkmenistan	10.0	10.1	10.3	11.3	12.6	12.9	2.2	0.5
Ukraine	82.5	74.3	68.8	70.6	68.5	65.8	—4.0	2.7
Uzbekistan	43.3	45.4	47.0	49.3	47.1	51.1	8.5	2.1
Other	7.9	8.8	9.2	7.9	7.6	8.6	13.4	0.4
Total	554.0	519.1	529.8	533.9	547.0	548.5	0.3	22.8
Asia-Pacific								
Australia	19.9	19.6	20.3	19.8	21.3	22.5	6.0	0.9
Bangladesh	7.6	7.6	7.8	8.3	10.0	10.8	8.2	0.4
China	17.7	19.3	19.3	21.4	24.5	27.7	12.9	1.2
India	20.6	21.3	24.2	24.8	26.0	26.3	1.3	1.1
Indonesia	31.4	31.9	27.8	31.8	30.6	29.7	—2.9	1.2
Japan	66.1	65.1	69.5	74.6	76.2	79.0	3.7	3.3
Malaysia	15.9	16.7	17.4	18.5	20.3	21.6	6.2	0.9
New Zealand	4.7	5.1	4.5	5.2	5.5	5.7	5.1	0.2
Pakistan	15.4	15.6	16.0	17.3	18.9	20.1	6.1	0.8
Philippines	--	--	--	--	--	0.1	>100.0	--
Singapore	1.5	1.5	1.5	1.5	1.7	2.5	43.3	0.1
South Korea	13.5	16.4	15.4	18.7	21.0	23.1	9.8	1.0
Taiwan	4.5	5.1	6.4	6.2	6.9	7.5	9.7	0.3
Thailand	11.8	14.6	15.9	17.4	20.5	21.1	3.0	0.9
Other	3.7	4.2	4.5	4.8	4.9	4.9	0.2	0.2
Total	236.0	246.6	253.0	273.0	290.8	305.1	5.0	12.7
Total World	2226.5	2213.5	2241.3	2298.7	2397.2	2404.9	0.3	100.0

World Natural Gas Imports and Exports

IMPORTS	1999			2000			2001		
	Pipelines	LNG	Total	Pipelines	LNG	Total	Pipelines	LNG	Total
	billion cubic feet						billion cubic meters		
North America									
Canada	31.08	--	31.08	66.28	--	66.28	4.87	--	4.87
US	3,399.39	161.39	3,560.78	3,791.95	220.33	4,012.28	109.67	6.59	116.26
Latin America									
Brazil	--	--	--	74.1	--	74.1	3.00	--	3.00
Chile	70.63	--	70.63	151.56	--	151.56	4.60	--	4.60
Mexico	63.92	--	63.92	115.81	--	115.81	4.28	--	4.28
Puerto Rico	--	--	--	-	-	--	--	0.63	0.63
Uruguay	--	--	--	1.49	--	1.49	0.07	--	0.07
Europe									
Austria	219.30	--	219.30	220.45	--	220.45	6.04	--	6.04
Belgium	418.13	142.67	560.80	439.41	148.3	587.71	13.22	2.40	15.62
Bulgaria	113.01	--	113.01	119.16	--	119.16	2.90	--	2.90
Croatia	39.20	--	39.20	41.33	--	41.33	1.08	--	1.08
Czech Rep.	321.36	--	321.36	317.27	--	317.27	9.20	--	9.20
Finland	148.32	--	148.32	160.12	--	160.12	4.54	--	4.54
France	1,089.81	362.33	1,452.14	1,206.88	396.53	1,603.41	31.14	10.45	41.59
Germany	2,586.10	--	2,586.10	2,859.88	--	2,859.88	78.75	--	78.75
Greece	52.97	--	52.97	59.58	10.59	70.17	1.48	0.50	1.98
Hungary	310.77	--	310.77	342.59	--	342.59	9.97	--	9.97
Ireland	74.16	--	74.16	96.82	--	96.82	3.42	--	3.42
Italy	1,642.13	100.29	1,742.43	1,966.17	168.78	2,134.95	49.55	5.25	54.80
Luxembourg	28.25	--	28.25	22.34	--	22.34	0.80	--	0.80
Netherlands	259.21	--	259.21	446.86	--	446.86	13.13	--	13.13
Poland	261.33	--	261.33	238.01	--	238.01	8.40	--	8.40
Portugal	67.10	--	67.10	81.92	--	81.92	2.20	0.26	2.46
Romania	120.07	--	120.07	126.61	--	126.61	3.00	--	3.00
Slovakia	261.33	--	261.33	294.18	--	294.18	7.90	--	7.90
Slovenia	34.96	--	34.96	37.61	--	37.61	1.04	--	1.04
Spain	283.93	252.85	536.78	315.41	299.08	614.49	7.76	9.84	17.60
Sweden	30.37	--	30.37	40.96	--	40.96	0.90	--	0.90
Switzerland	104.18	--	104.18	107.99	--	107.99	3.05	--	3.05
Turkey	310.77	112.30	423.07	383.55	130.65	514.20	11.04	4.83	15.87
UK	48.38	--	48.38	74.48	--	74.48	2.70	--	2.70
Others	--	--	--	--	--	--	1.97	--	1.97
Middle East									
Iran	113.01	--	113.01	98.68	--	98.68	4.20	--	4.20
Africa									
Tunisia	35.31	--	35.31	37.24	--	37.24	1.20	--	1.20
Asia Pacific									
Japan	--	2,446.60	2,446.60	--	2,558.56	2,558.56	--	74.07	74.07
Singapore	52.97	--	52.97	55.86	--	55.86	2.50	--	2.50
South Korea	--	618.71	618.71	--	694.90	694.90	--	21.83	21.83
Taiwan	--	188.93	188.93	--	208.33	208.33	--	6.30	6.30
Thailand	--	--	--	--	--	--	1.75	--	1.75
World Total	15,854.18	4,386.09	20,240.27	14,497.13	4,836.06	19,333.19	411.32	142.95	554.27

World Natural Gas Imports and Exports-cont'd

EXPORTS	1999			2000			2001		
	Pipelines	LNG	Total	Pipelines	LNG	Total	Pipelines	LNG	Total
	billion cubic feet						billion cubic meters		
North America									
Canada	3,344.30	--	3,344.30	3,785.62	--	3,785.62	109.02	--	109.02
US	95.00	58.27	153.27	182.09	58.26	240.35	--	1.79	1.79
Latin America									
Argentina	70.63	--	70.63	156.40	--	156.40	5.17	--	5.17
Bolivia	76.63	--	76.63	70.75	--	70.75	2.50	--	2.50
Mexico	55.09	--	55.09	6.33	--	6.33	0.65	--	0.65
Trin. & Tob.	--	--	--	--	123.94	123.94	--	3.65	3.65
Europe									
Denmark	93.94	--	93.94	134.06	--	134.06	3.10	--	3.10
France	30.02	--	30.02	29.79	--	29.79	1.20	--	1.20
Germany	121.84	--	121.84	132.57	--	132.57	4.48	--	4.48
Netherlands	1,236.01	--	1,236.01	1,363.66	--	1,363.66	42.20	--	42.20
Norway	1,606.47	--	1,606.47	1,824.66	--	1,824.66	50.50	--	50.50
UK	196.00	-	196.00	488.20	--	488.20	15.78	--	15.78
Middle East									
Iran	--	--	--	--	--	--	0.11	--	0.11
Oman	15.89	--	15.89	--	87.22	87.22	--	7.43	7.43
Qatar	--	287.11	287.11	--	495.75	495.75	--	16.54	16.54
UAE	--	--	--	--	244.70	244.70	--	7.08	7.08
Africa									
Algeria	1,190.11	909.71	2,099.81	1,315.25	929.36	2,244.61	32.15	25.54	57.69
Libya	--	32.14	32.14	--	28.25	28.25	--	0.77	0.77
Nigeria	--	--	--	--	198.09	198.09	--	7.83	7.83
Former Soviet Union									
Russian Fed.	4,433.41	--	4,433.41	4,853.23	--	4,853.23	126.86	--	126.86
Turkmenistan	113.01	--	113.01	98.68	--	98.68	4.20	--	4.20
Asia Pacific									
Australia	--	355.62	355.62	--	356.98	356.98	--	10.20	10.20
Brunei	--	297.00	297.00	--	310.37	310.37	--	9.00	9.00
Indonesia	--	1,370.56	1,370.56	--	1,260.57	1,260.57	1.00	31.80	32.80
Malaysia	52.97	725.72	778.59	55.86	742.57	798.43	1.50	20.91	22.41
Myanmar	--	--	--	--	--	--	1.75	--	1.75
Taiwan	--	--	--	--	--	--	--	0.41	0.41
World Total	13,299.87	4,285.79	17,585.66	14,497.13	4,836.06	19,333.19	411.32	142.95	554.27

Source: BP Statistical Review of World Energy 2002; Cedigaz
NOTE: Flows are on a contractual basis and might not correspond to physical gas flows in all cases.

PETROLEUM PRICES

Crude Oil Prices

YEAR	US refiner acquisition cost			US landed cost of imports							Avg. FOB price	
	Domes.	Imp.	Compos.	Total	Can.	Mex.	UK	Nig.	Saudi Arabia	Venez.	OPEC	Non-OPEC
	$/bbl											
1984	28.53	28.88	28.63	28.54	26.56	26.85	29.45	30.36	29.20	25.19	27.79	27.45
1985	26.66	26.99	26.75	26.67	25.71	25.63	28.36	28.96	24.72	24.43	25.67	25.96
1986	14.82	14.00	14.55	13.49	13.43	12.17	14.63	15.29	12.84	11.52	12.21	12.87
1987	17.76	18.13	17.90	17.65	17.04	16.69	18.78	19.32	16.81	15.76	16.43	16.99
1988	14.74	14.56	14.67	14.08	13.50	12.58	15.82	15.88	13.37	13.66	13.43	13.05
1989	17.87	18.08	17.97	17.68	16.81	16.35	18.74	19.19	17.34	16.78	17.06	16.72
1990	22.59	21.76	22.22	21.13	20.48	19.64	22.65	23.33	21.82	20.31	20.40	20.32
1991	19.33	18.70	19.06	18.02	17.16	15.89	21.37	21.39	17.22	15.92	16.99	16.77
1992	18.63	18.20	18.43	17.75	17.04	15.60	20.63	20.78	17.48	15.13	16.87	16.66
1993	16.67	16.14	16.41	15.72	15.27	14.11	17.92	18.73	15.40	13.39	14.78	14.65
1994	15.67	15.51	15.59	15.18	14.83	14.09	16.64	17.21	15.11	13.12	14.00	14.34
1995	17.33	17.14	17.23	16.78	16.65	16.19	17.91	18.25	16.84	14.81	15.36	16.02
1996	20.77	20.64	20.71	20.31	19.94	19.64	20.88	21.95	20.49	18.59	18.94	19.65
1997	19.61	18.53	19.04	18.11	17.63	17.30	20.64	20.64	17.52	16.35	16.26	17.51
1998	13.18	12.04	12.52	11.84	11.62	11.04	13.55	14.14	11.16	10.16	10.20	11.21
1999	17.90	17.26	17.51	17.23	17.54	16.12	18.26	17.63	17.48	15.58	15.90	16.84
2000	29.11	27.70	28.26	27.53	26.69	26.03	29.26	30.04	26.58	26.05	25.56	26.77
2001	24.34	22.01	22.96	21.92	20.72	19.39	25.38	26.64	21.20	19.87	19.73	21.04

Source: U.S. Energy Information Administration
NOTE: All prices are US average

Petroleum Product Prices

YEAR	Retail unleaded motor gasoline			Refiner resale (wholesale)								
	Reg.	Prem.	All	Home Htg. Oil	Motor Gasol.	Aviation Gasol.	Keros. Jet Fuel	Keros.	No. 2 Fuel Oil	No. 2 Diesel Fuel	Propane	Resid. Fuel Oil
	$/gallon			cents/gallon								
1984	1.21	1.37	1.20	109.1	83.2	116.5	83.0	91.6	82.1	80.3	45.0	65.4
1985	1.20	1.34	1.20	105.3	83.5	113.0	79.4	87.4	77.6	77.2	39.8	57.7
1986	0.93	1.09	0.93	83.6	53.1	91.2	49.5	60.6	48.6	45.2	29.0	30.5
1987	0.95	1.09	0.96	80.3	58.9	85.9	53.8	59.2	52.7	53.4	25.2	38.5
1988	0.95	1.11	0.96	81.3	57.7	85.0	49.5	54.9	47.3	47.3	24.0	30.0
1989	1.02	1.20	1.06	90.0	65.4	95.0	58.3	66.9	56.5	56.7	24.7	36.0
1990	1.16	1.35	1.22	106.3	78.6	106.3	77.3	83.9	69.7	69.4	38.6	41.3
1991	1.14	1.32	1.20	101.9	69.9	100.1	65.0	72.3	62.2	61.5	34.9	31.4
1992	1.13	1.32	1.19	93.4	67.7	99.1	60.5	63.2	57.9	59.1	32.8	30.8
1993	1.11	1.30	1.17	91.1	62.6	96.5	57.7	60.4	54.4	57.0	35.1	29.3
1994	1.11	1.31	1.17	88.4	59.9	93.3	53.4	61.8	50.6	52.9	32.4	31.7
1995	1.15	1.34	1.21	86.7	62.6	97.5	53.9	58.0	51.1	53.8	34.4	36.3
1996	1.23	1.41	1.29	98.9	71.3	105.5	64.6	71.4	63.9	65.9	46.1	42.0
1997	1.23	1.42	1.29	98.4	70.0	106.5	61.3	65.3	59.0	60.6	41.6	38.7
1998	1.06	1.25	1.12	85.2	52.6	91.2	45.0	46.5	42.2	44.4	28.8	28.0
1999	1.17	1.36	1.22	87.6	64.5	100.7	53.3	55.0	49.3	54.6	34.2	35.4
2000	1.51	1.69	1.56	131.1	96.3	133.0	88.0	96.9	88.6	89.8	59.5	56.6
2001	1.46	1.66	1.53	125.0	88.6	125.6	76.3	82.1	75.6	78.4	54.0	47.6

Source: U.S. Energy Information Administration
NOTE: All price s are US average

Comparative Energy Prices

YEAR	Natural gas delivered to consumers, $/1000 cu ft				Fossil fuels delivered to utilities, cents/million Btu			Retail electricity, cents/kWh			
	Res.	Comm.	Ind.	Util.	Coal	Petroleum	Nat. gas	Res.	Comm.	Ind.	All
1984	6.12	5.55	4.22	3.70	166.4	481.0	360.3	7.15	7.13	4.83	6.25
1985	6.12	5.50	3.95	3.55	164.8	424.4	344.4	7.39	7.27	4.97	6.44
1986	5.83	5.08	3.23	2.43	157.9	240.1	235.1	7.42	7.20	4.93	6.44
1987	5.54	4.77	2.94	2.32	150.6	297.6	224.0	7.45	7.08	4.77	6.37
1988	5.47	4.63	2.95	2.33	146.6	240.5	226.3	7.48	7.04	4.70	6.35
1989	5.64	4.74	2.96	2.43	144.5	284.6	235.5	7.65	7.20	4.72	6.45
1990	5.80	4.83	2.93	2.39	145.5	331.9	232.1	7.83	7.34	4.74	6.57
1991	5.82	4.81	2.69	2.18	144.7	246.5	215.3	8.04	7.53	4.83	6.75
1992	5.89	4.88	2.84	2.36	141.2	247.5	232.8	8.21	7.66	4.83	6.82
1993	6.16	5.22	3.07	2.61	138.5	236.2	256.0	8.32	7.74	4.85	6.93
1994	6.41	5.44	3.05	2.28	135.5	240.9	223.0	8.38	7.73	4.77	6.91
1995	6.06	5.05	2.71	2.02	131.8	258.6	198.4	8.40	7.69	4.66	6.89
1996	6.34	5.40	3.42	2.69	128.9	303.4	264.1	8.36	7.64	4.60	6.86
1997	6.94	5.80	3.59	2.78	127.3	278.8	276.0	8.43	7.59	4.53	6.85
1998	6.82	5.48	3.14	2.40	125.2	207.9	238.1	8.26	7.41	4.48	6.74
1999	6.69	5.33	3.10	2.62	121.6	243.6	257.4	8.16	7.26	4.43	6.66
2000	7.76	6.59	4.48	4.38	120.0	445.0	430.2	8.22	7.22	4.46	6.68
2001	9.63	8.45	5.16	4.51	123.4	431.1	634.1	8.36	7.68	4.99	7.07

Source: U.S. Energy Information Administration
NOTE: All prices are US average

Price History of Oil, Gas, and Gasoline

YEAR	Actual prices				Producer price index	Inflation-adjusted prices			
	Petroleum		Natural gas			Petroleum		Natural gas	
	US avg. wellhead $/bbl	Retail gasol. $/gal	US avg. wellhead $/1000 cu ft	Consumer avg. $/1000 cu ft		US avg. wellhead $/bbl	Retail gasol. $/gal	US avg. wellhead $/1000 cu ft	Consumer avg. $/1000 cu ft
1935	0.97	0.188	0.058	0.224	13.8	7.029	1.365	0.420	1.623
1940	1.02	0.184	0.045	0.217	13.5	7.556	1.364	0.333	1.607
1945	1.22	0.205	0.049	0.214	18.2	6.703	1.126	0.269	1.176
1950	2.51	0.268	0.065	0.266	27.3	9.194	0.980	0.238	0.974
1955	2.77	0.291	0.104	0.400	29.3	9.454	0.992	0.355	1.365
1960	2.88	0.311	0.140	0.500	31.7	9.085	0.982	0.442	1.577
1965	2.86	0.312	0.156	0.522	32.3	8.854	0.966	0.483	1.616
1970	3.18	0.357	0.171	0.550	36.9	8.618	0.967	0.463	1.491
1975	7.56	0.567	0.445	1.120	58.4	12.945	0.971	0.762	1.918
1980	21.59	1.191	1.59	2.800	89.7	24.069	1.328	1.773	3.122
1981	31.77	1.311	1.98	3.390	98.0	32.418	1.338	2.020	3.459
1982	28.52	1.296	2.46	4.150	100.0	28.520	1.296	2.460	4.150
1983	26.19	1.241	2.59	4.820	101.2	25.879	1.226	2.559	4.763
1984	25.88	1.212	2.66	4.850	103.6	24.981	1.170	2.568	4.681
1985	24.09	1.202	2.51	4.720	103.1	23.366	1.166	2.435	4.578
1986	12.51	0.927	1.94	4.130	100.1	12.498	0.926	1.938	4.046
1987	15.40	0.948	1.67	4.050	102.8	14.981	0.922	1.625	3.940
1988	12.58	0.946	1.69	4.090	106.9	11.768	0.885	1.581	3.826
1989	15.86	1.021	1.69	4.220	112.2	14.135	0.910	1.506	3.761
1990	20.03	1.164	1.71	4.190	116.3	17.223	1.001	1.470	3.603
1991	16.54	1.140	1.64	4.071	116.5	14.197	0.979	1.408	3.494
1992	15.99	1.127	1.74	4.173	117.2	13.643	0.962	1.485	3.561
1993	14.25	1.108	2.04	4.429	118.9	11.985	0.932	1.716	3.725
1994	13.19	1.112	1.85	4.509	120.4	10.955	0.924	1.537	3.745
1995	14.62	1.147	1.55	4.128	124.7	11.724	0.920	1.243	3.310
1996	18.46	1.231	2.17	4.668	127.7	14.456	0.964	1.699	3.655
1997	17.23	1.234	2.32	4.988	127.6	13.503	0.967	1.818	3.909
1998	10.87	1.059	1.96	4.600	124.4	8.738	0.851	1.559	3.698
1999	15.56	1.165	2.19	4.521	125.5	12.398	0.928	1.657	3.602
2000	26.72	1.510	3.69	5.732	132.7	20.143	1.138	2.773	4.371
2001	21.84	1.461	4.12	6.505	134.2	16.274	1.089	3.070	4.847

* 1982 = 100

Source: U.S. Energy Information Administration

NOTE: All prices are US average

INTERNATIONAL RIG COUNT

REGION	1996	1997	1998	1999	2000			2001			2002		
					Land	Off.	Total	Land	Off.	Total	Land	Off.	Total
North America													
Canada	270	374	259	245	339	5	344	262	4	266	273	4	277
US	779	943	827	625	776	140	916	866	134	1000	751	111	862
Subtotal	1049	1317	1086	870	1115	145	1260	1128	138	1266	1024	115	1139
Latin America													
Argentina	68	57	44	35	57	0	57	62	0	62	55	1	56
Bolivia	7	5	12	13	11	0	11	6	0	6	5	0	5
Brazil	22	21	20	19	9	14	23	11	16	27	12	14	26
Chile	2	2	1	1	0	1	1	1	0	1	0	1	1
Colombia	14	16	12	12	14	0	14	16	0	16	10	0	10
Ecuador	3	4	5	3	6	1	7	12	0	12	8	0	8
Mexico	47	49	55	43	38	6	44	54	9	63	59	21	80
Peru	7	9	5	2	3	1	4	2	1	3	1	0	1
Trinidad	4	4	6	3	1	3	4	0	5	5	0	2	2
Venezuela	108	110	82	57	48	15	63	44	16	60	25	5	30
Subtotal	281	277	242	188	187	41	228	208	47	255	175	44	219
Europe													
Croatia	3	1	0	0	0	0	0	0	0	0	3	0	3
Denmark	2	2	3	2	0	3	3	1	6	7	0	4	4
France	1	1	2	1	1	0	1	0	0	0	0	0	0
Germany	9	4	3	4	3	0	3	5	0	5	4	0	4
Hungary	2	1	2	1	1	0	1	1	0	1	2	0	2
Italy	13	12	9	9	7	1	8	6	1	7	3	2	5
Netherlands	11	11	8	4	1	3	4	2	6	8	2	3	5
Norway	15	17	17	17	0	22	22	0	24	24	0	15	15
Poland	18	16	15	14	15	0	15	12	1	13	12	0	12
Romania	3	2	2	2	0	0	0	2	0	2	2	0	2
Turkey	10	10	7	6	6	0	6	6	0	6	4	0	4
UK	31	34	28	18	0	17	17	1	22	23	0	23	23
Other	3	2	4	2	3	0	3	2	1	3	4	0	4
Subtotal	121	113	100	80	37	46	83	38	61	99	36	47	83
Middle East													
Iran	22	22	25	27	24	3	27	26	7	33	25	9	34
Kuwait	5	6	8	11	12	0	12	6	0	6	7	0	7
Oman	19	23	24	19	24	0	24	27	0	27	32	0	32
Qatar	8	13	12	6	2	4	6	3	11	14	2	7	9
Saudi Arabia	17	26	28	20	21	4	25	27	4	31	29	4	33
Syria	15	14	14	13	14	0	14	19	0	19	24	0	24
UAE	17	17	18	15	8	5	13	10	8	18	9	7	16
Yemen	3	3	4	4	6	0	6	7	0	7	11	0	11
Other	1	0	1	1	1	0	1	4	0	4	2	1	3
Subtotal	107	124	134	116	112	16	128	129	30	159	141	28	169
Africa													
Algeria	28	29	23	13	15	0	15	20	0	20	22	0	22
Angola	10	8	6	5	0	6	6	0	4	4	0	2	2
Congo (Braz.)	3	5	6	3	0	3	3	0	1	1	0	1	1
Egypt	19	21	22	17	12	6	18	17	7	24	19	5	24
Gabon	3	5	6	2	1	1	2	0	1	1	2	0	2
Libya	13	13	13	8	7	0	7	4	0	4	9	1	10
Nigeria	14	13	12	8	1	7	8	2	7	9	3	9	12
S. Africa	0	1	1	1	0	1	1	0	1	1	0	0	0
Sudan	--	--	--	--	--	--	-	--	--	--	5	0	5
Tunisia	2	2	2	1	1	0	1	1	1	2	1	2	3
Other	4	5	6	2	0	3	3	0	3	3	1	3	4
Subtotal	96	102	97	60	37	27	64	44	25	69	62	23	85

International Rig Count-cont'd

REGION	1996	1997	1998	1999	2000 Land	2000 Off.	2000 Total	2001 Land	2001 Off.	2001 Total	2002 Land	2002 Off.	2002 Total
Asia-Pacific													
Australia	14	14	15	10	5	5	10	7	3	10	5	5	10
Brunei	2	3	2	3	0	2	2	1	1	2	1	2	3
China	9	9	12	9	0	9	9	0	8	8	--	9	9
India	61	60	52	46	40	9	49	39	11	50	46	12	58
Indonesia	43	54	50	34	23	9	32	39	10	49	41	12	53
Japan	7	5	5	5	6	0	6	9	1	10	9	0	9
Malaysia	6	6	8	6	0	7	7	0	11	11	0	15	15
Myanmar	11	7	8	7	9	0	9	6	0	6	7	0	7
New Zealand	1	1	1	1	1	0	1	2	0	2	1	0	1
Pakistan	11	12	11	7	10	0	10	5	0	5	13	0	13
Pap. N. Guinea	2	2	2	1	0	0	0	1	0	1	0	0	0
Philippines	6	2	1	0	0	0	0	1	0	1	2	0	2
Taiwan	2	2	1	0	0	0	0	0	0	0	0	0	0
Thailand	7	7	8	7	1	5	6	1	7	8	0	7	7
Viet Nam	6	6	7	9	0	8	8	0	7	7	0	9	9
Other	1	2	2	2	0	1	1	0	0	0	0	1	1
Subtotal	189	192	185	147	95	55	150	111	59	170	125	72	197
Total World	1844	2126	1845	1461	1583	330	1913	1658	360	2018	1563	329	1892

Source: Baker Hughes Inc.
NOTES: US land count includes inland water rigs
China count includes offshore rigs only

IPE INFORMATION SOURCES

National Oil Companies and Energy Ministries

A

ALBANIA
Albpetrol (formerly DPNG)
c/o Geological Institute for Oil & Gas
Fieri, ALBANIA
Ph: 355 423-4542
Fax: 355 423-4204

National Petroleum Agency
Agjensia Kombetare e Hidrokarbureve
Rruga "Duressit," No. 83
Tirana, ALBANIA
Ph. & Fax: 355 423-1034

ALGERIA
Sonatrach Inc.
Immeuble Mauretania Pl. Perou
80 Ave. Ahmed Ghermoul
Algiers, ALGERIA
Ph: 213 21 54-80-11
Fax: 213 21 54-77-00
E-mail: sonatrach@sonatrach-dz.com
Website: www.sonatrach-dz.com

Ministry of Energy & Mines
80 Ave. Ahmed Ghermoul
Algiers, ALGERIA
Ph: 213 (0)21 65 22 22
Fax: 213 (0)21 65 19 04
E-mail: info@mem-algeria.org
Website: www.mem-algeria.org

ANGOLA
Sociedad Nacional de Combustiveis
de Angola (SONANGOL)
Box 1316
Luanda, ANGOLA
Ph: 244-2-334448
Fax: 244-2-391782
Website: www.sonangol.co.ao

ARGENTINA
Repsol YPF
Av. Roque Sáenz Peña, 777
C.P. 1364, Buenos Aires, ARGENTINA
Website: www.repsol-ypf.com

ARMENIA
Republic of Armenia
Ministry of Energy
Petroleum Project Implementation Unit
2 Government House
Republic Square
Yerevan 375010, ARMENIA
Ph: 3742 521964
Fax: 3742 151687

AUSTRALIA
Commonwealth of Australia
Department of Industry, Tourism and Resources
Resources & Energy
GPO Box 9839
Canberra ACT 2601, AUSTRALIA
Ph: 61 2 6213 6000
Fax: 61 2 6213 7000
Website: www.fed.gov.au

AUSTRALIA, NEW SOUTH WALES
Ministry of Energy and Utilities
PO Box 536
St Leonards 1590
New South Wales, AUSTRALIA
Ph: (02) 9901 8888
Fax: (02) 9901 8777
E-mail: energy@energy.nsw.gov.au
Website: www.doe.nsw.gov.au

AUSTRALIA, NORTHERN TERRITORY
Department of Industry,
Technology and Resources
Minerals and Energy
Ph: +618 8999 5511
Website: www.nt.gov.au

AUSTRALIA, QUEENSLAND
Department of Natural Resources and Mines
AXA Building
144 Edward Street
GPO Box 1401
Brisbane Queensland
4001 AUSTRALIA
Ph: (07) 3227 6626
(07) 3896 3111
Fax: (07) 32278758
Website: www.nrm.qld.gov.au

AUSTRALIA, SOUTH AUSTRALIA
Department of Primary
Industries and Resources
Office of Minerals and Energy Resources
Petroleum Group
7th Floor, 101 Grenfell Street
GPO Box 1671
Adelaide, South Australia
5001 AUSTRALIA
Ph: 61 8 8463 3204
Fax: 61 8 8463 3229
Website: www.pir.sa.gov.au

AUSTRALIA, TASMANIA
Department of Infrastructure,
Energy and Resources
GPO Box 936
Hobart, Tasmania
7001 AUSTRALIA
Ph: +61 3 6233 7503
+61 3 1300 135 513
E-mail: info@dier.tas.gov.au
Website: www.dier.tas.gov.au

AUSTRALIA, VICTORIA
Department of Infrastructure
GPO Box 2797Y
Melbourne, Victoria
3001 AUSTRALIA
Ph: +61 3 9655 6666
Fax: +61 3 9655 6752
Website: www.doi.vic.gov.au

AUSTRALIA, WESTERN AUSTRALIA
Office of Energy
9th Floor, Governor Stirling Tower
197 St Georges Terrace
Perth, Western Australia
6000 AUSTRALIA
Ph: 61 8 9420 5600
Fax: 61 8 9420 5700
E-mail: enquiries@energy.wa.gov.au
Website: www.energy.wa.gov.au

AUSTRIA
Federal Ministry for Economic Affairs
Supreme Mining Authority
Department VII/3
Landstrasser Hauptstrasse 55-57
A-1011, Vienna, AUSTRIA
Ph: 43/1/711 00-0
E-mail: service@bmwa.gv.at
Website: www.bmwa.gv.at

OMV Aktiengesellschaft
Otto Wagner-Platz 5
A-1090 Vienna, AUSTRIA
Ph: 43-1-40440-0
Fax: 43-1-40440-20091
E-mail: info@omv.com

AZERBAIJAN
State Oil Co. of Azerbaijan (SOCAR)
Neftchilar Ave. 73
Baku 370004, AZERBAIJAN
Ph: 799412 93-23-12
Fax: 799412 93-64-92
E-mail: eakhmedo@aiocaz.com
Website: www.azer.com

B

BAHRAIN
Bahrain National Gas
Company (Banagas)
P.O. Box 29099, BAHRAIN
Ph: (973) 756222
Fax: (973) 756991
Website: www.banagas.com.bh

Bahrain National Oil Co.
PO Box 25504
Awali, BAHRAIN
Ph: 973 75-46-66
Fax: 973 75-32-03

BANGLADESH
Bangladesh Oil, Gas & Mineral Corp.
(PETROBANGLA)
Petrocentre
3 Kawranbazar
Dhaka 1215, BANGLADESH
Fax: 880 2 9120224
Website: www.petrobangla.org

Minister of Energy & Mineral Resources
Dhaka, BANGLADESH
Website: www.bangladeshgov.org

BOLIVIA
Yacimientos Petrolifero Fiscales
Bolivianos (YPFB)
Box 401
La Paz, BOLIVIA
Ph: 591 2 374-468
Fax: 591 2 374-469

Vice-Ministry of Energy
and Hydrocarbons
Av. Mariscal Santa Cruz eaq. Oruro
Edificio Palacio de
Comunicaciones, Piso 12
La Paz, BOLIVIA
Ph: 591 2 374-050
Fax: 591 2 392-758
E-mail: enerhid@ceibo.entelnet.bo

BRAZIL
Ministry of Mines and Energy
Website: www.mme.gov.br

Petroleo Brasileiro SA (PETROBRAS)
Caixa Postal 809
Avenida Republica do Chile 65
20035-900 Rio de Janeiro, RJ, BRAZIL
Ph: 55 21 534-4477
Fax: 55 21 534-1939
Website: www.petrobras.com.br

Petrobras Internacional SA (BRASPETRO)
Rua Gen'l Canabarro, 500, 10th Floor
20.271-900, Maracana,
Rio de Janeiro, RJ, BRAZIL
Ph: 55 21 566-4477
Fax: 55 21 566-3400/3401
Website: www.petrobras.com.br

BRUNEI
Office of the Prime Minister
Petroleum Unit
Bahirah Building
Jalan Menteri Besar
BRUNEI Darussalam BB3910
Ph: 673 2 387 102
Fax: 673 2 383 004
Website: www.petroleum-unit.gov.bn

BULGARIA
Committee of Geology & Mineral Resources
22 Princess Marie-Louise Blvd.
Sofia 1000, BULGARIA
Ph: 359 2 981-8861, 981-8841
Fax: 359 2 980-5561
E-mail: cgmr-gf@cserv.mgu.bg

C

CAMEROON
Societe Nationale des Hydrocarbures (SNH)
Box 955
Yaounde, CAMEROON
Ph: 237 20-19-10, 20-98-64
Fax: 237 20-46-51

CANADA
ATTN: Canada Site
Communication Canada
Ottawa, Ontario
K1A 1M4 CANADA
Ph: 800-622-6323
Fax: 613-941-1827
E-mail: sitecanadasite@communication.gc.ca
Website: www.gc.ca

National Energy Board
444 Seventh Avenue SW
Calgary, Alberta
T2P 0X8 CANADA
Ph: 403-292-4800
800-899-1265
Fax: 403-292-5503
E-mail: info@neb-one.gc.ca
Website: www.neb-one.gc.ca

Natural Resources Canada
Energy Sector
Ottawa, Ontario
K1A 0E4 CANADA
Ph: (613) 995-0947
Fax: (613) 992-8738
Website: www.NRCan.gc.ca

CANADA, ALBERTA
Government of Alberta
Energy Ministry
2004 – 14 Street N.W.
Calgary, Alberta
T2M 3N3 CANADA
Ph: 780-427-7425
Website: www.gov.ab.ca

Department of Energy
Ph: 780-427-0265
Fax: 780-422-8731
E-mail: info.energy@gov.ab.ca
Website: www.energy.gov.ab.ca

Energy and Utilities Board
640 - 5th Avenue SW
Calgary, Alberta
T2P 3G4 CANADA
Ph: 403-297-8311
Fax: 403-297-7336
E-mail: eub.webmaster@gov.ab.ca
Website: www.eub.gov.ab.ca

CANADA, BRITISH COLUMBIA
Ministry of Energy and Mines
PO Box 9324, Stn Prov Govt
Victoria, British Columbia
V8W 9N3 CANADA
Ph: 250 387-5896
Fax: 250 356-2965
Website: www.gov.bc.ca/em

CANADA, MANITOBA
Industry, Trade and Mines
327 Legislative Bldg.
450 Broadway Avenue
Winnipeg, Manitoba
R3C 0V8 CANADA
Ph: 204-945-8029
Fax: 204-948-2403
Website: www.gov.mb.ca/itm

CANADA, NEW BRUNSWICK
Department of Natural Resources and Energy
Hugh John Flemming Forestry Complex
Floor: 3
1350 Regent Street
Fredericton, New Brunswick
E3C 2G6 CANADA
Ph: 506-453-3826
Fax: 506-444-5839
Website: www.gnb.ca/0078/index-e.asp

CANADA, NEWFOUNDLAND & LABRADOR
Department of Mines and Energy
Energy Office
P.O. Box 8700
St. John's, Newfoundland
A1B 4J6 CANADA
Ph: 709-729-2349
Fax: 709-729-2871
E-mail: bsaunder@gov.nf.ca
Website: www.gov.nf.ca

CANADA, NORTHWEST TERRITORIES
Resources Wildlife and Economic Development
PO Box 1320
Yellowknife, NT
X1A 2L9 CANADA
Ph: 867-669-2388
Website: www.gov.nt.ca/RWED

CANADA, NOVA SCOTIA
Department of Energy
Bank of Montreal Building Suite 400
5151 George Street
P.O. Box 2664
Halifax, Nova Scotia
B3J 3P7 CANADA
Ph: 902-424-4575
Fax: 902-424-0528
Website: www.gov.ns.ca/energy

CANADA, NUNAVUT
Department of Sustainable Development
Minerals, Oils and Gas
Iqaluit, Nunavut
CANADA
Ph: 867-975-5917
Fax: 867-975-5980
Website: www.gov.nu.ca/Nunavut/English/departments/DSD

CANADA, ONTARIO
Ministry of Energy
900 Bay Street, 4th Floor
Hearst Block
Toronto, Ontario
M7A 2E1 CANADA
Ph: 416 326 4483
877-818-2900
E-mail: write2us@energy.gov.on.ca
Website: www.energy.gov.on.ca

CANADA, PRINCE EDWARD ISLAND
InfoPEI Primary Resources
Oil and Natural Gas
P.O. Box 2000
Charlottetown, Prince Edward Island
C1A 7N8 CANADA
Ph: 902-368-4000
Fax: 902-368-5544
E-mail: island@gov.pe.ca
Website: www.gov.pe.ca/infopei/Primary_Resources/Oil_and_Natural_Gas

CANADA, QUEBEC
Department of Natural Resources
Energy Office
Website: www.mrn.gouv.qc.ca/english/energy

CANADA, SASKATCHEWAN
Department of Industry and Resources
2103 11th Avenue and
2101 Scarth Street
Regina, Saskatchewan
S4P 3V7 CANADA
Ph: 306-787-2232
Ph (Oil & Gas): 306-787-2528
Website: www.ir.gov.sk.ca

CANADA, YUKON TERRITORY
Department of Energy, Mines and Resources
211 Main Street, Suite 400
Box 2703
Whitehorse, Yukon Territory
Y1A 2C6 CANADA
Ph: 867-667-5466
Fax: 867-667-8601
E-mail: emr@gov.yk.ca
Website: www.emr.gov.yk.ca

CHILE
Empresa Nacional del Petróleo (ENAP)
Todos los derechos reservados
Av. Vitacura 2736 piso 10 Las Condes
Santiago, CHILE
Ph: (56-2) 280 3000
Fax: (56-2) 280 3199
Website: www.enap.cl

CHINA
Ministry of Petroleum Industry
PO Box 1411
Beijing
PEOPLE'S REPUBLIC OF CHINA

China National Offshore Oil Corp. (CNOOC)
Jingxin Bldg., Dongsanhuan Bei Rd.
Box 4705
Beijing 100027
PEOPLE'S REPUBLIC OF CHINA
Ph: 86 10 6466-9001, 6466-3696
Fax: 86 10 6466-2994, 6466-9007
Website: www.cnooc.com.cn

China National Petroleum Corp.
PO Box 766
Beijing
PEOPLE'S REPUBLIC OF CHINA
Ph: 86 1 201-5544
Website: www.cnpc.com.cn

China National Petroleum Corp.
One Westchase Center Suite 840
10777 Westheimer Rd.
Houston, Tex. 77042
Ph: 713 784-8598
Fax: 713 784-9197

COLOMBIA
Empresa Colombiana del Petroles
(ECOPETROL)
Carrera 13, No. 3624
Apartado Aereo 5938
Bogota, COLOMBIA
Ph: 57 1 287-9308
Fax: 57 1 288-0071
E-mail: prozo@infantas.ecp.com
Website: www.ecopetrol.com.co

Ministerio de Minas y Energia de Colombia
Centro Administrative Nacional
Avenida El Dorado
Bogota, COLOMBIA
Ph: 57 1 44-520, 44-525
Website: www.minminas.gov.co

CONGO (BRAZZAVILLE)
General Direction of Hydrocarbons
Ministry of Mines & Energy – S.P. 2120
Brazzaville, CONGO
Ph: 242 83-12-81, 83-58-73, 83-59-74
Fax: 242 83-62-43

Hydro Congo
B.P. 2008
Pointe Noire, Brazzaville, CONGO
Ph: 242 81-35-60, 81-40-23
Fax: 242 83-12-38

CONGO (FORMER ZAIRE)
Petrozaire
B.P. 7617
Kinshasa/Gombe, ZAIRE
Ph: 243 12 20-344, 20-492

CROATIA
INA-NAFTAPLIN
Subiceva 29
10,000 Zagreb, CROATIA
Ph: 385 1 4592-222
Fax: 385 1 4640-589
Website: www.ina.hr
E-mail: ina@ina.hr

CUBA
Ministerio de la Industria Basica
Empedarado 113
Havana, CUBA
Website: www.cubagob.cu

Comercial Cupet SA (CUPET)
km 37.5 Via Blanca
Caribe, Sta. Cruz del Norte
Havana 32900, CUBA
Ph: 53 7 62-40-11
Fax: 53 7 33-34-60, 33-80-27, 33-30-72

CZECH REPUBLIC
Ministry of Economic Policy & Development
tr. SNG 65 - 101 60 Prague 10
CZECH REPUBLIC
Ph: 42 2 712-1111
Fax: 42 2 73-13-57
Website: www.czech.cz

DENMARK
Ministry of Environment and Energy
Højbro Plads 4
1200 København K, DENMARK
Ph: 45 33 92 76 00
Fax: 45 33 32 22 27
E-mail: mim@mim.dk
Website: www.mim.dk

DUBAI
Dubai Petroleum Co. (DUPETCO)
Box 2222
Dubai, UNITED ARAB EMIRATES
Ph: 971 4 44-29-90

E

ECUADOR
Petroleos del Ecuador (Petroecuador)
Alpallana E8-86 y
Av. 6 de Diciembre
P.O. Box 17-11-5007
Quito, ECUADOR
Ph: (593-2) 2563060
(593-2) 2561589
Fax: (593-2) 2503571
Website: www.petroecuador.com.ec

Ministerio de Recursos Naturales y Energeticos
Direccion General de Geologia y Miras
Carrion 1016 y Paez
Casilla 23-A
Quito, ECUADOR
Ph: 593 2 23-12-04

Organizacion Latinoamericana
de Energia (Olade)
Av. Occidental N5863
Sector San Carlos
Quito, ECUADOR
Ph: (593)-(2) 2598122 / 2598280 / 2597995
Fax: 2539684
E-mail: olade@olade.org.ec
Website: www.olade.org.ec

EGYPT
Egyptian General Petroleum Corporation
(EGPC)
Palestine Street part 4
New Maadi, Cairo, EGYPT
Tel: 706 59 56
706 59 54
Fax: 702 88 13
703 14 57
E-mail: egpcinf@starnet.com.eg
Website: www.egpc.com.eg

Arab Republic of Egypt
Ministry of Petroleum
1 (A) Ahmed El Zomor str.
8th District Nasr City
Cairo, EGYPT
Ph: 00(202)670-64-35
00(202)670-64-05
Fax: 00(202)670-64-19

EQUATORIAL GUINEA
Ministry of Mines & Energy
Box 778
Malabo, C/12 de Octubre
EQUATORIAL GUINEA
Ph: 240 9 3567
Fax: 240 9 3353
Ph: (00 240) 7 7502
E-mail: d.shaw@ecqc.com
Website: www.equatorialoil.com
E-mail: mme@intnet.gq

F

FAROE ISLANDS
Faroese Petroleum Administration
Debesartro
100 Torshaun
FAROE ISLANDS
Ph: 298 12-306
Fax: 298 18-438
E-mail: oms@oms.fo

FINLAND
Ministry of Trade & Industry
Aleksanterinkatu 4
P.O. Box 32
FINLAND 00023 GOVERNMENT
Ph: +358 9 160 01
Fax: + 358 9 1606 3666
Website: www.vn.fi

FRANCE
Ministry of the Economy, Finance, and Industry
366 Avenue Napoleon Bonaparte
Rueil-Malmaison, FRANCE
Website: www.minefi.gouv.fr

G

GABON
Societe Nationale Petroliere Gabonaise
(PETROGAB)
Box 564
Libreville, GABON

Ministry of Mines, Energy, Oil and
Hydraulic Resources
PO Box 2199
Libreville, GABON
Ph: 241 76-10-71
Fax: 241 72-49-90
E-mail: mcn@internetgabon.com

GREECE
Ministry of Development
80 Michalakopoulo St.
Gr-101 92 Athens, GREECE
Ph: 301 7482762-4
Fax: 301 7772485
Website: www.bsrec.bg/greece

GUATEMALA
Ministerio de Energia y Minas
General Directorate of Hydrocarbons
Diagonal 17, 29-78 Zona 11
01011 Guatemala City, GUATEMALA
Ph: (502) 477-0382
Fax: (502) 476-8506
E-mail: informatica@mem.gob.gt
Website: www.mem.gob.gt

HUNGARY
MOL Hungarian Oil & Gas PLC
H-1502, PO Box 22
Xi, Schonherz Z. U. 18
1117 Budapest
HUNGARY
Ph: 36 1 166-2646
Fax: 36 1 186-8281
E-mail: sczenthe@mol.hu

INDIA
Oil & Natural Gas Corp. Ltd. (ONGC)
7th Floor, Bank of Baroda Bldg.
Parliament Street
New Delhi 110 001, INDIA
Ph: 91 11 331-7205, 371-5291
Fax: 91 11 331-6413
E-mail: endongc@del2.vsnl.net.in
Website: www.ongcindia.com

Director General of Hydrocarbons
Ministry of Petroleum & Natural Gas
Hindustan Times House
18-20, Kasturba Gandhi Marg.,
New Delhi 110 001, INDIA
Ph: 91 11 335-2650, 335-2617
Fax: 91 11 331-7081, 335-2649
E-mail: dspdi.png@sb.nic.in
Website: www.petroleum.nic.in

INDONESIA
Perusahaan Pertambangan Minyak Dan
Gas Bumi Negara (PERTAMINA)
Jalan Merdeka Timur No. 1A
Jakarta 10110, INDONESIA
Ph: 62 21 381-5111, 381-6111
Fax: 62 21 384-3882, 384-6865
Website: www.pertamina.co.id

IRAN
Petroleum Ministry of Islamic Republic of Iran
National Iranian Oil Co. (NIOC)
Ave. Takhte Jamshid
Box 1863
Tehran, IRAN
Ph: 98 21 6151
Website: www.nioc.org

IRAQ
Iraq National Oil Co. (INOC)
Al-Khullani Square
Box 476
Baghdad, IRAQ
Ph: 964 1 80-066, 80-069

Ministry of Oil
Box 6178
Al-Mansour, Baghdad, IRAQ
Ph: 964 1 541-0031

IRELAND
Department of Communications,
Marine & Natural Resources
Leeson Lane
Dublin 2, IRELAND
Ph: 353 1 6782000
Fax: 353 1 6618214
Website: www.marine.gov.ie

ISRAEL
Israel National Oil Co. Ltd.
Beit Gibor, 6 Kaufman St.
Tel-Aviv 68012, ISRAEL
PO Box 50199
Tel-Aviv 61500, ISRAEL
Ph: 972 3 514-2020
Fax: 972 3 514-2061

Ministry of Energy & Infrastructure
234 Yaffo St.
Box 1442/91013
Jerusalem, ISRAEL
Ph: 972 2 31-61-11
Fax: 972 2 38-14-44

ITALY
Ministero delle Attivita Produttive
Direzione Generale dell'Energia e
delle Risorse Minerarie
Via Molise 2
00187 Roma, ITALY
Ph: 06 47887835
E-mail: martino@mica-dgfe.casaccia.enea.it
Website: http://mica-dgfe.casaccia.enea.it

Ente Nazionale Idrocarburi (ENI)
1, Piazzale Enrico Mattei
00144 Rome, ITALY
Ph: 39 6 59002
Fax: 39 6 5900-2141
Website: www.eni.it

IVORY COAST

Director of the Office of the Minister
Ministere des Mines
Box V 50
Abidjan, COTE D'IVOIRE
Ph: 225 21-50-03
Fax: 225 21-53-20

Societe Nationale d'Operations Petrolieres de la Cote d'Ivoire (PETROCI)
Box V 194
Abidjan, COTE D'IVOIRE
Ph: 225 20-25-00, 20-25-10
Fax: 225 21-68-24

J

JAPAN

Japan National Oil Corp. (JNOC)
Fukoku-Seimei Bldg.
2-2 Uchisaiwaicho, 2-chome
Chiyoda-ku, Tokyo 100, JAPAN
Ph: 81 3 3580-5411
Fax: 81 3 3591-0172
Website: www.jnoc.go.jp

K

KAZAKHSTAN

KazakhOil
Republik Ave., 60
473000 Astana
KAZAKHSTAN
Ph: 7 3172 28 0609, 28 0270
Fax: 7 3172 28 0296
Website: www.o2.kz

Ministry of Fuel & Energy
Bakenbay-Batyj St. 142
480091 Almaty
KAZAKHSTAN
Ph: 7 3272 62-64-10, 62-60-80
Fax: 7 3272 62-66-30

KUWAIT

Kuwait Petroleum Corp. (KPC)
Box 26565
13126 Safat, KUWAIT
Ph: 965 245-5455
Fax: 965 246-7159
Website: www.kpc.com.kw

Ministry of Oil
PO Box 5077
Safat, KUWAIT

L

LIBYA

General Department of Technical Affairs
Ministry of Petroleum
PO Box 256
Tripoli, LIBYA

National Oil Corp. (NOC)
Box 2655
Tripoli, LIBYA
Ph: 218 21 46-181

M

MALAYSIA

Petroliam Nasional Berhad (PETRONAS)
Menara Dayabumi, Komplex Dayabumi
Jalan Sultan Hishamuddin
Box 12444
50778 Kuala Lumpur, MALAYSIA
Ph: 60 3 274-3833, 274-8011, 274-8022
Fax: 60 3 274-0217
Website: www.petronas.com.my

MEXICO

Petroleos Mexicanos (PEMEX)
Marina Nacional
#329 Col. Huasteca
Delegacion Miguel Hidalgo
Mexico C.D. 11311, MEXICO
Website: www.pemex.com

MONGOLIA

Mongol Petroleum Co. (MGT)
Uildverchnii Street 37
Ulaanbaatar, MONGOLIA
Ph: 976 1 38-34-83
Fax: 976 1 33-11-76

Petroleum Authority of Mongolia
Ulaanbaatar-37
Uildverchnii gudamj
MGT, MONGOLIA
Ph: 976 1 38-34-83
Fax: 976 1 33-11-76

MYANMAR

Burma Ministry of Mines
Minister's Office
Yangon, MYANMAR

Myanmar Oil & Gas Enterprise
Box 1049
74-80 Min Ye Kyawzwa Rd.
Yangon, MYANMAR
Ph: 95 1 21-394, 21-027
Fax: 95 1 22-964, 22-965

N

NETHERLANDS

Ministry of Economic Affairs
Agency for Energy and the Environment
Novem Utrecht
PO Box 8242
3503 RE Utrecht
NETHERLANDS
Ph: 31-30-2393493
Fax: 31-30-2316491
E-mail: info@novem.org
Website: www.novem.org

NEW ZEALAND

New Zealand Crown Minerals
Ministry of Economic Development
Box 1473
Wellington, NEW ZEALAND
Ph: 64 4 472-0020
Fax: 64 4 499-0968
E-mail: crown.minerals@med.govt.nz
Website: www.med.govt.nz/crown_minerals

NIGERIA

Ministry of Petroleum Resources
44, Eric Moore, Surulere
P.M.B. 12701
Lagos, NIGERIA
Ph: 234 1 60-31-00

Nigerian National Petroleum Corp. (NNPC)
Falomo Office Complex
Private Mail Bag 12701 (Ikoyi)
Lagos, NIGERIA
Ph: 234 1 261-4650, 4228, 4871, 4959

NORWAY

Den norske stats oljeselskap AS
(STATOIL)
Box 300
N-4035 Stavanger, NORWAY
Ph: 47 51 90 00 00
Fax: 47 5199 00 50
Website: www.statoil.com

Ministry of Petroleum and Energy
P O Box 8148 Dep
N-0033 Oslo, NORWAY
Ph: +47 22 24 61 07
E-mail: postmottak@oed.dep.no
Website: odin.dep.no/oed/engelsk

OMAN

Director General of Petroleum and Gas Affairs
Ministry of Oil & Gas
Sultanate of Oman
Box 551
Muscat, OMAN
Ph: 968 60-33-33
Website: www.omanet.com

Petroleum Development Oman (PDO)
Box 81
Mina Al-Fahal, OMAN
Ph: 968 67-81-11
Fax: 968 67-71-06
Website: www.pdo.co.om

P

PAKISTAN

Ministry of Petroleum and Natural Resources
3rd Floor, A Block, Pak Secretariat
Islamabad, PAKISTAN
Ph: 051-9208233
Fax: 051-9205437
E-mail: info@mpnr.gov.pk
Website: www.mpnr.gov.pk

PAPUA NEW GUINEA

Papua New Guinea Department of Mining & Petroleum (Petroleum Division)
Private Mail Bag
Port Moresby Post Office
PAPUA NEW GUINEA
Ph: 675 22 4200
Fax: 675 22 4222

PERU

Ministerio de Energia & Minas y
Direccion General de Hidrocarburos
Esq. Avenida Javier Prado Este y
Avenida Aviacion
Edifico Sol Gas, 5 Piso
Lima 34, PERU
Ph: 51 14 35-29-96, 36-62-36

Perupetro SA, San Borja
Av. Luis Aldana No. 320
Lima 41, PERU
Ph: 51 1 475-9590
Fax: 51 1 475-7722
E-mail: admweb@perupetro.com.pe
Website: www.perupetro.com.pe

PHILIPPINES

Department of Energy
PNPC Complex, Merritt Rd.
Fort Bonifacio
Makati, Metro Manila, PHILIPPINES
Ph: 63 2 844-1021
Fax: 63 2 812-4016

Ministry of Energy
7901 Makati Ave.
Box 1031 MCC
Makati, Metro Manila, PHILIPPINES
Ph: 63 2 85-88-56/9

Philippine National Oil Co.
PNOC Bldg.
7901 Makati Ave.
Makati, Metro Manila 1200, PHILIPPINES
Ph: 63 2 85-90-61, 85-88-56, 88-75-51
Fax: 63 2 815-3094, 810-6728

POLAND

Ministry of Environmental Protection,
Natural Resources, and Forestry
Ul. Wawelska 52/56
00-922 Warsaw, POLAND
Ph: 48 22 57-92-900
Website: www.mos.gov.pl

Polish Oil & Gas Co. (PGNiG)
6/14 Krucza St.
00-537 Warsaw, POLAND
Ph: 48 22 583 50 00
Fax: 48 22 583 58 56
E-mail: pr@pgnig.pl
Website: www.pgnig.com

PORTUGAL

Petroleos de Portugal EP (PETROGAL)
Edificio Europeia
Av. Fontes Pereira de Meio, No. 6-10
1100 Lisboa, PORTUGAL
Ph: 351 1 353-8821
Fax: 351 1 310-2950

Q

QATAR

Ministry of Finance & Petroleum
Department of Petroleum Affairs
Box 2233
Doha, QATAR
Ph: 974 41-41-23, 41-35-71

Qatar General Petroleum Corp. (QGPC)
Box 3212
Doha, QATAR
Ph: 974 49-14-91
Fax: 974 83-19-95

R

ROMANIA

Ministry of Industry and Resources
152 Victoriei Way
Bucharest, ROMANIA
Ph: 40 1 231.02.62
40 1 313.66.66
Fax: 40 1 312.05.13
Website: www.minind.ro

National Agency for Mineral Resources
36-38 Mendeleev St.
70169 Bucharest, ROMANIA
Ph: 40 1 613-2204
Fax: 40 1 210-7440

Petrom RA
109 Calea Victoriei, Sector 1
Cod 70176
C.P. 22-109
Bucharest, ROMANIA
Ph: 40 21-2125001
Fax: 40 3102213
Website: www.petrom.ro

RUSSIA

Ministry of Fuel & Energy of the Russian
Federation (Roskomnedra)
4/6 Bol'shaya Gruzinskaya St.
Moscow 123812, RUSSIA
Ph: 7 95 254-8277
Fax: 7 95 943-0013
Website: www.gov.ru

S

SAUDI ARABIA

Saudi Arabian Oil Co. (Saudi Aramco)
Administration Bldg. Room 2220
Box 5000
Dhahran 31311, SAUDI ARABIA
Ph: 966 3872-0115
Fax: 966 3873-8190
Website: www.aramco.com

SOUTH AFRICA

Department of Minerals and Energy
Pretoria, SOUTH AFRICA
Ph: 27 12 317-9121
Fax: 27 12 322-0810
Website: www.dme.gov.za

Petroleum Agency SA
PO Box 1174
Parow 7499
Cape Province, SOUTH AFRICA
Ph: 27 21 938-3500
Fax: 27 21 938-3520
E-mail: info@petroleumagencysa.com
Website: www.petroleumagencysa.com

SOUTH KOREA

Korea Petroleum Development Corp.
#1588-14, Kwanyang-dong, Dongan-gu
Anyang, Kyungki-do,
KOREA 431-711
Ph: 82 343-80-2114, 2932, 2933
Fax: 82 343 80-9321, 9322

SPAIN

Repsol YPF
Paseo de la Castellana, 278 – 280
28046 Madrid, SPAIN
Website: www.repsol-ypf.com

SYRIA

Ministry of Petroleum & Mineral Resources
Jadet Al Katib
Damascus, SYRIAN ARAB REPUBLIC
Ph: 00963 11 4451624
Fax: 00963 11 4463942
E-mail: mopmr@net.sy
Website: www.mopmr-sy.org

Syrian Petroleum Co.
Box 2849
Damascus, SYRIAN ARAB REPUBLIC
Ph: 963 11 221-7944
Fax: 963 11 223-2083

T

TAIWAN

Chinese Petroleum Corp.
83 Chung Hwa Road
Taipei 10031, TAIWAN
Ph: 886 2 361-0221, 371-7121
Fax: 886 2 331-9645
Website: www.cpc.com.tw

THAILAND

Department of Mineral Resources
Ministry of Energy
Rama VI Rd.
Bangkok 10400, THAILAND
Ph: 66 2 511-0215, 511-0749
Fax: 66 2 245-9855
Website: www.thaigov.go.th

Petroleum Authority of Thailand (PTT)
555 Vibhavadi Rangsit Rd., Ladyao, Chatuchak
Bangkok 10900, THAILAND
Ph: 66 2 537-2000
Fax: 66 2 537-3499
Website: www.pttplc.com

TRINIDAD & TOBAGO

Ministry of Energy & Energy Industries
Riverside Plaza
3 Besson Street
Port of Spain, TRINIDAD
Ph: 1-868-623-6708/6719
Fax: 1-868-625-0306
Website: www.energy.gov.tt

Petroleum Co. of Trinidad & Tobago Ltd.
(PETROTRIN)
Administration Bldg.
Pointe-a-Pierre, TRINIDAD
Ph: 868 658-4306
Fax: 868 658-1163
E-mail: lramyad@petrotrin.com

TUNISIA
Entreprise Tunisienne d'Activites Petrolieres (ETAP)
27 bis, Ave. Kherreddine Pacha
B.P. 367, 1002 Tunis de Belvedere
TUNISIA
Ph: 216 1 78-22-88, 78-32-33
Fax: 216 1 78-40-92, 78-61-42
E-mail: dexprom@etap.com.tn
Website: www.etap.com.tn

TURKEY
General Directorate of Mineral Research & Exploration (MTA)
Ankara, TURKEY
Ph: 90 41 287-3430
Fax: 90 41 222-2878

Turkiye Petrolleri Anonim Ortakligi (TPAO)
Mustafa Kemal Mahallesi
2. Cadde No. 86
06520 Ankara, TURKEY
Ph: 90 312 286-9100
Fax: 90 312 286-9000
E-mail: tpaocc@petrol.tpao.gov.tr
Website: www.tpao.gov.tr

TURKMENISTAN
Ministry of Oil and Gas
27 Bitarap Turkmenistan St.
Ashgabat, TURKMENISTAN 744000
Ph: 9 9312 39-38-34
Fax: 9 9312 51-04-43

U

UKRAINE
Counsel of Ministries
Vladimirskaya St. 34
252003 Kiev, UKRAINE
Ph: 7 044 226-2007

State Committee on Geology & Utilization of Mineral Resources
34 Volodymyrska St.
252601 Kiev, UKRAINE
Ph: 7 044 228-6051, 3243
Fax: 7 044 228-6051
Website: www.brama.com/ua-geology

UNITED ARAB EMIRATES
Abu Dhabi National Oil Co. (ADNOC)
Box 898
Abu Dhabi, UNITED ARAB EMIRATES
Ph: 971 2 602-0000
Fax: 971 2 602-3389
E-mail: adnoc@adnoc.com
Website: www.adnoc.com

Ministry of Petroleum & Mineral Resources
Box 59
Abu Dhabi, UNITED ARAB EMIRATES
Ph: 971 2 62810

UNITED KINGDOM
Department of Trade and Industry
Energy Division
Oil and Gas Directorate
DTI Enquiry Unit
1 Victoria Street
London SW1H 0ET
Ph: 020 7215 5000
Website: www.dti.gov/uk/energy

UZBEKISTAN
Ministry of Foreign Economic Relations
Buyuk Ipak Iuli 75
700077 Tashkent, UZBEKISTAN
Ph: 7 371 2 68-76-31
Fax: 7 371 2 68-72-31, 68-74-77
Website: www.gov.uz

State Committee for Geology & Mineral Resources
T. Shevchenko St., 11
70060 Tashkent, UZBEKISTAN
Ph: 7 371 2 56-19-98
Fax: 7 371 2 56-32-83

Uzneftegazdobycha
Amir Temir Ulitsa 66
700084 Tashkent, UZBEKISTAN
Ph: 371 2 35-83-23
Fax: 371 2 35-83-37

V

VENEZUELA
Ministerio de Energia y Minas
Edificio Parque Central
Caracas, VENEZUELA
Ph: 58 2 483-2033, 483-4033, 483-2317

Petroleos de Venezuela SA (PDVSA)
Avenida Libertador, Torre Este,
La Campina
Apartado 169
Caracas 1010-A, VENEZUELA
Ph: 58 2 708-4111
Fax: 58 2 708-4461, 708-4420
Website: www.pdv.com

VIET NAM
Ministry of Petroleum & Gas
193/6A Nam Ky Khoi Nghia
Ho Chi Minh City 3, VIET NAM

Vietnam Oil & Gas Co. (Petrovietnam)
Oil & Gas General Dept.
22 Ngo Quyen Street
Hanoi, VIET NAM
Ph: 84 48 25-25-26
Fax: 84 48 26-59-42
Website: www.petrovietnam.com.vn

REPUBLIC OF YEMEN
Ministry of Oil & Mineral Resources
Petroleum Exploration & Production Board
Sanaa, YEMEN
Ph: 967 2 263-246
Fax: 967 2 268-949

Yemen Petroleum Co.
YPC Bldg.
Box 3360
Al Hudaydah
YEMEN

GLOSSARY

Abbreviations and Acronyms

A

ACG	Azeri-Chirag-Guneshli (Azerbaijan oil fields)
Adma-Opco	Abu Dhabi Marine Operations Co.
Adnoc	Abu Dhabi National Oil Co.
AEP	American Electric Power Co.
A.G.A.	American Gas Association
Agip KCO	Agip Kazakhstan North Caspian Operating Co., formerly OKIOC
AIOC	Azerbaijan International Oil Co.
AMBO	Albanian Macedonian Bulgarian Oil Co.
AMPCO	Atlantic Methanol Production Co. LLC
ANP	National Petroleum Agency (Brazil)
ANWR	Arctic National Wildlife Refuge (Alaska, US)
API	American Petroleum Institute
ASEAN	Association of Southeast Asian Nations
ATHEER	Abu Dhabi Gas Co. (United Arab Emirates)

B

b/d	barrels per day
BALAK	national oil and gas agency's Indonesian-language acronym
Banagas	Bahrain National Gas Co.
Banco	Bahrain National Oil Co.
Bapco	Bahrain Petroleum Co.
bbl	barrels
bcf	billion cubic feet
bcfd	billion cubic feet per day
bcfe	billion cubic feet equivalent
bcm	billion cubic meters
bcmd	billion cubic meters per day
bcme	billion cubic meters equivalent
b/d	barrels per day
BGL	Bhagyanagar Gas Ltd. (India)
bo/d	barrels of oil per day
boe	barrels of oil equivalent
boe/d	barrels of oil equivalent per day
bscf	billion standard cubic feet
bscfd	billion standard cubic feet per day
BTC	Baku (Azerbaijan)-Tbilisi (Georgia)-Ceyhan (Turkey) oil pipeline
Btu	British thermal unit
bw/d	barrels of water per day

C

CAPP	Canadian Association of Petroleum Producers
Caricom	Caribbean regional economic union
CD	compact disk
CEO	chief executive officer
CERI	Canadian Energy Research Institute
cf	cubic feet
cfd	cubic feet per day
CGES	Centre for Global Energy Studies (London)
Chinaoil	China National United Oil Co.
CITIC	China International Trust & Investment Corporation
CNG	compressed natural gas
CNOOC	Chinaese National Offshore Oil Corporation
CNPC	China National Petroleum Corporation
CNR	Canadian Natural Resources Ltd.
CO	carbon monoxide
CO2	carbon dioxide
CPC	Caspian Pipeline Consortium
CPC	Chinese Petroleum Corp. (Taiwan)
CRE	Energy Regulatory Commission (Mexico)
CSO	Coflexip Stena Offshore
cu m	cubic meters (metres)

D

DDCV	deep-draft caisson vessel
DEPA	Greek Public Gas Co.
DIGP	Dauphin Island Gathering Partners
DOE	Department of Energy
DONG	Dansk Olie ag Naturgas AS (Denmark)
DUC	Dansk Undergrunds Consortium (Denmark)
dwt	deadweight tons

E

E&D	exploration and development
E&P	exploration and production
EC	European Commission
ECO	Economic Cooperation Organization (Central Asia)
EDP	Electricidade de Portugal SA
EGAS	Egyptian Natural Holding Gas Co.

EGPC	Egyptian General Petroleum Corp.
EIA	Energy Information Administration
ELNG	Egyptian LNG joint venture
ENAP	Empresa Nacional de Petroleo (Chile)
EniNI	Ente Nazionale Idrocarburi (Italy)
EPA	Environmental Protection Agency
ETAP	Enterprises Tuniseienne d'Activites Petrolieres
EU	European Union
EVE	Basque Energy Authority (Spain)
EWT	extended well test

F

FCC	fluid catalytic cracking
FCCU	fluid catalytic cracking unit
FERC	Federal Energy Regulatory Commission
FNA	Frigg Norwegian Association (Norway)
fob	free on board
FPC	Formosa Petrochemical Corp.
fph	feet per hour
FPSO	floating production, storage, and off-loading
FPU	floating production units
FSRU	floating storage and regasification unit
FSU	Former Soviet Union
ft	foot
FTC	Federal Trade Commission

G

GAIL	Gas Authority of India Ltd.
GASCO	Abu Dhabi Gas Industries Ltd. (United Arab Emirates)
GCC	Gulf Cooperation Council
GCDS	Gaqsoducto Cruz del Sur (Argentina-Uruguay)
GdF	Gaz de France SA
GDP	gross domestic production
GEA	Gas Energy Adria
GFU	Gas Negotiating Committee (Norway)
GNPOC	Greater Nile Petroleum Operating Co. (Sudan)
GTI	Gas Technology Institute
GTL	gas-to-liquids
GTLB	GTL Bolivia SA

H

H&P	Helmerich & Payne Inc.
HCU	Hydrocarbon Unit of the Bangladesh Energy and Mineral Resources Division
HDPE	high-density polyethylene
HDS	hydroskimming
hp	horsepower
HP	Hellenic Petroleum (Greece)
HPCL	Hindustan Petroleum Corp Ltd. (India)

I

IADB	Inter-American Development Bank
IEA	International Energy Agency
IEC	Israel Electric Co.
IET	integrated exploration technology
IGCC	integrated gasification combined cycle
IMF	International Monetary Fund
IMO	International Maritime Organization
INA	Industrija nNafte ddDD Zagreb (Croatia)
INAC	Indian and Northern Affairs Canada
INGAA	Interstate Natural Gas Association of America
IPAA	Independent Petroleum Association of America
IPO	initial public offering
IOC	Indian Oil Corp.

J

JDA	Joint Development Area (Malaysia-Thailand)
JIPA	joint investment production activity
JV	joint venture

K

KIO	Karachaganak Integrated Organization (consortium)
km	kilometer
KNOC	Korean National Oil Corporation.
Kogas	Korea Gas Corporation.
KVA	kilo-volt-amperes
KZM	Ku-Zaap-Maloob complex (Mexico)

L

LAB	linear alkylbenzene
LNG	liquefied natural gas
Logic (UK)	leading oil and gas industry competitiveness
LPG	liquefied petroleum gas

M

m	meter
Mcf	thousand cubic feet
MIDOR	Middle East Oil Refinery Ltd.
Migas	Directorate General of Oil and Gas (Indonesia)
MMBtu	million British thermal units
MMcf	million cubic feet
MMcfd	million cubic feet per day
MMS	Minerals Management Service
MOL Rt.	Hungarian Oil and Gas Co.
MSC	multiple-service contract
MTBE	methyl tertiary butyl ether
mtpy	million tonnes per year

N

NAM	Nederlandse Aardolie Maatschappij BV (Netherlands)
NCMA	North Coast Marine Area (Trinidad & Tobago)
NETA	New Electricity Trading Arrangements (UK)
NGO	non-governmental organization
NIOC	National Iranian Oil Company
NLNG	Nigeria Liquefied Natural Gas Co.
NNPC	Nigerian National Petroleum Co.
NOx	Nitrogen oxides
NPD	Norwegian Petroleum Directorate
NPR-A	National Petroleum Reserve-Alaska
NPRA	National Petrochemical & Refiners Association (US)
NWS	North West Shelf (Australia)
Nymex(YMEX)	New York Mercantile Exchange
NYSE	New York Stock Exchange

O

OCA	Overlapping Claims Area (Thailand-Cambodia)
OCP	Oleoducto de Crudos Pesados (consortium and pipeline, Ecuador)
OCTG	oil country tubular goods
OD	outer diameter
OECD	Organization for Economic Cooperation and Development
OGDC	Oil & Gas Development Co. Ltd. (Pakistan)
OGIP	original gas in place
OGJ	Oil & Gas Journal
OKIOC	Offshore Kazakhstan International Operating Company, re-named Agip KCO
OLNGC	Oman Liquefied Natural Gas Co.
ONAREP	Office National de Recherches et d'Explorations Pétrolières (Morocco)
ONGC	Oil & Natural Gas Corporation (India)
OPEC	Organization of Petroleum Exporting Countries
OPIC	Overseas Private Investment Corp., a US government loan agency
OPL	oil prospecting license (Nigeria)
OSA	operating service agreement

P

Parco	Pak-Arab Refinery Co. Ltd.
PDO	Petroleum Development Oman
PdDVSA	Petroleos de Venezuela SA
PEGASO	Excellence in Environmental Management & Operational Safety Program (Brazil)
PEP	Petroleum Exploration Permit
Pemex	Petroleos Mexicanos
Petrobangla	Bangladesh Oil, Gas & Mineral Corp.
Petrobras	Petroleo Brasileiro SA
Petroecuador	Petroleos del Ecuador
Petrogal	Petroleos de Portugal
Petronas	Petroliam Nasional Sdn. Berhad (Malaysia)
PetroSA	Petroleum Oil and Gas Corp. of South Africa
PetroVietnam	Vietnam Oil & Gas Co.
PG&E	Pacific Gas & Electric Co.
PGC	Potential Gas Committee
PGN	PT Perusahaan Gas Negara (Indonesia)
PGNiG	Polskie Gomictwo Naftowe I Gazownictow SA (Polish Oil & Gas Co.)
PIC	Petrochemical Industries Co. (Kuwait)
PIRINC	Petroleum Industry Research Foundation Inc.
PKN	Polski Koncern Naftowy (Poland)
POGC	Polskie Gomictwo Naftowe I Gazownictow SA (Polish Oil & Gas Co.)
PNSC	Pakistan National Shipping Corporation
Poas	Petrol Ofisis (Turkey)
PSA	production-sharing arrangement
PSA	pressure swing adsorption
PSC	production-sharing contract
PSO	Pakistan State Oil
PTEN	PT Exspan Nusantara (Indonesia)
PTT	polytrimethylene terephthalate
PTT PLC	Thai state-owned oil and gas firm (formerly Petroleum Authority of Thailand)
PTTEP	PTT Exploration and Production PLC

R

RasGas	Ras Laffan Liquefied Natural Gas Co. Ltd. (Qatar)
RFCC	residual fluid catalytic cracking
RFG	reformulated gasoline
ROSBOS	Rosetti Marino SPA-Bouygues Offshore joint venture
RR	railroad
RTO	regional transmission organization

S

SABIC	Saudi Basic Industries Corp.
SAGD	steam-assisted gravity drainage
SAPREF	South African Petroleum Refineries (Pty) Ltd.
scfd	standard cubic feet per day
SDFI	state's direct financial interest (Norway), re-named Petoro
SEEL	South-East European Line
semi	semisubmersible offshore production platform
Sinochem	China National Chemical Import & Export Company
SNH	National Hydrocarbon Corp. (Cameroon)
SNPC	Société Nationale des Pétroles du Congo (Brazzaville)
SO2	sulfur dioxide
SOx	sulfur oxides
SOCAR	State Oil Company of Azerbaijan Republic

SOEP	Sable Offshore Energy Project (Canada)
SOTE	Trans-Ecuadorian pipeline's Spanish-language acronym
SPE	Society of Petroleum Engineers (US)
SPLA	Sudan People's Liberation Army
SPP	Slovensky plynarensky priemysel (Slovakia)
sq km	square kilometer
sq mi	square mile
SPR	Strategic Petroleum Reserve
SRS	Schmierstoff Raffinerie Salzbergen (German refinery)
Statoil	Den norske stats oljeselskap AS (Norway)

T

TAC	technical assistance contract
TAP	Trans-Alpine Pipeline
tcf	trillion cubic feet
tcfe	trillion cubic feet equivalent
tcm	trillion cubic meters
tcme	trillion cubic meters equivalent
TCP	Trans-Caspian Pipeline
TD	total depth
TGS	Transportadora de Gas del Sur (South American pipeline operator)
tj	terajoules
TMD	total measured depth
toe	tons or tonnes of oil equivalent
TPAO	Turkiye Petrolleri Anonim Ortakligi (Turkey)
tpd	tons or tonnes per day
tpy	tons or tonnes per year
TVD	true vertical depth
TVDS	true vertical depth subsea
TVK	Tiszai Vegyi Kombinat Rt. (Hungary)
TWh	terawatt-hour

U

UAE	United Arab Emirates
UDS	Ultramar Diamond Shamrock Corp.
UES	Unified Energy Systems (Russia)
ULCC	ultra large crude carrier
UN	United Nations
Unipec	United International Petroleum & Chemicals Company Ltd. (China)
UOG	United Arab Emirates Offsets Group

V

VLCC	very large crude carrier
VP	Vice President
VSS	vortex separation system

W

WNTS	West Natuna Transportation System (Indonesia to Singapore)
WOPP	White Oil Pipeline Project (Pakistan)
WTO	World Trade Organization

North American Geographical Abbreviations

US STATES

AL	Alabama
AK	Alaska
AS	Arkansas
AZ	Arizona
CA	California
CO	Colorado
CT	Connecticut
DC	District of Columbia
DE	Delaware
FL	Florida
GA	Georgia
HI	Hawaii
ID	Idaho
IL	Illinois
IN	Indiana
IA	Iowa
KS	Kansas
KY	Kentucky
LA	Louisiana
MA	Massachusetts
MD	Maryland
ME	Maine
MI	Michigan
MN	Minnesota
MO	Missouri
MT	Montana
NC	North Carolina
ND	North Dakota
NE	Nebraska
NH	New Hampshire
NJ	New Jersey
NM	New Mexico
NV	Nevada
OH	Ohio
OK	Oklahoma
OR	Oregon
PA	Pennsylvania
RI	Rhode Island
SC	South Carolina
SD	South Dakota
TN	Tennessee
TX	Texas
UT	Utah
VA	Virginia
VT	Vermont
WA	Washington
WI	Wisconsin
WV	West Virginia
WY	Wyoming

CANADA PROVINCES

AB	Alberta
BC	British Columbia
MB	Manitoba
NB	New Brunswick
NL	Newfoundland & Labrador
NS	Nova Scotia
NT	Northwest Territories
NU	Nunavut
ON	Ontario
PE	Prince Edward Island
QC	Quebec
SK	Saskatchewan
YT	Yukon Territory

IPE INFORMATION SOURCES

Advertiser's Index/Index of Contents

ADVERTISER'S INDEX

INDEX OF CONTENTS

A

B

C

E

F

G

H

N

O

T

W

X

Z

FOLD-OUT MAP LEGEND:

Use this legend for all IPE 2003 maps.

Legend

Oil field

Oil sand

Gas field

Crude oil pipeline

Natural gas pipeline

Products pipeline

Pipeline planned
or under construction

600
Refinery in operation
Refinery capacity in
1,000 b/d

Tanker terminal

Cities

Capital

International boundary

Water depth

0 to 200m

200m and deeper